MAPS IN THOSE DAYS

MAPS IN THOSE DAYS

CARTOGRAPHIC METHODS BEFORE 1850

J.H. Andrews

FOUR COURTS PRESS

Typeset in 11pt on 14pt Caslon by
Carrigboy Typesetting Services for
FOUR COURTS PRESS LTD
7 Malpas Street, Dublin 8, Ireland
e-mail: info@fourcourtspress.ie
and in North America for
FOUR COURTS PRESS
c/o ISBS, 920 NE 58th Avenue, Suite 300, Portland, OR 97213.

A catalogue record for this title is available
from the British Library.

ISBN 978–1–84682–188–2

Printed in England
by MPG Books, Bodmin, Cornwall.

'Are you a scientific man? Can you make a map?'

(Minutes of evidence on the Dublin improvement bill
(local acts preliminary inquiries), House of Commons
sessional paper 1847 (124), xxvi, p. 21)

Contents

Illustrations

Preface

'HOW DID THEY MAKE maps in those days?' The speaker was an earnest young man, the time was a day in 1962, the place an exhibition room containing a large number of mainly pre-nineteenth-century maps. A memory with that amount of staying-power deserves to be acknowledged – and perhaps also to be exorcised by some process of re-enactment. So note first that in this case both questioner and respondent were Europeans speaking a European language and looking at European artifacts in a European country. To some people that might seem limitation enough, but it was not the only factor conditioning my answer. For such an issue to be worth raising, 'those days' must have been at least superficially different from our own. In the context of the maps then under inspection, they could well exclude the twentieth and twenty-first centuries and probably more, a convenient cut-off point being the advent of photography as an aid to map production around 1850. At the same time a degree of technical advancement was apparently being imputed to the products on display. Early medieval maps would hardly lead an uninstructed modern observer to ask the question 'how?' He would be more likely to assume without prompting that a thirteenth-century *mappamundi*, say, could be classed as idle speculation or flight of fancy, a criticism that leaves little room, in the present attempt at re-enactment, for several interesting varieties of early map. My companion also seemed to think that all the items we were looking at had been made in much the same way. He was therefore not referring to thematic cartography, where every subject requires a different mode of data-collection. Our concern was with the mainstream of map-making as charted by writers like Herbert George Fordham, I.J. Curnow, W.W. Jervis, Leo Bagrow, Ronald Tooley, Lloyd A. Brown, Gerald Crone, Eva Taylor and Peter Skelton.

So how *did* they make maps in those days? The past forty-five years have seen an explosion of books and articles on cartographic history. To the authors just mentioned the English-speaking reader must now add Paul Harvey, Tony Campbell, Brian Harley and David Woodward, Peter Barber and Chris Board, John Goss and many more. Besides the general histories written or edited by these and other scholars there have been catalogues and cartobibliographies, facsimile editions with their attendant commentaries, biographical monographs and directories, histories of individual survey departments and publishing houses, and studies of particular map-types such as town plans, estate surveys and marine charts. Although this literature contains useful information about how cartographers did their job, its technical content is not quite enough to meet our rather demanding initial specification. (The bias among recent writers towards the social and ideological significance of early maps has proved a mixed blessing in this respect.) The present book attempts to drop a few possible stepping-stones into the gap.

Apart from stating some facts I have a thesis – not to prove, or even to recommend, but to help save from the oblivion that threatens to overtake it. Confronted with an early map, many non-academics react, even today, with either disapproval, condescension or amusement to what they perceive as errors or deficiencies. When they start reading modern scholarly literature these vigilant observers find themselves berated for their naiveté. The past was another country, insists the map-historical establishment: they did things differently there. And one of the things done differently, according to this line of thought, was the assessment of cartographic merit. From which it eventually follows that historians should not complain or even comment if early map-makers worked to standards different from those of today. The fact is (we are given to understand) that these artists and their clientele did not really *want* their maps to be good in our sense of the word.

This view contains an element of truth, but surely not the whole truth. It ignores the possibility that some people in the past might have felt much the same about their maps as we do, a possibility that is at least consistent with many of the texts on which the present work depends. But to pronounce upon the longevity of cartographic principles we have to say what those principles are: hence the apparently incongruous mixture of historical narrative and classroom-style exposition attempted in this book. How well the principles in question were actually applied in 'those days' is of course a different matter, with its own historical interest. We must certainly be prepared to find less common ground between remote and recent periods in achievement than in aspiration, and this difference may well have been a cause of some misunderstanding in map-historical debate. The main point, however, is that when I comment 'judgementally' upon an early map it is in the belief that some, perhaps many, contemporaries would have been prepared to express themselves along similar lines.

The substance of the following pages has come from teaching elementary undergraduate courses in cartography and its history, from personal research into the cartographic record of one small but diversely mapped country, and from a perusal of published writings on the history of maps in general. This last experience suggests that the more abundant the secondary materials for any branch of knowledge, the more necessary is a kind of tertiary literature represented in our case by books like Norman Thrower's *Maps and civilization*, John Noble Wilford's *The mapmakers* and Jeremy Black's *Maps and history*. As readers of these works will agree, there is room for considerable variation within the tertiary sector, room perhaps for a treatment that is at once summary, critique, questionnaire and agenda – perhaps with a few subtextual elements of autobiography.

Tertiary writers have no hope of tracing every factual statement to its ultimate documentary roots. Indeed it might seem logical to stipulate that all their authorities should be secondary, though in practice I have often ignored this precept. However one classifies it, the literature of map history has grown too large to be digested by any one individual. The obvious remedy is sampling, with the inevitable disadvantage that what look like universal claims can actually have no more than limited reference. My choice of examples is systematic in depending almost entirely on British and Irish libraries and record repositories. Within these limits I have drawn from widely separate places and periods, often deployed in rapid succession. For some readers, 'cherry-picking' on this scale may induce a certain degree of disorientation. As a partial corrective the book begins with a thumbnail sketch of world cartographic history in roughly chronological sequence (chapter 1). It then follows what might conceivably have been the successive experiences involved in making a single map. An essential preliminary is to consider the cartographer's theoretical presuppositions about the world (chapter 2). Next comes the art of the sketch map (chapter 3), and then three essays on surveying by measurement which deal in turn with the surveyor's instruments and their use (chapter 4), the objects to which these instruments were directed (chapter 5) and the lay-out of lines and angles that resulted from their employment (chapter 6). After the plotting or protraction of ground measurements (chapter 7), perhaps preceded by the choice of a suitable map projection (chapter 8), the end-product finally materialises in three forms: the 'normal' topographical map (chapter 9), the reconnaissance or exploratory survey whether military or civil (chapter 10), and the marine chart (chapter 11). Two less essential kinds of fieldwork that seem to need separate treatment are the recording of relief features (chapter 12) and the collection of placenames (chapter 13). Then there are the presentational aspects of cartography. One of these is copying and reproduction (chapter 14), another is the more creative business of compilation from existing materials (chapter 15). To edit a map (chapter 16) is to change it in ways not involving an addition to its geographical substance, and this can entail the use of symbols and other conventions (chapter 17) as well as the addition of purely decorative features (chapter 18). Finally, a few social aspects of map-making history are referred to in a brief postscript (chapter 19).

The spelling, capitalisation and (in extreme cases) punctuation of quoted passages have been modernised. Editorial interpolations in published texts have been adopted without comment. Endnotes have been used almost exclusively for citations. Some of these locate quotations or verify facts; some point towards particular instances illustrating a general proposition, or even (on rare occasions) running counter to that proposition. Others adduce cases parallel to those chosen for explicit notice. Where more than one source appears in the same note, the first is the one most closely related to the corresponding passage in the body of the chapter. Where dates, names and map-titles seem in danger of overburdening the text, they have been endnoted after the appropriate source-reference.

* * *

Matthew Stout is the 'online begetter' of this book in the sense that without his interest and enthusiasm it would never have proceeded from disc to print. As well as reading every chapter with critical attention he has devoted much time to scanning, editing and where necessary redrawing the illustrations. As in so much of my previous work, Arnold Horner has been a generous source of encouragement and map-historical expertise. Others who have kindly read and commented on the text are Mary Davies, Paul Ferguson, Jacinta Prunty and Anngret Simms. Help with particular problems has been gratefully received from the late Brian Adams, Peter Barber, David Buisseret, Matthew Champion, Beatrice Coughlan, Raymond Frostick, Francis Herbert, Agustin Hernando, James Killen, Rob Kitchin, Andrew Bonar Law, Paul Laxton, Mark Monmonier, Angel Palatini, J.B. Post, Günter Schilder, Jeffrey Stone and Adrian Webb. Robert Towers gave crucial support at a difficult stage of my endeavours. Not least, I am deeply indebted to Martin Fanning of the Four Courts Press for his unfailing patience, wisdom and professionalism.

J.H.A.

CHAPTER 1

Map history miniaturised

Maps are more easily recognised than defined. Even within the narrow framework of this book, our specification must be something of a mouthful: a map, for the next five hundred pages, will be a partly schematised graphic representation whose signs are interrelated in ways that resemble or could reasonably be thought to resemble the horizontal relations connecting the objects represented. The first characteristic to note here is resemblance. Space relations can be described by words like 'adjacent to' or 'north of', but a sentence formed from these and other verbal utterances could never count as a map because it does not sufficiently resemble the objects it refers to. In this context the word 'resemble' serves better than 'imitate', 'replicate' or 'mirror' because resemblance admits of degrees and in a map the similarity between the image and its object can never be complete: if it were, all terrestrial maps would be life-sized globes or parts of globes. This means that 'mappiness' itself may be a matter of degree. Finally, the somewhat vague reference to schematicism in our definition is a way of acknowledging that the visual differences between a map and a vertical air photograph are deliberate – the result of 'scheming' – and not just a by-product of technical constraints.[1]

Charts and plans may be accepted as categories of map, but few people with no philosophical axe to grind would grant full cartographic status to views, panoramas or architectural elevations: hence the criterion of horizontality embedded in our definition. An infinite number of planes can be sliced at different angles through any group of objects. We favour one particular angle of representation by giving it a name of its own for the obvious reason that horizontal or nearly horizontal motion has a special significance for creatures who travel mainly by walking or riding rather than flying or swimming. Notice however that the horizontal 'planes' of normal experience are not flat but curved, maintaining an approximately constant distance from the earth's centre. (Note also for future reference that horizontal planes, though fewer than planes in general, are themselves infinitely numerous.) From the foregoing qualifications it follows that many planetary or lunar surfaces could in principle be represented by a map. Nor is that the end of the matter, because the word 'horizontal' can also be applied, legitimately if unusually, to the notional surface across which heavenly bodies appear to move when seen from the earth. A graphic representation of this 'celestial sphere' must therefore meet our initial specification and it has often seemed appropriate for terrestrial and astronomical maps to appear in the same atlas or even on the same sheet of paper.[2]

(There is no room for star maps in this book, though.) On the other hand, a diagram of the planets revolving round the sun would have to be excluded from our definition, though it does possess one negative map-like characteristic, namely that of not reproducing the exact contents of anyone's visual field.

Map-making is often seen nowadays as a manifestation of wilfulness, hope, fear, ethical preference, aesthetic sensibility, fantasy, greed, taciturnity, power-lust and various other more or less discreditable mental states.[3] In the following chapters these preoccupations will be given a rest. Instead, maps will appear as tokens of belief about what is considered to be the real world. 'Token' does not imply that the opinions expressed in a map must necessarily be held by its author. An Elizabethan Englishman accused the Spaniards of purposely falsifying their maps of South America 'that they might deceive strangers, if any gave the attempt to travail that way'.[4] So some maps are deliberate lies, while others, of which the most famous appears in Robert Louis Stevenson's novel *Treasure Island*, are accepted by producer and consumer alike as no more than harmless inventions.

Even if we ignore this element of insincerity, maps may still be made for various reasons. Their most common use has been for real or simulated determination of a route by land or sea, either by tracing exact lines of movement or merely by identifying the territories through which a traveller has to pass. Maps can also assist in the numerical reckoning of distance for the benefit of engineering, agriculture, estate management, or taxation. More simply, a map can act as a kind of non-quantitative inventory, with the advantage that the empty spaces representing the author's ignorance may be more easily noticed, and therefore less dangerous, than the omissions in a written list. Another function of cartography has been as a tool of science, including the science of theology, and particularly as a means of showing how spatial variations in different natural or social phenomena are interrelated. Finally, a map may be valued for its own sake as a true statement, useful or otherwise, within the sum of knowledge. In other words, not every map need have a practical part to play in human life. The more advanced a civilisation (if we are not too fastidious to admit the possibility of advancement in this connection), the more of these ends a map is likely to promote.

Primitivism in cartography

All the above-mentioned purposes can be served without the aid of maps, even if not very effectively. We cannot therefore be sure that cartography has been as widespread among mankind as, say, methods of getting food or looking after children. Its existence in any past community requires to be demonstrated by means of historical research. We can agree however that map-making has been practised more widely than writing. Many maps in illiterate societies would have been difficult if not impossible to understand without an oral commentary. Lacking the

1.1 Conventional and unconventional scale indicators. *A new mapp of the cittyes of London and Westminster* [1685].

words that accompanied it, only part of such a map can be said to remain for posterity. When that happens the survival may not be recognised as genuinely cartographic, especially if the places represented no longer exist. Some old maps can be seen for what they are by their resemblance to modern maps, with individual landscape features readily distinguishable. In other cases, among them numerous prehistoric rock carvings, we may have only irregular patterns for which the most that can be said is that in the absence of recognisable profile drawings they are capable of being given a planiform interpretation.[5] Some formerly authentic-seeming maps, like those on certain Greek coins, have been rejected by later scholarly opinion as wholly non-cartographic.[6] What this means is that unless an oral component was recorded at the time by literate visitors from another culture, historians investigating some populations may never become aware that pre-literate maps existed.

No society is known to have been debarred from cartographic activity by an absence of drawing materials, but many maps have been too perishable to survive, for instance those drawn on sand, snow, ashes, wood or bark. Perhaps surprisingly, relief models and globes are even more vulnerable to destruction than flat surfaces.[7] But through all the hazards of time and chance it is clear that realistic maps did exist in

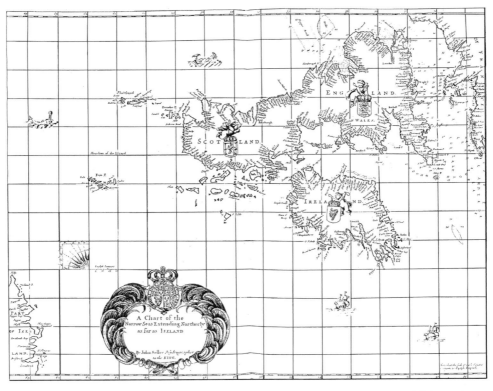

1.2 Graticule of meridians (longitude) and parallels (latitude). John Seller, British Isles and adjoining seas, from *The English pilot* (London, 1675).

prehistoric Europe, in early Egypt and Mesopotamia, in imperial China and some of its neighbours, and in parts of native North America, Mexico and the Pacific.[8]

Less widely demonstrable is the link between cartography and numeracy. Not every map benefits from being mathematically exact. Harry Beck's famous diagram of the London underground railway system is actually better for nowhere attempting to make its distances correct.[9] To distinguish 'true' from 'sketch' maps is to overlook this important fact, as well as to be guilty of political incorrectness in respect of certain non-European cartographers. In most cases, however, maps are improved by being amenable to precise measurement. For this purpose they must necessarily have a scale, that is a known relationship between map distances and ground distances (Fig. 1.1), and for territories of large extent there must also be a projection, which is a mathematical formula for transferring a sphere or part-sphere on to a plane, often expressed on the map itself as a graticule or network of geometrical-looking lines (Fig. 1.2).[10] Perhaps it is worth mentioning at an early stage that scales and graticules on otherwise impressive maps are quite often self-evidently incorrect. The point to note at present, however, is that before the European expansion these appendages, accurate or otherwise, seem to have existed only in parts of the so-called old world, so it has been natural for the concept of the sketch map to be somewhat depreciated by European writers.[11]

Especially, we may add, because there is a sense in which sketch maps do not have very much history. The ability to draw them may be developed by talent and personal experience but not to any great extent by formal training. For the drawing of a scale map, by contrast, a certain level of education is required, if not necessarily the education of a full-time cartographer: in the past, at any rate, other kinds of expert have sometimes proved equally competent with maps – a medical doctor, for instance, though hardly a witch doctor.[12] Eratosthenes of Alexandria (*c.*275–*c.*194 BC) was nicknamed 'Beta', meaning in this case jack of all trades and therefore master of none, but that did not prevent him from being one of the most successful map-makers of antiquity. Since then a good deal of first-class cartography has been embodied in once-in-a-lifetime achievements of relatively short duration. In this respect maps differ from, say, surgical operations or orchestral scores. The result of so much authorial diversity is a certain untidiness in the historical structure of our subject.

Maps and civilisation

Professionalism in map-making has to be paid for, perhaps from the taxes imposed by a monarch or other agent of statehood, perhaps by individuals economically reliant on an underclass of slaves, serfs or proletarians. In one slave-owning society, ancient Greece, cartography was impressively precocious, its exponents comparable with modern academics on almost every count. Their choice of working methods was recognisably scientific, a quality best demonstrated in this case by distinguishing the mathematical framework of a map from the individual features held in place by that framework.[13] As in a present-day university, interdisciplinary frontiers – notably the frontier between geography and astronomy – were clearly drawn but often transgressed. And like many modern scholars, Greek map-makers published under their own names with no hope of immediate personal reward apart from the approval of their peers. They were happy not only to conduct research but also to promulgate their results by writing books in which they criticised their predecessors without rancour while freely giving credit where it was due.[14] These were useful habits historiographically when we remember that there is only one Greek cartographer whose works survive in abundance. Of Claudius Ptolemy (AD *c.*90–*c.*168), royal librarian at Alexandria, it is impossible to say how high his reputation would rank today if his maps and cognate writings were comparable on level terms with those of Eratosthenes, Hipparchus, Marinus of Tyre and others who came very near to dying cartographically intestate. As things stand, Ptolemy must be placed near the top of any all-time map-makers' league table.

Greek academicism included a belief in progress, future as well as past. Its practitioners knew that their knowledge had its limits, and as geographers they

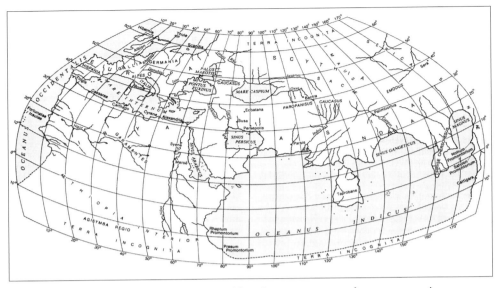

1.3 Claudius Ptolemy, the known world, 2nd century AD, a modern reconstruction.

invented a way of defining these limits with commendable precision. This was a world-wide system of numerical coordinates in which two numbers, latitude and longitude, were sufficient to locate any point on the surface of a spherical earth. Among scholars like Ptolemy who made use of such coordinates (Fig. 1.3) it could be argued, though nobody is known to have done so at the time, that for all its variety geographical science comprises only a single map, and that the output of any particular local or regional cartographer is no more than a fragment of one larger whole. This implicit recognition of terrestrial unity was the most important achievement in the history of map-making – in its way a comforting thought, because as it turned out the Greeks did little to apply their principles by actually measuring, plotting and drawing the earth.

Even in theoretical terms, not all classical cartography could be accommodated by a one-world schema. Some maps, 'niche maps' they might be called, were made to fit particular circumstances, and needed no academic-style support system except perhaps a set of practical instructions to would-be authors. Typical niche maps were the large-scale surveys of urban and rural properties conducted in the Roman empire.[15] There may also have been cartographic surveys of individual roads, canals and harbours. Then there was at least one schematic route map, perhaps originating in the fourth century AD, which anticipated the London underground 'journey planner' by showing the interconnections of the Roman road system without reference to exact distances or bearings (Fig. 1.4).[16] Harder to pigeonhole, especially as it no longer exists, is a map of the Roman dominions displayed in the emperor's capital city, considered by some scholars to have been no more than a diagram or perhaps just an enumeration of places and distances.[17]

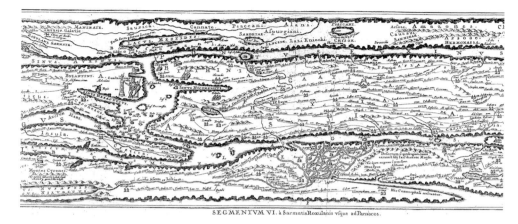

SEGMENTVM VI. à SarmatisRoxulanis vſque adParnacos.

1.4 Parts of Thracia, Sarmatia, Asia Minor, Crete, North Africa and Syria with diagrammatic roads.
The Peutinger table: a 12th- or 13th-century copy of a Roman original from the 4th century AD,
printed in Gorg Hoen, *Historis totius orbis* (Amsterdam, 1658).

Outside Europe the nearest contemporary equivalent to the Graeco-Roman
world was in China, whose earliest maps date from the fourth century BC. Here
scientific principles of map-making were codified for the first time, and several
innovations of cartographic significance gained currency earlier than in the west,
notably paper, printing and the magnetic compass. Most Chinese maps had some
governmental purpose, but the results varied a good deal in character: by and large,
portrayals of small areas looked remarkably like the regional maps of medieval
Europe, while those of the whole empire were more accurate than anything known
elsewhere (Fig. 1.5). Yet there was nothing in China to match Ptolemy's integration
of geography with mathematics and astronomy. In particular, early Chinese cartographers made no use of latitude or longitude. Some of their maps were gridded in
squares, but these followed no single numerical system and were incapable of being
extended to cover the whole earth. So, unlike the Greeks, Chinese geographers did
not allow the same intellectual status to the known and unknown worlds, and even
where a knowledge of foreign lands was available the scale of any wide-ranging
Chinese map would probably decrease outwards from the homeland. On the other
hand the progress of Chinese map-making was never arrested by barbarian
invasions. In eastern Asia the barbarians came from Europe and brought their own
maps with them,[18] by which time home-grown concepts of cartography had been
safely exported to Japan and Korea.[19] And now to strike a 'chauvinistic' note: once
east Asian maps begin reflecting European influence, they cease to be of crucial
importance for the present essay, which in any circumstances could never be more
than highly selective in its choice of examples.

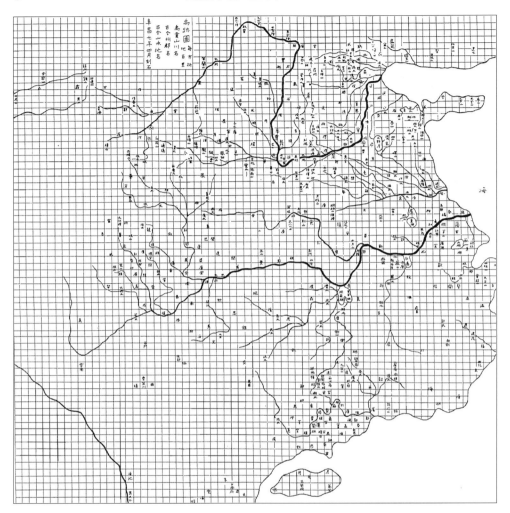

1.5 Chinese map of China, carved on stone, AD 1137. Each square is 100 li, about 50 kilometres. P.D.A. Harvey, *Medieval maps* (London, 1991), p. 17.

Medieval world views

Further west, map-making declined in parallel with the Roman empire. By the end of the fourth century, Ptolemy's library at Alexandria had been destroyed and a Christian church built on its site. Cartographically if not spiritually, this change is hard to see as an improvement. It was among Moslems rather than Christians that Greek science now proved most influential. In the ninth century much of Ptolemy's scientific work had become available to Arab scholars, who were soon making spectacular progress in astronomy, mathematics, metrology, mensuration, geodesy and the development of celestial and terrestrial coordinate systems. A characteristic

1.6 Richard of Holdingham, world map, *c.*1280. Hereford Cathedral.

achievement of Moslem geography was the listing of placenames and coordinates that were different from Ptolemy's and often closer to the truth. At the same time Moslem cartography had much in common with that of medieval Europe. Maps were made by and for an elite, and many of them were remarkable artistic compositions, drawn and coloured by hand on the newly discovered medium of paper. Some were regional, others aspired to full terrestrial coverage. The latter, for the most part chronologically ahead of their nearest counterparts in post-Roman Europe, and certainly different in style, anticipated western practice by depicting a disc-shaped old world surrounded by a narrow belt of ocean. Like all early medieval maps, they owed little to the science of mathematical geography, a fact more surprising in the astronomically-precocious Moslem world than it would have been anywhere else.[20]

From early medieval Europe, more than from ancient Greece, it was a literary scholar's concept of map-making that came down to later generations. The role of the academies and the royal libraries now fell to Christian monasteries endowed by some secular power and under no compulsion to make money from either prayer or work. Perhaps the monks' most novel accomplishment was to produce maps that survived long enough to be seen by modern historians, many as book illustrations, a few as *mappaemundi* exhibited in medieval palaces, castles and churches (Fig. 1.6). Like their lost Greek predecessors, these maps appear in retrospect as neutral contributions to knowledge rather than blueprints for any kind of earthbound action. They also purported to embrace the whole surface of the globe, though unknown areas, instead of being first defined by Ptolemaic coordinates and then left agnostically blank, were now brought to life with various kinds of verbal and pictorial misinformation.[21]

The *mappaemundi* had neither scale nor projection and made no acknowledgement to what would later be regarded as scientific method. They were usually circular in shape but did nothing to exploit the latitudes and longitudes recorded by contemporary astronomers. Time was treated as indiscriminately as space, with Biblical, classical and medieval themes all accompanied on the same map by a large admixture of undatable fiction. Clearly the authors' interests were encyclopedic rather than geographical in any narrow sense. Indeed centuries later an official custodian of the Hereford *mappamundi* would routinely warn visitors that 'this is not a map' – without explaining why in that case an adaptation of the word 'mappa' had eventually superseded 'cart', 'description', 'table', 'plate' and 'plot' as a label for the ultimate post-medieval cartographic experience in every English-speaking country.[22] Meanwhile, whatever may be said for or against the *mappaemundi* there were some monkish scholars, most notably Matthew Paris in thirteenth-century St Albans, who came as close to the modern concept of a regional map as could be expected from a non-surveyor working alone in any historical period (Fig. 1.7).[23]

There were also medieval niche maps associated with architecture, land holding and boundary delimitation, but these appear to survive less abundantly than the

1.7 Matthew Paris, Britain, mid-13th century. See note 23.

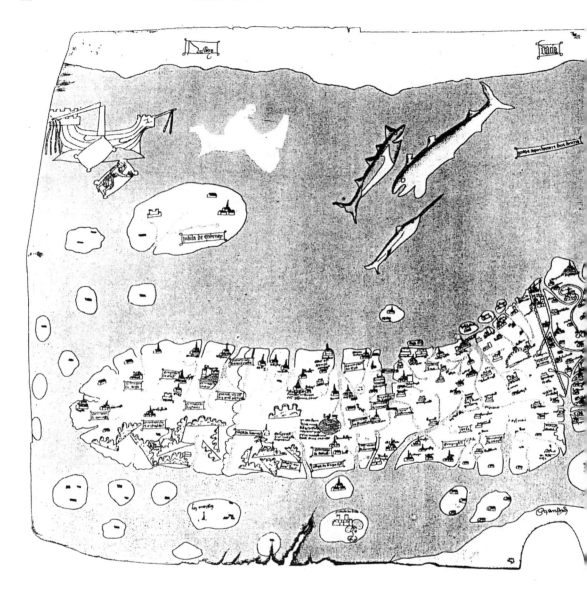

1.8 'Gough' map of Britain, mid or late 14th century. See note 25.

mappaemundi and are so varied in style that it is easier to consider them as historical accidents than as parts of any single tradition.[24] In general, modern scholarship has regarded feudal society outside the monasteries as by nature inhospitable to cartographic enterprise. Where the control of government, justice, taxation and economic activity was devolved among relatively small territorial units, the spatial knowledge requisite for everyday life could be held in people's heads, and in

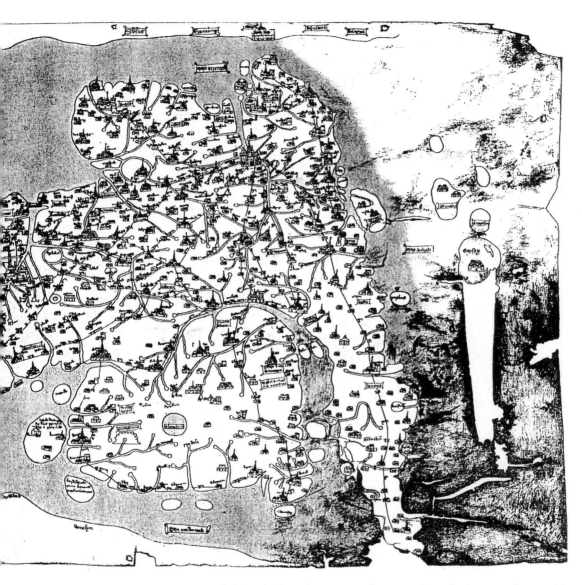

crossing a landscape well furnished with towns, villages and roads the traveller could find his way by word of mouth or even trial and error. This at least would be an orthodox historian's view of medieval space-consciousness. Unfortunately, it disregards such oddities as the remarkably 'modern' fourteenth-century Gough map of Britain (Fig. 1.8), an achievement not wholly explicable by postulating an unusual degree of centralisation in the English governmental system.[25] To repeat: map history has never been a model of tidiness.

On some journeys it was impossible to ask the way. Where land lay out of sight the navigator would have to consult instead the sun, the stars, the winds and, if such

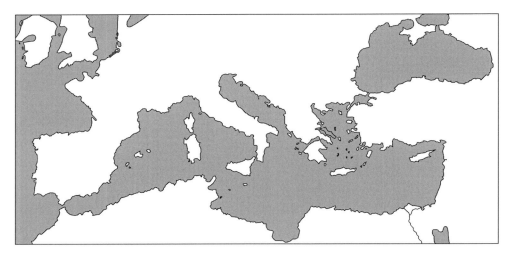

1.9 Typical portolan outline, Mediterranean, Black Sea and western Europe. Angelino Dulcert, 1339.

things existed, a chart. For all the self-sufficiency of the feudal system, there were merchant ships on Europe's marginal seas throughout the medieval period, but it is only in 1270 that light dawns for the map historian when King Louis IX of France was shown cartographic proof of being safely en route from Aigues-Mortes to Tunis. What relieved the king's anxiety on this occasion was almost certainly a portolan chart, a genre soon to become familiar in all the harbours of northern Italy and Catalonia.[26] Here was portrayed on vellum the coastline of the Mediterranean and Black Seas together with parts of Atlantic Europe and North Africa, usually with a wealth of named harbours and in most cases at a high level of accuracy (Fig. 1.9) signalled – for the first time in surviving maps – by the presence of a marginal scale line.

There is nothing odd about the existence of artifacts designed to facilitate the process of navigation: merchants were probably the element in medieval society best able to pay for an accurate map. The problem is why there are no charts of this kind before the thirteenth century. A common explanation is that the newly perfected magnetic compass was essential to the technique of marine surveying. Equally worth noting is the survival of so many specimens from the next two centuries. It would be pleasant, though entirely speculative, to credit King Louis with starting a fashion among royal and noble collectors in southern Europe. For not all such charts were wholly utilitarian: their almost empty continental interiors left ample space for decorative embellishments attractive to the connoisseur, and most of the survivals are evidently presentation copies.

By omitting geographical coordinates the portolan cartographers contrived to ignore the rest of the world. All the same, they cut a strikingly modern figure, though not quite post-modern: unlike some scholars of more recent times, they

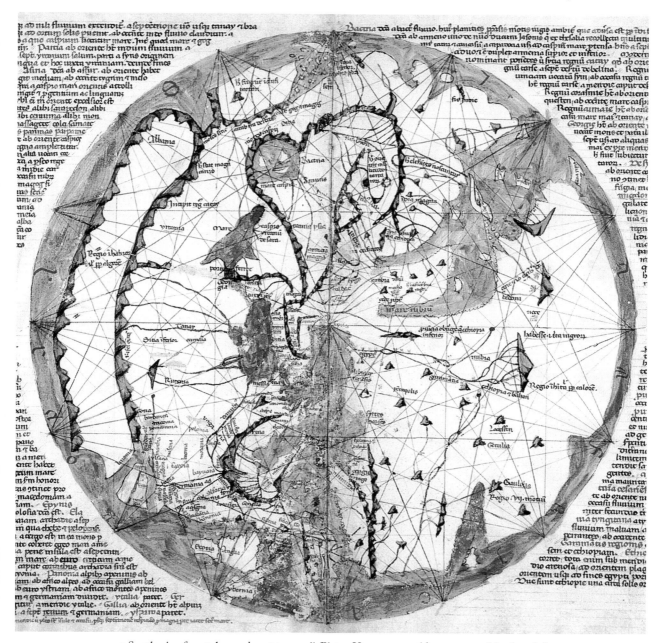

1.10 Synthesis of portolan and *mappamundi*. Pietro Vesconte, world map, *c.*1321. North to left.
John Goss, *The mapmaker's art: an illustrated history of cartography* (New York, 1993), p. 36.

would have refrained from characterising the Hereford parchment as 'not a map'. On the contrary, fourteenth-century compilers were happy enough to bring charts and *mappaemundi* together on the same drawing board. The earliest known fruits

1.11 Parts of Spain and north Africa. From the Catalan atlas, 1375. See note 29.

of such a marriage are maps by a Minorite friar, Paolino Veneto, in *c.*1320 and by a chart-maker, Pietro Vesconte, in *c.*1321 (Fig. 1.10). Here both secular and religious traditions were contributing to a new synthesis; it was world cartography in the spirit of Ptolemy, despite the participants' continuing ignorance of Greek precedent.[27] The difference between classical and post-classical map-making is that

1.12 Fra Mauro, *mappamundi*, 1459, modern reconstruction.

we can compare at least some of the medieval raw materials with their finished product and with each other. There can have been few occasions when the ingredients of a single cartographic compound were so different.

Not that the *mappaemundi* found universal acceptance as doing justice to the world beyond the portolans. Increasingly they were supplemented by reports from contemporary missionaries and merchants, of whom the best known was the thirteenth-century Venetian traveller Marco Polo.[28] The two most famous compilations showing Polo's influence were the Catalan atlas of 1375, a sectional view of the world thought to have been prepared for the king of France by the Majorcan Jewish compass-maker Abraham Cresques (Fig. 1.11), and an equally eclectic world map of 1459 by Fra Mauro, a monk of Murano near Venice, who enlisted the chart-maker Andrea Bianco to help with another royal commission, this time for the king of Portugal (Fig. 1.12).[29]

The European renaissance

With intellectuals, statesmen and merchants thus brought into collaboration, the stage was set for a reform of European cartography.[30] The change when it came was driven by a number of more or less concurrent forces. One early stimulus, in 1406, was the translation from Greek into Latin of Ptolemy's *Geography*. In the course of the next century this work became extraordinarily if belatedly influential, not only as a source of information about capes, rivers, mountains and cities but as a testimonial to the practical importance of latitude and longitude in mapping large territories. Remarkably – because it was by no means compulsive reading – the *Geography* now became one of Europe's earliest printed books, passing through numerous editions (one of them translated as rhymed verse) in Italy and later in Germany.[31] Ptolemy's maps, laid down from the supposedly original coordinates, were printed on paper from woodblocks or copper plates in issues of up to a thousand copies. Meanwhile these new duplicative processes were bringing added circulation to several other kinds of cartography, especially in Europe north of the Alps, where the art of manuscript copying had been less developed than in Italy. Eventually almost every form of word-printing would feature at least an occasional attendant map – textbooks, histories, scientific journals, newspapers, almanacs, advertisements, sales particulars, travel timetables, letterheads, even novels and poetry.

A post-medieval development with implications far beyond the world of publishing was the increased power of central governments at the expense of both ecclesiastical dignitaries and feudal lords. This was due partly to causes themselves associated with surveying and map-making, notably the advent of fortification techniques and weaponry that were too expensive for private ownership. As feudalism weakened, land became more of an economic commodity, to be measured, mapped and valued for the material benefit of its proprietors and tenants. Meanwhile taxation assumed greater importance as a source of government revenues and maps gained further currency as a method of surveillance by a bureaucratic state machine.[32] Cartography rose to these occasions without too much trouble, using new methods that could easily have been developed in the middle ages if anyone had seen the need for them. One was the adaptation of the magnetic compass to land surveying, another was the principle of triangulation, both to be discussed later in this book. By comparison, technical progress in the seventeenth and eighteenth centuries was to be unaccompanied by any great refinement of essential cartographic principles. Unusually by comparison with other industries, the effect of such progress was to make production more expensive rather than cheaper (this remained true until the post-Napoleonic era) but maps were now so much more useful that nobody objected.

As they grew stronger, Europe's nation states became increasingly prone to external conflict, both with each other and with what looked like a more anarchic

1.13 Giovanni Contarini, world map (Venice or Florence, 1506).

world beyond. At home, politicians took special care for the mapping of frontiers; abroad they used cartography for staking rival claims in territories that seemed to have no governments of their own. Hence the new interest by European statesmen in maritime discovery and exploration outside their own continent, an interest also pursued by officially-licensed trading companies and later by national geographical societies. Cartography thus received a powerful stimulus in the seaports of the Atlantic fringe; at the same time the Mediterranean had begun to lose its economic pre-eminence, especially as Africa was proving less commercially attractive than America and south-east Asia.

In 1578 it could be said that 'within the memory of man within these fourscore years, there hath been more new countries and regions discovered than in five thousand years before; yea, more than half the world hath been discovered by men that are yet (or might very well for their age be) alive'.[33] This meant that even as Ptolemy's cartographic influence approached its peak, his errors were becoming harder to ignore. At first sight these two strands of historical development might seem to have been incompatible; in reality, the appetite for maps was voracious enough to digest them both.

It was between the latitudes of forty degrees north and south that the world's land masses were being most quickly and most accurately outlined at this period. At first extra-European coasts were mapped from the observations of pilots and shipmasters, presumably using the methods of the portolan chart-makers sometimes accompanied by attempts at finding latitude and (more rarely and less successfully) longitude. From around the year 1500, in particular, there is a bewildering sequence of original world maps, each with its own interpretation of the new discoveries, some drawn for European heads of state, others seeking a wider market. Examples were by Henricus Martellus, Juan de la Cosa, Giovanni Contarini (Fig. 1.13) and Johannes Ruysch.[34] A cause of cartographic diversity within this group was the desire of national governments to monopolise profitable geographical knowledge for their own citizens. The impulse to secrecy varied in strength from one country to another, depending on how much profit was at stake. This is why the two pioneering nations of maritime exploration, Portugal and Spain, had less impact on world cartographic history than might be expected from the size of their empires.

Later the pace of discovery lost some of its impetus as only the world's less inviting seas and coasts remained to be made known. As a result, science gradually began to rival trade and politics as a motive for subsidising exploration – a trend that might be illustrated, if there were space to do so, from the contrasting careers of the only two seventeenth- or eighteenth-century navigators who deserve to be named in a chapter as short as this. Abel Tasman (1603–59) put New Zealand, Tasmania and much of coastal Australia on the map.[35] James Cook (1728–79) criss-crossed the Pacific, and without actually seeing Antarctica made it seem highly probable that any such land mass must be too cold for European settlement.[36]

1.14 Louis de Hennepin, Mississippi valley and lands to the east (Utrecht, 1697).

Contact with other cultures also taught the discoverers that theirs was not the only map-making society in the world. How much of either geography or cartography they acquired from such encounters is difficult to say. There is certainly little sign that European draughtsmen and engravers learned anything new about style, layout or symbolism.[37] Even the factual information that explorers obtained by tapping native knowledge was not spectacularly wide-ranging by comparison with what they collected themselves. This meant that continental interiors took much longer to map than coasts. Given the impressively realistic outlines that occupy the best mid sixteenth-century world maps, we do well to remember how late some famous inland sites were brought to the attention of Europeans: examples are Niagara Falls (1648), the Casiquiare Canal (1669), the Victoria Falls (1856), Angkor (1866), Zimbabwe (in the original sense of the word: 1868) and Ayer's Rock (1873).[38]

The exceptions to the foregoing generalisation were those continental interiors that could be most easily penetrated along navigable waterways in temperate latitudes. In the space of little more than twenty years, ending in 1682, French explorers travelled mainly by boat from Lake Superior to Hudson Bay, from Lake Ontario down the Ohio to Louisville, and from Lake Michigan along the Illinois, Wisconsin and Mississippi rivers to the Gulf of Mexico. Their findings were mapped in print before the end of the century (Fig. 1.14). At about the same time, and with comparable rapidity, the Rivers Yenesei and Lena began to appear in geographical descriptions of Siberia. Henceforth inland pioneer travellers were generally expected to return with some kind of sketch map, though even in North America it is not until the beginning of the nineteenth century that we find an interior journey being planned with as much care as had been given to the organisation of earlier sea voyages.

The cartographic specialist

However unevenly distributed, the achievements of the renaissance era had been enough to transform the pattern of European map-making. From now on the full-time cartographer was a recognisable member of the communications industry – although as it happened the word 'cartographer' did not become familiar until the twentieth century, one sign of Ptolemy's continued influence being that map-makers were known for so long as cosmographers or geographers. It was now also becoming accepted as normal for the same person to produce maps of many different scales and sizes. At St Dié in eastern France, for example, the output of Martin Waldseemüller (c.1475–c.1521) included maps of the world, the Americas, Europe, the upper Rhine and the province of Lorraine (Fig. 1.15).[39]

Waldseemüller also represents a growing number of specialist map-makers whose work survives mainly in print. Like the paper-maker, the early printer was restricted to sheets of no more than modest size. Above that limit there were two possibilities for physically expanding the cartographic medium, the multi-sheet wall map and the bound volume. In one vital respect these two formats stood opposed: wall displays were exceptionally perishable, books exceptionally durable. The word 'atlas' was first used by Gerard Mercator (1512–94) for a set of his own maps completed the year after his death.[40] There were a number of earlier publications, however, that modern historians would probably be willing to place in the same category. A literary treatise might be illustrated copiously enough for its maps to

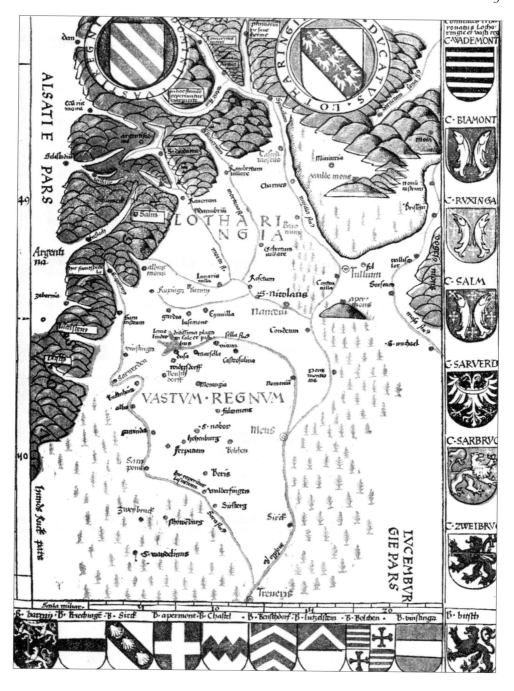

1.15 Martin Waldseemüller, Lorraine, original scale *c.*1:690,000. Claudius Ptolemy, *Geography* (Strasbourg, 1513).

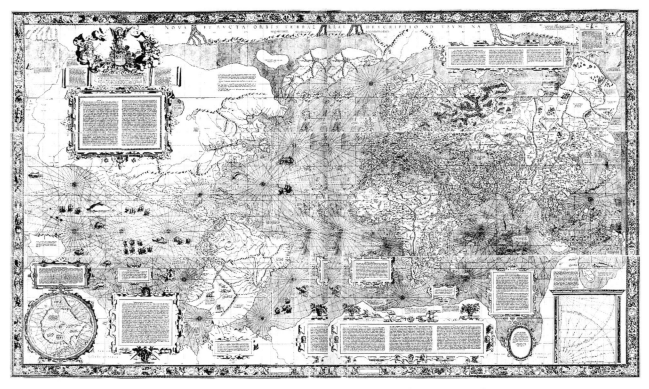

1.16 Gerard Mercator, *Nova et orbis terrae descriptio* (Duisburg, 1569).

seem the most important part of it. Sebastian Münster's *Cosmography* (1544) was an early case in point; another was Pietro Coppo's 'De summa totius orbis'.[41] Then in publishing Ptolemy's *Geography* some editors ignored the text and printed only the illustrations, perhaps augmented by more modern outlines of both old- and new-world countries, as in Waldseemüller's Strasbourg Ptolemy of 1513. Next came what have been called 'improvised atlases', which were collections of maps that had already been published individually, the atlas publisher contributing only covers and a title page. The pioneer of this not very demanding genre in *c.*1560 was Antonio Lafreri, a French engraver settled in Rome.[42] The true world atlas, with maps specially designed for publication *en bloc*, begins only in 1570 with the *Theatrum orbis terrarum* of Abraham Ortelius (1527–98).

From a modern perspective, the atlas format has strong claims to be adjudged the most important physical phenomenon in post-medieval cartographic history.[43] Most of the early maps familiar to present-day collectors were first made widely known through this medium, protected by substantial bindings against the hazards that so often proved fatal to single sheets. Among intending publishers, the preparation of an atlas naturally aroused a desire for completeness, which meant collecting, or if necessary constructing, maps of territories for which no generally accepted image was yet available. In these circumstances no atlas-maker could do more than a small proportion of his own fieldwork – so small that it was really not worth doing any, even if he had been trained as a surveyor. The atlas framework also helped to impose stylistic uniformity upon the output of professional cartographers. By aiming at global coverage it internationalised the practice of map-making at a time when art and culture were otherwise developing along national lines and when Latin was losing its monopoly as a medium for European scholarly communication, with the result that although world atlases appeared in different languages their cartographic content usually looked much the same. The similarity was increasingly a matter of substance as well as style. With Mercator's world map of 1569, in particular, a single view of the planet established itself more firmly than might have seemed possible in the free-for-all of previous decades (Fig. 1.16).[44] At the same time a major influence encouraging cartographic conservatism was the durability and adaptability of the reproductive medium. As copper plates accumulated in the publisher's store-room, it became correspondingly easier to reprint an old image than to engrave a new one. In a broad view of renaissance cartography, however, such economies were of little account.

So how were all these maps to be paid for? A monarch with funds to spare for a national map by one of his own subjects would almost certainly feel less generous towards a foreigner building up a foreign repertoire. The author of a large inter-national atlas would therefore probably have to publish it himself in the hope that his personal reputation would induce readers to pay an economic price. To establish that reputation required both commercial and scholarly acumen, and where motivating forces came into conflict it was usually the quest for profit that prevailed. A point of difference between the two philosophies was that the cartographic businessman unlike the true scientist would seldom let his readers penetrate behind the scenes. His maps displayed a bland, polished surface on which both contem-porary critics and future historians found it hard to secure a foothold. Thus although Mercator, for example, could claim an unassailably academic background, having studied at the University of Louvain with the mathematician Gemma Frisius (1508–55), he was not inspired by the example of his friend Ortelius's *Theatrum* to publish a list of the sources contributing to the *Atlas*. Nor did he explain the theory of his famous projection in which lines of constant bearing were always straight. Like most atlas-makers Mercator could also have blamed com-mercial pressures for his dependence on the least untrustworthy sources available, however dubious, as opposed to information satisfying some *a priori* criterion of reliability. Certainly cartographic entrepreneurs did not often employ their own surveyors or send out expeditions to distant lands.

Despite their limitations, early printed atlases gave the future discipline of map history plenty to think about, including a central cast of characters within which capital, expertise and goodwill were transmitted by inheritance or marriage from one generation to another through the medium of the family business. Publication was mainly located in a block of contiguous countries comprising Italy, Spain, Portugal, Germany, the Low Countries, France and England. In the seventeenth century, world atlases began to grow larger, partly as a result of absorbing national atlases. In Joan Blaeu's massive compilation of 1662, for instance, were incorporated copies of Timothy Pont's surveys of Scotland (1583–96), Johannes Mejer's maps of Schleswig Holstein (published in 1652) and a home-made fourteenth-century atlas of China recently edited for westerners by Father Martin Martini.[45] Blaeu's career ended with the fire that destroyed his Amsterdam workshop in 1672. His successor in the field of multi-volume atlas production was the Venetian Vincenzo Coronelli (1650–1718). It would be misleading to interpret this southward shift of activity as any kind of retrogression, and as it happened Coronelli spent a good deal of his time in France.[46] The fact remains that although his individual maps are full of interest the enormous *Atlante Veneto* that embraces them can only be described as, structurally, a mess: to this day there are few historians to tell us exactly when it was published or exactly how many volumes it contained. The truth is that by Coronelli's time there was too much geography to fit into a single book.

Such problems had done little to diminish the utility of the atlas as a vehicle for either regional collections or 'niche' maps. There were several volumes of charts and island maps that predated Ortelius's *Theatrum*, though none became widely known until Lucas Janszoon Waghenaer's *De Spieghel der Zeevaert* appeared in 1584.[47] Plans and views of towns were published as an atlas by Georg Braun and Frans Hogenberg at Cologne in 1572.[48] Christopher Saxton's county maps of England and Wales appeared in the same format seven years later, followed early in the next century by those of John Speed.[49] John Ogilby's strip road maps covering the same area were somewhat incongruously issued as a folio volume in 1675.[50] There were other atlases that posterity might almost consider labelling as thematic, such as Nicolas Tassin's plans of fortified places in France (1638)[51] and Augustin Lubin's distribution maps of Augustinian monastic houses (1659).[52] Meanwhile, in sixteenth-century manuscript property surveys as in printed literature, text and maps would often share the same pair of covers, though no historian of land tenure has yet nominated the earliest volume deserving to be known as an estate atlas.

Mapping in the age of Newton

If the typical atlas fell some way short of infallibility as a work of scholarship, just where, after *c.*1700, could genuinely scientific ideals like those of Ptolemy be expected to find cartographic expression? Nowhere, one is tempted to reply. And this despite the fact that by the late seventeenth century science in general, with its emphasis on the counting, weighing and measuring of natural phenomena, had begun to cross the limits laid down in ancient Greece. Its modern habitat was not

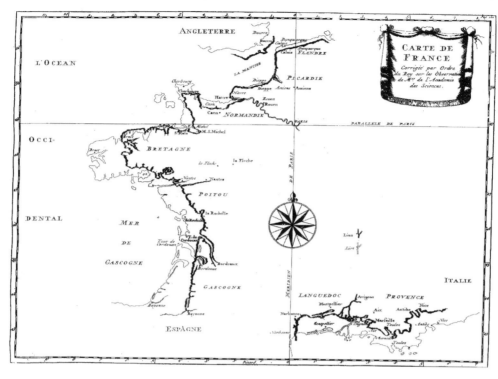

1.17 Coasts of France, by Nicolas Sanson (fine lines) and Jean Dominique Cassini,
Memoires of Académie des Sciences, Paris, 1729, first published 1693.

so much in any ancient university as among certain non-pedagogic seats of learning recently created with government patronage, such as the Royal Society in London and the Académie Royale des Sciences in Paris. Although Isaac Newton's *Principia* ranked first among the achievements emerging from this new milieu, the main focus of activity was in the France of Louis XIV.[53] In particular the French Académie, together with the Paris Observatory, promoted a series of longitude observations and other astronomical researches in coastal or near-coastal situations thought to have been seriously misrepresented on earlier world maps (Fig. 1.17). Among the non-European countries visited for this purpose between 1672 and 1724 were the Canary Islands, the Cape of Good Hope, Cape Verde, China, Egypt, Goa, Guiana, Panama, Siam, and various West Indian islands – a series of episodes still not adequately celebrated in the secondary literature of world cartographic history.[54]

Taken as a whole, however, regional cartography proved somewhat backward in responding to the new scientific challenges. There were certainly geographical hypotheses that needed testing – the existence of a southern continent, for instance – but the process of verification generally cost more than anyone was able to pay. Natural philosophers, like commercial publishers only more so, could not afford to survey whole countries with the accuracy attainable for selected points. The early eighteenth-century cartographer who won most respect from the French Académie, Guillaume Delisle (1675–1726), was better known for purging maps of unwarrantable assertions than for making them more abundantly informative. Such men were not so much discoverers as critics. J.B. Bourguignon d'Anville (1697–1782), for example, became Delisle's most eminent successor without ever setting foot beyond the environs of Paris. Neither Delisle nor d'Anville would risk their reputations with an atlas of the largest size. That was now a task for cartographers of the second rank, like the Vaugondy family in France and Thomas Jefferys in England.[55]

While atlases were continuing to flourish at their own level as a graphic version of the letterpress book, no map publisher had found an equivalent for an eighteenth-century learned society's proceedings or transactions. Yet the familiar modern characterisation of scholarly progress – 'learning more and more about less and less' – had already become applicable to cartography as the best maps continued to increase in scale and diminish in territorial coverage. It was this inflationary trend that degraded the bound volume to a less than universal medium. Pocket mapbooks served a useful if limited purpose for quick reference from around 1600,[56] but at the other extreme not many people had room for an atlas like the one with pages 'near 4 yards large' that belonged to King Charles II of England.[57]

A powerful force behind the enlargement of European map scales was the increasing articulation of the man-made landscape in an age of accelerated population growth. Among the land surveyor's new priorities were farms, tree plantations, coal mines, quarries, mills, kilns, furnaces, all kinds of public works and sometimes complete new villages and towns. Each major engineering project could be expected to generate its own survey, probably unpublished and unlikely to be preserved for future historians, who in any case were slow to take an interest in such maps. But the changing landscape could also claim attention from an eighteenth-century successor to Saxton and Speed. In England the response to geographical change was a new kind of county survey, published in several sheets at a scale (typically between 1:50,000 and 1:100,000) large enough to accommodate hills, water features and almost every visible object in the countryside other than field boundaries (Fig. 1.18).[58] As part of the same movement many growing towns were depicted at even more extended scales either in separate publications or as insets to a regional map.[59] This kind of mapping came to be described by the previously rather unspecific term 'topographical'. It was often performed by a local estate surveyor, civil engineer or architect and financed by collecting subscriptions in advance. Although most examples approximated to a common style, there were inevitably differences of scale, content and merit from one territorial division to another, so that no assemblage of independent regional surveys could be expected to constitute a full-size national map.

The most eminent engineers generally avoided topographical surveying, presumably because it was less profitable than furnishing the landscape itself with

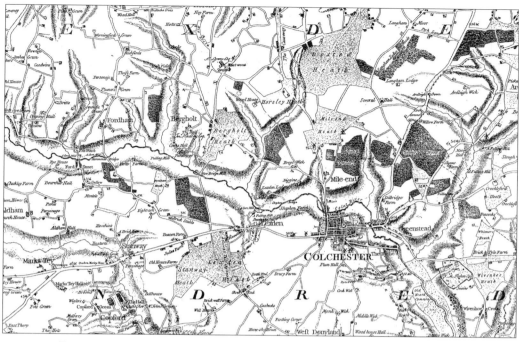

1.18 From John Chapman and Peter André, *A map of the county of Essex from an actual survey,* original scale 1:31,680 (London, 1777).

harbours, bridges and canals. What about their opposite numbers in the military sphere? Unlike the atlas-publishers of civil life, an army in the field could not make new maps simply by copying old ones: its interests were too specialised. So the eighteenth-century military cartographer, however far below later standards in other respects, could at least claim the merit of originality. In fact he and his predecessors had been accustomed throughout the post-medieval period to measuring and plotting individual forts and earthworks with the utmost precision. It was only in the seventeenth century that siege warfare yielded precedence to a more extensive kind of campaign, for which tacticians needed maps of surface relief and land cover stretching far across the horizon whether in their own homelands or in such hitherto unsurveyed territories as French and British North America. The problem in such cases was shortage not now so much of money – that came from the taxpayer – as of time. Eighteenth-century warfare is often seen with hindsight as a 'slow and learned game'; but the most sportsmanlike commander would hardly wait for his enemy to observe a set of latitudes and longitudes before launching an attack. To facilitate a mobile task force surveying had to be done quickly and without the refinement of detail necessary in delineating ravelins and bastions (Fig. 1.19).

A soldier familiar with both high and low standards of verisimilitude would surely be well placed to understand the level of medium-grade accuracy that separated these extremes. Nor was the military engineer's profession as isolated

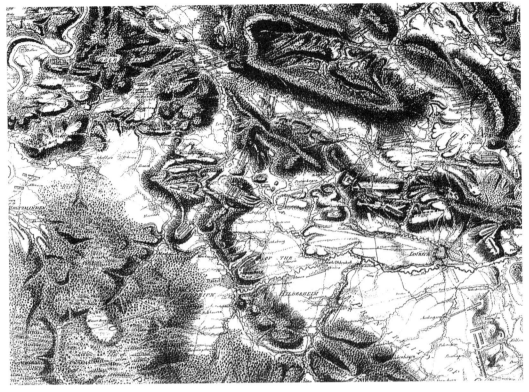

1.19 From William Roy, 'Plan shewing the movements of His Majesty's army in Germany …',
MS, 1761. Scale of original 1:63,360. The Royal Collection, RCIN 733040.
© 2008 Her Majesty Queen Elizabeth II.

from the rest of cartography as one might imagine from its distinctive repertoire of symbols and colours.[60] Security against 'industrial espionage' was not a pressing issue in the formalised gentlemanly warfare of the enlightenment. Military engineers could learn from each other, whether foreigners or compatriots, enemies or friends, and civilian map-makers could draw on the experience of soldiers – indeed some civilian map-makers had themselves not long been demobilised. Perhaps unsurprisingly then, it was an army officer, William Roy, who took a more comprehensive and realistic view of late eighteenth-century British cartography than anyone else;[61] and it was a member of the same profession, General Nicolas-Antoine Sanson, who in 1802 chaired a *commission de topographie* on the reform of French official maps.[62]

The advent of the national survey

Since at least the Elizabethan period cartographers had understood that provincial maps, besides serving a provincial purpose, could also be usefully scaled down and

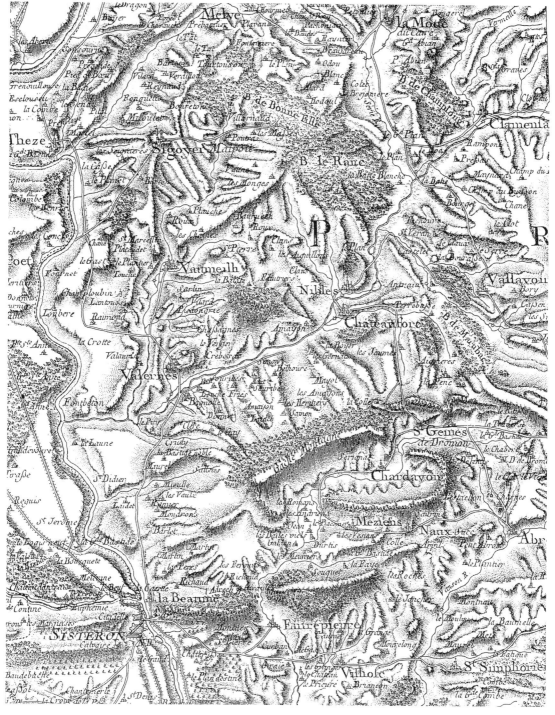

1.20 Europe's first modern national topographical survey. César François Cassini, printed map of France, original scale 1:86,400, 1779, sheet 152, Durance valley.

combined, an early example being the *Anglia figura* that prefaced Christopher Saxton's county atlas.[63] Paradoxically, by improving the quality of regional maps, eighteenth-century surveyors had made this kind of national union more difficult to achieve: for one thing, modern practice required all local operations to be held in place by a general skeleton survey of the highest quality. However, by degrees the effort came to seem worth making. The main consideration in mapping an entire kingdom to this standard was military defence but civilian interests would obviously benefit as well. Such a map should be accurate, uniform in style and content, and capacious enough to show all public roads. Most significantly in the long run, it should also be up to date. Accuracy and uniformity were unattainable without a single nation-wide organisation, and up-to-dateness meant keeping that organisation in existence for an indefinite period.

Here were the seeds of another major upheaval, for hitherto the cartographic efforts of the state had everywhere been uncoordinated and lacking in unity. Pre-eighteenth-century governments had acquired many of their maps as gifts from authors in search of future patronage. Others had been requested from administrators or commanders in outlying territories. Surveyors had been employed and paid by piecework on official projects of one kind and another at home or abroad. Map-making agencies had been created for particular tasks of long duration, as when surveyors-general were appointed to organise the allotment of farmland in new colonial possessions. Historically these arrangements are not always easy to distinguish, and their results present a variegated picture. But eventually in most countries the state can be found surveying its own national territory with its own employees and publishing its own results. By the nineteenth century a topographical map-making department had become as essential to modern statecraft as a police force, a post office or a customs service.[64]

This process first becomes recognisable in France, where a truly scientific map of the whole country had been advocated by the mathematician Jean Picard as early as 1681. Progress was facilitated by current geodetic researches on the size and shape of the earth, only to be obstructed by personnel changes in the administration of the survey and recurrent shortages of money at a time when the high cost of cartographic accuracy was not yet widely appreciated. In the end a national survey expected to occupy 180 sheets at a scale of 1:86,400 (Fig. 1.20) was made between 1750 and 1786 under the direction of César François Cassini de Thury (1714–84).[65] Most other European governments undertook their own topographical surveys in the course of the next hundred years. These are events too complex to be traced in detail, but their chronology shows the effect of military pressures: one cluster of foundation-dates comes in the 1760s, following the Seven Years War; another in the Napoleonic period of the 1800s and 1810s. In the present study the work of the British Ordnance Survey will be taken to represent this great historical theme.[66]

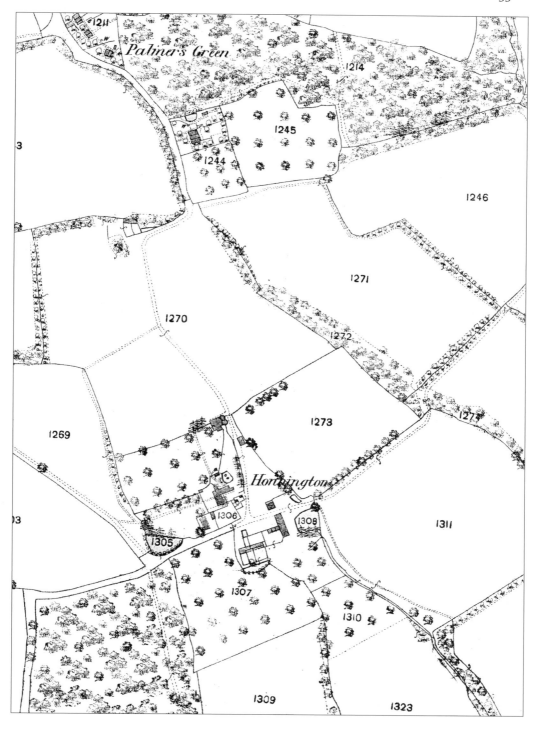

1.21 Ordnance Survey, reduced from 1/2500 county series, Kent, first edition, [1869], sheet 61–8.

Meanwhile the story of marine cartography was following a parallel course as national hydrographic services inherited the role of private chart-makers.[67]

Once a national survey department was at work it would naturally feel impelled to build its own cartographic empire, typically by reaching out along the scale-spectrum in one or both directions. Thus a small gridded map of a country or province, meant as an index to a mosaic of large-scale sheets, might later seem superior on its independent geographical merits to the output of contemporary private enterprise, surfacing after a few editorial changes as a marketable publication in its own right. Diversification into larger scales was not so easy, but there were several good arguments in its favour. In a freshly subjugated colony, law and order could be promoted by officially delimiting and recording the estates allocated to European settlers. In older countries land was coming to be seen by government planners as a profitable and conveniently immovable asset that could not be hidden from the tax inspector's scrutiny. This gave each state an interest in covering its entire territory with what came to be called cadastral maps, on which every property could have its area calculated with unchallengeable precision.[68] The best scale for such calculations, according to an international conference of 1853, was 1:2500 (Fig. 1.21). Topographical and other small-scale maps could then be made by a process of reduction. The results of nineteenth-century and later cadastral operations are accepted by most historians as a standard for the assessment of earlier maps, a practice adopted without hesitation in the following pages.

With official cartography cadastralised, the private land surveyor need be no more than a valuer, the private draughtsman no more than a copyist. In the event this was not quite how matters turned out. What restored the balance, from the mid nineteenth century onwards, was a huge growth in thematic cartography, mainly for scientific and educational purposes, using specialised data drawn from outside the purview of the national survey departments.[69] Such maps finally brought cartography into the world's universities as a respectable academic subject. The present book deals with a slightly murkier past.

CHAPTER 2

A science of facts?

MAPS EXPRESS BELIEFS about the surface of the earth. We should expect such beliefs to derive ultimately from sensory perception of the objects they refer to. A cartographer puts land in one place and water in another because he or some one else has seen them there. But seeing is a complex process, in which raw experience is notoriously difficult to separate from the products of habit, memory and anticipation. In the presence of a new landscape, much of what we perceive is what our brains have led us to expect. This platitude has sometimes been treated as an argument for philosophical scepticism, but in everyday life we accept the direct and indirect components of perceptual experience as a successful partnership. It is generally recognised, for example, that painters and sculptors can benefit from some acquaintance with the science of anatomy. A close cartographic parallel, in early nineteenth-century Britain, was the Ordnance Survey's insistence that its field staff should learn something of geology before attempting to map the form of the land surface.[1] Nobody seems to have said that the hills looked as they looked, that they might still bear the same appearance even if every single hypothesis of geology turned out to be untrue, and that all the surveyor needed to do in these circumstances was use his eyes. The implication here was that without the corrective power of science the sketcher's perceptions would be distorted by some other, anti-scientific influence from outside the realm of direct sensation.

Geographical beliefs of presumptively non-sensual origin are the main subject of the present chapter. Their philosophical status is often hard to define. The term 'theory', for instance, may well exaggerate the amount of mental effort and ability involved. According to an apocryphal book in the Christian Bible only one seventh of the earth's surface is occupied by water.[2] The author had clearly never authenticated the other six-sevenths from his own experience of dry land, so what was the source of his belief? In this case revelation would probably be as good a label as theory. Other opinions may be too unimportant, idiosyncratic or absurd to be dignified as theoretical. Others again may turn out not to have been opinions at all. Consider for instance the improbable boat-like shape of all the woods and forests shown on Samuel de Champlain's map of New England and Nova Scotia (Fig. 2.1).[3] It might plausibly be argued that these shapes, like the circles denoting railway stations on a modern Ordnance Survey map, were meant to be conventional rather than veridical. The most one can say is that diagrammatic-looking symbols may also have been influenced by a certain slight concern for realism, allowing

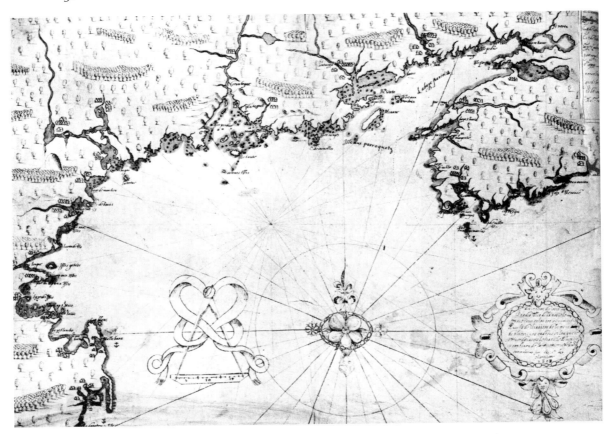

2.1 From Samuel de Champlain, *New England from the Cape Cod peninsula to south-west Nova Scotia*, MS, 1607. See note 3.

Champlain's 'boats' to count as a denial that most natural forests are bounded by straight lines.

Disregarding personal foibles, and concentrating on empirical rules of manifestly wide appeal, we can still try to evaluate cartographers' opinions with reference to philosophical status. Consider the assertion that well-defined peninsulas are more likely to point southwards than northwards. This seems a somewhat low-grade belief because there is no very obvious reason why it should be true. The most impressive and significant regularities of natural geography, we instinctively feel, are those that can be explained by an appeal to forces and tendencies familiar from other areas of human experience – those, in short, that eventuate from some kind of causal mechanism. At any rate map-makers have been at least as responsive as anyone else to opinions rooted in contemporary scientific doctrine. The effect of this doctrine has often been purely negative, for instance in ridding geography of freaks and wonders. More positively, and more challengingly, it could sometimes assert the existence of objects beyond the reach of normal perception.

Perhaps the most obvious contribution of science to cartography is in the thematic mapping of the nineteenth and later centuries, as when a knowledge of physics allowed climatologists to interpolate various kinds of 'isoline' between widely scattered weather stations.[4] In non-thematic cartography, general theories about nature seem harder to isolate. We might begin by seeking them in early scientific literature, but that still leaves the historian with the task of deciding whether the literature in question is likely to have been read by any particular map-maker. Alternatively, and no more satisfactorily, a cartographer's views may just have to be deduced from what he drew and wrote. Here, at the risk of seeming naively unhistorical, we must pay special attention to the errors on his map: non-erroneous opinions about the visible world are not very helpful for our present purpose, being just as likely to stem from observation as from theory.

Another obstacle to analysis is that some theories have been embraced unconsciously, and need a 'thought experiment' to reveal themselves. Take for instance what would generally be considered the most essential feature of any map, the line between land and water. Logically it would be possible for water to become progressively muddier, and land progressively marshier, until one medium merged into the other with no detectable boundary between them. In such a world there would be land and sea but no coastlines. That this never happens is itself a theoretical statement, as we can see by examining the contrary view advanced by the Greek explorer Pytheas in the fourth century BC. In arctic latitudes, according to Pytheas, 'neither earth was in existence by itself nor sea nor vapour, but instead a sort of mixture of these'.[5] Cartographers prefer to believe in a world expressible by lines and dots, and perhaps it is hardly surprising that Pytheas seems not to have made any maps. If he had, his 'mixture' would have presented some awkward problems of draughtsmanship.

The uniformity of nature

Let us now look more closely at a few of the early cartographer's favourite presuppositions. These can be identified at several levels of generality. Some of them concerned the entire universe, some the earth as a whole, others the pattern of the earth's surface either comprehensively or within individual regions. There is thus a hierarchy of theories diminishing in scope until near the bottom of the pyramid we cross a fairly definite logical boundary to reach beliefs concerning one particular geographical feature – beliefs which, given our present concern with error, must often be classified as delusions. The following paragraphs will take note of this progression from more to less general.

At the highest level it is not always easy to separate factual statements from methodological precepts. We may also experience some difficulty in distinguishing one rule from another. A case in point is the philosophical principle known as

Occam's razor, which warns metaphysicians against the unnecessary multiplication of hypothetical entities. One eighteenth-century cartographer modestly described his treatment of the north Pacific as 'no more than to connect together, according to probability, by points, the coasts that had been seen in various places'.[6] Here, on one interpretation, William of Occam had legitimised explaining detached fragments of coast or river as parts of one rather than several landmasses or watercourses.

Another fundamental postulate that may have some kinship with Occam's razor is the uniformity of nature, including human nature.[7] This sober doctrine came to override the more imaginative belief, influential among pre-modern geographers, that distance lends enchantment or, in more prosaic language, that widely separated places probably have less in common than those that are close together. It seemed more scientific in the long run to assume that unexplored localities would resemble those already known, and that familiar concepts – coasts, rivers, hills, forests, towns, roads, bridges and so on – would be transferable to unfamiliar areas.[8] This may help to explain why early maps of different countries so often look disappointingly alike. A difficulty here is that reasoning from known to unknown might not always harmonise with the kind of probability applied to the Pacific in the previous paragraph. A straight coastline could plausibly be assumed to maintain its course. A curving coast would probably continue curved, perhaps to the extent of completing a roughly circular or elliptical outline. This may be why peninsulas like Jutland, Yucatan, Korea, Lower California,[9] Kamchatka,[10] southern Greenland,[11] the North West Cape of Australia,[12] Alaska[13] and perhaps southern India have often been mapped as islands.

To take the same idea a stage further, certain more particular kinds of extrapolation may characterise the work of a single map-maker, for no two people were likely to have lived through precisely the same experiences and there would always be an individual element in the process of reasoning from past to future. A familiar cartographic manifestation of this phenomenon has been described elsewhere as 'personal curvature'.[14] Countless examples could be cited: a striking but little-known case is the idiosyncratically serrated pattern of capes and bays that appears with uniform emphasis all over the world in a map of 1566 by the Dieppe chart-maker Nicolas Desliens (Fig. 2.2).[15]

The uniformity of nature has to do with time as well as space, and here we need something like the metaphysician's 'postulate of quasi-permanence'.[16] If a surveyor's measurements gave two different sets of coordinates for the same mountain-top, it was more likely that he had made a mistake than that the mountain had changed its position between one sighting and the next. This postulate supplies a philosophical basis for what is perhaps the most notorious fact in the whole of cartographic history, namely the average map-maker's habit of borrowing from his predecessors. So might an early chart-publisher justify the copying of sandbanks and water-depths that were half a century out of date,[17] or a famous national survey

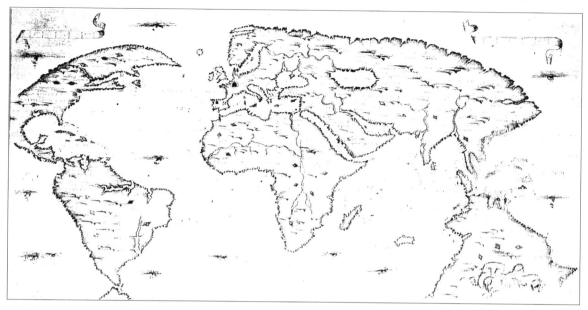

2.2 Nicolas Desliens, world map, MS, 1566. See note 15; also Figs 1.4, 1.7.

department revise its large-scale map of a maritime county without pausing to ask how much of the coastline had been washed away since the previous edition.

From metaphysics we pass to physics. Until modern times the most successful applications of this subject were in the field of astronomy or cosmography. Map-makers representing the whole earth have almost always assumed it to be round – though when latitude and longitude were omitted it might be impossible to say whether round in this context meant spherical or disc-shaped. Astronomical evidence for terrestrial rotundity fell into two categories. The first was the curvature of the shadow cast by this planet on other heavenly bodies, a phenomenon first explained by Aristotle.[18] The second, not firmly established until the invention of the telescope, was an analogy from the shape of those bodies as viewed by terrestrial observers. Other arguments for a spherical earth, such as the image of a ship crossing the horizon, depended on general physics rather than astronomy.[19] Sphericity brought its own reassurances. With an earth of limited size cartography had a reasonable chance of ultimate success: mapping an infinitely extensive surface might well have seemed a task altogether beyond contemplation.

One strange 'theory' that might at a pinch be classed as astronomical had to do with the size of the planets, or at least of our own planet. This was the belief in a simple relationship between linear and celestial units of measurement, as if the Almighty had ensured that in one respect, at least, mankind should escape the consequences of not being very good at arithmetic. Here we must tread carefully, for some terrestrial units have been invented on purpose to form simple fractions of the earth's circumference, as when Peter the Great of Russia defined a verst as one ninetieth part of an equatorial degree. However, such 'global' measures of linear distance have been comparatively rare. The coss can hardly have been chosen by the inhabitants of Hindustan with reference to the earth's dimensions, but eighteenth-century writers were nevertheless convinced that there were just 42 cosses to a degree of latitude.[20] Other traditional linear values for the degree were even rounder: 20 Dutch leagues, 50 Scottish miles, 80 Muscovite miles, 500 Attic stades, 600 Arabian stades, 250 Chinese lis and so on.[21]

Patterns of land and sea

Still in the domain of cosmography, many terrestrial attributes can be derived from the relative movements of earth and sun. In the ancient world continental and maritime climatic influences were not yet understood and air temperature was thought to depend more directly on the sun's elevation, in other words on latitude, than is actually the case. This allowed the earth to be divided into climatically uniform latitudinal belts, of which the tropical and frigid zones were long regarded as uninhabitable, a theory with obvious implications for the mapping of human settlement. Latitudinal zonation was a feature common to many medieval world maps; some of them showed little else.[22] At first sight this looks like a purely climatic issue: there was no manifest causal link that led from north-south temperature differences to the arrangement of the world's major land masses. But zonal concepts do seem to have influenced the pattern of continents and ocean as shown on certain early diagrammatic maps from both Europe and the Islamic countries,[23] and the same predilection for latitudinal thinking may also appear in the comparatively straight east-west coastlines attributed by sixteenth-century cartographers to Arctic North America and Asia. A development of the same idea was Gerard Mercator's hypothesis of a circular mountain range following the parallel of 78 degrees north 'like a wall' (Fig. 2.3).[24]

We now turn to the disposition of features on the earth's surface as affected by purely terrestrial forces. Much theocentric opinion grew out of teleological thought-processes in which the entire planet, if not the universe, was believed to have been made with no other purpose than to provide a home for man. Perhaps this helps to show why so many early cartographers agreed with the Book of Esdras in preferring land to water, and why their seas so often included convenient 'passages' leading between two continents towards a third land mass that was more attractive to human occupation. Since it has been easier at many periods to travel long distances by ship than by road, a benevolent supreme being also had a motive for connecting one sea with another,[25] though where the seas were small he would probably be less reluctant to make an exception: Ptolemy's enormous closed Indian Ocean thus did more violence to the divine plan than a land-locked Caspian or Dead Sea.

2.3 Circumpolar mountain ranges. Gerard Mercator, the Arctic,
inset to *Nova et orbis terrae descriptio* (Duisburg, 1569).

Another consequence of teleological thinking was that if the world had been
created by what human beings could recognise as an act of will, it might surely be
expected to show as much geometrical regularity as many human artifacts. A
famous example of physiographical symmetry is the kind of medieval *mappamundi*

2.4 Printed T-O diagram. Isidore of Seville, *Etymologiarum sive originum libri XX* (Augsburg, 1472).

known as a 'TO' map, in which the arms of the T were the Black Sea (or the River Don) and the Red Sea, and its stem the Mediterranean (Fig. 2.4). (The idea that the O and T shapes were created by a Latin-speaking deity to initialise the words 'orbis terrarum' is not so fully documented, but apparently dates from the fifteenth century.)[26] This would have seemed a fantastical conceit to travellers familiar at close quarters with the apparently random arrangement of capes and bays along the earth's coastlines; and strictly symmetrical representations of the continents were more likely to be regarded as non-cartographic at any historical period when people could see a difference between a map and a diagram. Exactly which periods are involved here is another matter. Herodotus in the fifth century BC may have failed to appreciate this critical distinction: he found it laughable for the earth's land masses to be depicted as a perfect circle, the first among innumerable occasions when cartographic error has been treated as a source of amusement.[27] Matthew Paris on the other hand knew he was wrong to represent the shape of thirteenth-century England by a circle or rosette, and made a point of saying so.[28] The merit of a diagram was to save trouble for the artist. It could also be claimed that geometrical outlines of this kind were not so much statements about the world as professions of agnosticism on points of geographical detail. To which it might be answered that if bold assertions are a way of expressing uncertainty they are not a very good way.

There was at least one attractive theoretical reason for acknowledging the distinction between ideal and reality as applied to patterns of land and sea.[29] For much of the seventeenth century a common belief in the science later to be known as geomorphology was that the earth existed in a state of ruin, the result, as one writer put it, of 'some tremendous convulsion' that had 'burst the strata, and thrown their fragments into … confusion and disorder.'[30] Whatever its early geological history, then, this planet had ceased to be a perfect work of art, and that was why its coasts, rivers, hills and other natural boundaries seldom followed the straight lines or regular curves of the designer's drawing board. It therefore seemed reasonable to consider islands as fragments of the mainland, so that the word 'broken', a morphological expression at least as old as Pliny, remained popular among geographical writers throughout the early modern era even in an apparently non-theoretical context.[31] Some cartographers seemed incapable of drawing a mainland coastline without peppering its offshore zone with islands.[32] Conversely, a catastrophist interpretation of the world's coastal outlines made it unlikely for a single tiny fragment of land to hide itself in the middle of an otherwise empty ocean: there ought to be some other fragments nearby.[33] Hence the diminutive Bouvet Island, sea-girt for almost a thousand miles in every direction, was for long regarded as a promontory of an unknown continent. This theory remained influential in the early nineteenth century.[34] Earlier it had sometimes been given a pseudo-biological twist, as when the Portuguese explorer Pedro Fernandes de Quiros described one large island as 'the mother' of smaller islands in the same area.[35] Once fragmentation and irregularity had become dogmas, the cartographic theorist might sometimes find himself caught out: compare the wavering course of Chesil Beach on Christopher Saxton's map of Dorset with the improbable straightness of the real thing as seen from the Isle of Portland.

The pursuit of symmetry

Breakages aside, perhaps the most striking case of a symmetrical relationship is the supposed resemblance between the northern and southern hemispheres, though in this case the arguments were physical rather than teleological.[36] It was long believed that for the earth to remain in stable equilibrium the proportion of land to sea must be nearly the same on both sides of the equator, and this seems to account for the appearance on many sixteenth-century maps of a huge antarctic continent that no human being had ever set eyes upon. In the same way the mass of Europe, Asia and Africa was thought to be balanced by one or more large land areas lying somewhere to the west, a doctrine especially associated with the Greek scholar Crates of Mallos in the second century BC.[37] When that same transatlantic world was made known by Columbus and others, the hypothesis of a southern continent seemed by analogy to become correspondingly more plausible. If northern and southern land areas did not quite match, other determinants of mass could readily be adduced, among them variations in the density of the earth's crust, in the height of its hills and mountains, and in the depth of the sea.[38] In its simpler form this view was vigorously if tortuously expressed by Mercator:

> Wherefore then the sentence standeth fast, that the machine of the earth is in itself equally balanced, without budging one way or other, and consequently also the sea, which is contained within the bosom of it. And also that the sea is in continual motion, lest it should be corrupted and infect the air, and kill the fishes. Moreover, it washeth the earth both within and without, that all things should be clean and wholesome: all corruption being consumed and dissipated, by motion and attrition. The constitution of the weight of the centre, and of the world importeth all these things: which if they had been known, and examined by the ancients, they had judged almost true, that which is of the situation and greatness of the continent of the new land, found out in our age: and of the meridional continent, not yet discovered, situated under the pole antarctic. For seeing that the lands known to the ancients, are comprehended in 180 degrees of longitude, that is to say, do only possess the one half of the sphere, it was necessary there should be also as much land in the other half. And seeing that Asia, Europe and Africa, for the greater part, are situated beyond the equinoctial, towards the north; it was necessary as great a continent to remain under the pole antarctic, which should be equivalent in the other lands, with the meridional parts of Asia and new India, or America.[39]

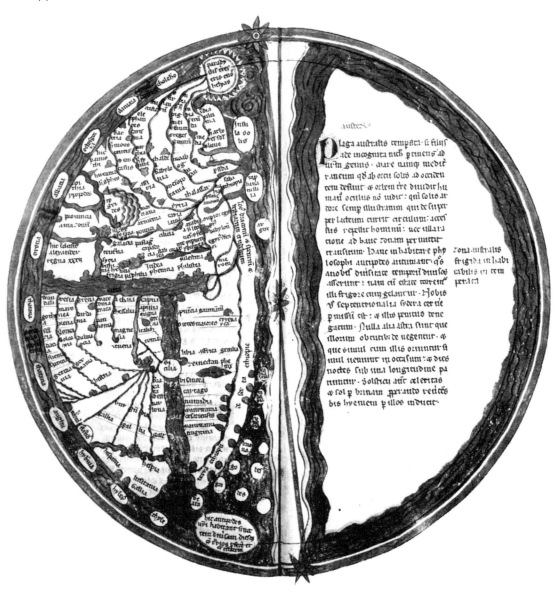

2.5 Near-symmetrical arrangement of marginal islands. Lambert de Saint-Omer, Liber Floridus, world map, 13th century. George Kish, *La carte: image des civilisations* (Paris, 1980), pl. 27.

Symmetry was by no means an exclusively western preoccupation, being well developed in at least one Islamic world map as early as the tenth century.[40] In Europe Mercator's view remained influential well into the age of enlightenment.[41] One late believer with good scientific credentials was James Cook's contemporary Joseph Banks, though unlike Mercator Banks wisely admitted being unable to substantiate his opinion. Even if inapplicable to entire continents, the argument for regularity might yet be adapted to some more localised scale. On many medieval world maps a tri-continental land mass is surrounded by small islands set at regular or almost regular intervals (Fig. 2.5).[42] In post-Columbian times there may have been a feeling that the longitudinal widths of North America and Eurasia should be more or less comparable: this at any rate is how the two land masses appear at high latitudes on world maps of the early seventeenth century.[43] Within the continents, some later students of North America looked for a Pacific-coast counterpart to the Gulf of Mexico[44] and a Canadian version of the Magellan strait,[45] thus acknowledging that not all such theories depended on a single world-wide physical force. Other appeals to symmetry for its own sake were the Pillars of Alexander (Bosphorus) balancing the Pillars of Hercules (Gibraltar),[46] a hypothetical lake in western Brazil to match the Ptolemaic lakes in central Africa[47] and an equally fictitious river in western Australia comparable with the American Mississippi.[48]

Heights and depths

In a vertical direction, early geomorphic science would seem to have been governed by a famous Latin maxim about nature not taking jumps. Thus land was assumed to slope gradually upwards from the ocean floor, and this helps to explain why depths were so assiduously plumbed by early navigators: where the water shallowed, a coastline was probably not far distant, even if no one could see it. One thinks of the ship's captain who deduced from a sounding of seventeen fathoms that he was little more than six leagues from the nearest coast, 'because that precise depth of water uses to be found in those seas at that distance from land'.[49] There was also a chart telling readers that however many fathoms they sounded, they were the same number of leagues from land.[50] By a similar mode of reasoning, the altitude of relief features was expected to throw light on the issue of continentality versus insularity: the higher the land, the more extensive the land mass. Eighteenth-century theorists sometimes forgot that the world's reputedly tallest mountain, the peak of Tenerife, was in an island no bigger than Cheshire.[51]

The early cartographer's propensity for interpolation and extrapolation was easily adapted to the portrayal of hills. There are linear mountain chains on what is often cited as the world's oldest extant regional map, showing northern Mesopotamia in c.2300 BC.[52] Similar ranges appeared in almost every kind of medieval cartography, including *mappaemundi*, portolan derivatives[53] and early editions of Ptolemy. Many subsequent maps featured lines of profile symbols, the width of the line being the same as the width of one hill. A good example for present purposes, because it must surely have been uncontaminated by empirical knowledge, was the network of Antarctic ranges shown in a world map of 1565 engraved by Ferando Bertelli (Fig. 2.6).[54] A similar linearity characterised the 'hairy caterpillars' that became commonplace on small-scale maps in the eighteenth century when hachures

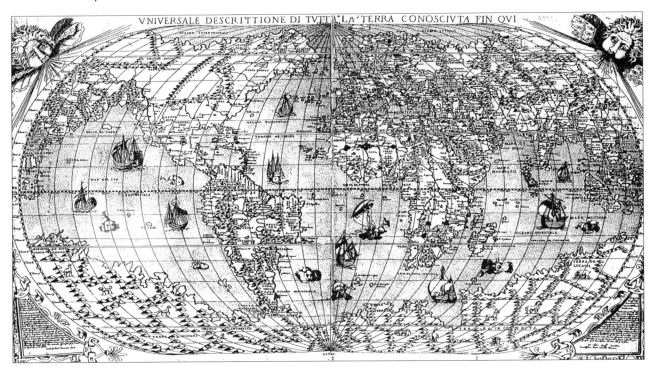

2.6 Ferando Bertelli, world map (Venice, 1565), based on Jacopo Gastaldi, showing
mountain ranges in Antarctica.

gradually superseded hill-profiles as a method of portraying relief. Sometimes these
inferences turned out to be correct, as in New Zealand when Cook combined his
fragmentary sightings of the Southern Alps into one almost unbroken range.[55]
Other chains were to prove delusory, like the Black Hills of Dakota on certain
nineteenth-century American maps.[56]

Apart from the metaphysical or aesthetic attractions of simplicity and singularity, long narrow hill formations may have reflected beliefs about the underlying
structure of the earth. Leonardo da Vinci compared mountain ranges with bones,[57]
and in the same spirit a non-existent range in southern Africa was given the
alternative name of 'Backbone of the World'.[58] Chinese geographers even found
terrestrial equivalents for the larger intestine and the bladder.[59] Like 'broken' islands,
bone-shaped uplands may also have been inspired by maps drawn from actual
observation in other regions, and particularly those depicting the mountain ranges
around the Mediterranean Sea. Their popularity gained strength from the opinion
that most river divides were mountainous rather than just gently elevated.[60] There
was also a teleological belief that the Andes and other linear mountain systems had
provided the nations of mankind with convenient political boundaries (Fig. 2.7).[61]
Even in the nineteenth century this comforting expectation could cause mountains
to be mapped in places where they did not exist.[62]

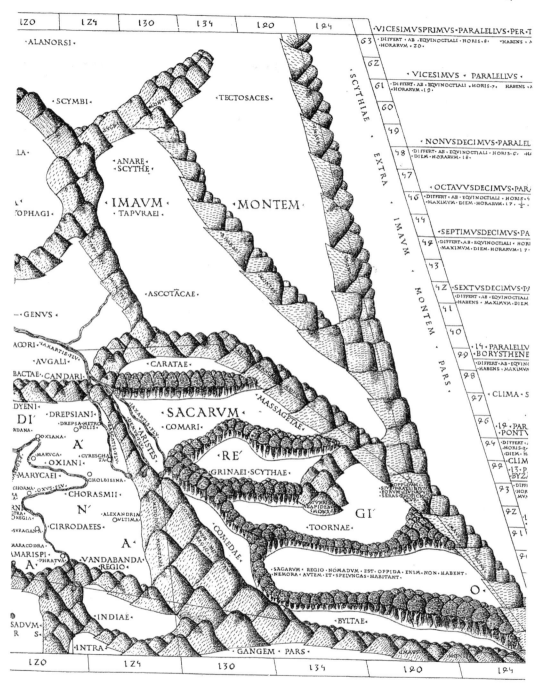

2.7 Mountains as political boundaries. Claudius Ptolemy, *Septima Asiae tabula*, from *Geography* (Rome, 1490).

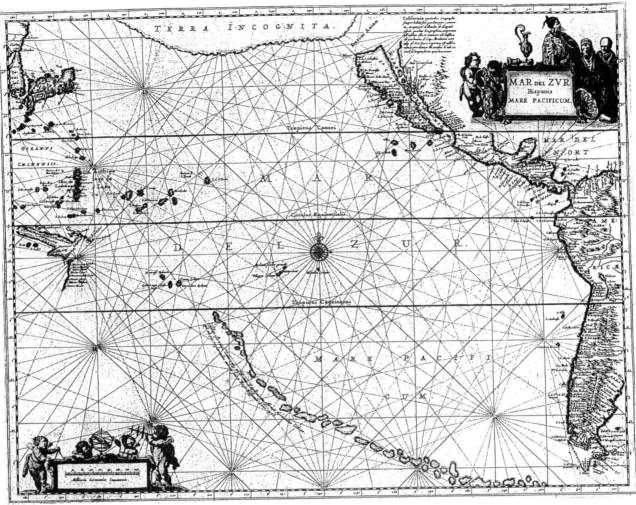

2.8 Chain of non-existent islands. Joannes Janssonius, sea chart of the Pacific (Amsterdam, 1650).

From which it must have seemed natural to infer that complete land masses, as well as mountains, were likely to be arranged in lines. The word 'chain' was applied to Indonesian islands early in the sixteenth century,[63] and in 1774 Cook used the same phrase when wondering if a succession of islands, sandbanks, and reefs might extend from New Caledonia to New South Wales.[64] A variant formula, 'range of islands', helped to make this orographic analogy explicit.[65] Examples abound on seventeenth-century maps: one of the longest was carried by Jan Jansson for about 4500 miles through a part of the South Pacific that happens to be almost devoid of real islands (Fig. 2.8).[66] Less often, a map-maker might show some inclination to arrange his woodlands as a chain.[67] Fra Mauro for example in 1459 caused a number of territorial boundaries to coincide with rows of trees. Similar lines appeared in central Asia as mapped some thirty years on by Ptolemy's editors at Rome[68] and later still in Olaus Magnus's representation of northern Europe.[69]

Atmosphere and hydrosphere

As east-west zones began to lose theoretical appeal, attention turned to the divergent meteorologies of land and sea. One natural laboratory for testing theories on this subject was the eastern end of the passage believed by many geographers to link Atlantic and Pacific along the Arctic edge of North America. Were all the inlets from Hudson Bay backed by a continental land mass, or did a large ocean lie within easy reach? It was here that Humphrey Gilbert conceived his 'infallible rules for the shortening of any discovery, to know at the first entering of any fret, whether it lie open to the ocean, more ways than one, how far soever the sea stretcheth itself, into the land'.[70] Explorers on the west side of the great bay were struck by the range of its tides and by the strength of tidal currents from the west, both suggesting the close proximity of open sea (presumably the Pacific Ocean) not far into the straits that they expected to lead them westwards across central Canada. Similar inferences were drawn from the seemingly anomalous duration of the ice-free period in this region – longer on the north of the Hudson Bay coastline than in the cul de sac of James Bay six hundred miles nearer the equator.[71]

A contrary process was to infer the presence or absence of nearby landmasses. Water free from strong currents was considered unlikely to adjoin a large continent.[72] Another guide to continental and oceanic influences was chemical analysis, as some navigators recognised by weighing samples of sea water to test its salinity.[73] Arguing along opposite lines, Cook admitted that a strong and persistent swell in calm weather was evidence against the proximity of a coast in the direction concerned.[74] From this premise George Bass deduced the existence of both Cook Strait and Bass Strait in Australasia before any European had sailed through them.[75] This was one of the few 'signs' that withstood Cook's scepticism, albeit in a negative sense. He was certainly reluctant to accept clouds and fog as indicators of land not far over the horizon. Another live issue from the same period was the significance of ice. For John Davis in 1595 'the experience of all that have ever travelled towards the north' (including himself) had shown that the sea never freezes.[76] Most of Cook's contemporaries thought the same, regarding any mass of ice encountered at sea as necessarily derived from the lakes and rivers of a nearby land mass. 'The more ice we find at sea', wrote one of them, 'the more land we expect to discover'.[77] Such expectations were usually disappointed. It was also considered possible to deduce the existence of banks and soundings by taking the temperature of the water above them.[78]

On land, where river floodplains had nurtured so many ancient civilisations, the basic properties of a fluvial drainage system must have been understood at an early date. If a traveller encountered flowing water he knew that, unlike the surface of a

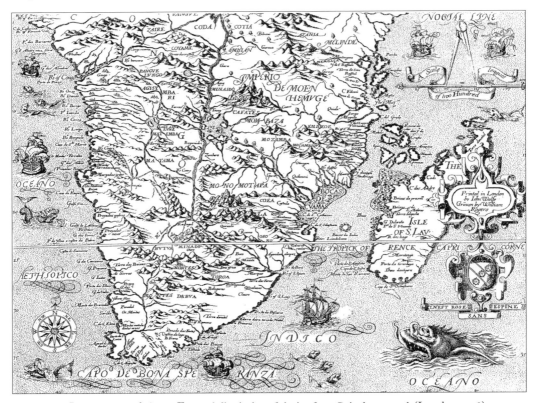

2.9 Interconnected rivers. From *A discription of Aegipt from Cair downwards* (London, 1598), based on Filippo Pigafetta, 1590.

lake or pond, it could be expected to continue for some distance outside his visual field, its direction of movement guiding him towards a river's upper and lower courses and ultimately to its source and mouth. If drainage channels grew steadily wider in a downstream direction, it seemed to follow that, other things being equal, the approximate size of a land mass might be deduced from the breadth of its rivers in their lower courses.[79] The fresh waters at the mouths of the Orinoco persuaded Columbus that his third voyage had brought him to a large continent and not just another West Indian island.[80] Inverting this argument, some eighteenth-century geographers deduced from the apparent lack of large rivers in Australia that the continent must be made up of several islands or that it possessed its own mediterranean sea.[81]

It would also seem reasonable at any historical period for a map-maker to join up fragmentary river observations into a single uninterrupted stream. Such tidy-mindedness had its dangers, especially in the presence of right-angled bends (the geomorphologist's 'elbows of capture'), but at least it was safer than interpolating along a coastline that might turn out not to exist. On maps showing relief as well as drainage the force of gravity had to be respected. One of its consequences was that except on very flat land rivers were more likely to branch upstream than

2.10 Inferential dendritic river systems east of the Caspian Sea. From Abraham Ortelius, *Asiae nova descriptio* ([Antwerp], 1570).

downstream, though maps of authentic deltas in the tideless Mediterranean may have encouraged early cartographers to be more generous with distributaries elsewhere than subsequent experience would warrant (Fig. 2.9). Sometimes the generosity was deserved, as when a seventeenth-century geographer counted the mouths of the River Amazon and found them to number eighty-four.[82]

Collectively, as individually, rivers were a never-failing subject for theoretical argument. A favourite comparison was between river systems and the veins and arteries of an animal's body, or the twigs and branches of a tree.[83] Thus tributaries were generally believed to enter a main stream at an acute horizontal angle, and a river's lower branches were expected to be longer than those nearer the source. Such 'dendritic' drainage networks gave a plausible appearance to many early maps, though where the actual pattern was 'trellised' rather than dendritic nature's truth could prove stranger than the map-maker's fiction: examples of plausible but incorrectly branching tributaries occur in Abraham Ortelius's maps of the upper Nile, the upper Amazon, and the rivers flowing westwards into the Caspian Sea (Fig. 2.10).[84] It was also often thought that drainage to a more or less straight coast would follow evenly-spaced parallel streams of uniform length and sinuosity. In land masses considered as a whole, according to what has been called the 'pyramidal height of land theory', radiating rivers were expected to rise within a short distance of each other.[85] If not, watersheds would probably run through the middle of a land mass rather than along one side.[86] And if this assumption failed, the same reasoning might still be true of individual rivers: in the Mississippi basin, the Missouri was assumed by some cartographers to rise as far to the west as did the Ohio to the east.[87]

Another less than universally valid theory was that large rivers rise in the highest mountains of their watershed, a belief manifested in Christopher Saxton's otherwise staunchly empirical maps of the English counties.[88] Nor was it true that even the weariest river wound somewhere safe to sea: it might dry up instead, or lose itself in a swallow hole. Still less credible was the belief that rivers normally rise in lakes, or at least spring fully-formed from well-like orifices big enough to be mapped with their own conventional sign.[89] This may have been connected with the apparent inability of the earth's rainfall to feed all the streams on its surface, and the resultant need for supplementary underground water to make good the deficiency. Until it was refuted in the late seventeenth century[90] this doctrine would allow the same lake to send rivers into two different oceans – an attractive possibility for early settlers in Virginia hoping to navigate their way due westwards from Atlantic to Pacific.[91] It was only in the late eighteenth century that such two-way drainage systems came to seem unusual – unusual enough for 'Lake of the Two Discharges' to be worth adopting as a placename in the vicinity of Hudson Bay.[92]

Life-forms and life-styles

A major influence of biological theory on early map-making was the idea of limits. Seas and oceans were accepted as the only serious barrier to the movement of many animals (including primitive man) who could be expected to occupy a single habitable land mass from end to end without leapfrogging further. This was Humphrey Gilbert's reason for accepting the separation of Asia and North America: if the two continents were joined, why had no Scythians or Tartarians been found in Canada?[93] Climatic limits were equally restrictive. Ptolemy believed that 'all animals and plants that are on the same parallels or parallels equidistant from either pole ought to exist in similar combinations in accordance with the similarity of their environments'. The animals he had in mind were the rhinoceroses, elephants, and dark-skinned humans reported from both northern and southern hemispheres.[94]

2.11 Crooked representation of straight Roman road (Watling Street), original scale 1:195,000. From [William Smith], *Warwici comitatus descriptio* (London, 1603).

Latitudinal limits had a long history ahead of them: 'no grass beyond 65 degrees' appears on a map of 1775.[95]

More interesting were possible signs of a ship's proximity to an unseen coast in the form of plants and animals whose dietary or other environmental needs prevented them from straying more than a certain distance from the shore.[96] In particular, different bird species were thought to range over varying mileages by sea or land; they might show mariners the direction as well as the proximity of the nearest coast by the simple act of flying towards it. The same argument was reversed along the western side of Hudson Bay, where seals and whales were thought to prove a close connection with the Pacific.[97] Like their meteorological equivalents, biological signs of land became less plausible as navigators gained more experience. As so often, Cook was ready with an apt comment, inspired by West Indian beans cast up by the sea in western Ireland: to Pacific explorers, he said, such plant fragments would have proved land to be just over the horizon, 'so apt are we to catch at everything that may in the least point out to us the favourite object we are in pursuit of'.[98] Biological inferences were not confined to the issue of land versus sea. In the 1820s the explorer of Guiana, William Hilhouse, deduced from the fish species present in the Corentyne River that its waters would not lead him to the main divide of the Sierra Accarai.[99]

In the realm of human geography, man's competence as a planner has given many landscapes a degree of regularity not to be expected in the physical world. At

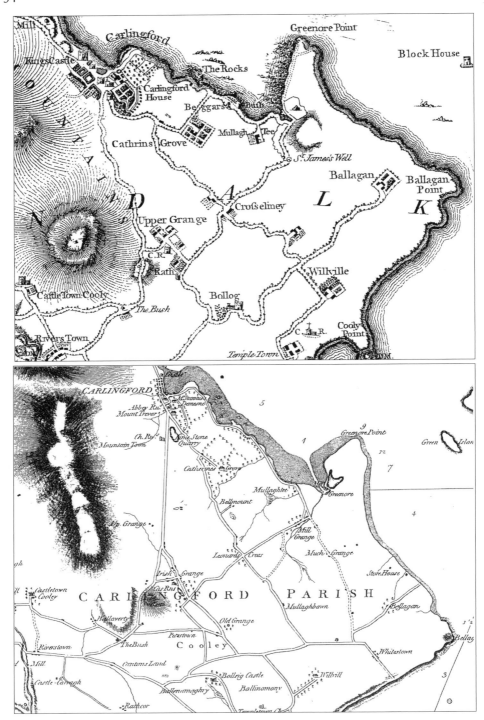

2.12 18th-century road, Carlingford to Grange, Co. Louth, Ireland, on two printed county maps. *Above*, Matthew Wren, 1766, original scale 1:36,206. *Below*, George Taylor and Andrew Skinner, 1778, original scale 1:42,240.

2.13 Diagrammatic political boundaries. Gregor Reisch, world map (Strasbourg, 1515).

cadastral and architectural scales this is most obvious in the graphic representation of buildings, yards, gardens and streets. In small-scale maps of Europe it appears in the alignment of what appear to be Roman roads. Other route networks presented a different problem. Marinus and Ptolemy, when reckoning straight-line distances, both made a considerable deduction from road mileages to allow for bends and turns, sometimes as much as a half.[100] Later it could prove easy to take the wrong decision on this point. In Warwickshire William Smith's Watling Street was less geometrical than it ought to have been (Fig. 2.11), a mistake rectified half a century later by a cartographer with superior antiquarian credentials.[101] By that time the distinctive character of England's Roman roads was well known; but with no comparable historical tradition to authenticate them, the new roads of the early eighteenth century were sometimes interpreted too 'naturalistically': a localised but effective example from Ireland is the genuinely straight minor road due south of Carlingford as shown with varying accuracy on maps of County Louth (Fig. 2.12). At this period, changes of direction were thought more likely to curve gently than to bend sharply, as one can see by comparing the shapes of partly interpolated roads in a typical eighteenth-century English county map with those in a more fully-surveyed contemporary estate plan of the same area.[102]

The mapping of political and administrative boundaries had a history not unlike that of roads. They appeared as straight lines on a number of medieval world maps that made a point of showing coasts and rivers with plausible irregularity – the Cotton map, the Hereford map, the Aslake map and the maps of Henry of Mainz

2.14 The legendary island of Frisland (Rome or Venice, *c.*1565).

and Ranulf Higden, to mention only those illustrated in a recent standard work.[103] Familiar enough in modern colonial boundaries outside Eurasia, such straightness was neither typically Roman nor typically medieval (Fig. 2.13). Perhaps it was derived from some remote archetype on which political geography had been conceived in diagrammatic rather than cartographic terms. Whatever their origin, by the sixteenth century such artificialities had come to look improbable.[104] Human creations of the largest size were now generally expected to resemble natural features in their geometrical untidiness, and political divisions not yet surveyed in detail were increasingly drawn as 'cellular' rather than rectangular in outline, a shape recognised in the historian's term 'fish-scale' for the administrative diagrams of Ming-dynasty China and also exemplified in milder form by many early maps of Japan.[105] There were limits to the impulse that produced the cells: as with the majority of coasts and rivers, the typical interpolated boundary was cautiously kept free from tight recurving salients or embayments – despite the probability that these would turn up sooner or later on the ground.

There is less to say about the *a priori* mapping of towns and villages, partly because geographical awareness of human habitation seldom assumed a theoretical character until the twentieth century. By the time latitude began to appear on post-Ptolemaic maps, climatic barriers to human occupation had lost some of their credibility: Mercator's contemporaries placed urban settlements within five degrees of the equator (including Manoa in Guiana, said to be the largest city in the world)[106] and at least three degrees inside the arctic circle. This trend was encouraged by a general disposition among map-makers to exaggerate the prevalence of urban life. Interpreting physical or territorial names as town names was a common mistake, well illustrated in maps derived from the Irish surveys made by Robert Lythe in 1567–71. In Jodocus Hondius's separate map of Ireland (1591), habitation sites were invented for 'Suillivant [Beare]', 'Suilivant more', 'Coner Donne' and 'Bagnal', though on any of Lythe's own drafts these would have been correctly shown as family names with no symbol attached.[107] Further afield the same tendency was exemplified by the seven fictitious cities of Antillia and Cibola and by the twelve towns recorded in the imaginary island of Frisland (Fig. 2.14).[108] In 1593 Cornelis de Jode named half a dozen towns in the interior of modern Canada.[109] There was even a sixteenth-century settlement deep inside northern Greenland.

Within a narrower spatial framework, the idea of repeated patterns occasionally found an echo in urban morphology. A possible example is the representation of different towns as identical arrangements of identical buildings. Thus Jerusalem in some sixteenth-century Bibles was indistinguishable from contemporary views of Lübeck and other north European cities.[110] Elsewhere different towns were made to look exactly similar even within the covers of a single book. A notable offender was Hartmann Schedel, who in 1493 made do with seventeen woodblocks in

2.15 Stylised town view, Würzburg, Bavaria. Hartmann Schedel, *Liber cronicarum* (Nuremberg, 1493).

separately representing a total of fifty-seven towns (Fig. 2.15).[111] (An alternative view would be to classify these symbols as very large conventional signs of the type discussed below in chapter 17.) When town plans became more locationally specific, with individual public buildings recognisable, they could still be distorted, even in the early seventeenth century, by a process of idealisation in which streets became unrealistically straight and wide, and perimeter walls unrealistically circular.[112]

The cartographer as theorist

Having reviewed a number of theoretical beliefs, we can now summarise their influence on map-making in more general terms. In most maps, as in most individual acts of perception, fact and theory are intermixed. Cartographic theorising seems to have been especially active in two contrasted sets of circumstances. On the one hand it could help to relieve a shortage of raw geographical data, as in pre-modern Europe where the cognitive vacuum was too large to be filled by any process with claims to empirical authority. A later incentive to theory-construction came from the progress and prestige of Baconian scientific method during and after the seventeenth century. In its general outlook the new spirit was cautious enough, but there had now accumulated a large mass of evidence, some of it conflicting, on which many geographers felt obliged to adjudicate. This could never be a mechanical operation. Thus reported signs of land would not necessarily in themselves persuade a post-medieval map-maker to show a coastline just across the observer's horizon. More probably, inference from meteorological or botanical phenomena would be combined with other kinds of testimony, for example a dubious sighting of the land itself on some previous occasion. Such inferences would then affect the direction in which a fragment of authentic coastline was extrapolated, and perhaps the amount of land that might reasonably be assumed to lie behind it.

The progress of natural science inspired a new spirit of audacity in some writers who might have done better to follow the herd. This happened particularly in eighteenth-century France, where Philippe Buache and other theoretical geographers gave special attention to the coastlands of the north Pacific, linking Mexico, Alaska and a peninsular Greenland with a single almost unbroken chain of mountains that separated mainly non-existent seas, bays and lakes of various shapes and sizes (Fig. 2.16).[113] Among individuals, however, the role of theory was largely a matter of temperament. Cook for instance was ready to approve a world chart marking oceanic observations of tree fragments, seaweed, pumice stone, icebergs, turtles and various bird species, and on occasion even allowed a draughtsman to write the phrase 'signs of land', but he would never countenance the interpolation of an unobserved coastline, insisting that nothing should ever be mapped where it had not been seen.[114] Louis de Bougainville in 1767 made the same point when

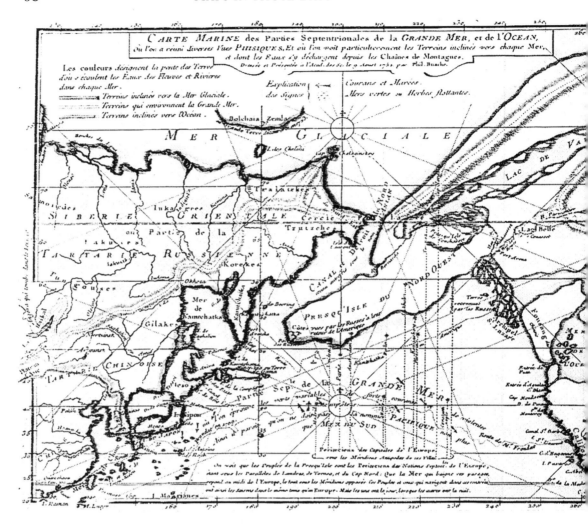

defining geography as a science of facts.[115] Eventually theorising was driven into retreat by the pressure of accurate observational records, but it never died out altogether. Even in twentieth-century Britain the wrong colour could be put on a layered Ordnance Survey sheet in the mistaken belief that a ring contour must necessarily represent an elevation and not a depression.[116]

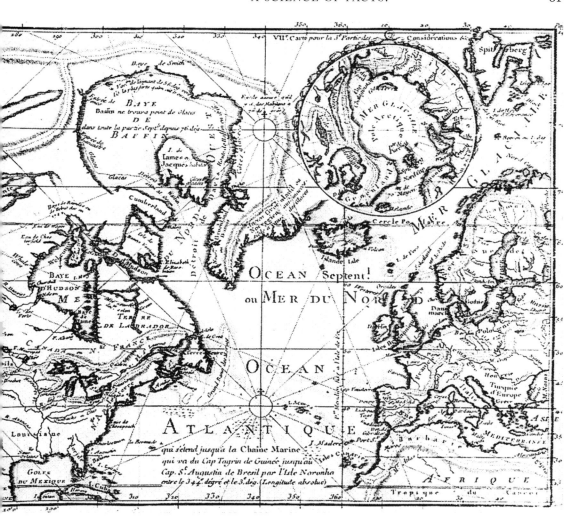

2.16 Actual and theoretical mountain ranges and water bodies. Philippe Buache,
Carte marine from *Considérations physiques et géographiques* (Paris, 1753).

CHAPTER 3

So many guess plots

UNTIL THE ADVENT OF photography, maps had to be drawn by coordinating eye and hand, and the easiest way for the draughtsman to represent external reality was by imitating what he saw. Unfortunately what he saw was not a map but a view. Consider some of the differences. In commonsense terms, a view is less abstract, less intellectualised and on the face of it more realistic than a map. It is what people apprehend directly as distinct from what their minds build out of more primitive experiences. However, it is the process of 'building', one hopes, that ultimately makes the map more reliable. This for instance is what John Byron saw off the coast of Brazil in 1764:

> All the people upon the forecastle called out at once land right ahead, I looked under the foresail and upon the lee bow, and saw it to all appearance as plain as ever I saw land in my life. It made at first like an island with two very scraggy hummocks upon it, but looking to leeward we saw the land joining it and running a long way to the south east. We were then steering south west. I sent officers to the mast head to look out upon the weather beam and they called out immediately they saw the land a great way to windward. I brought to and sounded and had 52 fathom. I now thought I was embayed and as it looked very wild all round I wished myself out before night. We made sail and steered east south east. All this time the appearance of the land did not alter in the least. The hills looked very blue as they generally do at some little distance in dark rainy weather, and many of the people said they saw the sea break upon the sandy beaches.

But now compare the sentences he wrote next.

> After steering out for about an hour, what we took for land all at once disappeared to our great astonishment, and certainly must have been nothing but a fog bank. Though I have been at sea now twenty-seven years and never saw such a deception before, and I question much if the oldest seaman breathing ever did, except it was some in that ship when the master made oath of seeing an island between the west end of Ireland and Newfoundland, and even distinguishing the trees upon it, and which since has never been heard of though ships have been sent out on purpose to look for it. And had the weather come on very thick after the sight we had for some time of this imaginary land so that we could not have seen it disappear as we did, I dare say there is not a man on board but would have freely made oath of the certainty of its being land.[1]

Such illusions, when not checked by any further sighting, could easily find their way on to a map – sometimes, embarrassingly, alongside the names of eminent persons unaccustomed to ridicule.[2] At least two secretaries to the British navy have suffered this indignity. Pepys Island was 'discovered' in 1683 by William Cowley in 47 degrees 40 minutes south, supposedly 80 leagues east of South America – a convenient place to take in water and wood with harbour-space for a thousand ships, said a reliable source, though in truth there were neither trees nor shelter where Cowley made his discovery and all the water was salt.[3] Far to the north, in 1818, non-existent mountains named after Samuel Pepys's successor John Wilson Croker were thought to block the north-west passage through Lancaster Sound.[4] Errors of this kind were later to be acknowledged in placename elements like 'False', 'Doubtful' and 'Mistaken'.[5]

So much for the perils of immediacy. Another difference between plans and views is that a picture, however vivid, omits various immaterial features commonly found on maps – placenames, political boundaries, the indoor uses of buildings and nearly all the subject-matter of thematic cartography. On the other hand, the viewer cannot help seeing many objects that few readers wish to see mapped, among them not just individual people, animals and vehicles, but also minor examples of buildings, roads, water features, fences and trees, which may in any case be incapable of representation at a manageable scale. To that extent the contents of a view are largely wasted.

Again, while maps are indefinitely extensible, views are interrupted by hills and other obstructions as well as by the curvature of the terrestrial spheroid. However, on paper it may be possible for more than one visual fixation to be amalgamated, an especially common practice in the iconography of towns. Thus by turning the observer's head the view is widened into a panorama. Other depictions of landscape can be understood as combining an infinite number of views taken from every point along a straight line of continuous forward movement, usually following a horizontal course and with a uniform angle of vision. If the line of sight were vertical, the result would be a map. If oblique, it would be a 'bird's flight' view – as opposed to a single 'bird's eye' view in which all features are seen from the same point.[6] Compared with individual observations, map and bird's flight view are equally remote from immediate experience.

Increasing the area covered by a view is not an easy task. Either its size becomes intolerably large, or its scale has to be made so small that the resulting picture soon loses whatever advantages it may possess over a map of the same region. In any case, there are unlikely to be enough elevated points, natural or artificial, for broad sweeps of an artist's vision to embrace an extensive area of average topographical complexity. With the coming of photography and powered flight, this last problem could be overcome. Before that, the long-distance view presented so many difficulties that there was, and is, little point in discussing just how a number of examples could be brought together.

From view to map

Perhaps the most serious difference between cartography and scenography is that a map of limited territorial coverage is expected to maintain a constant scale whereas in a view, however restricted, the scale of each feature varies with its distance from the observer. The steeper the angle of vision, the more closely will a view approximate to a plan. It will never actually become a plan, because even if we contrive to look vertically downwards, all places not straight below us must still be perceived more or less obliquely, so that in a drawing of what is seen from any one viewpoint the scale will decrease from the centre outwards. Again, this argument is hardly worth discussing except in relation to small areas. Beyond a certain size it soon becomes unmistakably clear that the earth's curvature must prevent even the most accurate of flat maps from achieving a uniform scale.[7] Only in a celestial context does the distinction between the two media begin to break down: an image of the moon as seen from the earth is mathematically almost identical with a map – though admittedly on an unusual (orthographic) projection.[8]

Within narrow territorial limits, a steeply inclined view – of a small islet sketched from a ship's masthead, for example[9] – could serve many of the purposes of a planiform representation. Some early surveyors acknowledged this possibility by also practising as landscape artists. The oblique *Landtafel* was especially characteristic of sixteenth-century southern Germany, a place and time in which views of interestingly patterned terrain were perhaps a more popular form of topographical record than any kind of vertical projection (Fig. 3.1).[10] Later subjects much in favour for an oblique portrayal were the towns of the nineteenth-century United States, where buildings had sprung up recently enough for their facades to inspire both pride and wonder (Fig. 3.2). In areas inaccessible to a surveyor it would also have seemed worth trying to use pictures of this kind as the raw materials of a true map – or, for that matter, of an imaginary oblique view taken from a different altitude. Such transformations are difficult to execute, as modern students can discover by attempting to conjure up a sketch map of an area unknown to them with no other guidance than a low-level ground photograph. Some relationships between oblique and vertical are admittedly not difficult to understand. Straight lines in the view remain straight on the map. Horizontal lines in the view are parallel on the map. Verticals in the view radiate from the observer's location on the map. The real problem is the diminution of scale with distance, because here the rate of change depends on circumstances undiscoverable within any single visual field.[11]

By way of illustration, imagine an extensive plain as seen from a moderately elevated vantage point. In the distance is a patch of woodland that looks like an ellipse with its long axis running parallel to the horizon. Every viewer knows that on a map the wood will have a different shape from the one he is looking at. But will that shape be a broader ellipse with the same general alignment, or a circle, or

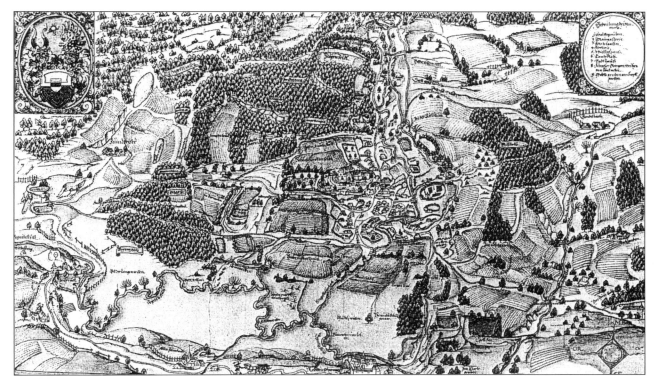

3.1 German Landtafel, combining elements of map and view. Paul Pfinzing, Hennenfeld, 1585. David Buisseret, 'The estate map in the old world' in David Buisseret (ed.), *Rural images: estate maps in the old and new worlds* (Chicago, 1996), p. 17.

an ellipse whose long axis follows the line of sight? And whatever we call the figure in question, exactly how will its length compare with its breadth? This problem first attracted attention among fifteenth-century artists attempting to portray floors and piazzas patterned with square tiles or flagstones. A related process was that of deriving a landscape view from a true plan. This could be done by drawing two networks of lines, a square grid on the plan and a perspective grid on a blank sheet intended for the view, and then copying the contents of each square with appropriate adjustments in the corresponding foreshortened quadrilateral. The perspective cells would naturally get narrower and shorter towards the horizon, though not in accordance with any simple formula (Fig. 3.3).

There was however a significant difference between plan-making and view-making by the method of grids. For a view there were many possible frameworks, each with a different angle of vision, and nothing much to choose between them, whereas for a map there was only one option. This made life more difficult for the cartographer than for the topographical draughtsman. Observing and copying the landscape from a single vantage point, he could probably insert a horizon and a

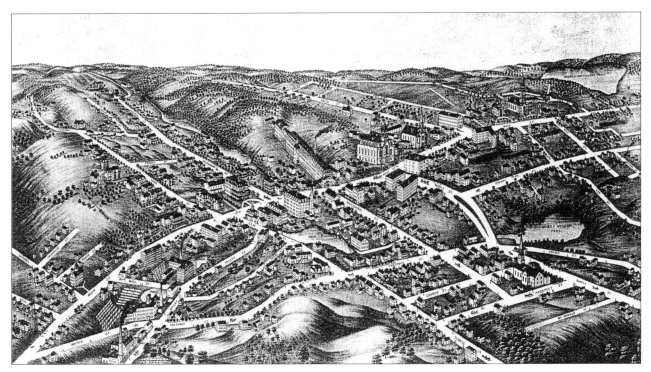

3.2 Panoramic view. O.H. Bailey and J.C. Hazen, Spencer, Massachusetts (Boston, 1877).

vertical axis (also called a principal plane), but would have no way of knowing where to draw the perspective grid lines unless his map already included a number of 'control points' in their true positions. If for example the map showed three identifiable points one behind the other, the distances between them could be compared with their equivalents in the view. A perspective grid could then be chosen with its horizontal lines spaced in the same proportions. But unless a map was already available the necessary controls could not be located without taking measurements on the ground, which by definition is exactly what our hypothetical topographer would wish to avoid.

Another obstacle to the geometrical 'rectification' of landscape views is the difficulty of making the observer's original freehand drawings accurate enough to justify the requisite measurements and calculations. One familiar problem is that the artist may involuntarily reduce his scale as he works his way across the paper and finds himself running out of space.[12] Even Leonardo da Vinci is known to have fallen into this trap.[13] A possible remedy is to view the landscape through a gridded transparent screen of the type found useful by Albrecht Dürer in the early sixteenth century.[14] This is unlikely to have been easily managed out of doors, especially in rough weather, and many observers would have found it hard to keep their eyes in the right position relative to the apparatus. A more elaborate sixteenth-century

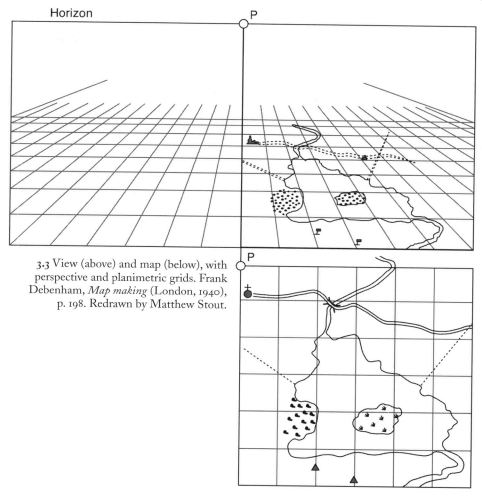

3.3 View (above) and map (below), with perspective and planimetric grids. Frank Debenham, *Map making* (London, 1940), p. 198. Redrawn by Matthew Stout.

device was the camera obscura, in which an image of the landscape could be projected on to a vertical surface by light-rays passing through a small aperture in a suitably placed screen. With the help of a lens and a mirror the image could be recreated at a more convenient scale and in a horizontal plane. Satisfactory results were obtainable from such a non-photographic camera only by eliminating extraneous light, but this could be done by enlarging the whole appliance to form a small room – or, more practicably, tent – and placing the artist inside it.[15] For a portable version of this device, the camera lucida, artists had to wait until the early nineteenth century.[16] Terrestrial surveys were supposedly made with a camera obscura by the astronomer Johannes Kepler in *c.*1620, but on the whole – and not surprisingly – this kind of transformation seems to have won little favour until views could be provided ready-made and in abundance by means of photography.[17]

Nothing in the above discussion excludes the possibility that maps of a sort might be made non-geometrically from views drawn freehand or 'by eye', with the shift from perspective to planiform imaging achieved by a process of estimation. This operation will henceforth frequently be referred to as 'sketching', though the same word may sometimes denote the freehand copying of extant maps, accurate or otherwise, and it must also be remembered that 'sketchiness' as a visual characteristic, like its converse, neatness, is not necessarily a guide to what has been happening in the field. Sketching in the outdoor sense must have been one of the commonest 'surveying' techniques ever used, both in the making of totally new maps and in the filling out of a surveyor's line diagrams. Its practitioners might have justified themselves by quoting Ptolemy, for whom chorography as opposed to geography dealt non-mathematically 'above all with the qualities rather than the quantities of the things it sets down'.[18] On at least one occasion, this liberal-minded doctrine of Ptolemy's has been held to justify a chorographer in drawing views rather than maps.[19] As a rule, however, most of Ptolemy's successors, including modern cartographic historians, have found it easier to write about the quest for planimetric rigour. In early literature on survey methods one must watch closely for any recognition that unless a landscape contains nothing but regular geometrical figures it cannot be mapped without a certain amount of sketching. A typically roundabout formulation is Robert Norton's: 'Note, that if there should be between any two of the [measured] angles some small crookedness or bending, the form thereof may well be represented on the paper of notes, the same being present to the observator's eye'.[20]

Estimation, rather than being recommended or even mentioned in cartographic textbooks, is most likely to betray its existence when a surveyor through no fault of his own was in too much of a hurry for anything more ambitious. This was especially liable to happen at brief stops on long voyages of discovery. An example is the plan drawn by a member of Christopher Middleton's crew in 1742 of everything he could see from the highest mountain near Cape Frigid in Hudson Bay.[21] The same predicament might occur in military surveying, where William Roy, a highly conscientious observer, admitted that some detail might have to be 'taken by the eye'.[22] Even in a properly triangulated survey (as described below in chapter 6) the most careful of Roy's contemporaries could only 'go along the coast, and in the way, join the several points protracted, imitating on paper the small curvatures that may be betwixt the corresponding points in the land' – in effect the same advice that Norton had given nearly a century and a half earlier.[23]

Like many cartographic processes sketching could sometimes be expected to reveal itself through characteristic omissions and errors in the resultant map. Among European surveyors such faults were perhaps commonest in the sixteenth and early seventeenth centuries, not so much in pure sketches as where sketching had been used to fill the gaps in a hasty and open-textured instrumental survey. Elizabethan county maps, in particular, have provided a salutary challenge to students trying to reconstruct the course of an author's fieldwork. In surveys like those of Christopher Saxton the use of sketching from a single station is best

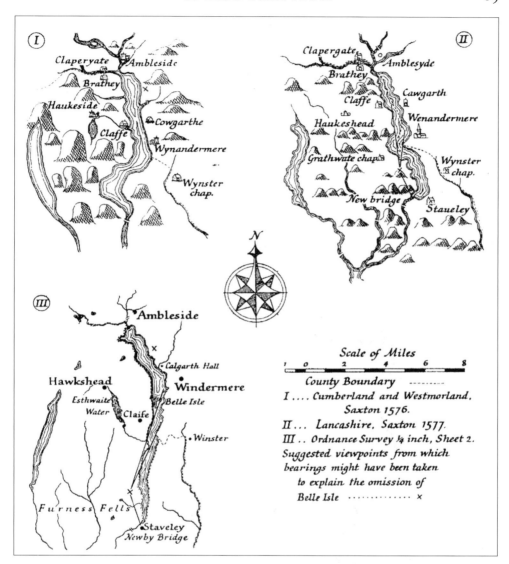

3.4 Windermere, from Christopher Saxton's published county maps of Westmorland, 1576, and Lancashire, 1577. Redrawn in Gordon Manley, 'Saxton's survey of northern England', *Geographical Journal*, lxxxiii (1934), pp. 312, 315.

demonstrated when the sketcher took no account of important detail that happened to be invisible from his viewpoint. Thus in Durham Saxton followed the River Tees upwards as far as Holwick Fell. From here he could see the tributary valley of the Harwood Beck but not the uppermost course of the main river, which was consequently omitted from his survey.[24] Perhaps the most serious kind of omission occurred at sea when a break of slope some distance inland was mistaken for a coastline, the true coast below it remaining unnoticed. The same thing could happen on

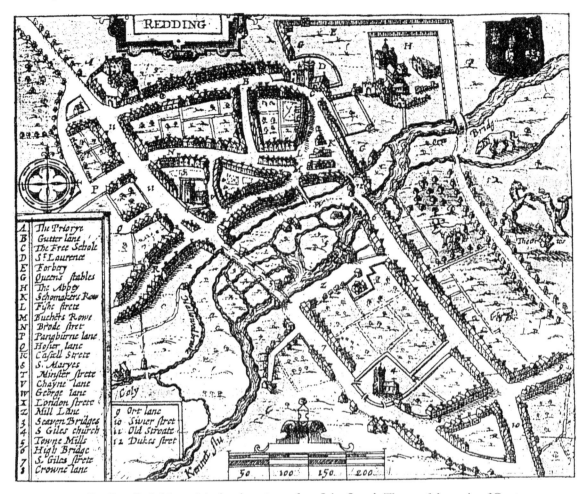

The Priorye
Gutter lane
The Free Schole
St Laurence
Forbery
Queens stables
The Abbey
Schomakers Row
Fishe strete
Buchers Rowe
Brode stret
Pangburne lane
Hosier lane
Castell Strete
S. Maryes
Minster strete
Chayne lane
George lane
London strete
Mill Lane
Seaven Bridges
S Giles church
Towne Mills
High Bridge
S. Giles strete
Crowne lane

9 Ort lane
10 Siluer stret
11 Old Streate
12 Dukes stret

3.5 Reading, Berkshire, original scale *c*.1:8000, from John Speed, *Theatre of the empire of Great Britaine* (London, 1611–12).

shore where a distant ridge-top might simulate the junction between land and sea.[25] Even a gifted surveyor like James Cook could occasionally be misled in this way.[26]

Where a feature did appear on the map its outline could suffer by not having been observed from more than one direction. The foreshortening of the Lleyn peninsula as viewed from Snowdonia has been held to explain the malformation of this coastline in Humphrey Lhuyd's small-scale maps of Wales in 1573.[27] Lhuyd made no claim to cartographic professionalism, but even a respected field worker might take the same kind of short cut in wild and unproductive country; thus Saxton is thought to have exaggerated the curvature of Windermere because he viewed the lake obliquely from somewhere near its southern end without actually riding along the shore (Fig. 3.4). A similar situation could arise in the mapping of towns. In John Speed's plans of Canterbury and Reading (1610) an attempt has been

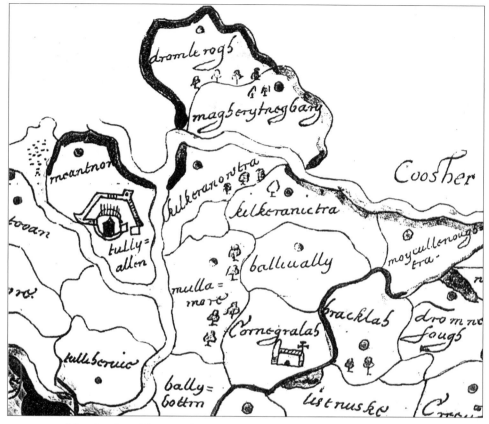

3.6 Mapping from 'abutments', original scale *c.*1:60,000. From Josias Bodley et al.,
barony of Orier, County Armagh, Ireland, MS, 1609–10. See note 29.

made to identify the surveyor's viewpoint from the premise that proportionate
distance-errors along his probable line of sight are greater than those measured in
other directions (Fig. 3.5).[28] Little research of this kind appears to be on record: even
the view from Snowdonia is not known to have been described at first hand by any
map historian. At present one can only suggest, 'intuitively', that most observers
of the landscape have underestimated the geometrical differences between a view
and a map.

Sketching en route

In short, looking at the world from one position is unlikely to produce a satisfactory
map. Let us therefore set our informant in motion between two of the places
selected for recording, and let his route be represented by a line. Other features of
interest could be placed, in order of appearance, either on this line or to the right or
left of it, depending on their observed position. He would no doubt also make some

attempt, if only unconsciously, to keep the distances laid down on the map proportionate with those he was covering on the ground. In Europe such lines of movement
were probably roads and the points of interest towns or villages. A traveller's map
of this kind could be sketched easily enough when the sequentiality of perceptual
experience conformed to the stringing of point-like locations along a visible thread,
as with harbours on a coast, towns on a road or tributary-junctions on a river.
A special case of this itinerary arrangement was the perambulation of a boundary,
either by naming and describing successive landmarks as they were encountered or
simply by listing the territories that abutted on to a given territory (Fig. 3.6).[29]

One attraction of roads for the early map-maker was that they were usually
interconnected. Some of course were dead-ends, but many of these led to a coastline
which from a constructional standpoint could be treated as part of the same
network. Such a skeleton held the different elements of a map together, prevented
duplication and reduced the incidence of gaps or holes. The nodal points in a route
system would help to prevent the map-maker's scale from varying too widely,
because in real life, and on the map, the length of a route was likely to bear some
relation to the number of nodes along its course. In the plotting of a line without
nodes, such as the coast of a large desert island, this check would not be available.
As an alternative framework, territorial boundaries had the advantage of being
entirely without loose ends, assuming that seas and lakes were allowed to count as
territories. Here again was a factor working (not very effectively, it is true) towards
a consistent scale, because the number of neighbours possessed by a territory was
likely to depend upon its size. Rivers, on the face of it, would provide a less
satisfactory network than roads or boundaries, because a drainage system would
always have more terminals than junctions and there was normally no water
connection between neighbouring basins and therefore no ready means of
establishing a correct relationship between them (Fig. 3.7). But since rivers have so
often been regarded as essential to cartography, an itinerant map-maker might still
choose roads that gave a view of running water. In Timothy Pont's surveys of
sixteenth-century Scotland, for example, the settlement sites appear to have been
sketched with reference to nearby stream courses, and the resulting district maps are
often bounded by watersheds.[30] Of course, some geographical patterns cannot be
treated as networks – islands in an archipelago, clearings in a jungle, trees in a
savannah, oases in a desert, tumuli on a tract of downland, craters on the moon –
and these are consequently more difficult to map.

The kind of information we have been considering was not necessarily collected
for cartographic purposes. Some of it was not collected at all but remembered
involuntarily, like the ten islands sketched on paper by Samoan informants at the
request of the eighteenth-century explorer La Pérouse.[31] Beyond a certain radius the
quality of such local information would naturally deteriorate. La Pérouse
experienced this difficulty when the people of mainland China 'corrected' his belief

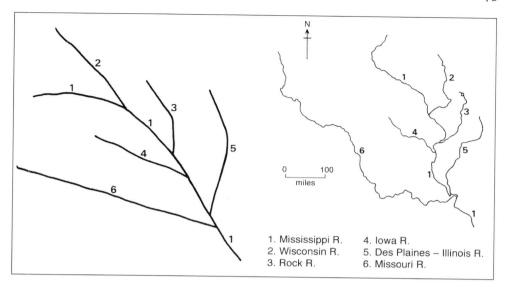

1. Mississippi R.
2. Wisconsin R.
3. Rock R.
4. Iowa R.
5. Des Plaines – Illinois R.
6. Missouri R.

3.7 Native American mapping of a river system (left) with modern equivalent. G.Malcolm Lewis, 'Indicators of unacknowledged assimilation from Amerindian maps on Euro-American maps of North America: some general principles arising from a study of La Vérendrye's composite map, 1728–29', *Imago Mundi*, xxxviii (1986), p. 27.

(true, as we now know) that Sakhalin was an island and not a peninsula.[32] The accidental element in this kind of memory-based data-collection can sometimes show itself at global level: a fourteenth-century world map known to emanate from Worcestershire gives many more names in western and south-western England than in the east and north.[33]

As for relative distances, the experience of William Parry among the Inuits may speak for itself:

> [G]reat caution is requisite in judging of the information these people give of the distances from one place to another, as expressed by the number of seêniks (sleeps) or days' journeys, to which in other countries a definite value is affixed. No two Esquimaux will give the same account in this respect, though each is equally desirous of furnishing correct information; for besides their deficiency as arithmeticians, which renders the enumeration of ten a labour, and of fifteen almost an impossibility to many of them, each individual forms his idea of the distance, according to the season of the year, and consequently the mode of travelling in which his own journey has been performed. Instances of this kind will be observed in the charts of the Esquimaux, in which they not only differ from each other in this respect, but the same individual differs from himself at different times.[34]

One might expect, however, that where knowledge from different witnesses could be pooled such differences would eventually tend to cancel each other out: in Ptolemy's culture-circle, the more people's experience went into a map the more correct it was likely to be.[35] Here, not for the last time, we must acknowledge the difference between original and secondary research. Collecting geographical data from local inhabitants in their place of residence may reasonably count as a kind of surveying. If the same data come in written form from the same location the recipient is not so much a surveyor as a compiler. It is, after all, a significant difference, prompting the question how far, territorially, a map can be extended by accumulating the notebooks of a single itinerant interviewer. In theory, a topological framework could be carried forward until eventually land gave way to sea. There was a limit, nevertheless, to how much travelling could be crowded into a human life-time. In what sense could a large and complex *mappamundi*, for example, have been the product of one person's experience? Not a totally futile question: the answer may be indirectly relevant to such profound issues as whether medieval European world maps were necessarily derived from Roman prototypes.

Maps without measurements

No special expertise is needed to recognise purely topological relationships such as junctions, enclosures and overlaps. This is why mankind is sometimes said to be universally capable of acting as if in possession of a map, a skill also thought to be widely distributed through the animal kingdom. The difference is that humans are assumed to be not just spatially cognisant but also able to express their knowledge graphically when equipped with suitable drawing materials. However, if everyone could draw a map, the term 'map-maker' would be no more useful than 'walker' or 'grasper' in distinguishing different members of the human race. For a clearer picture of reality, the historian must admit that whatever population is under scrutiny, some individuals had acquired more geographical experience than others; some were blessed with a better memory for details; some had more patience; and some showed more skill in reproducing the contents of their imagination – a faculty accompanied, one assumes, by a gift for drawing what was actually before their eyes. Beyond all these general capacities lay one specific talent: certain people were congenitally more able by a process of psychological 'rotation' to translate their essentially horizontal experience into something like a vertical view.

Natural abilities aside, the early sketch-mapper differed in one important respect from the subjects of a modern psychological experiment: we cannot ask him about his state of mind. Was he aiming at correctness, and under the impression that he had achieved it? Was he simply unconcerned about certain types of error, neither knowing nor caring whether he had committed them? Was he conscious of his fallibility and disappointed at being unable to do better? Perhaps all these attitudes were involved at one time or another. What we can say is that many maps must have been produced without the aid of measurement, in Europe before about 1600 and in certain non-European countries until more recently. Some of these maps are famous and some look surprisingly accurate. One seventeenth-century writer was willing to dismiss the work of even Ortelius, Mercator, Blaeu and Sanson as 'so many guess plots'.[36] However that may have been, a pre-modern society could exist for centuries without map-producers becoming recognised as specialist craftsmen on the same level as blacksmiths, potters or carpenters.

It is odd that so few surviving maps of any size are known to have been authoritatively described by their contemporaries as the product of topological experience alone. Perhaps this suggests that the majority of such maps were made by recording journeys undertaken incidentally to other business. A traveller activated by a definite cartographic purpose would probably expect some reward for his efforts, and requests for payment were more likely to be expressed in writing, and therefore to survive for posterity, than many other kinds of communication. It is also notable that history has failed to preserve the names of any individuals eminent in the art of topological reconstruction. However there are many allusions, especially outside Europe, to maps being drawn from memory – meaning in this case memory of the landscape rather than of some other map. An early example was a sketch of Walter Raleigh's settlement on Roanoke Island made by an Irish seaman, Darby Glande, who was rightly undeterred by not knowing 'how to draw'.[37] Perhaps the most startling of these references is a British Ordnance Survey officer's admission in 1834 that he had depended on his memory in sketching part of Lundy Island.[38]

How then do we recognise a memory map? In part from purely negative evidence, as where a cartographically active and literate society has left no documentary trace of any other method being used. Then there is the appearance of the map itself. Mere inspection of the ground, unaccompanied by measurement, is unlikely to yield more than a certain level of planimetric accuracy, a point that will be taken up again below. In fact attributing someone else's map to memory may sometimes have been a sarcastic way of finding fault with it.[39] Particularly interesting here is the question of consensus. Ideally, each verbal description of a traveller's experience should have been too detailed to bear more than one graphic interpretation. In practice such unanimity was improbable, and a set of independent word-based maps depicting the same area would almost necessarily contradict each other as well as differing from reality. Thus where authors worked independently, sketch maps would vary more among themselves than would true admeasurements of the same landscape. In 1616 Captain John Smith, collecting information on New England, anticipated William Parry's experience further north when he wrote of 'six or several [sic] plots of those northern parts, so unlike each to other … as they did me no more good, than so much waste paper, though they cost me more'.[40] The difference was that Smith's source-maps were presumably drawn by Europeans.

A more idiosyncratic clue to non-quantitative origins would occasionally be available if a cartographer's source had consisted of abutments and nothing else. His map might then turn out to have left and right reversed, as if the earth's surface were being viewed from underground, with east placed ninety degrees anticlockwise

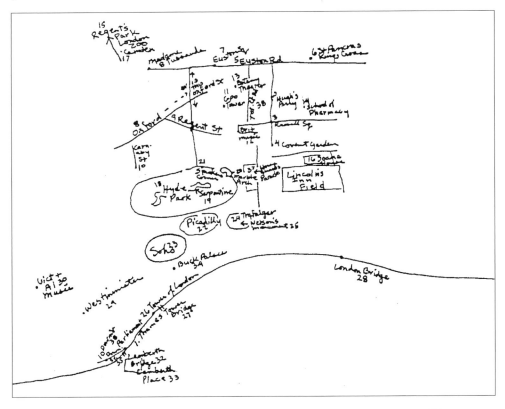

3.8 Student's mental map of central London. Denis Wood and Robert Beck, 'Janine Eber maps London: individual dimensions of cognitive imagery', *Journal of Environmental Psychology*, ix (1989), p. 20, Fig. 17.

from north instead of clockwise. This hazard would disappear if his data included at least one spatial fact that would suffer visible derangement by the process of mirror reversal, for instance a relationship with a point outside the map such as the earth's geographical or magnetic pole. It is the abundant scope for such external checks that has made the 'mirror-image' map a rare curiosity.

Despite differences of detail, maps deriving from their authors' memories will probably show certain broad structural resemblances. Here we can usefully refer to modern psychological research on cognitive or 'mental' maps. When committed to paper these often look like standard maps that have been deliberately kept simple. Their line-work and shading may appear crude and hastily executed (Fig. 3.8). Profiles or oblique views will probably be drawn to represent hills, water-waves, trees and buildings even in an era when more professional artists would prefer some kind of miniature plan. Apart from their recognisability, the main advantage of such profiles is to let the size of each symbol depend upon importance rather than extent, in the manner of young children who draw their mother larger than their father.[41]

Thus the sketcher simplifies the work of real cartography by doing as little of it as possible.

Other aspects of simplification can best be approached by first redefining the term 'feature', which in the next few paragraphs will mean anything on a map that can be treated without inconvenience as a point. Features in this sense include not just settlements, hill summits, bridges and all kinds of upstanding landmark but also bends, intersections and terminals in the linear elements of a landscape.[42] 'Mental' cartography proceeds by selecting fewer features than exist either in the real world or in the traveller's perception of that world – fewer, certainly, than would be recorded by a professional surveyor. Let us now assume, caricaturing reality for purposes of exposition, that apart from random variations the geographical distance between any mental-map feature and its nearest neighbour is always the same. Continuous gradation is therefore absent from the realm of mental mapping and the length of each line is determined by the number of features located on it. This explains a number of peculiarities common in mental maps, one of which is the underestimation or exaggeration of distances according to whether they are long or short.[43] In particular, length is more likely to be overstated in areas of close detail such as city centres or on routes with many bends and intersections.[44] For practical recognition of the same tendency we may consult the eighteenth-century French hydrographer Jacques Nicolas Bellin.[45]

> The dwelling too minutely on particulars has produced this error [on his own map of New France], and will always have the same effect. For by attending too nicely to express the form of the ports, and the windings of the capes and islands, it is scarce possible, if the scale be small, but they must appear larger than they are in reality.

A more famous geographer than Bellin had made much the same point in the second century AD:

> In the case of an undivided map [wrote Ptolemy], because of the need to preserve the ratios of the parts of the oikoumene to each other, some parts inevitably become crowded together because the things to be included are near each other, and others go to waste because of a lack of things to be inscribed. In trying to avoid this, most map-makers have frequently been constrained by the shapes and sizes of the planar surfaces themselves to distort both the measures and the shapes of countries, as if they were not guided by their research.[46]

An especially important kind of intersection is between a route and a territorial boundary, causing distances between points in different countries to be judged longer than similar distances within a single country.[47] The same kind of relative enlargement occurs in regions that are especially well known: the more familiar the terrain, the more of its features are likely to be chosen for mapping.[48] According to some psychologists, this also applies to places with happy rather than unpleasant associations. On the other hand a more sophisticated mapper will sometimes break

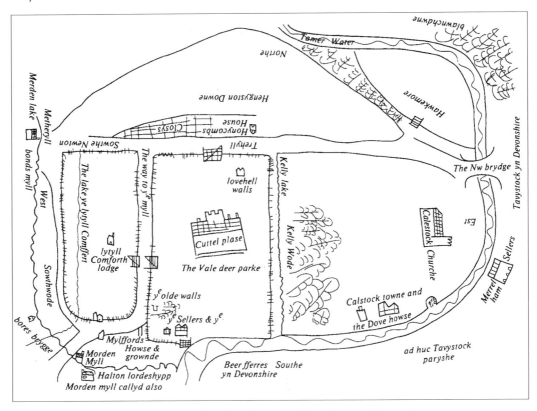

3.9 Cotehele and district, Cornwall, mid-16th century. Sketch-map with angular errors averaging about 50 degrees. Redrawn in William Ravenhill, 'The plottes of Morden Mylles, Cuttell (Cotehele)', *Devon and Cornwall Notes and Queries*, xxxv (1984), p. 166.

the rule of constant distance: then, by way of compensation, distances with one or more unfamiliar terminals are liable to be overestimated, too much allowance being made for hypothetical intermediate features.[49] Such a cartographer, in the vocabulary of this book, has passed from the realm of sketching into that of compilation. In normal uncorrected sketching, the standardisation of terrestrial distance is not without advantages; for instance, it allows the mental mapper's cartographic distances to deputise for other more or less closely related variables, including cost-distance (in money or energy), social distance (in which length is equated with infrequency of travel) and time-distance as reckoned in hours or days.[50]

Next consider the simplification of bearings as opposed to lengths. Just as all pairs of nearest neighbours on our postulated mental map are the same distance apart, so there are (roughly speaking) only two mental angles between connecting or intersecting lines: one is approximately 180°, the other approximately 90°. The first is illustrated by the common practice in modern sketch maps of aligning North and South America on a single straight axis[51] and by the belief that the Caribbean entrance to the Panama Canal lies eastward of the Pacific entrance.[52] The second appears in town maps that misrepresent oblique street-intersections as rectangular;[53] in a rural setting it can be illustrated by a mid-sixteenth-century map showing the country round the Cornish village of Cotehele (Fig. 3.9). No doubt this preference for ninety-degree junctions is caused ultimately by the arrangement of limbs on the human body, and more proximately by the man-made significance of the right angle in modern life, whether on sheets of paper or in the lay-out of many buildings, streets and fields – suggesting that rectangularity might have been less important in the mental maps of previous generations.[54] How far these and other psychological dispositions are exemplified in surviving early maps must be left for others to investigate.

Quantifying the qualitative map

A cartographer's methods, we have already seen, might be expected to declare themselves in the planimetric quality of his results. This raises the question of how the accuracy of a map can be measured and recorded. Most methods require the identification of as many points as possible that are common to the test map and a reliable standard map. One can then compare the distances between pairs of common points or the angles between pairs of common lines. A distance can be taken from the scale statement of the test map or expressed as a ratio of the corresponding modern distance. An angle can be related to the direction of north shown on the test map, or stated as a ratio of the corresponding modern angle in an equivalent triad of points. (If the test map lacks a scale and north point, there is obviously no choice.) In other methods the position of each point is defined by rectangular coordinates. These have the advantage of containing within themselves both distances and angles, which therefore no longer need to be treated as separate variables. For test maps with a graticule the most readily available coordinates are latitude and longitude but of course these are not always present. In the absence of a graticule no ready-made linear framework offers itself, but this is where W.R. Tobler's technique of bidimensional regression comes into its own.[55] Here the maps being compared can each be given their own coordinates, with their own origin, their own orientation and their own units of measurement, after which the Tobler programme eliminates variations of scale and alignment between the two point-patterns, and then subjects them to statistical comparison.

All these methods yield a single number expressing the degree of geometrical resemblance between the two maps, a relationship which given our initial premises measures the accuracy of the test map. In Tobler's system perfect accuracy, in other words total conformity between the test map and a modern map, is represented by an index of 100 per cent. A value of nought per cent would be harder to interpret: one can only define this rather vacuously as the score of a map whose author has succeeded in maximising its inaccuracy. In practice, of course, neither of these extreme values is ever encountered. Few map historians have reported using Tobler's method, and the present book makes no attempt to fill the gap: the sample values given below are approximations mainly designed to challenge readers commanding better facilities for experimentation. However, two preliminary hypotheses may be worth considering even at this stage. First, percentage values for nearly all kinds of map are higher than the uninitiated reader would expect. Secondly, other things being equal, such values seem generally higher for small-scale than for large-scale maps.

How much accuracy can we expect from the map-types described in the present chapter? Perhaps the best case to start with would be not a map at all but a perspective view taken from a single vantage point. Such oblique projections may be interpreted, a trifle perversely, as genuine maps that happen to be incorrect. (This is a reminder that some reference to authorial motives may be necessary to the definition of the word 'map' – a condition satisfied in most definitions by the word 'represent'.) In such cases the Tobler percentage will diminish with the observer's angle of vision and will also depend on the distribution and altitude of the points chosen for comparison. Planimetric values for oblique landscape photographs vary widely, but in general will probably fall short of 90 per cent. Where the viewer's perceptions were intended to appear under the guise of a map, he may have tried to convert oblique to vertical impressions by making an appropriate allowance for perspective. When performed freehand, without taking measurements, this operation was unlikely to be completely successful. In one experimental case, the score was 56 per cent for an uncorrected view and 76 per cent (still very low) for a sketch map based partly on the same view and partly on the author's common-sense attempt at transformation.[56]

Another kind of map that can easily be fabricated for experimental purposes is the 'abuttal'. Here the researcher's source is a list of territorial divisions, each accompanied by the names of its immediate neighbours, from which is put together, jigsaw fashion, a map that can then be made to undergo the Tobler test. For boundary networks mapped in this way at about 1:50,000 the percentage values seem likely to fall between 60 and 90, the effect of scale differences being illustrated by a similar map of French departments at c.1:12,000,000 which scored no less than 97 per cent.[57] Where abutments are deliberately plotted 'in mirror-image' (a result which in real-life cartography would count as an unfortunate accident) scores may be less than 30 per cent.

It remains to consider the much larger class of sketch maps thought to have been based mainly on memories or on other records of travel not involving actual measurement. The problem here is a shortage of satisfactory data. Most modern cognitive maps are liable to contamination from pre-existing cartographic knowledge; and most early maps are unaccompanied by any evidence as to whether they were sketched or surveyed, so that in practice a sketch may be identifiable only by a general appearance of inaccuracy and by the antecedent historical improbability of a precise survey. A sample of fifteenth- and sixteenth-century regional and local maps chosen for these qualities has yielded percentages ranging from about 60 to 75. Areas of nation-wide extent can also return low values. For the whole of Great Britain Matthew Paris's mid thirteenth-century map scores 83 per cent, a map of early sixteenth-century Ireland (much of it previously cartographic *terra incognita*) 66 per cent.[58]

The sheer magnitude of these percentages calls for comment. In elementary statistical textbooks a correlation coefficient of 0.8 (equivalent to a Tobler percentage of 64) is often described as 'high' or 'strong'.[59] Yet to the eye of common sense a map registering in the sixties would appear as almost absurdly erroneous. It seems difficult to explain this apparent contradiction, and for the time being only one oblique judgement will be offered. At first sight the high Tobler values seem to demonstrate the effectiveness of human sensori-motor competence in all its fallibility as a machine for discovering geographical truth. This characteristic might emerge more vividly if we could compare the cartographic percentages quoted above with the success rate of (for instance) fifteenth-century medicine in curing diseases. There might conceivably be a moral here. Historians with an interest in planimetric accuracy are often criticised for their naiveté in judging pre-modern maps by inappropriate criteria. It is wrong to describe an old map as poor, runs the argument, when it was never intended to be good. Perhaps we should say instead that it is wrong to describe maps as poor when they are not poor. But this is too paradoxical a note on which to end: the question of planimetric accuracy will have to be resumed in a subsequent chapter.

CHAPTER 4

Strict dimensuration

IN CARTOGRAPHY THE term 'surveying' has come to mean field work performed at levels of exactitude unattainable without some kind of apparatus. A surveyor is therefore among other things a person who takes measurements. The history of mensuration can be, and often has been, studied in its own right without reference to the purposes for which the measuring was carried out. Some degree of historiographical alienation between methods and motives may indeed be unavoidable. Early maps are seldom accompanied by documents that describe the techniques and instruments used to produce them. The same applies in reverse to early surveying instruments, which now exist mainly as uncontextualised museum pieces, so that there are very few surviving pre-eighteenth-century appliances known to have been used in the making of any particular map. The problem was exacerbated when the invention of aids to surveying took on a life of its own, an eventuality not made more acceptable by the popularity among inventors of erudite Greek or Latin names. (Many of these never penetrated ordinary language, driving authors with a non-professional readership back to the equally unhelpful word 'instrument'.) Genuine map-makers, apart from a few eccentrics, were traditionally reticent about matters of technique, and happy to leave the description of their equipment to authors unversed in practical cartography.

So the truth is that no one can be sure of knowing which instruments were most widely used at any given period: indeed the mere existence of a technique is no proof that it was ever put to the test. It would take a bold historian to defy tradition by simply ignoring all instruments for which he cannot find a clearly dated and located cartographic performance, and in the following paragraphs readers will be left to find a path between what is written and what has been left unsaid. There is one further complication that has tended to widen the divergence between theory and practice among surveyors. A map is always smaller than its subject, but no reader with normal eyesight expects to depend on a magnifying glass, and *minutiae* of terrain-measurement are cartographically unproductive if the results are incapable of being legibly plotted at any of the scales in general use. The practical surveyor, then, does not necessarily seek the highest level of accuracy that other kinds of expert might be trying to make feasible.

Measurement can be defined as any method, other than mathematical calculation, that enables spatial quantities to be accurately represented by numbers. It is therefore sharply different from topology, in which such relationships can be expressed entirely by words. On a theoretical level, admittedly, there may be intermediate stages between these two concepts. For instance, all the points in a region could be arranged in pairs to yield a series of linear distances, which could then be ranked from shortest to longest without assigning them any 'absolute' lengths in miles or kilometres. In primitive cartography it might have been acceptable for distances to count as either shorter or longer, but any non-numerical systematisation of this ranking process would make intolerable demands on the surveyor's capacity for visual judgement, as well as precluding quantitative comparisons between maps of different regions. In fact the only use of ordinal numbers in cartography that comes readily to mind is the mapping of 'command' by military surveyors, as noticed below under the heading of relief representation.

The principal subject of this chapter is the more familiar process of characterising spaces by the number of standard metrical units that they contain. Here the main difference is between measurement, which we have already defined, and estimation, which was implicitly excluded from that definition by the word 'accurately'. Estimation stands somewhat apart from John Ogilby's antithesis between 'strict dimensuration' and the making of 'guess plots'.[1] It involves visualising a unit of distance and mentally comparing it with the ground. This is not very reliable as a guide to the truth, but estimates can still contribute to all the survey methods which in a later chapter we shall find making use of measurement. The difficulty is that estimating was more likely to be associated with an extensive dependence on sketches, in combinations that even the closest contemporary observer (let alone a modern historian) might well have found it impossible to analyse.

Surrogates for distance

Measurement itself involves both seeing and counting. Though far from infallible, counting was simpler than estimation; the process of seeing, however, could in some circumstances be extremely difficult and for this reason the measurement of distance was often achieved by substituting time for space. The registration of time, which could be done in a state of immobility, had the advantage of being less energetic, as we can tell from the association of laziness with 'clock-watching'. It also presented fewer problems with units – at least until the highest levels of precision were required. One natural standard for time was given by the period between consecutive noons as indicated by the sun. Other units were then obtainable by subdivision of the day, and eventually these subdivisions produced the concept of equal hours, wholly independent of seasonal and geographical differences in the actual duration of daylight and darkness: no one expects to read of 'French hours' or 'German hours'. Outside modern Europe the day's journey has been a common unit of distance; even within Europe, maps were accompanied by

4.1 Some 17th-century units of distance. A.H. Jaillot, *Partie de la Nouvelle France* (Paris, 1685).
See also Fig. 18.7.

a scale of travel-times as recently as the eighteenth century.[2] If it comes to that, the
word 'league', though etymologically signifying no more than 'division', is said to
have once meant the distance that could be travelled in an hour, an origin preserved
in the seventeenth-century French 'lieue d'une heure de chemin' (Fig. 4.1). This
latter unit was regarded as equivalent to about three miles, but since different people
walked at different speeds, no single translation from time to space could expect to
command universal assent.

To pursue chronometric distancing further, the surveyor had first to identify
processes that take place at nearly uniform speeds, after which he could time the
movement of somebody or something over a given distance, calculate the speed of
that movement, and thereafter concentrate on recording the passage of minutes and
seconds. This method has been most important in navigating seas and rivers, where
more direct methods of distance measurement would be physically difficult or
impossible. Many sixteenth-century records of sea distance, even those specifying
hundreds, tens and units of leagues, turn out to have been based on judgement – 'to
the nearest estimation that possibly he can give', 'as near as he can guess', and so on.[3]

These appear to have been derived from estimates of a ship's speed, made by watching bubbles or weed in the water. An experienced seaman who knew his ship could hope to get within one mile per hour, or even half a mile, of the truth.[4] This could imply an error of 15 to 20 per cent at medieval sailing speeds. Later, a portable clock would help to make such measurements more accurate, as when Charles Marie de la Condamine surveyed his way down the River Amazon in 1743-4 with the help of a compass and a watch.[5]

For greater precision there was the log line, a mid sixteenth-century English invention first described by William Bourne.[6] Its essence was a block of wood (the 'log ship') at the end of a cord or rope. The block was thrown overboard and a sand-glass used to measure the time taken for the line to be paid out. Later, knots were tied in the line at intervals calculated to equate the number of knots passed out in thirty seconds with the speed of the ship in miles per hour. A seventeenth-century alternative was to time the interval between the throwing of an object (the 'Dutchman's log') from the bow and its movement past the stern of a ship whose length was known. A change in speed would obviously necessitate a fresh measurement. The Dutchman's log was less accurate than the normal log line, because a given error of timing would have a proportionately greater effect for short than for long distances. Both kinds of log measurement depended on the false assumption that the water itself was motionless. In fact, a ship could easily move backwards when the wind seemed to be carrying it forwards, a possibility which in the wrong circumstances might cause a single stretch of coastline to appear twice on the same chart.[7]

Another opportunity for time-distance measurement was the propagation of sound waves, recommended to surveyors by Edmond Halley as early as 1702[8] and in the late eighteenth century believed (almost correctly) to occur at a speed of 1142 feet per second. When a gun was fired, the time-difference between seeing and hearing the explosion could be measured at some other station and the distance between gunner and listener calculated arithmetically. Unlike the log, this method could also be used in land surveying, but it had obvious disadvantages for people unaccustomed to handling explosives[9] and despite needing two widely separated stations it seems to have been mainly used by marine surveyors without enough time to find a site suitable for conventional linear measurement.

One other kind of surrogate for distance remains to be considered, albeit practicable only in a vertical direction. This is atmospheric pressure.[10] In the mercury barometer, invented by Evangeliste Torricelli in 1643, a liquid was squeezed to varying degrees by the pressure of the air overlying it, which varied in turn with the height of the instrument above sea level, a rise by one inch of the liquid in a vacuum tube corresponding to an altitudinal difference of about 900 feet. One source of inaccuracy in this process was that mercury, a high-density medium chosen to keep the size of the barometer within manageable limits, proved insufficiently responsive to small altitudinal variations. Another difficulty was that pressure depended not just on altitude but on weather conditions in general, and adjustments for these extraneous influences could themselves be subject to serious error. In 1682 Snowdon was measured barometrically at 3720 feet, a result some 160 feet higher than the modern value.[11] By the early nineteenth century a careful application of the same method might hope to come within little more than one per cent of the truth.[12] From about 1845 mercury barometers were gradually replaced by the more convenient (but no more accurate) aneroid, in which atmospheric pressure was applied to the lid of a box largely exhausted of air.

Units of linear reckoning

The direct measurement of distance brought its own problems. Linear and areal units of some kind have existed in every historic culture of Asia, North Africa and Europe. Some originated independently, others evolved by diverging from a common source, requiring Italians, for example, to choose among Ancona miles, Bologna miles, Fermo miles, Ferrara miles, Firenze miles, Perugia miles, Ravenna miles and Roman miles.[13] Even within a single community, what was ostensibly the same unit might exist in several forms, as with the large miles, common miles and small miles often distinguished in seventeenth-century maps of England.[14] Different areal units might be used for measuring different kinds of surface, an English woodland acre being appreciably larger than an agricultural acre. Then there was the need to recognise higher and lower orders of distance, and thus to avoid contemplating the number of inches between Land's End and John o'Groats. Of course many units were calculated from other units either by division or aggregation, as when an inch was fixed at a twelfth of a foot or a mile at a thousand paces, but this did not prevent the definitions from outnumbering the terms to be defined, with perches for example composed of many different numbers of feet.[15] However, in any system of metrology at least one unit must be defined in terms imported from outside the system. As a matter of logic this external standard might be a single real-life distance, between two well-known landmarks for example, but such 'geographical' definitions have been understandably rare. More often measurement is based on organs or movements of the human body such as thumbs, hands, feet and paces, with anatomical variations among individuals helping to explain how a unit with a single name can vary from one place or time to another.

When measurement was internationalised these differences could become a serious cause of confusion, especially if no country possessed a standard that was absolutely invariable. The problem is obscured by an assumption among some European historians that one unit, the foot, happens to have been exactly the same all over the world.[16] To mention only a single counter-example, in the seventeenth century a Paris foot was estimated at 1.067 London feet.[17] Translations between pre-metric units differed considerably from one authority to another, and this has made it inadvisable to include a list of obsolete linear measures in the present work. Theoretically the conflicts could be resolved by a direct comparison of different measuring devices, though it seems hard to find evidence that this was done. Translated values would, except by an unlikely coincidence, almost certainly be burdened with awkward fractions: around the beginning of the Christian era Strabo defined a Persian parasang as either sixty, forty or thirty Greek or Roman stadia, but all these figures seem too round to be probable.[18] Once a unit has become obsolete historians may be unable to determine its value. This happened with the original stade or stadion of ancient Greece, which began as the length of a certain race track that nobody now has any means of measuring.[19]

A favoured escape route from this morass of uncertainty was to base one's standard on a single physical object, such as a particular metal bar kept in a certain laboratory. Since bars could become damaged by wear and tear or altered by a change in atmospheric conditions, all definitions ought strictly speaking to have specified both time and place at which a given length was to count as valid – a self-evidently useless stipulation in the case of time. It should at least have been feasible to state the temperature at which a bar might be expected to assume its standard length. One attempt to evade this problem took place in Revolutionary France, where the metre was officially defined in 1791 as one ten-millionth part of the distance from equator to pole on the earth's surface. Dividing a distance of this magnitude was not a task to be undertaken lightly, and in practice the metre became a bar of platinum kept in Paris.[20]

The scaling of length

Once a unit was chosen, lengths could be measured by repeatedly laying down a standard containing a given number of units. Pacing was perhaps the simplest way of doing this.[21] Tested against a known distance, the pace of an individual adult is remarkably consistent, yielding an accuracy of 99 per cent. So says the modern textbook, but although such a standard might work well enough on the surface of a present-day street, in the 1790s a town plan made by 'striding' could be dismissed as self-evidently unacceptable.[22] On rougher ground even less could be expected from a pedestrian's estimates. Otherwise the counting was a more probable cause of error than the stepping. Mistakes of enumeration could be avoided by the pedometer or passometer, a portable watch-like device invented in mid sixteenth-century Germany, in which each jolt of a walker's footsteps moved an index a small distance around a dial in units proportioned to the length of his stride.[23] The navigational equivalent of pacing was the counting of oar-strokes to determine the distance covered by a galley.[24]

A more accurate use of dials was in the perambulator, odometer (alias hodometer), cyclometer or way-wiser, where the clock measured rotations by a

4.2 Pedestrian measuring wheel, 19th century. J. Wartnaby, *Surveying: instruments and methods* (London, 1968), no. 5. Courtesy of Science Museum, London.

wheel of known circumference as it travelled the distance to be measured (Fig. 4.2). Way-wising could be made less energetic, though not it seems more accurate, by adapting one of the wheels on a horse-drawn vehicle (Fig. 4.3).[25] Reported from both ancient Rome and ancient China, the measuring wheel came into its own in the sixteenth century and grew more popular as roads became increasingly common on topographical maps, including the earliest British Ordnance Survey maps, losing favour only when such maps began to be made by reduction from larger scales.[26] In England its highpoint was to be recommended in 1759 by the Royal Society of Arts.[27] Joseph Lindley was still expressing a preference for the wheel in 1793: in Surrey his results generally agreed with a network of widely-spaced control-points to within ten yards and were never out by more than twenty yards.[28] Unlike the linear methods to be discussed below, clock-face mensuration could be practised on sea as well as land. For instance a pointer and dial could record the number of rotations of a propeller drawn through water, the first successful rotator of this kind being invented by Edward Massey in 1802.[29]

Wheel revolutions were generally a less accurate guide to distance than the setting out of linear standards in the form of rods, bars, tapes, ropes, cords, wires or chains.[30] The wheel-pusher might find it difficult to proceed in a straight line and his rather complicated apparatus was more likely than a measuring chain to develop faults that were difficult to correct. On the other hand, laying down a line required

4.3 Measurement by coach wheel, from Paul Pfinzing, *Methodus geometrica*, 1598.

two people, except in the case of plumbing water-depths as considered in a later chapter, whereas wheel surveys could be conducted by a single operator, though in practice chain-bearers were paid so little that this may have been a minor consideration. Before a straight line of any considerable length could be measured, it would have to be defined on the ground by placing visible marks along its course. On this alignment the chain was stretched taut between two pegs or arrows, the second peg in the first chain-length becoming the first in the next chain-length. As with pacing, it was surprisingly difficult to count how many times a given distance had been laid down: by the late seventeenth century, and probably earlier, there was an accepted routine in which each of ten arrows was successively passed from one chainman to the other until all ten were held by the same person.[31]

The word 'chain' is used here in a generic sense. Poles or rods were more suitable for very short distances. They benefited from firmness and stability, but were awkward to handle and laborious to carry. Cords (treated with wax and rosin to help maintain their length)[32] or wire lines were commonly used in the sixteenth century. Linked chains were recommended in 1579 but had probably been used earlier; their rods and eyes re-introduced an element of rigidity to the surveyor's standard, and by the early seventeenth century they were considered preferable to cords and wires.[33] The ascendancy of chain surveying was an achievement for which map historians have provided few clear punctuation-marks, though research might yield more instances similar to a Swedish ban on taping in any official survey of arable land after 1725.[34] Eventually the tape by general agreement was used only for short branches perpendicular to a longer chain line.

One disadvantage of any flexible measuring standard was that it might not always lie perfectly straight, tending to sag under its own weight and thus cause lengths to be over-stated. On the other hand a flexible medium measuring the horizontal equivalent of a hillside could be kept at least approximately level by being set out in short segmented lengths. But however it was used the chain could never be infallible. It might be shortened if the rods became bent or the links clogged with grass and mud, or accidentally lengthened by the spreading of the links when taut, which meant that each working chain had to be regularly checked against a duplicate kept for no other purpose than to serve as a standard.

Chains were made in various lengths. Four perches or twenty-two yards became the most popular choice among English surveyors, being the traditional width of an arable strip in their country's midland field system. This measurement became even more practicable when Edmond Gunter decimalised its length with a division into 100 equal links.[35] By the 1660s the chain had become not just a physical object but a purely abstract unit of length comparable with yards or feet. In military mapping, where areal calculations were not required, a 100-foot chain was sometimes recommended.[36] For the shorter distances of an urban survey 50 feet was a common choice.[37]

With suitable precautions chaining could achieve an accuracy of about 1 in 500 on a good surface. For any linear method to do better, a heroic struggle was needed against variations of alignment or tension, and against the expansion or contraction of the standard in response to changing temperatures. This necessitated more expensive equipment, more manpower and more skill, as well as a slower rate of progress. To accommodate such refinements the whole process of surveying had to be restructured, so that for any given map not more than a few measured lines, perhaps only one, could expect to achieve the highest level of precision. In late seventeenth-century France, rods of glass and (a surprising choice) wood were tried not very successfully as alternatives to the chain. But the most ambitious expedient before *c.*1850 was to counter the effect of temperature change by combining two metals with different coefficients of expansion that could be made to cancel each other out. This was the principle of Thomas Colby's compensation bars, first used in 1827.[38] In general, however, a steel chain could yield results as accurate as most surveyors would require.

The measurement of direction

In one sense the transition from linear to angular measurement is already behind us, because to measure a straight line in effect entailed laying out two right angles. Also meriting special treatment is the single right angle. In subtense or tacheometric methods a horizontal bar of known length was set up at one end of the survey line and at 90 degrees to it; from the other end of the line the surveyor took the angle

4.4 Vertical angles, altazimuthal instrument. Thomas Digges, *A geometrical practise, named pantometria* (London, 1571).

subtended by the two ends of the bar.[39] Simple geometry or trigonometry then gave the distance between bar and observer. This method was proposed at the beginning of the seventeenth century and was presumably the rationale behind a succession of later claims to measure distances from a single station. In the eighteenth and nineteenth centuries tacheometry was applied to surfaces unsuitable for chaining, for instance in mountainous country or over water, sometimes by using a ship's mast as a subtense bar in determining horizontal distances from ship to shore.

The superior status of right angles was also in evidence when the surveyor's measurements involved a third dimension (Fig. 4.4). This could happen for two reasons: first when the elevations of heavenly bodies were used in fixing positions on the earth's surface, and again in the mapping of terrestrial altitudes and relief features. Vertical angles could be determined only by reference to a horizontal plane, which on a spherical earth meant laying off a right angle from a straight line that passed through the centre of the globe. In a calm sea, nature provided a horizon. Away from the sea, a star could be reflected in an artificial surface such as a trough of mercury.[40] On land, for most of human history, a vertical line could be adequately defined in terms of gravitational force. A plummet or plumb bob attached to a T-square or to the apex of a wooden isoceles triangle was one possibility. Another was the instrument known simply as a level.

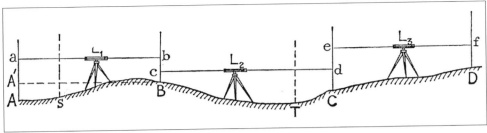

4.5 Height measurement, level with graduated staves. James Glendinning, *Principles of surveying* (2nd ed., London, 1960), p. 145.

In early levels the horizontal 'line of collimation' was obtained by filling part of a circular or U-shaped tube with a liquid whose two inter-communicating surfaces, though some distance apart horizontally, would stand at the same vertical height. Water levels of this kind were mentioned in ancient times by Hero of Alexandria and by Vitruvius. In the bubble level, first described in 1666 by Melchisédech Thevenot, the two surfaces were reduced to one: the tube, which could now be straight instead of U-shaped, was horizontal when a bubble in the liquid stood centrally between its two ends. A spirit such as alcohol was more mobile than water and yielded more accurate results. The spirit level was given its modern telescopic form by Jonathan Sisson in *c*.1725, but levelling remained a laborious procedure because on steep slopes the lines of sight were necessarily very short and each observation had to be made by two people (Fig. 4.5).[41] Until the mid-nineteenth century staff and level were used mainly in architectural and engineering work.

The next kinds of angularity for discussion are the obtuse and the acute. It was a brilliant idea to treat space as a sphere of indefinite size with the observer at its centre. The relation between any two points on the inner surface of the sphere was then expressible as the angle subtended by these points at the observer's instrument. Units of reckoning could now be defined not as aggregations of small quantities but simply by subdividing one enormous circle. The modern sexagesimal division of a complete surround into 360 degrees is among the oldest ways of doing this, having been devised by Babylonian astronomers early in the first millennium BC. Their Greek successor Eratosthenes was so impressed by this system that he altered his estimate of the earth's circumference for no other reason than to make it divisible without remainder by sixty.[42] In the ancient world, degrees co-existed with a classification of winds in which there were unequal intervals between one wind direction and another. A more symmetrical system, sometimes attributed to the Emperor Charlemagne, was for the circle to be successively bisected, ultimately giving thirty-two points (with non-quantitative designations such as north, north by east, north-north-east and so on) at a rather awkward interval of 11.25 degrees.[43] Not that the circle was easily susceptible to decimalisation: a thousand angular units were too many, a hundred too few. A hundred 'grades' instead of ninety degrees

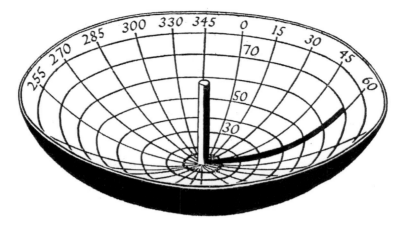

4.6 Measuring the sun's elevation by length of shadow in a gnomon.

would have 'privileged' the right angle to excess, though it was briefly attempted after the French Revolution.[44] Most surveyors preferred to live with 360 degrees in a circle, 60 minutes in a degree and 60 seconds in a minute, an arrangement whose user-friendliness was not improved by applying the words 'minute' and 'second' indiscriminately to time and space.

In angular measurement the necessary travelling was done not by the surveyor but by rays of either natural or artificial light, so that a measurement involving three widely separated points could take place at a single station. As a result, observation was possible not just to places on the earth's surface but also to far-off targets like the sun, moon and stars, a major advantage in cartography as we shall see. Another benefit of angular as opposed to linear units was that they were all interchangeable, the circle being common to every system. On the other hand bearings could never be directly converted into yards or miles because there was obviously no single linear distance separating the limbs that converged to form a given angle.

A typical angle-measuring appliance included two sights – fine strands formed by pins, strings, wires, hairs or even spider's webs – that could be brought into line with each other and with a distant object. The simplest such apparatus was the groma, recorded in the first century BC, in which the sights were set permanently at an angle of ninety degrees.[45] This served less for mapping in the conventional sense than for imposing rectangular streets and property boundaries on the landscape itself, as was done in the Roman system of land-allotment known as centuriation.[46] In later methods of surveying, more convenient fixed-angle instruments – the surveyor's cross and the optical square – were useful for setting out perpendicular offsets connecting chain-lines with individual objects on the ground, though in this situation, where the lateral distances were always kept relatively short, it was probably more common for the right angles to be estimated than measured.[47]

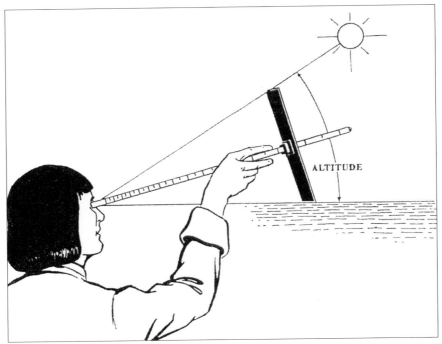

4.7 Vertical angles, cross-staff. H.O. Hill and E.W. Paget-Tomlinson, *Instruments of navigation: a catalogue of instruments at the National Maritime Museum* (London, 1958), p. 9. See also Figs 5.6, 11.6.

As a further logical step in instrumental design, the angle between two sight lines could be made adjustable. This was done in the Roman diopter or dioptra, but here too (though with less justification) the resulting angle seems to have usually been a matter of guesswork.[48] More serviceable, to say the least, were those instruments in which the sights accompanied some kind of index pointing to divisions on a numerical scale. One group of instruments used the subtense principle in which some elements of a right-angled triangle can be deduced from the others without having to be measured, one side of the triangle in question actually forming part of the instrument – perhaps the whole of the instrument. Thus by measuring the shadow of a vertical rod as cast by the sun on flat ground, one could find the sun's angular elevation, given a table of tangents and the length of the rod. But the angles themselves could be read directly if the rod were mounted in the centre of a hemispherical bowl like the gnomon used for astronomical observations by the Chaldeans and introduced to ancient Greece by Anaximander in the sixth century BC (Fig. 4.6).[49] Aristarchus (310–230 BC) is credited with the idea of graduating the sides of the bowl.

The progress of medieval European navigation brought a need for more portable instruments. Two of these had sights located on cross-pieces that could slide along a graduated axial rod.[50] The cross staff (Fig. 4.7) was known around 1350 and

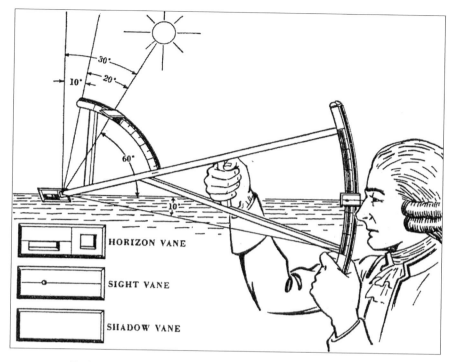

4.8 Back staff. Hill and Paget-Tomlinson, *Instruments of navigation*, p. 11.

became common in the sixteenth century.[51] Though capable of measuring horizontal angles, it was mainly used to find the sun's elevation above the horizon. To protect the observer's eyes, John Davis invented the back staff, first described in 1594, in which the sun's shadow acted as a substitute for the sun itself.[52] In the final version of this instrument there were two arc-shaped cross-pieces, one for the viewer and one for the sun (Fig. 4.8). Since the arcs defined two angles amounting to a quarter of a circle this appliance was often called a quadrant. It may be regarded as transitional between the staff instruments and those that depended on a single circle. Despite their awkwardness and fragility, staff instruments remained in service for vertical angles throughout the seventeenth and early eighteenth centuries. They were read to single minutes and according to Davis the cross staff was more effective than either the astrolabe or the single-arc quadrant, both described below.

In another group of angular devices the sights were aligned on two or more different targets in succession and rotated around the centre of a plane circle or part-circle whose circumference was calibrated in degrees and fractions of a degree. By turning the sights the angular distance between any pair of objects could be determined in relation to the observer's viewpoint. The oldest such instrument was a circular disc known as an astrolabe, originally used as a calculating device but simplified in the mid fifteenth century for use in measuring angular distances to

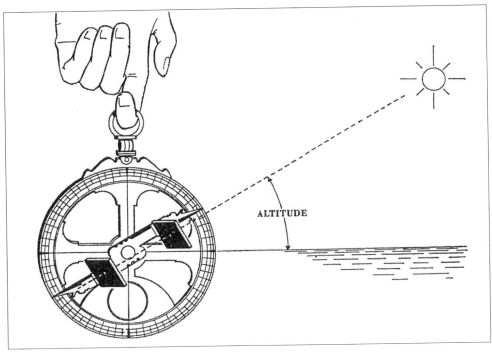

4.9 Vertical angles, astrolabe. Hill and Paget-Tomlinson, *Instruments of navigation*, p. 7. See also Fig. 11.6.

remote objects (Fig. 4.9).[53] Like the staff instruments, it could measure horizontally but in practice was used mainly for elevations.[54] Supported only by a handle at the top, it hung in correct alignment under its own weight to allow the sighting of vertical angles. The circle was usually quite small – a large astrolabe would have been too heavy to hold at eye level – and this made it difficult for scalar units to be read with much precision, but by the early seventeenth century a skilled observer could read to single minutes and achieve an accuracy of perhaps twenty minutes.

The idea of the astrolabe was adaptable in various ways. A circular scale could be mounted on a stand or tripod for measuring horizontal angles, in which case it might be called simply a 'circle' or sometimes a 'theodolite' (Fig. 4.10).[55] Alternatively the apparatus could be made lighter, less cumbersome, and presumably cheaper by leaving out a large segment of the circumference. In horizontal observations, an angle of more than two right angles could be subtracted from 360 degrees and annotated accordingly. Many eighteenth-century surveying instruments were based on this principle: an early French example was a kind of half-circular compass known as a graphometer, invented in 1597 and used later for fixing local detail in Cassini's survey of France.[56] For vertical measurement a quarter-circle or quadrant would be sufficient, since no angle of elevation can exceed 90 degrees. In its original form the quadrant had sights at the ends of one of its radial edges, while a plumb

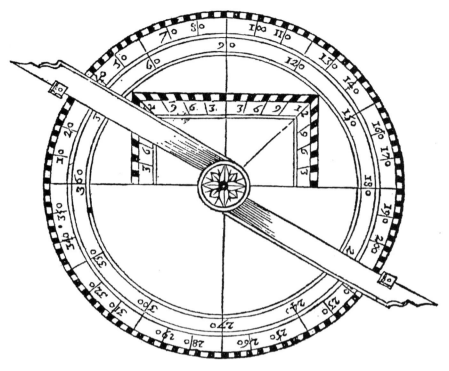

4.10 16th-century theodolite with open sights. Digges, *Geometrical practise.*

bob fixed at the centre gave a vertical alignment crossing the circumferential scale at a point determined by the angle of tilt (Fig. 4.11).

The accuracy of an early quadrant depended among other things upon its size. As Davis put it, 'those instruments whose degrees are of largest capacity are instruments of most certainty'.[57] For portable use a radius of two feet was considered sufficient. More than that, and the advantages of fine subdivision were outweighed by the difficulty of keeping a heavy weight under control, but in an astronomical observatory, where each quadrant could be mounted on a permanent base, the only size-limit was the dimensional stability of the materials from which the instrument was constructed: six feet, ten feet, even twenty feet were not considered excessive. Where the maximum angle required was known in advance, even smaller parts of a circle might constitute the entire apparatus, as in the sextant or sextile and the octant. Vertical instruments of greater accuracy but narrower range were known as zenith sectors.[58]

In 1731 a new variation on the quadrant was introduced by John Hadley, vice president of the Royal Society of London (Fig. 4.12). It was one of the few well-documented technical innovations that brought a sudden and large advance in the accuracy of maps, or in this case charts. Hadley laid claim to future celebrity by

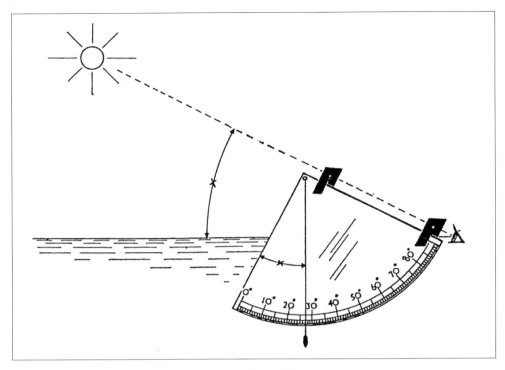

4.11 Vertical angles, portable quadrant. Hill and Paget-Tomlinson, *Instruments of navigation*, p. 6.
See also Fig. 11.6.

announcing his invention not through the usual instrument-maker's newspaper advertisement but as part of the Royal Society's *Transactions*, and later by writing about it in Latin.[59] One merit of his invention was to bring the images of the horizon and the sun or star into direct contact, making it easier to observe from the deck of a moving ship. Another was that by redirecting mirrored light rays within its own framework a small instrument could achieve levels of precision otherwise attainable only by placing an awkwardly long distance between back sight and fore sight. By 1766, according to Benjamin Martin, the largest quadrants had been 'quite laid aside'.[60]

The greatest distance measurable by a quadrant was ninety degrees. For observing wider arcs, a more useful instrument of similar design was the reflecting sextant, whose mirrors would enable a 60-degree reading to represent a real-world angle of 120 degrees. This extension of Hadley's method was introduced by Captain John Campbell in 1757. It became the seaman's favourite instrument for observing the moon and stars and was also applicable with limited accuracy to horizontal angles in land surveying.[61] Campbell's invention was a counter to the reflecting circle, devised in 1753 by Johann Tobias Mayer, which Campbell considered excessively cumbersome.[62]

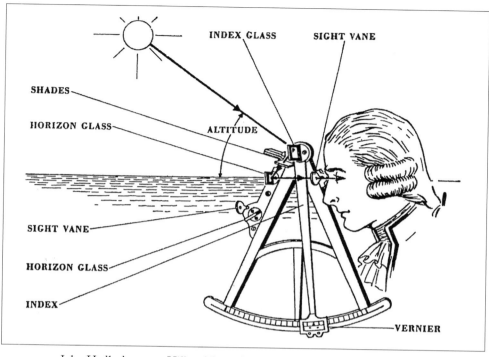

INDEX GLASS **SIGHT VANE**

SHADES

HORIZON GLASS **ALTITUDE**

SIGHT VANE

HORIZON GLASS

INDEX

VERNIER

4.12 John Hadley's octant. Hill and Paget-Tomlinson, *Instruments of navigation*, p. 14.

Next for discussion are what may be called 'altazimuthal' devices in which the line of sight could be rotated both horizontally and vertically, an advantage for levelling and field astronomy as well as for ordinary land surveying. In its early form this kind of instrument was sometimes called a 'polymetrum', first illustrated at the beginning of the sixteenth century, sometimes just a 'topographical instrument' (Fig. 4.13).[63] The term 'theodolite' appears with the same meaning as early as 1596, though as we have seen a theodolite could also be a simple horizontal circle. Theodolites in the later sense were heavy, cumbersome, fragile and expensive, but they met the special needs of cartography by measuring horizontally between two stations that were not on the same level, in other words by determining the angle between two vertical planes, whereas the angle taken between the same points with a quadrant or sextant would have been too large if plotted straight on to a flat map. The theodolite attained its later status as all-purpose angular instrument in the 1730s thanks to the ingenuity of Jonathan Sisson (Fig. 4.14).[64] By then, optimists put its limit of accuracy at one minute, though more realistic estimates were several times larger.[65] As with early quadrants the precision of a theodolite depended on its size: on Jesse Ramsden's three-foot diameter of 1787 readings could be taken to one second.[66]

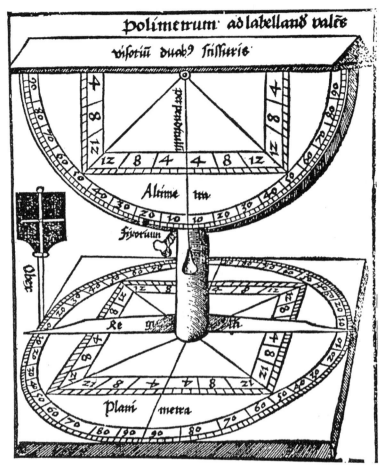

4.13 16th-century polymetrum, ancestor of the modern theodolite. Gregor
Reisch, *Margarita philosophica* (Strasbourg, 1504). See also Figs 4.4, 6.1.

The magnetic compass

An important medieval discovery was that magnetic needles, if free to rotate,
would point approximately northwards. The use of the compass for navigation was
known in eleventh-century China and mentioned by the English scholar
Alexander Neckham in 1187. During the next two centuries it was refined in
various ways – afloat in mercury or water, rotating on a pivot, mounted in gimbals
and a binnacle, and combined with a 'fly' or card distinguishing different directions
as compass points or (later) degrees. The first unambiguous description of the
compass in its box dates from *c*.1315.[67] This epoch-making device had two special
merits. One was to save labour. In non-magnetic angular measurement, sights were
separately aimed at two distant objects, and two separate readings taken on the dial

4.14 Theodolite by Utzschneider and Liebherr, Munich, 1820.
http://www.astro.helsinki.fi/museo/ laitekuvat/3_teodoliitti_valk_1.jpg.
Information from Matthew Stout.

of the instrument. But in every magnetic angle there is one limb for which nature takes aim and the instrument reads the result, thus reducing the surveyor's workload by half. Admittedly a compass survey might be interpreted as a series of figures each with one corner at the earth's magnetic pole, a remote and invisible object of no practical importance to most map readers. But if we have a compass bearing for each side of a more localised triangle its internal angles can easily be calculated and the pole forgotten. In any case it was the very invisibility of magnetic north that made the compass so useful to the navigator: earth and sky might both be totally obscured by darkness, cloud or fog, but with a compass he still knew where he was going.

Unfortunately the merits of the compass were balanced by several inherent weaknesses.[68] First, magnetic and geographical north seldom coincided exactly, a fact that was common knowledge in western Europe as early as the mid-fifteenth century. The difference could be easily measured by observing the compass direction of a line that was known to run from geographical north to south. The variation could then be recognised by marking it on the edge of the compass card and showing true north at the correct distance right or left of the needle. The trouble

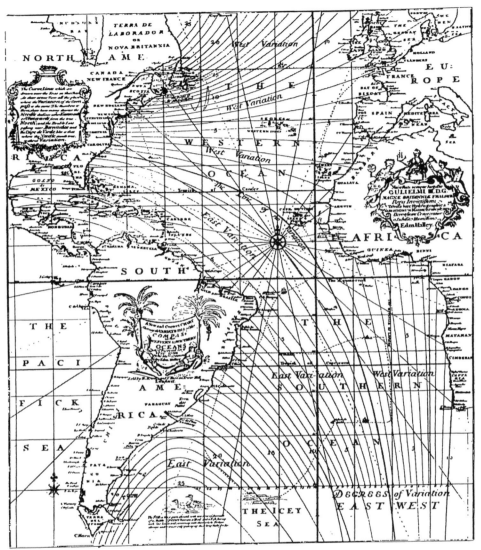

4.15 Geographical versus magnetic north. Lines of equal compass variation by Edmond Halley, *A new and correct chart shewing the variations of the compass in the Western and Southern Oceans* (London, 1701).

now was that magnetic variation differed considerably from place to place, a discovery made by Christopher Columbus in the course of sailing to America. In the Atlantic of Columbus's lifetime the variation was easterly in the east, westerly in the west, and even within a small country like England it could differ regionally by one or two degrees. In these circumstances, correcting the instrument for variation might do more harm than good, and a useful purpose could still be served by 'meridional' compasses that had *not* been adjusted in this way. The alternative

4.16 Circumferentor or surveying compass. J.C. Ludewig, Dresden, *c.*1720, Jagiellonian University Museum. Information from Matthew Stout.

was an 'azimuthal' compass with sights or a shadow vane through which the sun or stars could be observed to establish true north.

At this point an historiographical warning needs to be interpolated. An apparent advantage of magnetic variation for the student is the possibility of drawing inferences from early maps and charts in which north, as defined by a margin or a meridian line, turns out to be a contemporary magnetic compass direction rather than a reference to the earth's geographical pole. Such maps are often thought to have been plotted entirely from compass bearings but this argument falls some way short of total validity: a surveyor could take all necessary bearings with a non-magnetic theodolite and then at the last minute furnish his map with a magnetic north-point, presumably for the benefit of readers who might use a pocket compass for way-finding.

Regional differences in terrestrial magnetism could themselves be surveyed and mapped like any other fact of geography, an operation that was famously performed by Edmond Halley in 1700 (Fig. 4.15).[69] By that time the behaviour of the compass had been shown to vary in a temporal as well as a spatial sense. To complicate matters further, the time sequences also varied geographically in ways that were difficult to predict, though it was safe to say that at any given place the direction of

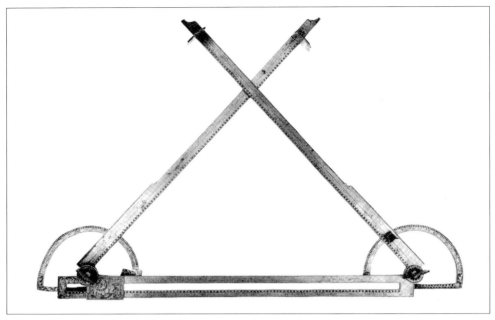

4.17 Late 16th-century trigonometer by Philippe Danfrie. J.A. Bennett, *The divided circle: a history of instruments for astronomy, navigation and surveying* (London, 1987), p. 45.

compass north might change by as much as one degree in seven years.[70] To such secular changes should be added a number of smaller short-term variations, perhaps amounting to as much as thirty minutes per day. Finally a compass needle could be disturbed by local magnetic attraction, whether from nearby metal artifacts or natural iron in the earth's crust. It is hardly surprising that no one tried to improve compasses by making them very large, as was done with quadrants and theodolites. Instead it soon became generally agreed that for all its advantages this was a measuring tool with a lower order of accuracy.

Nevertheless many azimuthal instruments were designed to incorporate a compass, either as a means of getting a rough preliminary orientation before attempting fine adjustments or as a last resort when visibility was poor. The most popular instrument intended for compass surveys and nothing else was the circumferentor, illustrated by Niccolo Tartaglia of Venice in 1546 (Fig. 4.16). This differed from the theodolite in that it was not the sights but the whole instrument, including the compass box, that rotated horizontally on its stand, making it more convenient for the circumferentor dial to be graduated 'in reverse', with easterly bearings counted from right to left. The circumferentor was used in colonial environments throughout the eighteenth century.[71] After that it fell out of favour, as exploratory surveyors came to prefer the pocket-sized hand-held prismatic compass introduced in 1806.[72]

4.18 Modern plane table with alidade and box compass.

Finally, there were instruments that measured angles by simulating them. Instead of one pair of sights they had two pairs, which could be rotated to make the surveyor's lines of vision converge upon his standpoint.[73] The sauterelle, recorded in 1567, was a hinged ruler used to measure turns in a road by taking alignments along both the limbs. By adding a third ruler Sebastian Münster made it possible to measure two angles. Other such instruments were the 'Zürich triangle' (1602) and the trigonometer (1684: Fig. 4.17). One ruler, or part of it, was treated as a base line corresponding to a specified distance on the ground. An angle was observed from one end of the ground base to a distant station, and the rulers defining that angle were clamped in position. A second angle was next observed to the same station from the opposite extremity of the measured distance, using the third ruler at the other end of the instrumental 'base'. The two rotary rulers now defined a point of intersection on the map corresponding to the station that had been intersected in the landscape. The terrestrial triangle had thus been miniaturised in wood or metal. Its corners could now be measured, and so could the relative lengths of its sides. In Abel Foullon's holometer of 1551 the rulers swung across a table which also incorporated a compass and a pair of vertically rotating sights for measuring altitude. In the familiar modern expression 'plane table', historians have interpreted the word

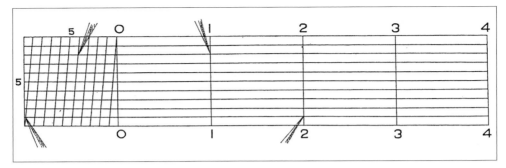

4.19 Diagonal scale. Of the two distances shown, the upper is 1.42 units, the lower 2.99 units.

'plane' as 'plain' in the sense of simple or straightforward, indicating that triangles could be reconstructed equally well without putting anything on Foullon's table except a sheet of paper, a pencil, and one free-standing pair of sights incorporated in a ruler or alidade (Fig. 4.18).[74] This however takes us from pure measurement to the subjects of later chapters on surveying and plotting.

The pace of improvement

It is usual in histories of surveying to proceed from one named measuring device to another, but this piecemeal approach may obscure some interesting broader issues. In particular, certain historical changes of great significance were common to a number of instruments. First there was the question of physical strength and stability. Metal was more durable than rope, cord or cloth, while steel was lighter than iron and just as strong. For azimuthal instruments brass was increasingly preferred to wood, especially from the middle of the eighteenth century. Scales engraved on metallic as opposed to wooden surfaces were capable of greater precision, and in the eighteenth century much care was given to dividing an arc by repeated graphic bisection of angles. A careful surveyor could hope to reduce errors of graduation by taking different observations of the same angle, each on a different part of a circular instrument, and then adopting an average of the results. After 1775, however, such precautions were rendered less important by Jesse Ramsden's dividing engine: this involved the use of a micrometer, in which the requisite angles could be set out by turning a fine screw through a definite number of rotations.[75] These new dividing techniques greatly improved the accuracy of small portable instruments like the sextant.

Another problem for the surveyor was the weakness of his own eyesight, whether in observing at long distances or in distinguishing scale division marks that lay very close together. An ingenious remedy was to expand the measurer's linear frame of reference into a second dimension. The oldest such device was the diagonal scale, familiar on modern school rulers (Fig. 4.19). Others were the nonius, a radial

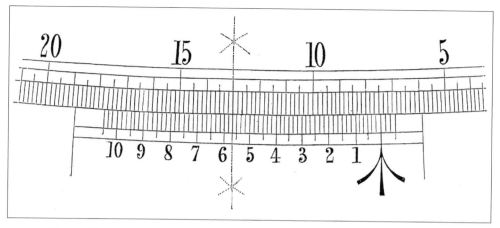

4.20 Vernier. The upper scale-intervals (reading downwards: 5 degrees, 1 degree, half a degree, 10 minutes) show a value at the arrow-head of 7 degrees, 20 minutes plus an unspecified fraction of 10 minutes. The lower, vernier scale-intervals (reading upwards: 1 minute, 30 seconds, 10 seconds) are counted from the arrow-head to the point where the two smallest sets of scale-divisions coincide, i.e. 5 minutes, 40 seconds. Value for the whole angle: 7° 25' 40".

development of the diagonal scale that was never very popular,[76] and, from the mid-eighteenth century onwards, the more effective vernier (Fig. 4.20).[77] At the highest level of precision, microscopes and micrometers, together with verniers, could enable a theodolite to measure with a *prima facie* accuracy of two seconds.

In observing far-off landmarks the cure for poor vision was the telescope, invented around the end of the sixteenth century and first used in terrestrial surveying instruments some fifty years later.[78] The replacement of open by telescopic sights extended an observer's command of distance, and fine cross hairs could be protected inside the cylinder of the instrument, thus increasing the precision with which the sight could 'cut' a narrow station-mark. Early telescopes were subject to chromatic aberration, which blurred the image of a distant object by dispersing light of different wavelengths, but in 1757 it was discovered by John Dollond that if two different lenses were combined 'achromatically', one could be made to correct the other in this respect.[79]

Perhaps what emerges most clearly from the foregoing paragraphs is the pre-eminence of the sixteenth century as a time of revolutionary change in the practical application of geometrical knowledge. In particular this period saw the advent of various measuring instruments not previously used in cartography. Designing such appliances was a matter of concern among independent scientists and mathematicians, and many practical surveyors sought credit by introducing new techniques rather than by drawing good maps. From the mid-seventeenth century, progress occurred chiefly by improving instruments that already existed. This was a cumulative process. Some inventions brought no immediate benefit because their advantages were nullified by apparently irremediable defects. Only when those defects were eliminated could the previous improvement come into its own. In the mid-eighteenth century enough of these changes had matured to constitute another revolution. Finally, the years immediately after 1815 brought an era of comparative peace in which generously funded national survey organisations could exploit a new range of advantages: precision engineering; further scientific progress in mathematics, physics and chemistry; a combination of military and industrial discipline in the work force; the appearance of able leaders whose energies might otherwise have been diverted into more warlike pursuits. More than ever before, it was now worth adopting technical improvements as soon as they became available, so the best maps of this age were as good as the art of mensuration could make them.

CHAPTER 5

Remarkable objects

NO MEASUREMENT CAN BE quite precise unless its terminals are geometrical points in the Euclidean sense of possessing neither size nor shape. Such entities are too abstract to appear on paper but the map-maker must still do his best to represent them. In practice the smallest mark he can draw with a pencil is usually reckoned to be a hundredth of an inch across.[1] So on the ground all map points possess a definite magnitude that depends on the cartographer's chosen scale; at one inch to a mile each dot corresponds to a circle about fifty feet in diameter, so a medium-sized house could be treated as a point by surveyors preparing a map at this scale.

There is another complication to be addressed before we go any further. Distances scaled from the surface of a normal map are usually thought to be horizontal: in deference to this belief, a surveyor with cartographic intentions should either hold his chain level or else determine the gradient of each chain line and reduce it to a horizontal equivalent by trigonometry. For the most accurate work, however, the concept of horizontality is indeterminate. Until some height-datum has been specified the points of cartography are not points at all, but straight lines of uncertain length passing through the centre of the earth. With increasing distance from that centre, the horizontal gap between any two such radial lines will naturally grow longer. Whatever may be done to keep the surveyor's measuring apparatus level, there is no definite distance between any two points unless we state the exact plane in which the distance will be reckoned. Clearly this plane should be the same for all the lines in any given survey, and by the end of the eighteenth century measurements were accurate enough for the position of a datum surface to be worth debating among aspirants to scientific exactitude. The obvious location for such a datum was sea level.[2] In some ways this was hardly an issue of the utmost importance. When the Ordnance Survey laid out a seven-mile base on Salisbury Plain in 1794, its reduced or corrected length was shorter than the measured length by about two inches.[3] As for exactly what one meant by sea level, it was another fifty years before this problem came to a head.

Terrestrial survey stations

A surveyor's choice of what the old textbooks called 'remarkable objects'[4] would depend not just on size but also on visual prominence and recognisability, practical importance to the map-reader and, in the most accurate surveys, ease of access for the field-worker. In the real world these virtues were not always combined. Hill summits, headlands, small rocks and islets, towers, spires, windmills, lighthouses and even isolated tall trees might be both visible and interesting, but they were not always within easy reach. Nor did they necessarily provide an unambiguous landmark. Hills and headlands might look less angular from one viewpoint than from another. False summits would occur on a convex slope, and false promontories on an outward-curving coastline. Cape Egmont in New Zealand, for example, was too 'round in form' for different observations of it to be comparable.[5] Caves, springs, river-mouths or confluences might appear as points on maps of the appropriate scale; so too might the junctions of roads, fences and ditches, but in the field such features were often hard to see at a distance. Objects visible from far away could still be difficult to identify: one Irish surveyor identified a landmark in his field book as 'Something on Roskey hill'.[6] Even the best modern survey departments have suffered in the same way. In Ireland two points on the Bills Rocks in the sea off County Mayo were noted as suitable marks for theodolite observations, but then the Ordnance surveyors of the 1830s mistook one of them for the other, shifting the rocks on the published map too far to the west by about 1200 feet.[7]

Landmarks could be too few as well as too many: in thick jungle, where the traveller's visual field was often devoid of anything that could serve as a point, a bearing might have to be taken instead to a sound produced by an invisible colleague. But most man-made survey stations were optically if not tactually accessible. They might be poles or flags, lamps observable at night, or mirrors that diverted the sun's rays towards the observer. If sights were needed from such places as well as to them, some kind of artificial observation platform might have to be provided. Complete stations were sometimes built from ground level upwards as an artificial mound or perhaps a wooden tower, remaining identifiable after demolition by the burial of a marked stone at a site described in the surveyor's written records.[8] This happened in two kinds of situation. As we shall see, some surveys required one or more specially chosen lines to be measured with the highest possible accuracy. The best sites were on flat, smooth, firm ground, typically in an area of well-drained alluvial land like Romney Marsh in Kent or Sedgemoor in Somerset, both used for this purpose by the British Trigonometrical Survey before the end of the eighteenth century.[9] Then there were sparsely populated tracts of lowlying forest or open prairie or even frozen lakes devoid of appropriate stations. In the Indo-Gangetic plains of India surveying was seriously curtailed by the expense of building towers.[10] At sea, the mast of an anchored ship might function as a survey point. Of course these artificial markers had no cartographic value in themselves. Separate measurements were needed to link them with more durable landscape features, but that could be a small price to pay for the accuracy of the survey as a whole.

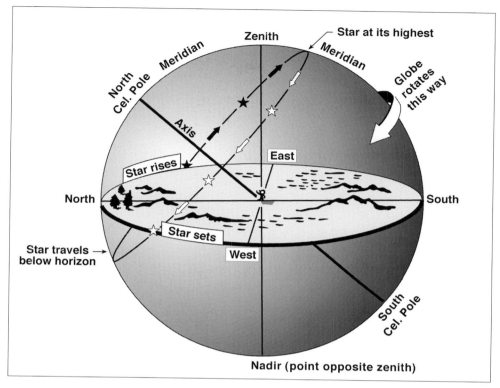

5.1 Path of a star across the celestial sphere. Redrawn by Matthew Stout.

The celestial sphere

It was the lack of observable points on so much of the earth's surface, especially at sea, that from time immemorial directed the surveyor's gaze towards the heavens. In a cloudless sky, at least one located object – sun, moon, star – was always visible, and these objects either looked like points or had identifiable centres that could be treated as such. Their movements, though sometimes highly complex, were usually predictable. Unfortunately the sky was not always clear: this is why the science of astronomy originated in areas of very low rainfall.

In astronomical observation the sky is treated as a hemispherical bowl with the observer at its centre and the various heavenly bodies attached to its inner surface (Fig. 5.1). The practical convenience of this assumption was unaffected by Copernicus's discovery that the earth moves round the sun and not vice versa, though early astronomers doubtless regretted that the bounty of Providence did not extend to calibrating the surface of the bowl. A second hemisphere of the same kind is out of sight, hidden by the bulge of the earth. Visible and invisible hemispheres together form the celestial sphere. (The earth itself is treated as a smaller sphere at the centre

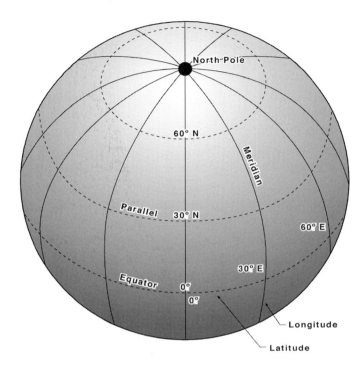

5.2 Parallels and meridians on the earth's surface with latitudes and longitudes in degrees.

of the celestial sphere: the effects of its spheroidal shape can be considered later.) The stars all follow parallel circular courses round the bowl, each circuit taking twenty-four hours. Some remain above the horizon at all times, others rise and set. A straight line connecting the centres of these circles will pass through the earth's north and south poles and will meet the celestial sphere at two points known as the celestial poles. For navigators the merit of a celestial pole consists in seeming to be always in the same place: in fact the earth wobbles under the gravitational forces from the sun and moon (a phenomenon known as precession), causing the celestial pole to follow a circular course around the sky but for purposes of cartographic history this phenomenon can be disregarded.[11]

It would be expecting too much for the celestial pole to coincide precisely with any visible object, but in the northern hemisphere one star is close enough to have served as a convenient substitute. This is the appropriately named Polaris or Pole Star. Its distance from the true pole has varied considerably. Today that distance is less than one degree, but in the fifteenth century it was 3 degrees 30 minutes, sufficient reason for mathematicians to produce tables showing how the height of Polaris should be corrected at different times and different seasons to pin-point the pole itself.[12]

The position of the celestial pole was important to early geographers in two ways. First, it defined a plane, known as a meridian, which passed through both the pole and the observer and which met the earth's surface along a line (also called a meridian, this time in a geographical rather than an astronomical sense) that followed a semicircular north-south course along one of the shortest surface distances between the two terrestrial poles (Fig. 5.2). Meridians introduced an element of absolutism into a system of horizontal angular measurement that would otherwise have been wholly relativistic. Thus 'A lies between B and C' means nothing unless we already know the whereabouts of B and C, but for understanding 'A lies north of B' no third point is necessary; this simpler proposition is intelligible everywhere on earth, and becomes seriously unhelpful only if B happens to coincide with the south pole. Meridians thus provide convenient lines of reference for reckoning the direction of one terrestrial point from another, a variable which when quantified is generally known as an azimuth.

Meridians could also claim a special status among the infinitude of circles on the earth, a privilege conferred on them by the apparently unchanging position of the celestial pole. 'Privilege' in this sense was an almost unavoidable feature of map-making, even in its least mathematical form. Readers expected a cartographer to avoid wasting paper or parchment by keeping the borders of a rectangular image at least approximately in line with the edges of its sheet; meridians drawn on a map could help to enforce this requirement.

Positions north and south

The celestial pole does more than show that one point is north of another: it can also indicate how much further north. This possibility is best approached from the viewpoint of a single observer. If the north celestial pole is vertically above our head we are at the earth's north geographical pole. If the celestial pole is on the horizon we are on the earth's equator. If it is half way between zenith and horizon we are half way between equator and pole, and so on. Clearly our north-south position on this quadrant of the earth's circumference can be expressed by an angle between nought and ninety degrees, the angle subtended at the earth's centre by the observer and the plane of the equator. The same applies in reverse to the southern hemisphere. The angle in question is terrestrial latitude, and it can be seen from the simplest of geometrical diagrams (Fig. 5.3) that the latitude of any point is the same as the elevation of the celestial pole at that point.

Traditionally both azimuth and latitude have been measured by observing the pole star. But any other star could be used instead. Its horizontal bearing from some arbitrary fixed point could be measured both at its westernmost and easternmost position. The mean of the two bearings would give the direction of the meridian. Similarly, the altitude of a star could be measured at its upper and lower crossings of the meridian, provided it remained above the horizon, and the mean of these two values would be the altitude of the celestial pole. We can go further: in theory, the

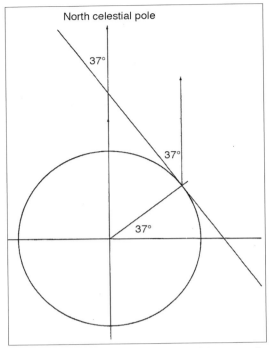

North celestial pole

37°

37°

37°

5.3 Terrestrial latitude as the angle of elevation of the celestial pole above the horizon.

elevation of any star at any moment could give the observer's latitude if we knew its angular distance or 'declination' from the pole, the time at which it crossed the meridian, and the time of our observation.

In latitude observations the sun was a daytime alternative to the pole star, with the further advantage that its shadow could act as a pointer in certain measuring instruments such as the gnomon and the sundial. The shadow could also be used to set out a meridian line, because the sun at noon, if not directly overhead, stood always either due north or due south of the observer. But now two complications had to be faced. First, the sun was not a point but a disc with a semi-diameter of about 16 minutes. Usually its edge or limb was observed and a correction made for this amount. More serious was the need to allow for seasonal variations. If the earth's axis were perpendicular to the plane of its annual revolution, the noonday sun at the equator would always be overhead, and the latitude in degrees of any point could be found by subtracting the sun's angle of elevation from ninety. In fact the earth's axis is inclined at an angle of about 23.5 degrees, so that the northern hemisphere leans towards the sun for half the year and away from it for the other half. The noonday sun is overhead on the equator twice a year, at the equinoxes, on about 21 March and 23 September. In June (about the 21st) it is overhead on the Tropic of Cancer at latitude 23.5 degrees north, and in December (about the 22nd) on the Tropic of Capricorn at 23.5 degrees south.[13] The June and December dates are known as the solstices, so called because the sun seems to stand still before moving in the opposite direction. At other times, the overhead sun falls somewhere between the two tropics. Its position north or south of the equator is known as the sun's declination. To find latitude from the noonday sun, we must know the declination on the day its height is measured.

Next comes another important sense in which latitude is related to time. At the equinoxes of March and September, as their name suggests, nights and days are of equal length. Otherwise the daily duration of sunlight increases from equator to pole in summer and decreases in winter. The arctic and antarctic circles, at 66.5 degrees north and south, delimit those polar regions that experience at least twenty-

5.4 'Climates' as numbered latitudinal belts defined by the duration of daylight. Joannes Myritius, *Universalis orbis descriptio* (Ingolstadt, 1590).

four consecutive hours of darkness in winter and daylight in summer. At other latitudes, since clocks of a sort were already available to the Greeks and Romans, ancient cartographers could in principle determine the maximum duration of continuous daylight wherever they chose and eventually such readings were to provide a world-wide system of latitudinal reference.[14] Strictly speaking, the necessary measurement for any given place was possible on only two days in the year, but as a metrological unit day-lengths were well adapted to a sedentary way of life where latitude was worth studying for scientific reasons rather than as an aid to navigation, and where the alternative idea of angular measurement in fractions of a circle was still not very familiar.

The concept of time-latitude was not without its difficulties, however. Perhaps the most fundamental of these was a certain failure to disentangle continuous variation from regional identity. The Greeks wanted definite zonal boundaries, but the only such lines provided by nature were the polar circles, tropics and equator. Another complication was the non-linear relationship between hours and degrees: a given increase of latitude lengthened the day by three times as much at 60 degrees as at 10 degrees. Matters were not improved by the practice of numbering zones from a point some way north of the equator, apparently in deference to the regional notion of the *oikoumene* or habitable earth. Finally, there was the problem of zonal names and zonal widths. Which would be equalised in the definition of latitudinal bands, hours of time or units of terrestrial distance? The answer was neither, as can be seen from the temporal width of Ptolemy's zones, which varied from fifteen minutes to a full hour. In the event two systems were to coexist on each side of the equator – 'climates', traditionally numbering seven, and 'parallels', of which Ptolemy recognised twenty-one.[15] Why not be content with 'hours', one wonders.

Within the body of a map, it is true, renaissance graticules usually specified degrees rather than hours. But latitudinal time-zones, however strange and unrewarding from a modern standpoint, remain an inescapable part of cartographic history. They appear in the margins of most Ptolemaic maps, both general and regional, as well as in numerous original world maps by sixteenth-century cartographers (Fig. 5.4), often generating more script than any other subject and sometimes dominating the entire composition, as for instance in Johannes Ruysch's *Universalior cogniti orbis tabula*. Only under the influence of the great modern atlas-makers did climates begin to relax their grip. Mercator re-tabulated the time-units, reckoning two parallels to one climate,[16] but except when editing Ptolemy he did nothing to recognise these divisions in his maps. Ortelius was equally unenthusiastic, on one occasion pointedly reserving the word 'parallel' for a reference to angular measure rather than hours.[17] Thenceforth the climates steadily fell out of use. Changing fashions in projection (discussed in chapter 8) may have had some influence here, because marginal captions to lines of latitude would be hard to fit in if these lines were steeply inclined to the horizontal. The main factor, however, was the ease and frequency with which angular latitudes were now being measured and recorded.

Whatever one's method of time-keeping, it was fortunate that, all over the earth, a single celestial observation at a single point could prove the observer to be standing on just one of an infinite number of parallel circles drawn across the terrestrial surface. One of the puzzles of map history is that so little was done in ancient times to seize this opportunity. Massilia, Rhodes, Alexandria, Syene and Meroe are the only well-authenticated cases of astronomical latitude determination in antiquity, and of those Syene was a lucky accident: its midday sun at the summer solstice happened to be visible as a reflection at the bottom of a vertical well-shaft.[18] By the time of the European renaissance things had changed. Egnatio Danti in 1577 seems to have thought that any ordinary tourist should be capable of finding his or her latitude.[19] 'Easily kept by any indifferent seaman' was a rather more modest judgement of 1638[20] and at about the same period a traveller to India could report how:

> Every day, about the hour of noon, the sun's altitude was infallibly observed, not only by the pilots, as the custom is in all ships, and the captain ... but ... at that hour twenty or thirty mariners, masters, boys, young men, and of all sorts came upon the deck to make the same observations: some with astrolabes, others with cross staffs, and others with several other instruments ... [21]

An early tribute to the status of the latitude concept was Francis Drake's plan for honouring Queen Elizabeth I in Chile with a monument to be located at exactly 52 degrees, the southern equivalent of her own position in the opposite hemisphere.[22] A more ephemeral mode of celebration customarily marked a ship's passage across the equator, the attendant ceremonies being described as 'ridiculous' rather than as novel even in the seventeenth century.[23] Where security was thought to be an issue, these freedom-of-information policies might seem less desirable. Henry Hudson in 1610 forbade his crew to keep 'account or reckoning', confiscating any instruments that might have been used for such a purpose.[24]

Ptolemy's latitudes were mostly recorded in degrees or half degrees. Many sixteenth-century navigators worked to the nearest minute, but they were often wrong by as much as twenty or thirty of these units. By the late seventeenth century the best measurements could give less than two minutes; in Cook's day, less than a minute. In the 1800s latitude was being quoted to the nearest second, and the British geodesist Henry Kater located a room in his house to the nearest tenth of a second.[25] By this time, however, the dominance of gravitational forces in the surveying process had encountered a new challenge. It had been suspected since Newton's time that plumb lines, rather than always pointing vertically downwards, might sometimes be deflected by the gravitational influence of a nearby mountain mass. In 1774 this possibility was tested when Nevil Maskelyne compared positions given by astronomical and terrestrial measurement on either side of the Scottish mountain Schiehallion. The plumb line of his zenith sector, he found, had been deflected by an angle of 5.8 seconds.[26] At first surveyors in quest of astronomical latitudes tried to solve this problem by avoiding mountainous country, but gravitational anomalies soon began to appear at quite low elevations, for example in the Isle of Wight. Such anomalies could be detected and measured by timing the swings of a pendulum, but it is not clear how many latitude values were actually corrected by this method. It seems to have been more common to identify a reliable astronomical observatory from which other latitudes could be deduced by means of terrestrial surveying methods.[27] Though important to geodesists, such refinements would make little difference to a normal map: at one inch to a mile, two seconds of latitude were represented by no more than a millimetre.

Positions east and west

Today everyone accepts that the similarity between latitude and longitude should be expressed by names with two final syllables in common, by being reckoned in the same units, and in general by enjoying the same geographical status. This symmetrical view is understandable if latitude and longitude are equated with width and length. Such an association seemed natural to the Greeks, who conceived the known world in essentially rectangular terms (as far as this was possible on a spherical earth) with boundaries that ran approximately parallel or perpendicular to one another and with an east-west length or longitude that considerably exceeded its north-south width or latitude. This quasi-rectangular conception obscured the profound difference between latitude and longitude as they appear on the globe.

Off the map, lines or parallels of latitude are circles of varying size, largest at the equator and smallest at the poles, where they become reduced to points. There are no longitudinal equivalents to the poles because the earth does not spin on an east-west axis. But a second series of coordinates could be set out (if anyone felt the urge to do so) from an arbitrarily chosen circle passing through the north and south poles to form a longitudinal version of the equator. Circular 'longitudinal' parallels of varying length could then be drawn on a map to enclose an east pole and a west pole. A certain tidiness would result from making parallels and pseudo-meridians as similar as they could be, but in fact no such system of geographical reference has ever been seriously considered: it would be intolerably awkward for each line of longitude to be constantly changing its azimuth and therefore intersecting the parallels of latitude at every possible angle. Instead, the meridians of longitude seen on maps have little in common with parallels of latitude apart from being circular. One difference is that all meridians are the same length. Their numerical designations range through a complete circle (two semicircles, if longitude is reckoned both eastwards and westwards from its point of origin) instead of through no more than a quarter-circle as with latitude.[28] Another difference is that whereas each unit of latitude represents the same linear distance along the surface of a spherical earth, the corresponding length of a longitudinal unit diminishes from the equator to the poles.

Despite these complications each line of longitude had one characteristic of great practical significance, which was that all points on its course experienced noon at the same moment – the moment when it was midnight at corresponding points on the opposite side of the earth. In short, the key to longitude was time. It has accordingly often been measured in chronometric units, with twenty-four hours representing a complete east-west circuit of the earth. On the other hand longitude

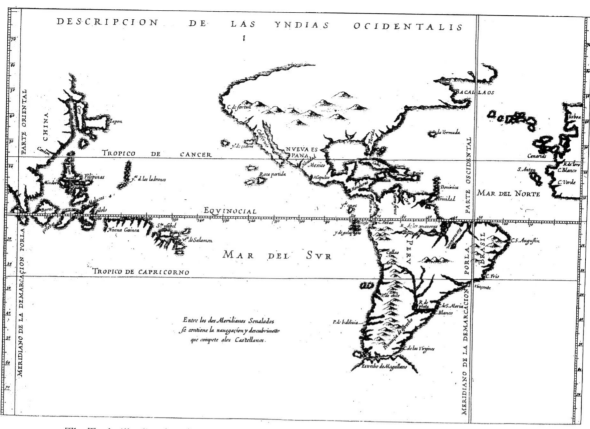

5.5 The Tordesillas line (1494) separating areas of Spanish and Portuguese control in the new world. Antonio de Herrera, *Descripcion de las Indias Occidentales* (Madrid, 1601).

could also be given an angular interpretation in which each meridian formed a plane, the longitudinal difference between two meridians being the angle subtended by these planes along the earth's axis, with 360 degrees as the equivalent of twenty-four hours. The difficulty with angles of longitude was to measure or calculate them. Amerigo Vespucci, discoverer of Brazil, said he had shortened his life by ten years trying to solve this problem.[29] A century later, in *c.*1588, Alvaro de Mendana was complaining about 'degrees there have been none that could ever measure'.[30] For early geographers to insist on reckoning longitude in degrees instead of miles or leagues can only be described as an act of heroism.

By the early seventeenth century, it is true, the concept of the terrestrial degree was so familiar that sea distances running obliquely to meridians and parallels were sometimes quoted in degrees not of latitude or longitude but simply of a great circle.[31] But in the world of practical navigation such statements must surely have been made partly in a spirit of bravado. Most masters in Mendana's time and for long afterwards avoided degrees and minutes when expressing east-west distance, preferring to estimate their position from rough measurements in linear units. Sure enough, the first historically important meridian of longitude, the Tordesillas line separating Spanish and Portuguese claims to the new world, was officially defined in 1493–4 as so many leagues – not degrees – west of the Cape Verde Islands (Fig. 5.5).[32] Genuinely non-linear determinations were seen to present a more serious problem, as governments came to recognise with offers of prizes for a method of obtaining them – the Spaniards in 1598, the Dutch in 1600, the French in 1668, the British in 1714.[33] The official British board of longitude was not dissolved until 1828.[34]

For a time it was hoped that magnetic variation could serve as a measurable surrogate for longitude. This seemed especially probable to early transatlantic voyagers like Sebastian Cabot and Christopher Columbus who found the variation changing as they sailed westwards. Humphrey Gilbert devised a spherical instrument with a built-in compass for the 'perfect knowing' of the longitude.[35] The idea was made plausible by the practice of counting degrees eastwards from the Azores, which as well as being appropriately placed outside the old-world land mass lay on the contemporary divide between eastern and western magnetic variation. The same view persisted until the early seventeenth century, eminent believers being Samuel de Champlain[36] and Abel Tasman, and it was subsequently reconciled with the fact that the variation differed in time as well as in space. As late as 1755 a pamphlet advocating the longitudinal theory of magnetic variation was ghost-written by Dr Samuel Johnson.[37] But magnetic longitude had been subverted by cartography itself when Edmond Halley's lines of equal variation were seen to differ considerably from geographical meridians.[38]

Alternative methods of longitude determination varied in merit. One of them appears written on the wall of a lunatic asylum in a famous picture by William Hogarth.[39] Some were less impracticable than they seemed. A project for firing rockets whose high-level detonations could be seen and timed at different places has been dismissed as fantastic, but eventually such rockets did help to determine the relative longitudes of observatories at Dublin and Armagh.[40] All in all, however, longitude remained a puzzle until late in the eighteenth century. Before that time it was more common for maps and charts to show parallels than meridians and when they did show both, the results were of very different quality. In Ivan Kirilov's instructions of 1721 to Russian surveyors, latitudes were to be found with a quadrant, longitudes to be 'as stated in old maps and catalogues'.[41]

A solution for longitude

Time on the earth's surface could most easily be measured by observing the sun. On any meridian the interval between one noon and the next was for most practical purposes constant, and this period could be subdivided by observing physical processes with a shorter but more or less regular rhythm, such as the movements of a sand glass, a pendulum, or a slowly uncoiling spring. In principle the rest was easy. Wind up two clocks, one at A and one at B; start each clock going at its own local noon; carry one clock from A to B, and compare it with the clock that had remained at B. For every minute of chronometric difference, the distance in longitude from A to B was a quarter of one degree. This possibility had been known to the ancient Greeks and was suggested as a means of longitude determination by Gemma Frisius in *c.*1530.[42] But the clocks of Gemma's time were vulnerable to changes in temperature, humidity and barometric pressure, as well as being easily upset by the motion of any vehicle used to transport them. Hence the rejection of chronometric longitude-measurement by all 'learned men', as Robert Hues put it in 1592.[43] Prospects improved with the advent in the 1670s of a spring-driven watch that worked reasonably well at sea. Mechanical time could now be relied upon at least from one noon to the next. But forty years later (when most new latitudes were correct to a minute or two) the British government was still offering £10,000 for any clock that would measure the longitude to within one degree of a great circle on a voyage from Britain to the West Indies.

The problem of chronometric longitude was finally solved by John Harrison in 1761–2.[44] But as long as man-made clocks were unreliable, the only alternative had been for the heavens themselves to be treated as a clock which did not need winding and which was large enough to be read by two widely separated terrestrial observers at the same time. The celestial clock-face was formed by the stars: these admittedly moved round the sky, but they did so in a simple and predictable manner. All the navigator needed was that the 'hand' of the clock should change its position on the face by moving with equal predictability along a different course. This function was discharged by the sun and moon. So what kind of celestial change could be observed and related to local noon in different parts of the world? Preferably this should be an event whose time on one arbitrarily chosen meridian had already been forecast and published by astronomers. That meridian could then be treated as a standard. The use of lunar eclipses for this purpose was proposed by Hipparchus in 150 BC, and Ptolemy gave one – but only one – example of such an observation.[45] Many renaissance geographers and explorers followed suit, but never as a matter of routine. Lunar eclipses occurred only about twice a year; solar eclipses, though equally suitable, were still less frequent. Another problem was that, however good the observer's clocks, the beginnings and endings of eclipses were difficult to define with precision.

The earth, sun and moon were not the only celestial bodies that eclipsed each other. Jupiter had four bright moons, two of which were eclipsed every two days by passing into the planet's shadow. These events were seen through Galileo's astronomical telescope almost as soon as he had invented it, but it was not until 1676 that their movements were reduced to order by the French geodesists Jean

5.6 Use of the cross-staff. Left: measurement of lunar distance as a means of determining longitude. J. Werner and P. Apian, *Introductio geographica Petri Apiani in doctissimas Verneri annotationes* (Ingolstadt, 1533).

Picard and Jean-Dominique Cassini.[46] From 1690 this information was regularly published, and in precise terrestrial surveys Jupiter's satellites now became a possible device for finding longitude, but only with the aid of a six-foot-long telescope unsuitable for use at sea. Even on land the method would always be too slow and cumbersome to provide a dense network of stations: 'impracticable in ambulatory survey' was Thomas Jefferson's verdict as late as 1816.[47]

It may seem odd that the earth's own satellite should have been less useful to its inhabitants than those of a remote planet, but so for a time it proved.[48] One reason was that the movements of our moon are extremely complicated. It travels annually around the sun within a system of planet-plus-satellite and also follows its own elliptical path around the earth (once in 27.3 days) in a plane that differs from that of the earth's revolution. On the other hand the moon suited the convenience of a terrestrial observer by moving through the sky comparatively fast. Its transit across the meridian provided a possible means of longitude-determination in which the time of this event could be compared with that of the corresponding transit at some standard meridian as recorded in a reference table known as an ephemerides. The only other information required was the moon's speed of movement relative to the earth.[49] This method was recommended by Johannes Werner of Nuremberg in 1514 with other sixteenth-century astronomers following his lead.[50]

More generally, longitude could be found from the angle between the moon and the sun or any star as subtended at the point of observation (Fig. 5.6). This value was the same everywhere on the earth's surface. It could be measured at a local time obtainable from the altitude of the sun. The astronomer then needed to know the local time when the moon would occupy the same position at some predetermined reference point. In modern textbooks the reference point is Greenwich Observatory and the appropriate Greenwich time is taken from a published almanac. These 'uncertain and ticklish' methods (as Hues called then) were seldom mentioned by navigators and cartographers of the early seventeenth century or before. The exceptions were recorded with an abundance of detail befitting a rare novelty and generally gave incorrect results: William Baffin's use of the moon's meridian transit in 1612 in Davis Strait was wrong by nearly eight degrees.[51]

Two developments brought lunar distances within the realm of practicality. One was the invention of Hadley's quadrant, the first effective means of taking angular measurements from the deck of a moving ship. The other was the calculation and tabulation of the moon's movements by Tobias Mayer of Göttingen, who differed from astronomers in maritime countries by showing more concern for land cartography than for navigation.[52] Mayer's figures were revised in Nevil Maskelyne's *Nautical almanac*, an annual publication which first appeared in 1766[53] and which gave distances between the moon and a selection of stars at three-hour intervals throughout the year.[54] Before Maskelyne took this initiative, it has been said, 'probably not more than a score of navigators of any nationality had succeeded in measuring their longitude when out of sight of land'.[55]

A difficulty with the lunar method was that for any given error in a measured angular celestial distance the corresponding error of longitude on the ground was thirty times larger. Here was the root cause of longitudes being so much harder to measure accurately than latitudes. The requisite observations therefore had to be taken with extreme care, and apart from needing good instruments this meant making laborious corrections too small to be thought worthwhile in other kinds of measurement associated with cartography. An example was the error known as parallax, caused by assuming an observer to be located at the centre of the earth instead of on its surface, a displacement accentuated by the relatively short sight-lines connecting earth and moon. Another such error was refraction, or the bending of light in its passage through the non-homogeneous medium of the earth's atmosphere. In mid eighteenth-century longitude determinations it would take four hours of arithmetic to process a single instrumental reading. For a practised observer, the *Nautical almanac* helped reduce this delay to fifteen minutes, but

longitude was still more difficult than latitude.[56] Before setting out across North America in 1803 (a successful journey in most other respects) Meriwether Lewis and William Clark made elaborate attempts to prepare themselves for determining longitudes en route, but in the end their results were dismissed by a more expert judge as unusable. At sea James Cook's longitudes for New Zealand were correct to half a degree, but even he did considerably better with latitude. Nevertheless by this

5.7 World map emphasising lines of astronomical significance. H.A. Chatelain, *Mappemonde ou description generale du globe terrestre* (Amsterdam, 1718–30). See also Fig. 8.12.

time any serious map, whatever the circumstances of its construction, was expected to include meridians as well as parallels.

By chance the successful exploitation of lunar distances happened to coincide with the advent of an accurate chronometer, after which the only astronomy needed in determining longitude was for the establishment of local time. Chronometers gave a narrower margin of error than lunars – 3 to 6 miles instead of 30 to 60 miles[57] – but the earliest models were unlikely to be used at sea except on special occasions like Cook's voyages, hardly a matter for surprise when each of them cost thirteen times as much as a Hadley quadrant.[58] From the 1780s, however, ships' records began to include daily references to chronometric longitudes, and for a marine surveyor to be without a good timepiece had become a cause for complaint.[59] After 1791 there was a separate column for the appropriate entries in the printed logbooks issued by the British East India Company. As a check on the chronometer the more economical lunar distances were still used to a diminishing extent, less so after sea travel was expedited by the advent of steam navigation. Robert Fitzroy, whose ship the *Beagle* carried twenty-two chronometers, is credited with showing that accurate time-based longitudes were obtainable without recourse to any earlier method.[60] By circumnavigating the globe he was able to check the position of Greenwich against itself. The difference was 33 seconds of time.[61]

Astronomy on the world map

For the average user of terrestrial cartography the main interest of celestial bodies and their movements comes from the help they can give in measuring and mapping the earth's surface. A distantly related development was for astronomical paths and boundaries to be shown in purpose-drawn diagrams illustrative of mathematical geography. The appearance of such graphics in the margins of ordinary world maps is a topic that historians have seldom considered worth writing about. Nor is it often remarked that any point on the celestial sphere can be connected by a straight line with the centre of the earth, and that wherever such a line crosses the earth's surface there could conceivably be a subject for the geographer. Mercator for instance did a service to navigators by projecting star positions on to the surface of a terrestrial globe.[62] Better-known features of this kind are the north and south poles, parallels of latitude (including the Arctic and Antarctic Circles and the Tropics of Cancer and Capricorn) and meridians of longitude. Beginning with reconstructions of Ptolemy, these lines have been a commonplace of world cartography since the fifteenth century, bringing benefits that are almost too obvious to need stating: together with their numerical values in degrees, they provide a convenient reference system for the reader, as well as helping to identify the projection on which a map is drawn.

Other astronomical lines are less familiar. An extreme case in this respect is the demarcation of areas from which an eclipse of the sun could be seen, a geographical phenomenon first treated as mappable by Edmond Halley. Such maps should probably be considered as part of thematic cartography irrelevant to the present book.[63] The same is true of maps highlighting the world's astronomically fixed locations, like that of Johann Gabriel Doppelmayr in *c.*1722.[64]

An intermediate level of currency is represented by the ecliptic or solar circle. In a terrestrial sense, this is probably best defined in practical terms. On a world map mark the point where the sun is overhead at any given moment. Do the same twenty-fours later, and again every twenty-four hours until, after a year, the noonday sun returns to its original position. A line connecting all the marked points is a great circle connecting one tropic with the other, in a plane inclined by 23.5 degrees to the plane of the equator. Clearly there are as many such ecliptic circles as there are moments of time within a twenty-four-hour period. Take for instance the two earliest ecliptics illustrated in a well-known cartobibliography: they show a considerable difference between the Bologna Ptolemy of 1477 and the anonymous Nuremberg globe-gores of 1535.[65] In the latter map the sun is overhead on the equator at longitude 180 degrees: this is the version of the ecliptic most often mapped in the sixteenth and seventeenth centuries, though its position relative to the continents and oceans naturally depended on the point from which longitude was counted, a subject to be taken up in chapter 7. Ecliptic lines were commonest on maps showing the earth as a western and eastern hemisphere, crossing the equator at the junction between the two. There was no accepted rule as to whether the upward curve to the June solstice should be in the west or the east. The lines also appeared occasionally in hemispherical maps based on the north and south poles as well as in non-hemispherical world maps.[66]

Like other lines on the earth's surface, the ecliptic has its equivalent on the celestial sphere. Here it is the circular path followed by the sun (neglecting the effects of daily rotation) as seen from the revolving earth. In this apparent journey the sun passes successively in front of twelve constellations, known collectively as the zodiac, famous from their associations with astrology, each of which gives its name to a 30-degree arc of the ecliptic-circle. Of these twelve, the sun enters Aires at the March equinox and Libra at the September equinox. On world maps the line of the ecliptic was usually divided into 360 degrees, the sun's apparent movement from one sector to another often being marked alongside the line by the names of the appropriate constellations, by astrological signs or even by miniature star maps.[67] Accordingly such lines were often labelled 'zodiac' rather than 'ecliptic'.

Some early world maps came surrounded by enough information about the sun, moon and planets to fill a textbook. In such company it is hardly unexpected to see the ecliptic snaking its way across the middle of the sheet. But most of its appearances were on ordinary maps of land and sea that made no other contributions to astronomical knowledge (Fig. 5.7). What purpose was being served here? As a cartographic device the line was inappropriate and misleading, or at any rate hard to understand, and there seems to be no reason why any reader should wish to see it. Perhaps there was simply something attractive about the zodiac's double curve: if so, this would not be the only case of map-makers putting aesthetics above utility.

CHAPTER 6

A survey in its literal sense

THE CARTOGRAPHIC FIELD worker must do more than pick out suitable observation points and know how to measure between them. He must also combine these operations to make a framework for a truthful and practicable map. Thus it would be useless to determine a series of unconnected distances AB, CD, EF and so on, in which no point was common to more than one pair. Equally, given a scatter of twenty locations it would be wasted effort to measure each of the 190 different lines that can be drawn from any one point to any other. The aim is rather to minimise the amount of measurement while at the same time satisfying certain fairly obvious conditions. First the survey must 'plot'. In other words it must be expressible on paper as one and only one self-consistent network of lines that is not manifestly incompatible with the corresponding network on the ground. Beyond that, the resulting map should be as correct as its users require. For a far-sighted and public-spirited surveyor, this should mean as correct as the scale allows, for there is no way of predicting how future readers may decide to use a map. A prerequisite of correctness, first stated by Ptolemy and applicable to nearly all survey methods, is that map-making should where possible proceed from the whole to the part.[1] In other words, a small number of widely separated points should be fixed with special precision, setting limits to the error that can accumulate in the spaces encompassed by those points.[2] From all this it follows that accuracy should not only exist but be seen to exist, a requirement satisfied by making each part of every survey subject to verification by some other part.

Among the documentary sources for the study of early map-making there are two that happen to cover much the same historical time-span: one is literature on the art of surveying (Fig. 6.1),[3] the other comprises those early maps that look as if they were based on actual measurement. The synchronicity between these two kinds of evidence does not mean that successful cartographers always worked by the book. A surveyor might be happy enough to augment his income by publishing an instructional text, but less ready to describe his own procedure on any given assignment. Descriptions of particular surveys, like those of particular instruments, were mainly written by would-be scientists who devoted much of their lives to non-cartographic pursuits. Most of these amateurs omitted the humdrum routines of map-making and concentrated on procedures that brought out their own originality. A similar reticence was generally exhibited by the maps themselves. Usually a cartographer chose to cover his tracks, if only to avoid confusion between

6.1 Title page, surveying textbook. Aaron Rathborne, 1616, implicitly stating a preference for the theodolite over the plane table.

the surface features of the ground and the invisible lines and points that were used
in plotting those features. If he did lift the veil of secrecy, it would probably be to
publicise some innovation on which he had lavished particular care.

Of course, if a map is accurate beyond a certain point it may plausibly be attributed
to what James Cook called 'a survey in its literal sense'[4] (as opposed to a sketch), but
we may still be unable to deduce a surveyor's choice of any particular technique from
cartographic evidence alone. At this stage a distinction must be drawn between
geometry and cartometry. Circles of latitude and longitude on a map are not
necessarily a product of its author's astronomical research; nor do scales of distance
imply the use of rods or chains. The same is true of 'construction lines' not normally
meant to be seen by the reader. When examining such lines it is easy to forget that
shapes on paper can be 'surveyed' in the same way as landscapes, by measuring lengths
and angles across them. Sometimes there were good practical reasons for subjecting a
map to this invasive treatment. In determining the area of a polygonal field or farm,
for example, distances could be taken either from the land itself or from a plot that
showed the same parcel at a prescribed scale. If in the latter case the area was found
by dividing the map into pencilled triangles, the sides of those triangles would appear
to posterity as very much like representations of lines traced out on the ground by a
surveyor's chain. In the same way a square grid could conceivably tell us something
about the author's surveying methods, but is more likely to mean that the map had
been or would be copied, enlarged or reduced; or simply that an alphanumeric
reference guide had been added as a convenience to the reader.

Surveying in squares

Returning for want of any better expedient to literary sources, we may next attempt
a classification of the survey methods prescribed in early textbooks. The possible
combinations were: (1) all lines; (2) all angles; (3) a mixture of the two. One
difficulty for this approach is posed by the peculiar status of the right angle
(mentioned in an earlier chapter) which so often reduced mensuration to a matter
of length and breadth. Right angles certainly seemed more amenable to estimation
than other angles, so much so that some authorities could see no need for
instrumental assistance. One of these was Richard Benese, who in 1537 became the
first English writer on surveying to discuss the art of land measurement (though
not, unfortunately, the making of maps) as well as the processes of examining and
valuing tenements and estates.[5] His title page promised 'the manner of measuring
of all manner of land', which he recommended should be done with rods or cords
of known length. He favoured the dividing of irregular shapes into rectangles and
like many contemporaries he seems to have regarded acres and other abstract units
of area as themselves inherently rectangular. But he never explained how to
determine whether any given parcel actually was a rectangle or some other figure.

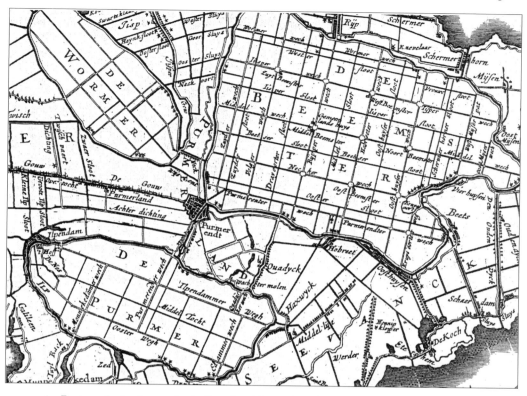

6.2 Rectangular landscapes in the Dutch polders, original scale 1:105,000. Gerard Mercator, Henricus Hondius and Joannes Janssonius, *Atlas or a geographicke description of the world*, ii (Amsterdam, 1636), p. 252.

In fact the right angle had won its privileged status many centuries earlier by being easy to set out on the ground. The ancient Babylonians and Egyptians, anticipating Pythagoras's theorem, had known how to do this by constructing a triangle with sides in the proportion of 5, 4 and 3;[6] for those who wanted something quicker there were appliances like the Roman groma. Right angles were also of practical convenience in architecture, in street planning and in measuring enclosures. This explains their popularity in land settlement schemes, even at periods when instruments quantifying other angles had reached a high level of precision. Famous examples of rectangular land allotment were the Roman system of centuriation in Italy, North Africa and elsewhere; the Dutch polders in the sixteenth and seventeenth centuries (Fig. 6.2); and the lands settled in the United States and Canada from the 1780s onwards.[7] In many such schemes the boundaries were intended to run from north to south, directions easily obtainable by the sun, with transverse links that could be laid out with equal facility from west to east. As a method of land allocation, this 'squaring' process had the well-known geometrical disadvantage that true north-south lines were not parallel but convergent towards

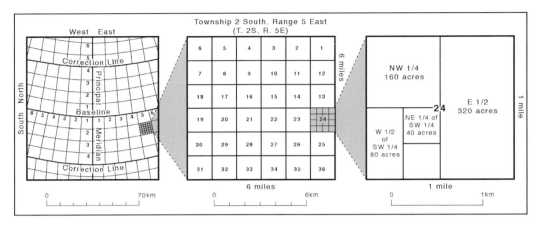

6.3 United States rectangular surveys according to the north-western ordinance of 1785: adaptation to the earth's sphericity. Redrawn by Matthew Stout from Hildegard Binder Johnson, *Order upon the land: the U.S. rectangular land survey and the upper Mississippi country* (New York, 1976), p. 58.

the pole. So either land allotments would get smaller as the network was extended polewards, or else their boundaries would cease to be meridians and their corners cease to be rectangular. In practice each grid was confined to a block of holdings small enough to be treated as a plane surface, but this left discontinuities between one such block and its neighbours (Fig. 6.3), a fact made plain to subsequent travellers in the American mid-west wherever the straight course of a north-south boundary road was interrupted by a dog-leg bend.[8]

A landscape that had already been squared could be put on the map by chaining distances along its roads and boundaries, provided that the grid had been correctly set out. The danger with this approach was that instead of checking the angles by measurement the surveyor might be tempted simply to make assumptions about them. Sometimes, inevitably, his bluff would be called. Christchurch Cathedral in Dublin was mapped by the Ordnance Survey with nave and chancel on the same alignment but had not actually been built like that. In an irregular landscape, such risks could be avoided by laying out a new and independent grid of survey lines, if necessary across open country, and locating detail through perpendicular offsets measured from these axes. It would still be possible for error to pass undetected, however. The angles of a quadrilateral could not show conclusively whether it was a rectangle rather than a square, nor could its linear dimensions prove that it was not a rhombus. The only way to discover the errors in a rectangular survey was to take all the measurements again. What this meant was that such surveys could never be entirely legitimate except when restricted to short offsets from the main lines of an otherwise non-rectangular network capable of being more easily checked. Such networks will now claim our attention.

Trilateration

Any plane triangle could be mapped with a ruler and compasses if the lengths of its three sides were known. (Strictly speaking this rule applies not to single triangles but to pairs of indirectly congruent triangles, one the mirror image of the other; however, of any two such possible triangles it should be obvious on the ground which one is correct.) The only one-word name for this ancient technique is the modern term 'trilateration', and to avoid any appearance of anachronism it may be slightly preferable to substitute the expression 'chain survey', this being the method of distance measurement adopted by most trilateralists in the period under consideration, though we should not forget that the trilateral principle can govern any method of determining lengths, including the roughest of estimates. To map a measured triangle, one of its sides was shown as a ruled line, and the other two set off to their proportionate lengths with dividers or compasses from the ends of the first, to form intersecting arcs at what could now be identified as the opposite corner.[9] There was no other plane figure that could be drawn as easily as this. In the ideal trilateral survey all the angles would be greater than about 30 and less than about 90 degrees: outside these limits it would take only a small error to cause an unacceptably large displacement of the intersected point.[10]

Most surveys required the measurement of more than one triangle, and for a single map these had to form a continuous network. In surveys of no more than a few square miles, one or more lines should if possible run straight through the whole area, provided the ground was smooth enough for them to be measured accurately. In some places, unavoidably, an otherwise convenient line might be interrupted by an impassable obstruction such as a river or marsh, but then the missing portion could be 'displaced' to a more accessible position by laying out a simple geometrical construction. The main survey triangles could be divided if necessary by 'split lines' creating new sub-triangles. The chain should pass as closely as possible to features that would appear on the map but, as in our hypothetical gridded survey, the exact position of these features had to depend on perpendicular offsets taken from prescribed points along the main lines (Figs 6.4, 6.5). In a survey of roads, fields and houses, one chain or one and a half chains was often considered a maximum length for offsets. Without some such rule, trilateration could easily begin to degenerate into squaring.[11]

Oddly enough there is no record of trilateration being used by practical surveyors until well into post-medieval times. This silence is best interpreted as history's tribute to the revolutionary impact of another system soon to be discussed, namely triangulation. At first, trilateration probably served for plotting lengths that were already known in the form of road distances, either through the emplacement of Roman-style milestones or in less well regulated societies by a consensus among successive generations of travellers. To measure along every line in an extensive

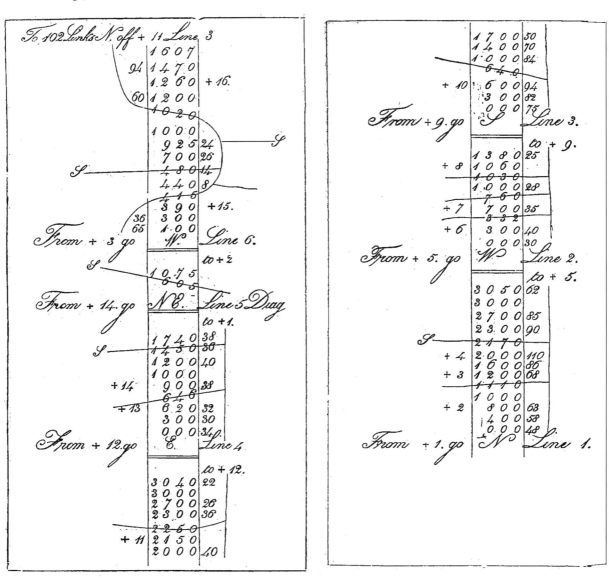

6.4 Pages from a chain surveyor's field book, with offsets on either side of the double column recording principal measurements in links. A. Nesbit, *A treatise of practical mensuration* (London, 1847), p. 106.

network for the sole purpose of making a map would have been regarded at most historical periods as excessively laborious. Only in the seventeenth century was chaining recommended as better than any other survey method, or at least as no worse.[12] It went on to earn praise from eighteenth-century writers, and in England may have received something of a fillip from contemporary enclosure legislation, which provided for the survey of openfield strips on farm land presenting few physical obstacles to the passage of the chain. But this was always in the context of

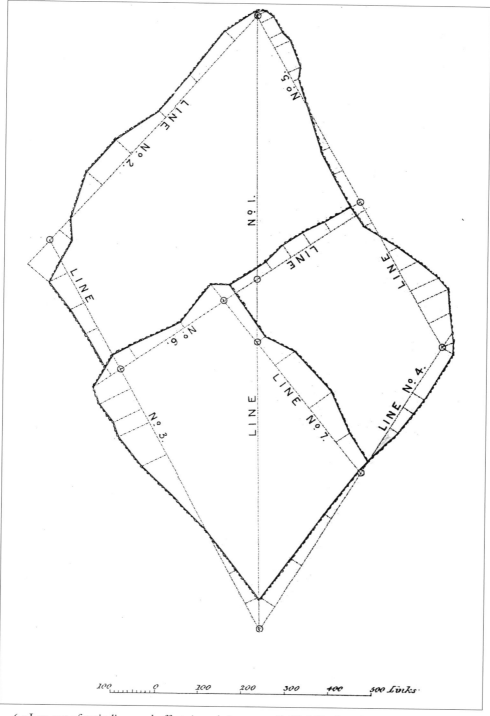

6.5 Lay-out of main lines and offsets in a chain survey. 'A Civil Engineer', *An introduction to the present practice of surveying and levelling* (London, 1846), pl. 2.

single parishes or estates. It was not until the 1830s, in the Ordnance Survey of Ireland, that anyone thought of employing hundreds of chain surveyors to trilaterate a whole country.[13]

Triangulation

Next consider a survey made up entirely of angles. In principle the simplest way to avoid chaining might seem to be a determination of latitudes and longitudes for all the points to be shown on the map. This was the method proposed for mapping France by Oronce Finé, who published not only the map itself (in 1525) but also a list of latitudes and longitudes for 124 French towns with instructions for the reader to make his own plot – using 'the places observed by Ptolemy himself, or by others, or by thyself, or by us'.[14] In practice there would never be enough of these points to provide a tolerable outline. It is clear for instance that not all the detail on surviving Ptolemaic maps could have been plotted from Ptolemy's coordinates as given in his *Geography*, which might include no more than three points to define the course of a 300-mile river.[15] (In fact hardly any of these points were genuinely astronomical, but that is another story.) It was in the spirit of Ptolemy that Finé admitted the need to 'adjust such places as one pleases, by means of comparison and distance from two or three places noted and situated on the map'. The same point might appropriately have been conceded by William Cuningham, whose specimen map of England in 1559 was based on latitudes and longitudes for just nineteen places.[16]

No doubt these writers saw the futility of pursuing latitude and longitude to a logical conclusion. The objections to doing so are easily stated. Non-astronomical measurements across the earth's surface could be adjusted to the complexity of the landscape by choosing from a variety of instruments and techniques, quick or slow, cheap or costly. By contrast a surveyor who depended on celestial observations would have virtually no way of matching method to scale. In the late eighteenth century, for instance, an observation for latitude might be wrong by as little as half a minute, but this was equivalent to about 3000 feet on the ground: an estate surveyor who tried to map a farm from stellar observations might well finish by moving his employer's house on to somebody else's property. For latitude errors to fall below plottable limits, the map-scale would have to be less than about 1: 4 million, say that of Spain in a school-atlas map. In short, latitude and longitude observations were useful not as a comprehensive system of measurement but rather as a way of pegging some other kind of survey on to the globe.

Which brings us finally to triangulation in a horizontal plane. This was the same as trilateration except that angles were measured instead of sides. Of course it was not always practicable for every angle in the network to be visited with a theodolite but, wherever circumstances allowed, measurements needed checking against the fact that the corners of any plane triangle amount to exactly 180 degrees. There was

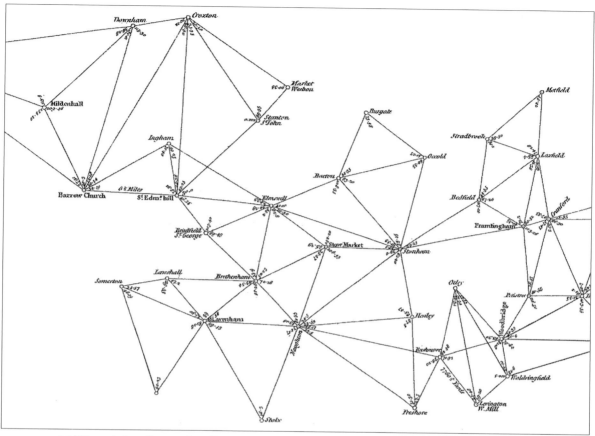

6.6 Triangulation diagram from Joseph Hodskinson's survey of Suffolk, 1783. The base was evidently the line between Rushmore and Levington W. Mill for which a length is stated in yards.

however a complication, which may be best explained by once again visualising the points of cartography as lines that radiate from the centre of the earth like the spines of a spherical hedgehog. At the corners of any theodolite triangle, each plane of observation was perpendicular to a different vertical, and the surveyor who took the three angles was measuring parts of not one but three different figures. (Alternatively, he may be pictured as observing a spherical triangle in which each side is part of a great circle.) Because the verticals diverged upwards, the angles would be in a sense opened out, causing their sum to exceed 180 degrees by a 'spherical excess' whose magnitude depended on the size of the triangle. The usual solution, proposed by Adrien Marie Legendre in 1787, was to deduct one third of the excess from each of the measured angles.[17]

In the government surveys of the nineteenth century, rays connecting high mountain-tops were sometimes prodigiously long. The distance observed in 1880 from Mount Shasta to Mount Helena in California, for instance, was no less than

192 miles.[18] In triangles of that size the spherical excess could amount to more than a minute. But in earlier surveys, like those of an eighteenth-century English county, the largest triangles might have sides of about twelve miles,[19] giving a surplus of less than 0.8 seconds. In a broad view, correcting for spherical excess was no great hardship compared with the practical benefits of triangulation.

Although the distances in a triangulated network were correct in relation to each other, none of them could be stated in absolute linear units. To make good this deficiency the survey network had to become a mixture of angles and lines, though admittedly in very unequal proportions, with dozens of bearings measured in the field as against only one distance, a strategically placed line known as the base (Fig. 6.6). From base and angles the length of every line in the system could be determined in either of two ways. One was graphically, by measurement from the triangles as plotted on paper. The other, more exact, was by trigonometrical calculation. The mathematical relations between the sides and angles of a triangle, familiar in principle to the ancient Greeks, were first numerically formulated by the scientists of the renaissance, and in 1533 Gemma Frisius could advise serious students of triangulation to consult a 'table of sines', at the same time warning them that trigonometry was 'too difficult for the common man'.[20] Such tables were much improved in the course of the sixteenth and seventeenth centuries. Otherwise the main refinement in modern triangular surveys was the practice of measuring two or more bases in different parts of the network so that every side of every triangle could be calculated at least twice and any one base could act as a check on any other. Other operations involving bases and angles will be treated in the following chapter. For the present we may simply note that some modern writers would probably restrict the word 'triangulation' to a survey that makes use of trigonometry, but there is no particular merit in this usage except that it avoids the awkward phrase 'trigonometrical triangulation'. The fact is that a triangulation could be conducted at any level of accuracy.

The principles outlined above were understood in fourteenth- and fifteenth-century Europe and China but not, it seems, very widely, for Gemma could announce the subject as one that was 'never seen before'.[21] For his contemporaries, at any rate, it was an idea whose moment had arrived: other descriptions were by Joachim Rhaticus in 1540, Sebastian Münster in 1544,[22] and (the first from an English writer) William Cuningham in 1559,[23] followed shortly afterwards by William Bourne[24] and Leonard Digges.[25] Not all these people actually made their own maps: the Flemish example quoted by Gemma gave the game away by directly connecting named stations that were too far apart to be intervisible.[26]

The great advantage of triangulation in all its forms was that shapes could now be surveyed and plotted with far less effort than had to be put into chaining. The difference is hard to quantify with any precision. Consider, nonetheless, the case of Christopher Saxton, whose surveys of Elizabethan England and Wales are generally thought to have been based on angular observation. To judge from the dates given on his maps he got through 37 million acres in five years, or one county every five weeks.[27] This is more than some historians can accept: hence the suggestion that Saxton may have copied other people's surveys. But compare the slightly earlier maps, not much inferior in quality, made by Robert Lythe in Ireland, an undeveloped country where the prior existence of ready-made field data was less likely. Lythe's rate of progress was almost the same as Saxton's, the equivalent of an English county in less than six weeks.[28] British county surveys of the Georgian era, while owing much to triangulation, often made considerable use of linear measurement, so eighteenth-century speeds were naturally slower. According to the Society of Arts in 1760, a good county map might require anything up to two years.[29] Long enough by Elizabethan standards, but how quickly could a pair of chain surveyors have covered the same area without measuring any angles? To judge from the admittedly narrow experience of an Ordnance Survey staff-member many years later, the answer is fifty years.[30] The difference is surely wide enough to swallow up all the uncertainties inherent in such a calculation.

However fast and however precise, triangulation involved one handicap that textbook writers seldom took very seriously. How did the surveyor pass finally from his own abstractions to the features placed by natural and human agencies on the earth's surface? In chain surveying it was the offsets that provided this connection, branching from the nearest main line at a point whose position was automatically given by the number of links counted along the chain. In a theodolite or compass survey the only equivalent process was to make the triangles smaller and smaller. In the end, either sketching or linear measurement would be needed to locate the detail of the landscape. There was one application of the triangular principle, however, that went a long way towards closing the gap.

Plane tabling

In a normal triangulated survey the angles were read from a graduated circle, recorded as numbers in a field book, and plotted at the surveyor's leisure. An alternative procedure was for intersections to be drawn in the field on a paper-covered plane table by taking sights with an alidade, so that surveying and plotting became a single operation (Fig. 6.7). Once an identifiable point had been intersected, the table could be set up at that point, new rays plotted to create new intersections, the table moved again, and so on, with the map gradually taking shape by the addition of further points. Three-ray intersections gave a constant check on accuracy. This led to the *pons asinorum* of plane tabling, the process known as resection, meaning an intersection that had to be performed more than once. The table was taken to a new position that could not conveniently be intersected from previously determined points. The surveyor estimated his position on the paper as

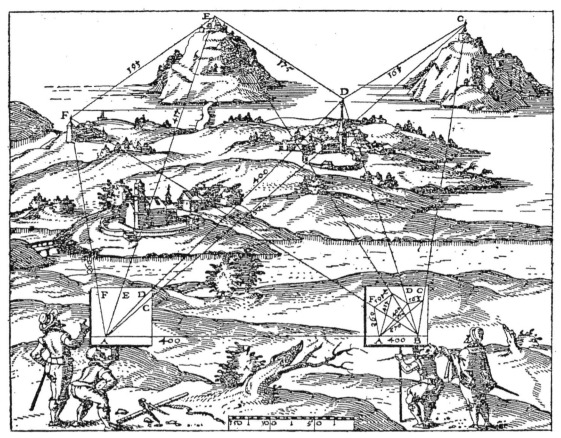

6.7 Successive plane-table stations at each end of a measured base. Leonhard Zubler, *Fabrica et usus instrumenti chorographici* (Basel, 1607). See also Fig. 4.18.

carefully as he could. Provisional rays were drawn backwards from three established points to this assumed location, forming a small 'triangle of error' that was then reduced to nothing by successive retrials.[31] Points on a plane-table sheet would usually be encircled (and also annotated in some way) to prevent them being lost. This doubtless influenced the use of a dot within a circle to symbolise a settlement on many medium-scale published maps, leaving the reader to infer that the site in question had been fixed by precise measurement.

Apart from saving labour, the plane table allowed a map to be put together on its own ground under the eye of its own surveyor, avoiding the 'clerical' errors inevitable in any complicated arithmetical operation. On the other hand a separation between fieldwork and plotting made any survey more amenable to checking by an independent examiner as long as the numerical measurements had been properly recorded. Otherwise the effectiveness of plane tabling depended on environmental conditions: it was unsuited to a wet climate, or to surveying the

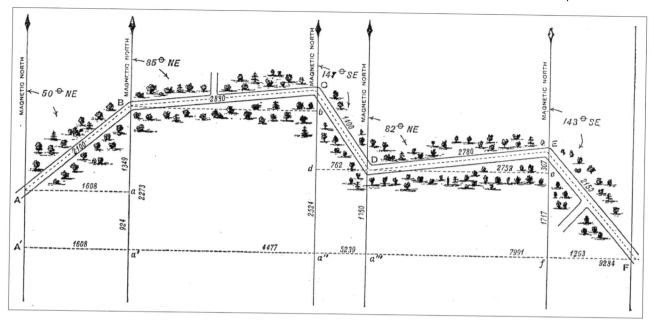

6.8 Lay-out of traverse survey, 1898: distances by chain and angles by compass bearings, with rectangular coordinates calculated from these measurements.

streets of a busy town. It was also subject to all the uncertainties of manual and graphic operations, and to its own forms of instrumental error, such as a disturbance to the horizontality or correct orientation of the table while in use.[32] Finally, being dependent on what was visible from a single point this method to be effective required a fairly large scale, ensuring that a substantial proportion of what the observer saw was likely to appear on his map. Most plane table surveys were for maps of 1:20,000 or more. On the other hand, the larger the scale the smaller the area that could be shown as a unit on a platform that seldom seems to have exceeded 15 inches in length[33] and the more often it became necessary for separate sheets to be combined, always a likely cause of error. (For historians there is the advantage that an apparently needless proliferation of sheets in a finished map may suggest that plane tabling had been the surveyor's method).[34] In general, the plane table was less popular among British surveyors than might have been expected. One eminent writer on cartography put this down to simple ignorance.[35]

Traverse surveys

Chain surveying and triangulation may be regarded as extreme cases. Between them lie systems in which linear and angular measurement were combined on more equal terms. These can be grouped together as 'traversing'. In the classic traverse survey the principal points were connected by a single journey in which the observer had no need to retrace his steps. His route formed a series of straight 'legs' each of which was measured in two ways, its length by a chain or wheel and its bearing by a compass or other azimuthal instrument (Fig. 6.8). The results were usually entered as numbers in a field book, though in some circumstances it might be thought desirable to plot the bearings on a plane table. The need for numerous separately determined distances made traversing a more laborious process than triangulation, but these linear measurements facilitated the taking of short offsets to detail on either side of the main lines. In a compass traverse the magnetic bearing of any given line AB, as taken at A, could be checked by a back bearing from B to A, which would have to differ from the forward bearing by 180 degrees. A closed traverse was distinguished from an open traverse by finally returning to its first station. This was presumably the 'polygon' method of surveying employed by Georg Friedrich Meyer in 1678.[36] The advantage of closure lay in introducing an element of self-verification: in fact there was almost always a slight discrepancy between the first and last positions of the initial point, an error that could be adjusted by 'distributing' it around the survey polygon. An open traverse was less reliable, because its termination had been left hanging without support in empty space – though in practice a traverse that did not return to its origin would if possible be made to close on a point that had been surveyed by some more accurate method.

Occasionally a cartographer might give himself away by allowing traverse lines and stations to remain visible on what otherwise purported to be a finished map, as seems to have been done for instance by Nicolaus Person in c.1690 (Fig. 6.9).[37] Even more rarely he might fail to eliminate the gap in what was meant to be a closed traverse.[38] In other cases the evidence for traversing is literary or documentary. An early description of the method was that of Sebastian Schmid, who recommended it in 1566 as a suitable means of surveying the River Rhine.[39] He might also have been expected to advocate its use for roads, especially as a road unlike a river would often return to its starting point. Earlier still, an authenticated example of a traverse closure is Leonardo da Vinci's survey of the city wall at Cesena.[40] Territorial boundaries were another likely subject for this method, as was well illustrated by successive admeasurements of confiscated lands in seventeenth-century Ireland.[41] Essentially the same procedure might be followed in setting out survey lines between the street-blocks within a town, though in this case many of the figures were likely to be quadrilaterals rather than complex polygons. Traversing by ships was generally of the unclosed variety known to navigators as 'dead reckoning', distances being estimated or measured with the log line and bearings taken by compass.

In short, traversing suited a wide variety of situations, scales and levels of precision, but it was never as accurate as triangulation, and it lost much of its appeal as angle-measuring instruments were improved. Richard Norwood notoriously depended on a traverse to find the distance from London to York in 1634 – as it happened, with surprisingly good results – but it is hard to imagine any later

6.9 Traverse lines on a finished map. From *Gericht Katzenberg in Hessen* in Nicolaus Person, *Novae archiepiscopatus Moguntini tabulae* [Mainz, 1690].

surveyor trying to emulate this feat.[42] In Swedish land surveys after 1780 lay-outs that were not self-checking were prohibited.[43] In the British Ordnance Survey traversing, 'though not absolutely forbidden, was discouraged' from the 1820s onwards.[44] Everyone agreed however that where visibility was restricted, and especially where only one survey-line could be seen at a time, triangular networks might be so impracticable as to make traversing the only option. Thick woodland and deeply incised valleys were perhaps the most likely obstacles to triangulation. Town surveys constituted another difficult case, for unless the streets were very wide there might be no room to set out well-conditioned triangles.

It should be clear from the foregoing examples that textbook classifications of survey techniques are unlikely to cover every eventuality. In reality, the surveyor's methods had to acknowledge the varying character of each landscape, even within a small area. The result might be an amalgam of trilateration, triangulation and traversing. Perhaps such mixtures are best seen as extensions of the traverse, designed to strengthen it with an element of verification or control. For instance in surveying along a route it might be easy to take a succession of angles to some conspicuous landmark well away from the main axis. Whether or not such landmarks were worth mapping in their own right, they had the effect of converting at least part of the survey into a local triangulation, because when three side-bearings to the same feature were plotted they were obviously required to intersect at the same point. In the survey of a tightly winding linear feature, like a meandering river, there might be in effect two connected traverses, the first made up of a few long lines remote from the watercourse, the other following the individual curves more closely. Each station in the short lay-out was also a station in the long one, which was thereby converted from a single open traverse into a series of closed traverses.[45]

Another way of combining lengths and angles was in what might be called a radial survey, where every point could be defined by a distance and a bearing taken from one central station. There would then be a simple check on the bearings, because the sum of all such angles must be 360 degrees. In the mid-fifteenth century Leon Battista Alberti recommended radial angles (expressed in forty-eighths of a circle) as a method of mapping Rome from a viewpoint at the Capitol, but he did not say how the distances were to be determined.[46] Some scholars have given an Albertian interpretation for the famous plan of Imola attributed to Leonardo da Vinci, which certainly shows a number of radiating lines, but it seems that Leonardo's ground distances were actually laid out along the streets, as indeed would be unavoidable where most other lines of sight were blocked by buildings.[47] In due course radial surveys were to be recommended by a number of textbook writers, but in areas of any size they must often have been impracticable.

Control and detail

Surveyors had another reason for varying their methods and this was to extend the idea of control, or proceeding from whole to part. The precautions thus imposed were an essential feature of many survey systems, among them chain surveying and plane tabling as distinct from open traversing. The tendency in modern practice has been towards sharpening the antithesis between control and detail, each characterised by different methods, different instruments, and sometimes different personnel. The more independent the methods the more authoritatively could one operation check the other. Then by a further refinement one set of controls could be controlled by another set. The received wisdom was that no purely graphic method of surveying could achieve the highest accuracy; nor could any system that made more than minimal use of linear distance measurement. This left two candidates for the role of control survey: triangulation and field astronomy. Both provided a network of points within which detail could be inserted by traversing, plane tabling or chain surveying.

Astronomical controls had one unique advantage. The unreliability of terrestrial measurement was clearly proportional to the linear distances involved: common sense suggests that the more angles or chain lines were required to get from one point to another, the greater was the chance of at least one link in the sequence being incorrect. The value of such cumulative measurements accordingly decreased in proportion to their total length. But in observing the heavens at two different places the risk of error was much less affected by the distance between terrestrial stations. The more extensive the survey, then, the greater the corrective power of astronomical intervention. This principle gained force in the seventeenth century when Jupiter's satellites began to play a part in longitude-measurement. Its first major application came when points were fixed around the periphery of France, inspiring Louis XIV's famous lamentation that the astronomers had cost him more territory than an unsuccessful military campaign.[48] A comparable development was the determination of longitudes as well as latitudes in east and south-east Asia by Jesuit missionaries.[49]

The lasting appeal of an astronomical control system was illustrated in 1816 by Thomas Jefferson's suggestions for a map of Virginia. His chosen point of origin was a hill summit from which both terrestrial and celestial readings could be taken. At each other principal station he recommended two sets of observations, one for latitude by field astronomy, the other a theodolite angle between the local meridian and a prominent mountain, thus enabling all other principal longitudes to be calculated. Within this framework, further astronomical fixes would be needed (the longitudes by lunar distances) for river-junctions, rapids, falls, ferries, mountain tops, towns, courthouses and 'angles of counties'. If these landmarks were more than a third or half of a degree apart, additional control points would have to be inserted in the network.[50]

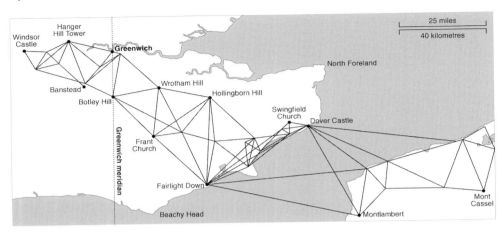

6.10 South-east England and the Strait of Dover with part of the chain of triangles connecting Greenwich and Paris, 1784–90. *Philosophical Transactions of the Royal Society of London*, lxxx (1790), Tab. XIII. Redrawn by Matthew Stout.

Once gross errors had been rectified, the utility of a latitude and longitude control was limited by the same factors that made this kind of observation unsuitable for a detail survey. As terrestrial measurements gained in accuracy, so astronomical methods lost much of their advantage, especially when, from the late eighteenth century onwards, the influence of gravitational anomalies began to put their geodetic effectiveness in doubt. The main credit for the change belonged to the instrument-makers of the middle eighteenth century and after, another important factor being a new facility for the carriage of delicate appliances to remote observation points. For Englishmen the breakthrough came in 1783 when Cassini de Thury persuaded them that an essential element in determining the correct latitude of Greenwich was to extend a line of triangles between Greenwich and his own observatory in Paris. Here, for a change, terrestrial would be checking celestial measurement rather than vice versa (Fig. 6.10).[51]

Before the era of government-supported national surveys the most accurate forms of triangulation had been, to coin a phrase, 'para-cartographic', not so much a framework for any particular map as an indirect contribution to all serious maps requiring a knowledge of how linear mileages might be equated with degrees of latitude. In the nineteenth century relations between geodesy and run-of-the-mill cartography became rather more intimate. The conceptual bridge was built by distinguishing different orders of triangulation – three of them in a survey of Languedoc as early as 1730.[52] In high-order networks, the triangles were larger, as also were the instruments that measured them. Each order was independent of those below it and adjusted separately as a self-contained system. 'Adjusting' meant distributing small errors so that they became undetectable, a practice familiar in all kinds of map-making since time immemorial. The procedure in a scientific triangulation was first to observe each angle several times – six according to Thomas Breaks in 1771, though he himself thought this was carrying precision too far[53] – and then to strike an average. The answers would not yield exactly two right angles in every triangle or four right angles at every point; nor would every line maintain a constant length however it was calculated. To produce a self-consistent result the averages would have to undergo some slight alteration, and there were clearly various ways of doing this. Mathematically the most satisfactory method was the one that gave a minimum value for the sum of the squares of the corrections. The appropriate theory – 'least squares' – is usually attributed to Carl Friedrich Gauss but was first publicised by Adrien Marie Legendre in 1805.[54] Its earliest cartographic application, reported in 1838, was in the triangulation of Prussia by Friedrich Wilhelm Bessel and Johann Jacob Baeyer.[55]

One final aspect of the relationship between different orders of surveying deserves more comment than it seems to have received. On the principle that prevention is better than cure, most experts assumed that control should precede detail in temporal succession as well as in scientific status. If higher and lower systems seriously disagreed about a given distance, the control points should be treated as correct and the relevant ground measurements repeated – preferably before the map had been drawn and the survey parties disbanded. In practice, however, the two admeasurements were sometimes done in what Ferdinand Hassler described as an 'inversion of the order, which nature, and science, dictates', usually because the control survey had some ulterior purpose and because the corrections required by it were unlikely to be cartographically significant at the surveyors' chosen scale.[56]

But what if these corrections *were* significant? In that case the error might be eliminated by changing the main detail survey lines to make them fit the control and then readjusting the minor lines in sympathy. The method would doubtless be some two-dimensional version of a text-book traverse adjustment, perhaps involving the construction of two grid systems, 'before' and 'after', as one map was replaced by another with a slightly different shape. (No advice on this subject has been discovered in the literature.) Such a procedure could perhaps be described as 'fudging', but it was not so irresponsible as leaving boundaries, roads and rivers at odds with the skeleton that allegedly supported them. This issue became a matter of debate in early eighteenth-century Russia where the visiting French expert, Joseph Nicolas Delisle, wished to lay down a preliminary control, while his more pragmatic native colleague, Ivan Kirilov, favoured starting at once on the detail survey.[57]

Kirilov was not alone. In Britain William Roy expected his forthcoming triangulation to be partly 'filled in' from an 'original map' made some time earlier.[58] Among the sources for Aaron Arrowsmith's map of South America, the 'surveys' were all completed by 1806, while the astronomical observations that 'corrected' them apparently went on for another four years.[59] Half a century later, a similar situation confronted Roy's successors in the Ordnance Survey. The final adjustment of the Survey triangulation was not complete until 1858, by which time Ordnance maps had already been published for the greater part of England and Wales. In the offhandedly dismissive words of an official historian, 'the principal triangulation … was not used as a framework for the mapping of the United Kingdom'.[60] So what exactly *was* it used for? In the context of this book we need not stay for an answer: it is time to start discussing maps rather than surveys.

CHAPTER 7

Laid down in our drafts

To 'PROTRACT' OR (in more modern terminology) 'plot' a survey means to convert numerical field data into a graphic composition that forms the basis of a map, without yet adding the kind of information that is collected by non-quantitative methods. Not every map had to go through the process of plotting in this sense. Some were drawn in the field from direct observation, either as freehand sketches or within a network of rays reconstructed by pencil and alidade on a plane table. In such cases surveying and plotting became impossible to separate. Other maps were made by 'graphicising' verbal data. No doubt trial and error contributed to this latter operation, in which case the word 'plot' as distinct from 'map' could perhaps denote a preliminary outline later replaced by something more correct. However, in the following paragraphs the plotter's raw material is taken to have consisted essentially of numbers, which could be either astronomical latitudes and longitudes or terrestrial bearings and distances, accompanied by placenames that might at this stage be only provisional. His end-product was a skeleton of lines and points to which were later added the flesh and blood of roads, rivers, boundaries and symbols. Some early map-users seemed to acknowledge this distinction by their habitual use of the expression 'lay down', as in 'laid down in our drafts', rather than 'draw' or 'delineate'.[1]

Whether the plotter was a different person from the surveyor, as the foregoing definition may suggest, would depend on extraneous circumstances. A surveyor laying down his own measurements might be tempted to eliminate intolerably large misclosures by 'distributing' them through the system. The alternatives would be either to repeat the field work until no inconsistencies remained, or simply to let the errors stand. At this point there is one obvious comment: although many finished maps are planimetrically inaccurate to varying degrees, their errors hardly ever exhibit the kind of stark discontinuity that must sometimes betray itself in a surveyor's field book. It therefore seems advisable to illustrate a phenomenon that may lie outside some readers' experience. In an early nineteenth-century plan of Chepstow, Monmouthshire, a fictitious dog-leg bend diverts the course of the medieval town wall between the main street and the castle (Fig. 7.1). The quantities involved are far from negligible at the large scale of a town plan: length of non-existent displacement, 80 feet; non-existent angle with main wall, about 100 degrees at each end. The true alignment, a smooth curve, appears on both earlier and later maps as well as surviving conspicuously on the ground.[2]

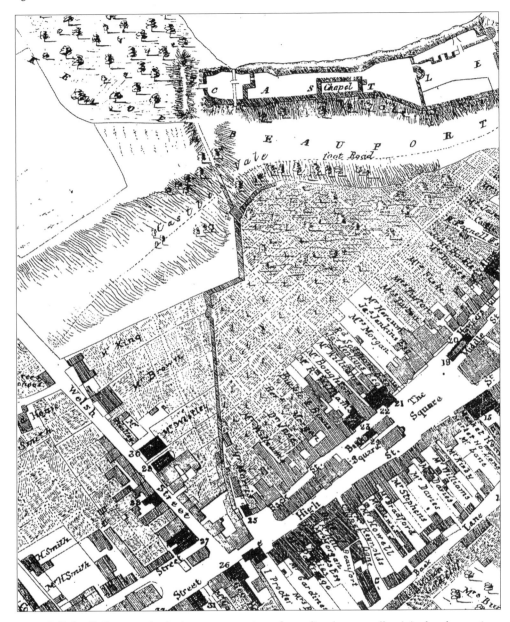

7.1 A 'fudged' alignment in the incorrect mapping of a medieval town wall, original scale 1:2376.
[John Wood], plan of Chepstow, Monmouthshire, 1835.

Segregating the two processes of survey and protraction would set limits to the possibility of 'distribution' by allowing an impartial critic to check the measurer's results. The value of such checks may have been understood by early surveyors,[3] but it was only in a large organisation like the British Ordnance Survey that the

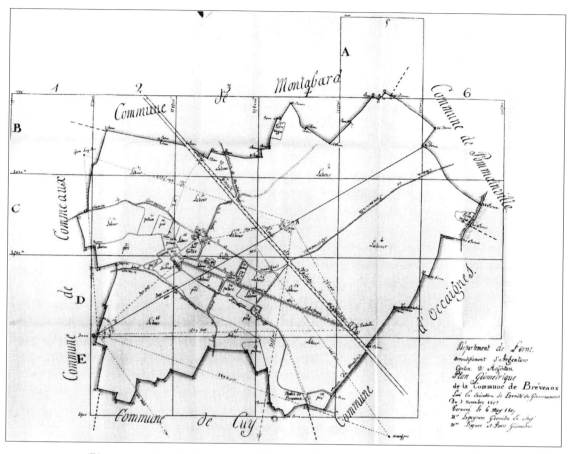

7.2 'Plan geometrique de la commune de Bréveaux', 1803, with measured lines shown for
instructional purposes, original scale 1:5000. See note 4.

distinction could be rigidly enforced. For a small firm, separate surveyors and
plotters would probably be an unaffordable luxury.

Despite its importance, the work of plotting could hardly be presented as
glamorous. Unlike surveying, it was untouched by any spirit of open-air adventure
and had none of the artistic appeal of drawing or engraving. This may explain why
historians have never shown much interest in it. Nor does it figure prominently in
the unpublished records of bygone cartographers. The plot itself, being an
essentially derivative statement that added nothing of substance to the surveyor's
field notes, was unlikely to be preserved for future reference. If it did survive, its
essential character was probably obliterated by having the 'construction' lines erased
and the fair drawing penned on top of them. It was only when a surveyor expected
a challenge from other surveyors – and felt capable of meeting such a challenge –
that his plotted lines might be deliberately left visible in the final draft (Fig. 7.2).[4]

Plotting tools and media

Most of the instruments and methods used in plotting were already well known to architects and draughtsmen before the revival of European cartography at the end of the middle ages.[5] A trilateral survey could be laid down with dividers or compasses and a scale of equal parts. A triangulation or traverse required a scale of chords or, after *c.*1580, a protractor.[6] A surveyor's protractor was a semi-circle or circle of brass or horn with its circumference graduated in degrees and parts of a degree, used for setting off angles from a given point on a given line. Its purpose was to put the operation of a circumferentor or theodolite into reverse, and some authorities recommended that a survey should be plotted with a protractor of the same radius as the instrument used for taking the angles in the field.[7] A map based on latitudes and longitudes with a simple rectangular projection made even fewer demands: the plotter began with a quadrilateral graduated in degrees both upwards and sideways and then fixed positions inside it with the aid of a straight edge or a piece of thread.

Plotting could be improved in three ways: equipment, media, and method. One possibility was to increase the effectiveness of the material components, organic or inorganic, from which an instrument was assembled. In this respect the histories of plotting and surveying have followed a similar course. From the beginnings of modern cartography, measuring scales of every kind have been made stronger, larger, and more precisely graduated. Wood and horn were replaced first by brass and then, from the end of the eighteenth century, by a mixture of nickel and silver. For the protractor, large diameters (say eight inches) were increasingly preferred to small (say four or five inches) but, as with surveying instruments, the best substitute for size was to apply the simple geometrical principles of the nonius and the diagonal scale, by which the smallest practicable linear intervals could be further subdivided by scaling along another dimension.

Many of the instrumental advances of the early modern period affected the relative motion of two or more components within the same appliance, not always fitting together as accurately as the textbook writers assumed but still setting new kinds of limit to the scope of human clumsiness. The invention of parallel rulers was claimed by Gustav Mordente in 1584.[8] A complementary item of equipment, surprisingly late in origin, was a nineteenth-century device for plotting offset measurements by a graduated cursor that gave distances at right angles to a full-length scale along which it could be made to slide. Another group of devices was for the measuring or laying-off of angles. The 'new protractor', incorporating a rotary arm, was already being manufactured in the early eighteenth century.[9] The three-legged station pointer was an aid to plotting the position of a survey ship from angles taken to the coast: it appears to have been first described in 1774 by Murdoch Mackenzie senior.[10]

Then there was what one instrument-maker described as 'drawing lines tending to a centre at a great distance', in the manner required by several kinds of map projection or by the protraction of very large triangles.[11] The beam compass, tracing arcs with a radius of three feet or (more awkwardly) even longer, had been known to Leonardo da Vinci, and by the end of the eighteenth century had been supplemented by a variety of other aids that were capable of drawing ellipses and parabolas as well as 'arcs from an infinite radius … to those of two or three inches diameter'.[12] By the time George Adams the younger made this last claim, most kinds of portable drawing instrument had reached the highest pitch of accuracy that human hands and eyes could hope to achieve.[13] In many situations, however, it was easier to use ready-made rulers of appropriate shape. Rulers with a variety of different radii became especially useful when railway surveys had to be plotted;[14] in some of them, called 'French curves', the radius varied from one part of the instrument to another;[15] in others, known as splines, a flexible strip could be set to the right curvature and held in place by weights.[16] These were commonplace in the late nineteenth and the twentieth centuries: their previous history has received little attention.

The usual choice among plotting media was between vellum or parchment and paper.[17] Vellum was recommended for its dimensional stability and for the large sizes in which it was available: about 50 inches by 40 inches per sheet, as compared with 22 inches by 30 inches which was quoted as a maximum for eighteenth-century hand-made paper and which approximated closely to the sheet-size recommended in 1802 by the French *commission de topographie*.[18] However, vellum was an expensive substance to be damaged by perforation with a plotting needle; many cartographers saved it for the final copy and did their plotting on paper. The manufacture of paper in the larger sheets made possible by nineteenth-century machinery, up to a 'double-elephant' size of 40 inches by 30 inches, reduced the errors caused by joining different sheets. Expansion and contraction were still a problem, and could amount to as much as two per cent, though diminished to some extent by mounting paper on some other substance.[19] A possible remedy was to draw on a more rigid medium. Plotting on slate was not meant to be more than a way of calculating areas on a stable surface, and seems to have been seldom adopted in practice even for that purpose.[20] The use of copper plates was taken more seriously, and had the effect of protecting measurements against instability for at least a part of their active life. The printing on paper of ready-made grids as a framework for the plotter's angles or coordinates was already current in the seventeenth century.[21] The next and more expensive step was to plot triangulation points in reverse on a printing plate, perhaps simply as a basis for manuscript drawing, perhaps as the first stage in engraving a complete map.[22] This last idea was still thought of as an innovation when adopted by the Ordnance Survey in the nineteenth century.[23]

Most suggestions for the improvement of plotting techniques depended on the premise that a protractor ('the least precise of all the map-makers' instruments', according to one authority)[24] was more difficult to use than a ruler. In general, once a surveyor had determined his base and angles there were two possible courses of action. In the first, a scale was chosen for the map and the base drawn to that scale. The survey could then be laid down angle by angle using a protractor and ruler without any direct reference to the lengths of the other lines. The trouble with this simple technique was that drawing and measuring on a map were precariously dependent on the quality of pencil, ruler and paper and the steadiness of the plotter's hand and eye. A safer alternative was for all linear distances except the measured base to be calculated by simple trigonometry, using for instance the constant relationship that exists between the sine of each angle in a plane triangle and the length of the opposite side. From one side (the base), and two angles, it was now possible for the lengths of the other two sides to be first computed and then fixed on the plot as an intersection of compass arcs. This procedure was advocated by Arthur Hopton in 1611[25] and by John Holwell in 1678.[26]

For plotting a traverse, the protractor was regarded by many writers as indispensable, but to avoid repeated settings of the instrument some authorities advised laying off all angles from a single point near the centre of the proposed map and then transferring each line from that centre to its correct position by means of parallel rulers.[27] However much or little was gained by shifting the burden of accuracy from the protractor to the ruler and compasses, plotting from a centre resembled the ordinary traverse plot in allowing each successive error to be carried forward through the rest of the survey.

One way of checking this propagation of traverse errors was the kind of numerical table described by Richard Norwood in 1637, in which each bearing and distance could be read off as a pair of rectangular coordinates.[28] In a closed traverse, Norwood's 'eastings' and 'northings' made it possible to distinguish errors committed outdoors and indoors, because before starting to draw his map the plotter could ascertain from numbers alone whether the accumulated positive and negative coordinates cancelled each other out when his calculation returned to its point of origin. Like several other ideas developed on the higher intellectual levels of the cartographic fraternity this was by no means an immediate success, being still seen as new-fangled towards the end of the eighteenth century.[29] But the value of rectangular coordinates was not confined to traversing: the points of a triangulation could be laid down in the same way. In general, the appeal of coordinate systems grew stronger as angular readings became more accurate and as different orders of triangulation began to be differentiated.

Sheet lines

The plotter also faced a number of other decisions before he took up his pencil. In particular he had to make a choice of sheet lines, orientation, scale and projection.

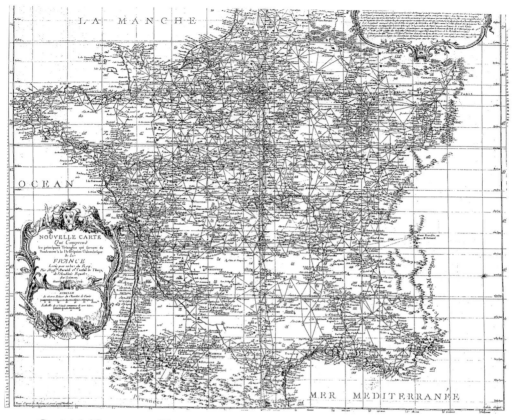

7.3 Cassini de Thury, map of France: triangulation diagram and index to sheet lines (Paris, 1744), with enlarged extract. See also Fig. 1.20.

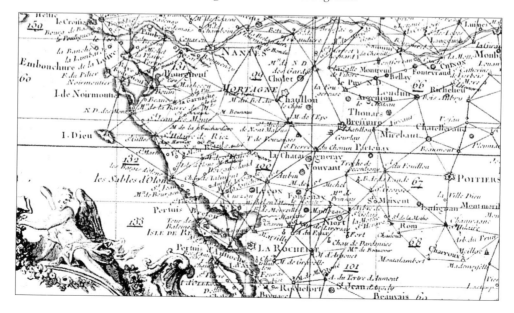

At this stage an important distinction must be introduced: which came first, the sheet lines or the map? Once plotted and drawn, a map might be re-divided, re-orientated, re-scaled and re-projected in different ways, and for different reasons, and these reasons – however worthy of notice in themselves – had no necessary connection with the original act of plotting. Sheet lines first came to attention in the thirteenth century when Matthew Paris apologised for making a map no bigger than the page on which it was drawn.[30] A century later his problem was solved in the Catalan atlas by dividing one map into four sections. With the build-up of geographical information from the sixteenth century onwards, a detailed map of almost any subject, from the whole world down to a single city, became likely to extend over several sheets. Jacopo de' Barbari's view of Venice needed six, Waldseemüller's world map of 1507 twelve, Philipp Apian's map of Bavaria forty,[31] the Cassini map of France a hundred and eighty (Fig. 7.3).[32]

Sheet lines raise interesting questions of a kind that might be classed as epistemological, perhaps almost as metaphysical. Suppose a map to be divided among a number of rectangular non-overlapping sheets, each bordered by four finely drawn neat lines. Suppose too that a small point-symbol, say for a church, lies within a millimetre of a right-hand sheet edge. The neat line itself obviously has no existence in the real world: it is a mere contrivance for making a large map easier to handle. Nevertheless the same line is stating, in the plainest terms, that the church has not been misplaced by the ground-distance equivalent to a millimetre; because if the survey was repeated, and the original church symbol found to be slightly west of its revised position, the difference would become glaringly obvious when the symbol in question jumped on to an adjacent sheet. If the whole map was drawn on one large undivided expanse of paper, this difficulty would not arise and the error could hardly be said to exist: without the neat line to guide him, no critic would hope to identify a difference of a millimetre. By subdividing a large map, in other words, the author increases its claim to accuracy. Of course this kind of reasoning probably assumes that all sheets of a divided map are identical in size and shape. Where this is nearly true we can at least assume that such was the author's intention. But suppose the same sheet to be two or three millimetres wider at the top than at the bottom. Has a correct map been misleadingly, but not 'incorrectly', divided into trapezia instead of or as well as rectangles? Or has a rectangular piece of territory been wrongly plotted as a trapezium? (Or have the sheet's original dimensions been subsequently altered by changes in temperature and humidity?) The answers to these questions could determine whether the artifact in question is catalogued as a single map or as several maps – but only if the cataloguer is fanatically conscientious.

The foregoing remarks may well be dismissed as futile, but it still seems worth asking how such composite maps were produced. There was no theoretical limit to the number of geometrical figures that a plotter could build outwards from his baseline; or the number of sheets that could be pasted together; or the capacity of whatever ladders, pulleys, cranes and scaffold-poles might be needed to give an artist access to his more ambitious constructions. So, were very large maps made in one piece and then cut into rectangles of convenient size? This is another no man's land of cartographic history. In early surveying literature the only relevant discussions are those referring to the plane-table, an apparatus that imposed severe practical limits to the size of any individual plot. Contemporary instructions for extending the plane-tabler's survey from one sheet to another, circumstantial as they are, have little to say about constructing a really large mosaic. At any rate it is clear that, whatever the medium, a one-piece map of more than pocket-book size would probably be subject to perceptible distortion. The need to hold error in check by working from whole to part had been implicit since the sixteenth century in the idea of triangulation. At first sight, the same need was given further recognition when surveyors of manors and estates were repeatedly urged to space their stations as widely as possible.[33] But that was in the interests of speed as much as correctness. Maps of larger areas were unashamedly pieced together from independent local surveys until far into the seventeenth century. Only later did English-language textbooks begin to insist that the 'principal and most eminent stations' should be plotted first.[34]

The best way to regulate the plotting process was developed in France by Jacques Cassini and his son César François Cassini de Thury. In the Cassini map of the whole kingdom at 1:86,400 each control point was defined by two rectangular coordinates reckoned from a pair of great circles crossing at right angles in the heart of Paris. One of these circles was a meridian of longitude. Given the dimensions of the 180 rectangles that composed the map, each point could then be laid down within the appropriate sheet, and the size of that sheet (65 x 95 centimetres) would determine the length of the longest single line that the plotter was required to measure rather than to calculate.[35]

Orientations

'Orientation' is a comparatively modern word, applied originally to the building of churches along an east-west axis and then more generally to the alignment of any object in relation to points of reference like those of the compass. Its use in cartography developed too late to justify any historical inferences about east-facing maps. The most we can say is that, other things equal, the 'orientation' of a finished map might simply depend on the plotter's original choice of coordinate-axes. But this was not the only factor at work. East did stand at the top of many medieval world maps to focus attention on the earthly paradise, and perhaps to allow the rivers issuing from this source to flow downwards.[36] But in any astronomically-conditioned view of geography it was natural for scholars domiciled in the northern hemisphere to visualise the pole star as hanging above the 'top' of the earth. Ptolemy for instance was characteristically explicit in identifying 'up' with 'north', though he gave no particular reason for doing so.[37] The pre-eminence of a north-south axis among the cardinal points may have been strengthened by the advent of the magnetic compass.

At all events, for large and well-known territories that were frequently mapped, an image with north pointing upwards eventually became fixed in the public mind, though both practical and psychological considerations delayed this process more for some countries than others as cartographers sought to retain what Jean Jolivet called an 'advantageous viewpoint'.[38] Thus for an Englishman in contemplation of the Isle of Man, Ireland or the colony of Virginia the advantageous viewpoint was obtained by putting east at the bottom, nearest the viewer on the page as it was in reality. For areas with a limited readership, whose members brought no particular geographical preconceptions to a map, lay-outs could be varied to fit the available paper or parchment, but by the late seventeenth century a northerly alignment was being recommended even for estate maps.[39] Consistency and practicality were to some extent reconcilable by putting north nearly at the top but not quite, as in Saxton's map of Warwickshire.[40] Some scope for this kind of approximation was acknowledged by the many sixteenth-century cartographers who omitted a separate north-arrow and instead wrote the names of the cardinal points half-way along the map borders, a practice that could cheerfully accommodate deliberate 'errors' of up to 30 degrees.[41]

This brings us to the kind of meridian line that was drawn as an aid to orientation and not as part of a graticule. Such lines made their first appearance on marine charts, but became rapidly more popular on land maps after about 1590. At sea it was desirable to distinguish all thirty-two points of the compass, and indeed some of the first north indicators can be interpreted as pictures of a real mariner's compass – just as some early cartographers' scale-statements can be seen as facsimiles of a wooden plotting scale or ruler. This convention also spread to land maps, more as a symbolic claim to professional status than because a plethora of compass directions had any great practical value. The principal points were often identified on these compass roses by means of initial letters (with 'S' sometimes representing the Latin 'septentrio', meaning north, and with 'O' not very usefully doing service for both 'oriens' and 'occidens'), but elsewhere an arrow or fleur-de-lys was left to speak for itself, though there were some late sixteenth- and early seventeenth-century maps, like Ortelius's Friesland (1568), with arrowheads indicating south and not north (Fig. 7.4).

The distinction between true and magnetic north was slow to manifest itself on maps. Some early cartographers were doubtless unaware of the difference, and if they gave any thought to the matter may not have considered it particularly important. For a cartographer engaged in plotting, the easiest course was to adopt

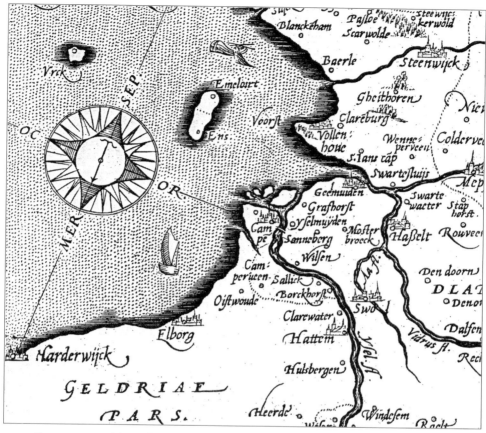

7.4 Compass rose with south-pointing arrow. Abraham Ortelius, *The theatre of the whole world* (London, 1606), p. 48.

whichever meridian he had used in his survey. This is presumably why John Norden recommended magnetic north.[42] His advice became less helpful, however, when the difference between true and compass north was found to vary in time as well as space. The problem was especially obtrusive in western Europe where compass and pole star continued to diverge throughout the later seventeenth century and for most of the eighteenth. From now on magnetic north began losing ground, to be unequivocally rejected in the setting of county maps by able practitioners like the Warwickshire surveyor Henry Beighton as early as 1721.[43] This change was evidently related to the declining popularity of the magnetic compass as a survey instrument. It was a slow and irregular trend, however: even in 1780 a competent English surveyor could alternate between true north and compass north, without warning, from one page to another of a single estate atlas.[44]

A possible escape from the orientation dilemma was to show two kinds of north on the same map, for example by a pair of intersecting arrows, one or both clearly labelled. It was a solution obvious enough to have been adopted by one precocious

7.5 Compass north and true north distinguished but not identified. Abraham Ortelius, *The theatre of the whole world* (London, 1606), p. 39.

surveyor at the impressively early date of 1532,[45] but for some reason the difference between the two lines proved difficult to express (Fig. 7.5). A number of sixteenth-century cartographers simply drew a compass in which the needle was inclined to the north-line on the card, but both angle and direction were so often incorrect that many such drawings are best interpreted as a vague nod towards the existence of declination rather than as a precise record of its numerical value.[46]

Other maps frustrate the reader by defining the relation between true and magnetic north in words and numbers without adding any lines to show their relation to other geographical features. Even in the eighteenth century, when the best map-makers were beginning to make themselves clear on this subject (Fig. 7.6), it was still common for the traditional compass star and fleur de lys to be crossed, without explanation, by a single oblique line ending in a plain arrow-head. Which was true north, the arrow or the lily? At first it was often the former (Fig. 11.13), in the nineteenth century the latter, a change that was perhaps to be expected from the steadily diminishing authority of the compass itself. To show magnetic north by half an arrowhead, as on some twentieth-century maps, was an appropriate reflection of its geodetic inferiority, but not necessarily intelligible to a wide readership.[47] The dating of meridian lines was never more than a matter of personal taste. A date could demonstrate the author's ability to keep abreast of events, as in Lewis Morris's chart of the Welsh coast,[48] and it might also be deliberately omitted to hide his failure in this respect: not many cartographers were as honest as the Society for the Diffusion of Useful Knowledge, whose map of the Falkland Islands admitted to lagging more than sixty years behind the times in the matter of magnetic declination.[49]

Choosing a scale

Like several other claims to exactitude, the scale line made its debut on the stage of late medieval marine cartography, invading various kinds of land map in the course of the sixteenth century. Before its intricacies are confronted, something should be said about the motives influencing a plotter's choice of scale-ratio. His most reasonable course was to save paper, and storage space, by adopting the smallest size

7.6 Magnetic meridian, original scale 1:84,480. From Richard Budgen, *An actual survey of the county of Sussex* (London, 1724).

that could convey the cartographer's message in a legible form. A rather less straightforward issue is the relation of scale to planimetric error. Two opposite tendencies might seem relevant here. In some cases, error could be expected to vary inversely with scale. The process of graphically reducing a map was usually accompanied by generalisation, and this involved omitting certain curves or angles and exaggerating others, with a consequent shift in the position of at least some identifiable points. However, replotting a survey was a different operation from redrawing a map and at present we are concerned only with what happens when the same field data are laid down at different scales. The contrary effect is that with protraction on smaller scales more of the surveyor's errors will fall below the threshold of plottability, and fewer will remain detectable in the final product.

Perhaps this is related to the frequency with which small-scale maps have beaten large-scale maps in tests of planimetric merit.

There could also be more practical reasons against minimising the scale of a map. One was the intrinsic attractiveness of non-fractional numbers and well-known linear standards. In a measured survey of a small area, laid down with ruler and compasses from a field book, there was often a simple relationship between ground units and paper units, with a round number of chains or perches to the inch. This could save both author and reader from elementary arithmetical errors. In early triangulation surveys, however, it was quite possible for plotting to precede scaling, and the results were then less likely to be definable in simple terms. Another complication was that when a map formed part of a series such as an atlas, its scale might justifiably be selected not for any individual advantages but to meet the requirements of the whole collection. Also worthy of attention, to judge from the tone of certain eighteenth-century map advertisements, is the habit of cultivating size as an end in itself for reasons of prestige. The sixteen sheets comprising John Rocque's plan of London would seem to be a case in point: having decided not to distinguish the individual houses in his street blocks he could easily have chosen to work within narrower limits (Fig. 7.7).[50]

The best known kind of scale statement is the numbered line in its various forms of divided map-margin, grid squares, circles of specified ground radius, or (most commonly) the separate 'trunk' or bar first seen in the portolan charts of the fourteenth century. The more scientific-looking device of the 'representative fraction' came much later, being ascribed by most historians to the year 1806.[51] Its subsequent popularity was due in part to the influence of metrication as adopted by the French authorities fifteen years earlier and recommended for cartography in 1802 by the *commission de topographie*.[52] It was only under the metric system that an arithmetical fraction could be expected to achieve simplicity both as a single number and in the alternative form of a ratio between different ground and paper units.

Miles versus degrees

Where the plotter's raw material consisted of latitude and longitude values it was natural for him to start by setting out a graticule. Ptolemy's data provided one source of coordinates for his fifteenth-century successors; others, at the same period, were the first-hand astronomical observations that were beginning to supplement compass bearings as chart-makers emerged from the Mediterranean into the Atlantic; others again were the records kept by observatories and scientific expeditions which from the late seventeenth century were increasingly used as a corrective to earlier maps of the world and of the larger countries.

On any map that carried both scale and graticule the author necessarily committed himself, if only by implication, to offering a comparison between

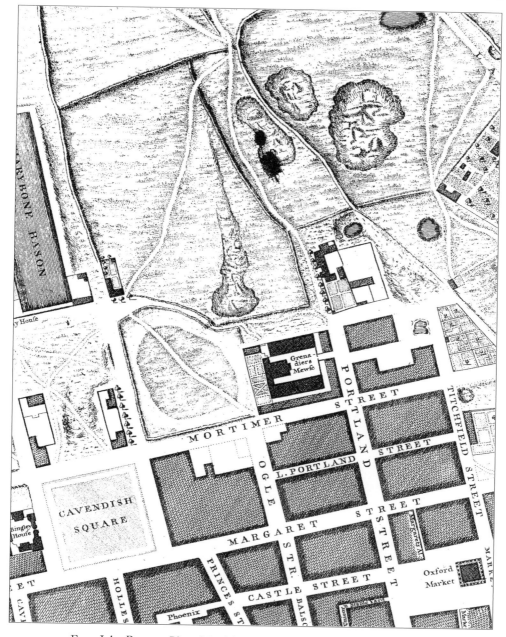

7.7 From John Rocque, *Plan of the cities of London and Westminster* (London, 1746).
Original scale 26 inches to 1 statute mile (1:2437).

astronomical and terrestrial measures. This relationship was unlikely to be very harmonious when one unit was defined by aggregating the human foot and the other by subdividing the earth's circumference. As we have seen, however, it was common until the seventeenth century to quote suspiciously simple formulae identifying one degree with a round number of national or regional units. Some of these equations were much older than any of the maps in which they appeared. Thus the Greek geographer Strabo had defined the mile as eight and one-third stades and a stade as 1:180,000 of the earth's circumference or 1:500 of an equatorial degree – a definition that subsequently found its way to England, through early sixteenth-century German sources, in such naturalised forms as Hans Woutneel's 'sixty English miles answerable to a degree'.[53] Not every Englishman accepted the authority of either Greek or German. William Cuningham claimed to have found sixty miles in a degree by his own efforts,[54] and William Bedwell in 1631 attributed the same value to 'experience'.[55] Of course it was quite permissible to *define* a mile as one-sixtieth of a degree, and this unit did indeed enjoy a respectable life of its own in the eighteenth century and afterwards as what was variously titled a nautical, geographical or geometrical mile. The mile as reckoned along the roads of England was something different, becoming stabilised in the seventeenth century as an aggregate of aggregates to give a total length of 5280 feet, now known as a statute mile after an Elizabethan act of parliament regulating the expansion of London.[56] According to the test conducted by Richard Norwood between London and York in 1634, the number of statute miles in a degree was not 60 but 69 and a half plus 231 feet.[57]

Norwood's version of the degree was still being quoted on English maps towards the end of the eighteenth century. However, some of his earlier successors – among them Francis Lamb, Joel Gascoyne and John Seller – continued to treat the minute of latitude as identical with the statute mile.[58] This elementary error caused trouble when one unit was substituted for another, because if degrees were converted into miles the map would be too small and vice versa. Shape, as distinct from size, was unaffected by such errors, provided that latitude and longitude had been treated in the same way; but since latitude was so much more easily measured than longitude, north-south distances miscalculated from astronomical data must sometimes have been combined with east-west distances taken from a topographical survey, in which case the map would have the wrong shape as well as the wrong scale. Differential error of this kind has been reported more than once from the sixteenth century, for example in Martin Waldseemüller's map of the Rhine valley and in Humphrey Lhuyd's map of Wales.[59]

In the eighteenth century this danger receded as current linear interpretations of angular measure became sufficiently correct for any residual disagreement to be cartographically meaningless. The improvement in accuracy made it worth asking how the length of a degree varied with latitude, a variation foreseen by Isaac Newton and later confirmed by experiment in Lapland (Fig. 7.8) and Peru between 1735 and 1744.[60] No sooner was the Peruvian operation complete than John Elphinstone announced his map of Scotland as the first to recognise the earth's spheroidal form.[61] But at Elphinstone's scale of 1:253,344 the resultant difference in

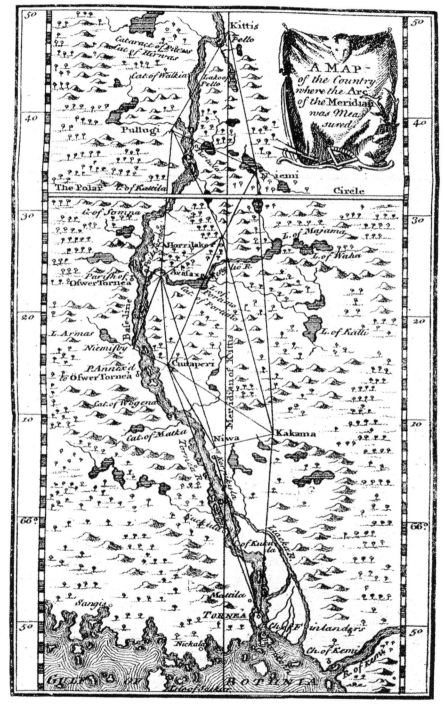

7.8 Determining the length of a degree of latitude. Pierre-Louis Moreau de Maupertius, Lapland, 1736–7.

spacing between the northernmost and southernmost degrees of latitude in mainland Scotland would have been no more than 0.05 inches. Further south, at least one eminent practitioner in Europe's most cartographically successful nation was prepared to condemn such refinements as a waste of time.[62]

Divided margins became customary at an early date as a convenient way of showing latitude and longitude. Unfortunately they were by no means free from ambiguity. Experience suggests that where a number was placed half-way between two marginal scale divisions (a common habit in the sixteenth century) it probably referred to the line further from the equator or further from the central meridian as the case may be. A more serious problem for the reader was how to run his own meridians and parallels across a map that showed marginal divisions but no graticule. In such cases a slope in the invisible lines of latitude and longitude was often signalled by a corresponding inclination in the short marginal cross-strokes within the map border, but the curvature of the missing lines would still remain uncertain. Otherwise most modern historians would probably interpolate the meridians and parallels of an early map as straight and hope for the best, but apart from a hint with the same implication in Mercator's *Atlas*, it is hard to find any contemporary pronouncement on the matter.[63] The situation was not made easier by maps with asymmetrically graduated margins that showed one end of each meridian or parallel but not the other end.[64]

Meridians, prime and central

The most problematical line in many early graticules is the prime meridian. Here, as with several other parameters considered in this chapter, the frankness of the eighteenth century was preceded by a less communicative period in which cartographers' intentions could be hard to deduce from whatever results they might have hit upon by accident. Many prime meridians are difficult to identify because they fell outside the limits of the map in question, and even where a zero line existed it could bewilder posterity by passing through two or more well-known places that did not in fact lie north and south. Not that the first meridian necessarily had to coincide with any named geographical feature. A cartographer could simply rule a meridian line at random, write a nought against it, and then insert the other meridians according to his chosen scale. If it came to that, there was no strictly logical reason why longitude should be reckoned from an origin at all: meridians could simply announce themselves as being so many degrees apart.[65]

It seems clear from contemporary literature, however, as well as from historical common sense, that on many maps the choice of prime meridian was influenced by a number of definite if conflicting principles.[66] One was the advantage of a single world-wide system. Another was the desire to avoid a mixture of positive and negative numbers by reckoning longitude from one extremity of the mapped area.

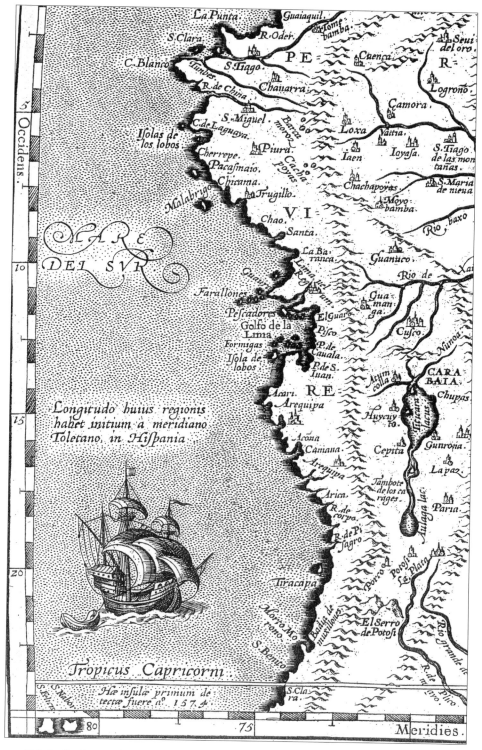

7.9 Toledo, Spain, as prime meridian for Peru. Abraham Ortelius, *The theatre of the whole world* (London, 1606), p. 9.

It also seemed natural for users of Greek or Roman script to count from left to right. This was why Ptolemy chose the Canary Islands and why, after the discovery of the Azores (*c.*1427) and the Cape Verde Islands (1461), it seemed appropriate for the zero of longitude to be shifted westwards, some cartographers believing that it would be more convenient to pass between the archipelagoes than to divide them.[67] For a time in the sixteenth century the line of zero magnetic declination was favoured in the belief that it ran from north to south. But it was thanks to Ptolemy's prestige that the Canaries held sway for so long.[68] (Nobody seems to have thought that longitudes should be counted from the Tordesillas line, although this feature was occasionally mapped as an historical phenomenon.)[69] Perhaps Louis XIII was reflecting on Ptolemy's example when he adopted Ferro, the westernmost of the Canaries, as prime meridian for French maps in 1634.

Another possibility was to choose some important town in the map-maker's own country (Fig. 7.9), making this key geographical datum accessible to his compatriots at work in subjects like astronomy and geophysics. Hence the preference for London in England, Cadiz or Toledo in Spain, Copenhagen in Denmark, Philadelphia or Washington in the United States and so on. In France ambiguity threatened when the Paris meridian began to find favour for national maps, a situation not much improved by placing the city 20 degrees east of Ferro as a matter of definition.[70] Of course it remained possible for both versions to appear on the same map, sometimes along with one or more others, and this was still happening after 1800,[71] though once Paris had been chosen for Cassini's survey it was inevitable that Ferro should enter a long period of decline. 'We could make a large catalogue', wrote Robert Sayer in 1775, 'of meridians which caprice and national pride have led different authors to make use of. There is hardly a considerable town in Europe which has not had this honour, as have almost all the islands of the Atlantic Ocean in their turn'.[72] The choice of large cities as origins may have helped to inspire a new fashion for reckoning longitude in both directions up to a maximum of 180 degrees each way, instead of counting eastwards in a complete circle.[73]

Prime meridians should not be confused with central meridians. The former have a purely numerical significance; the latter are associated with map projections and may also refer to whichever line of longitude lay half-way between the western and eastern margins of a particular map. In Ptolemaic world graticules the western limit normally stood at zero longitude in the Canary Islands and the central meridian at 90 degrees east. In fact the 90-degree meridian, usually passing through or near the Persian Gulf, remained central to most world maps, whatever their source of information about lands and seas, until around 1550, the only important exception being a certain preference for 80 degrees on the part of Sebastian Münster. In this connection a case that may be worth a passing glance is that of North America's Pacific seaboard in the first half of the sixteenth century. If a world map had a central axis at 100 degrees east, the new continent might, according to

7.10 Straight coast of Pacific North America with conventional irregularities. Sebastian Münster and Hans Holbein, world map. Simon Grynaeus and John Huttich, *Novus orbis regionum* (Basle, 1555).

current information, be divided between its right-hand and left-hand margins. Such a map might show eastern North America near its western edge but fail to pick up the remainder of this land mass at the far right, perhaps deliberately as a confession of ignorance. Imagine a copy of the same map with the central meridian moved westwards to 80 degrees. Part of the Pacific Ocean would now appear on the left, and the former intra-continental map margin would be reclassified as an unnaturally straight American west coast running from north to south along a meridian – which is just how the feature in question was shown on maps by Johannes de Stobnicza (1519),[74] Johann Honter (1530), Sebastian Münster (1532) and Joachim von Watte (1534).[75] Fortunately this phase did not last long (Fig. 7.10). It soon became desirable to accommodate the full width of an emerging American continent on the left-hand side of the world map in what a European navigator would regard as its proper direction. Accordingly, from the 1560s onwards prime and central meridians would often coincide, either as the north-south axis of an oval or rectangular map or else as the boundary between a pair of hemispheres. However, there were enough exceptions to all these statements to make central meridians nearly as confusing as prime meridians.

At scales where the longitudinal extent of built-up areas was large enough to be measurable, the most appropriate object from which to define a prime meridian was a tall building. One writer in 1696 proposed to make it an act of sacrilege for Englishmen to reckon longitude from anywhere but St Paul's cathedral.[76] Not all map-makers were intimidated by this threat: in Emanuel Bowen and Thomas Kitchin's *Large English atlas* (1760) the central meridian lies nowhere in particular about one minute eastwards from St Paul's. By that time, however, thanks to a new campaign for finding the astronomical longitude at sea, the idea of a world-wide standard was due for a revival, with a centre of scientific activity rather than a westerly island as reference point. The observatories of Greenwich (5' 37" east of St Paul's), Paris and Washington remained in contention for a long period, but fortunately it was now becoming customary for a point of origin to be specified in any map that gave the longitude. Greenwich remained a popular choice; its formal adoption at an international conference of 1884 (by a large majority but with France as one of the abstainers) lies outside the scope of this book.[77]

CHAPTER 8

So strangely distorted

A MAP PROJECTION IS a mathematical formula that converts the earth's curvature into a flat surface. More particularly, it translates the numerical values of latitude and longitude into two rectangular or Cartesian coordinates. Although there are infinitely many ways of doing this (a new projection appears as soon as one of the terms in such a formula is multiplied by any rational number not previously used for a similar purpose), as a matter of historical fact only a few hundred projections have until recent times been adopted for the portrayal of geographical reality.[1] A graticule of visible meridians and parallels is by no means essential to the idea of a projection: the existence of a particular formula may be implicit in the drawing of geographical features even without a gridded outline. From which it seems to follow that a map may be scientifically projected without its author's knowledge: for example, if the bearings and distances of a radial survey (as defined in chapter 6) are plotted to a uniform scale the resulting projection will automatically be classifiable as 'zenithal equidistant' – though in real life such a survey could hardly extend far enough to make this a matter of any practical significance. Where the cartographer's field data include two or more independent latitude and longitude values the situation is different: now he is forced to choose some formula for conversion.

Whether every existing map can be mathematically analysed in this way seems doubtful. A distinguished scholar describes a well-known marine chart as 'not based on a projection'.[2] Another states with equal conviction that 'any map (of a part of the world large enough for its sphericity to be taken into account) that is drawn on a flat plane, even the crudest of sketches, involves some sort of projection'.[3] Before trying to settle this apparent conflict of opinion let us ask what is meant by 'large enough' in the second of the foregoing quotations. Perhaps surprisingly, not more than a single author has been found to answer this important question, and that only in a somewhat roundabout way.[4] But according to one unpublished contribution, the largest area showing no detectable difference between a spherical and a flat earth at a scale of 1:10,000 would have a radius of about 42 statute miles.[5] Inside that limit, different projections would be equally effective, except for those designed in a spirit of perversity with the intention of proving this statement wrong.

Most of the maps known to history have covered a wider radius than 42 miles. Imagine then a world or continental map, a medieval *mappamundi* for instance, that was made by a mathematical ignoramus from estimation, hearsay or guesswork. Using Waldo Tobler's technique of bidimensional regression or some equivalent

8.1 British Isles, derivative of George Lily (1546) with parallel meridians. Longitude of Thule 26–27 degrees. Giovanni Valvassore, *Britanniae Insulae … nova descriptio* (Venice, 1556).

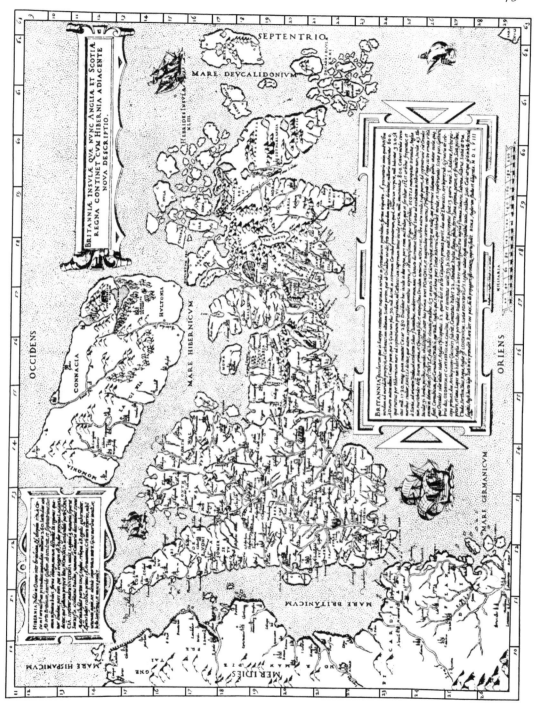

8.2 British Isles, derivative of George Lily (1546) with convergent meridians. Longitude of Thule 29–30 degrees. Sebastiano di Re, *Britanniae Insulae … nova descriptio* (Rome, 1558).

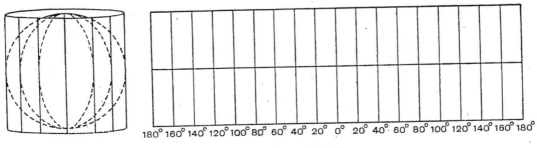

CYLINDRICAL

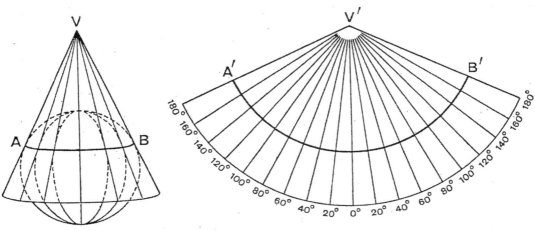

CONICAL

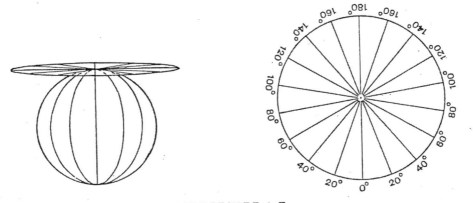

ZENITHAL

8.3 Three common varieties of map projection.

method, let such a specimen be compared with a series of maps showing the same area on every possible projection. We might now suggest that if our map resembles a given projection, p, more closely than p resembles any other projection (let us call this condition 'C') then this map must be 'on' p and that if there is no value of p for which C is true, then the map is not on any single projection. Such reasoning obviously has its difficulties. But we can accept as a matter of common sense that the statistical links between a map and its supposed projection may sometimes prove too weak to be taken very seriously. In other words, the idea of a projection-less map is not necessarily without value.

If some maps have no projection, others must be credited with two or conceivably more. This would occur if within the same border one formula was used for meridians and parallels and another for coasts, rivers and other geographical detail.[6] Consider for example George Lily's map of the British Isles (1546) and the various maps based on it. On what seem to be identical geographical outlines, some of the derivatives have convergent meridians and others parallel meridians, causing the same feature to vary in longitude from one Lily-type map to another by as much as a whole degree (Figs 8.1, 8.2).[7] Later, a famous map of Abel Tasman's discoveries is thought to have inadvertently combined two projections to be dealt with more fully below – Mercator's for the coastlines and the plat carrée for the graticule.[8]

A single projection is definable by its formula, but in modern geography projections have been grouped into broader categories according to several different criteria (Fig. 8.3). Thus some graticules have been likened to a cone which touches the earth along a parallel of latitude and which is then unrolled or 'developed'. Slide this line of contact down to the equator and the cone becomes a cylinder unrolling into a rectangle. Slide it in the other direction to the north or south pole, and the conical projection becomes circular or zenithal, starting and finishing its career as a plane. Many projections can therefore be classed as either conical, cylindrical or zenithal. In the most common cylindrical varieties the equator forms a line of minimum error running horizontally across the middle of the map.

With no change of the underlying projection-formula, this line may be replaced either by a meridian of longitude or by any other great circle. Similarly in a central projection the centre may be moved from the pole to the equator or to any other point. From such manoeuvrings are derived the 'cases' or 'aspects' of the modern textbook – polar, equatorial, transverse, oblique. Finally, projections may preserve or sacrifice certain properties of the spherical earth: for instance they may be correct for either relative areas (equivalence) or directions (conformality) or neither. All these ideas will reappear in the following paragraphs, but in a historical treatment it may help to lay more stress initially on a simple two-fold distinction between astronomical and global mapping. From a mathematical standpoint the story can begin in the heavens, even though it was not until post-medieval times that astronomy came to have much effect on the projections chosen for terrestrial maps.

Star maps and earth maps

In the sky, what made most impression upon the viewer was the circular movement of the stars around the pole. So, although in theory any geographical projection might be applied to the celestial sphere, in practice nearly all astronomers chose from the group that is variously known as central, polar, zenithal or azimuthal, in which the parallels are concentric circles from whose centre at the north or south pole the meridians radiate as straight lines (Fig. 8.4). Applied to the earth, some of these zenithal networks are 'perspective' in the sense that the meridians and parallels could be literally projected by rays of light on to a plane touching one of the poles. An important variable factor in this arrangement is the supposed location of the light-source. In the gnomonic projection the light is at the centre of the earth. In the stereographic, it is on the earth's surface, antipodal to the point of tangency. In the orthographic it is an infinite distance away, so that the lines of projection are parallel rather than divergent. The zenithal equidistant, already mentioned, is a non-perspective projection without an imaginary light: its circles of latitude are correctly spaced along the meridians, which (unlike those of the perspective zenithal projections) could be extended if necessary to take in the whole sphere.[9] The gnomonic, most familiar from its use in sundials, is as old as the earliest recorded Greek mathematics. The orthographic and stereographic were both known to Hipparchus in the second century BC.[10] The zenithal equidistant was used for a number of early European star charts and may have been derived from classical or Arab models, though among geographers it is especially associated with Guillaume Postel and other sixteenth-century map-makers.[11]

In geography the zenithal systems proved most effective for maps of the arctic and antarctic, not surprisingly when it was in these regions that other kinds of projection, to be considered later, were at their least useful. One factor encouraging maps with a polar focus was the post-medieval European obsession with discovering new routes to China through either a north-east or a north-west passage. But zenithal projections could also be drawn with the plane of tangency passing through some point other than one of the poles. Perhaps the most probable choice for non-polar use was the type most likely to occur by accident, namely the zenithal equidistant, which among other advantages could show correct distances to any-where on earth from the cartographer's or customer's home town, or from any other point that might be chosen as origin.[12]

Of the three perspective zenithal projections, two have been relatively unimportant for most of their history. The orthographic, recommended nowadays for simulating a view from outer space, attracted few enthusiasts in the world of early cartography, though one of them was the charismatic Albrecht Dürer.[13] The spectacular scale-variations of the gnomonic had a certain curiosity value.[14] Much more familiar was the stereographic, particularly in its equatorial aspect. This was

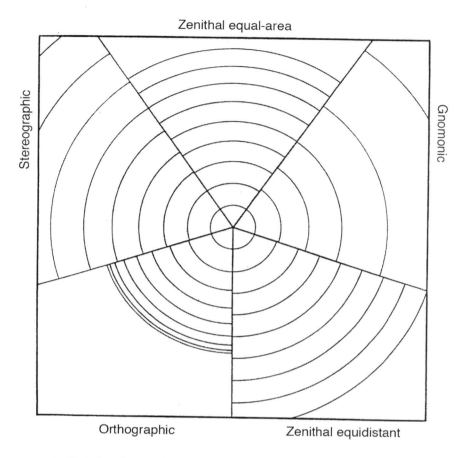

8.4 Varieties of zenithal or azimuthal projection, parallels at equal intervals.

fairly common in renaissance world maps, the oldest surviving example being by Walther Ludd in 1507,[15] but the period of its greatest success began eighty years later when Rumold Mercator produced a version soon to be made famous in his father's atlas (Fig. 8.5). Thereafter the best-known application for the stereographic was in pairs of hemispheres, each with the north pole at the top.[16] These may be recognised from latitudinal intervals that widen with increasing distance from the equator and longitudinal intervals that widen correspondingly with increasing distance from the central meridian. The result is a conformal projection, preserving the correct shapes of small areas; and whether or not contemporaries fully understood the mathematics involved,[17] they must have been glad to see the parallels and meridians crossing at right angles. At any rate the popularity of the stereographic in the seventeenth century rivalled that of Mercator's projection among the Victorians: far from seeming impossibly difficult and esoteric, it was sometimes used for world maps printed on the backs of playing cards.[18] Like Mercator, the stereographic was

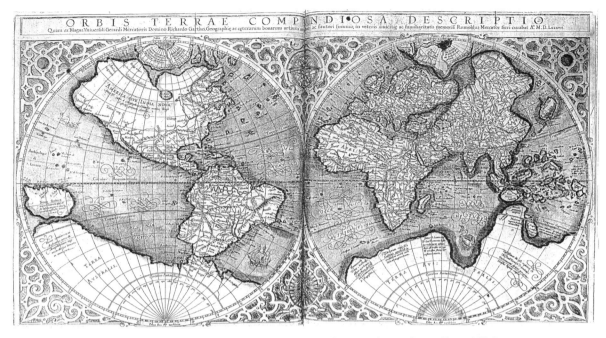

8.5 Equatorial stereographic projection, western and eastern hemispheres. Rumold Mercator, *Orbis terrae compendiosa descriptio* (Geneva, 1587). See also Fig. 5.7.

eventually to suffer a slump: by the early twentieth century it was 'scarcely found in use at all'.[19]

Two of the four best-known zenithal projections can show more than half of the terrestrial surface, but in practice most early zenithal maps confined themselves to a single hemisphere, a fact that may have encouraged the development of a similar-looking non-zenithal projection known as the equatorial globular. Here, predictably, the equator and central meridian were straight lines of equal length passing at right angles through the centre of a hemispherical circle. Globular parallels of latitude could be added in one of three ways: (1) equal division of the outer circumference, with straight lines of latitude that necessarily cut the central meridian at unequal intervals (Fig. 8.6);[20] (2) equal division of the central meridian by straight perpendiculars, as in Peter Apian's world map of 1524;[21] (3) equal division of both marginal and central meridians, causing non-equatorial parallels as well as non-central meridians to be curved.[22] This third category must be subdivided, the meridians being made elliptical by Georges Fournier in 1643 and circular by Giovanni Battista Nicolosi in 1660.[23] It was Nicolosi's method which eventually came to monopolise the name 'globular' and which superseded the equatorial stereographic as the most popular way of mapping the earth in hemispheres, probably because it seemed easier to understand. With Aaron Arrowsmith it even achieved the rare distinction of having its name included in the title of a published map.[24]

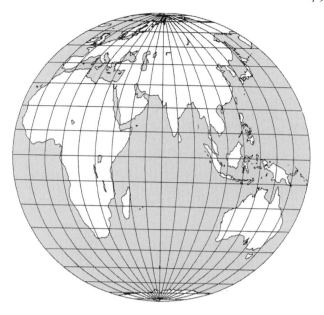

8.6 Roger Bacon's globular projection, *c.*1265, reconstructed with modern coastlines in John P. Snyder, *Flattening the earth: two thousand years of map projections* (Chicago, 1993), p. 15. Redrawn by Matthew Stout.

Ptolemy on projections

Having followed the idea of circular projections as far as it will take us, we must now revisit the ancient world in quest of map-types originating with terrestrial rather than celestial knowledge. For earth-dwelling travellers with no scientific education, the most important spatial concepts associated with locomotion are right, left, forwards and backwards. These relationships are best shown by lines crossing at right angles, and from here it is but a short step to the idea of a rectangular or cylindrical projection. The simplest rectangular network is the plat carrée, in which the scales of degrees along the meridians and the equator are the same (Fig. 8.7). For present purposes the history of this perhaps under-rated projection may be thought to have begun before AD 100, when mathematical geography was studied by Marinus of Tyre. Unfortunately Marinus is now known only through the *Geography* of his critic Ptolemy, but like other ancient map-makers he seems to have been concerned not so much with the whole earth as with the *oikoumene* or known world, an area covering about 80 degrees of latitude and 180 degrees of longitude. It was natural for this region to be conceived as a rectangle and mapped on a rectangular projection.

If Marinus had thought of representing the entire globe he might well have preferred some other shape. As it was, he had good reason to build his graticule from straight lines, making the meridians parallel to each other – 'just as most [cartographers] have done', added Ptolemy, casually telling modern historians something important about ancient geography that they would have found it hard to learn from any other source.[25] Presumably Marinus's latitudes and longitudes

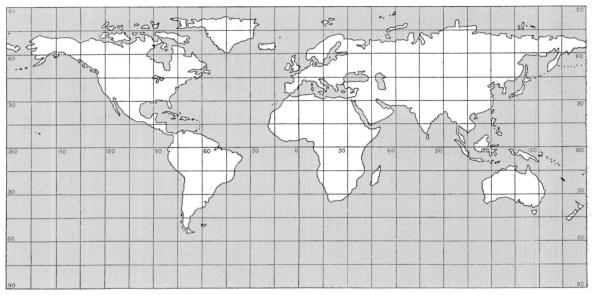

8.7 Simple cylindrical projection, also known as square projection, plat carrée or plane chart, with modern coastlines.

intersected at right angles. He can perhaps be pictured as beginning with a plat carrée and then noticing how grossly its east-west scales became too large northwards and southwards from the equator, a serious problem considering that the lands and seas best known to the Greeks all lay well inside the northern hemisphere. This was why Marinus matched his north-south scales not with the equator but with the 36-degree parallel cutting through the island of Rhodes in the heart of the Graeco-Roman world. The result was a modified cylindrical projection.

Even after this adjustment Marinus's east-west distances at the northern edge of the *oikoumene* were about eighty per cent too long, as Ptolemy was quick to point out. The solution was to abandon the idea of a rectangle. Perhaps Ptolemy expected some resistance from his readers on this point, because he introduced his new curvilinear projection by drawing a rectangle that had nothing to do with the ensuing argument. In fact most modern students will find the whole of his exposition difficult to follow, and the same may be true of recent historical writings covering the same ground, which generally fall back on quoting verbatim from the *Geography*. It may be better to begin with a modern-style account of a later projection, the simple conic, to which Ptolemy's first graticule is closely related (Fig. 8.8). A conveniently placed parallel of latitude is chosen as line of contact between the map cone and the terrestrial globe. Tangents to this line meet at the tip or vertex of the cone, which is a point outside the sphere somewhere 'above' the pole. The 'slant length' of the cone is the radius of the standard parallel, which appears on the map as part of a circle. (Like Marinus, Ptolemy chose the island of

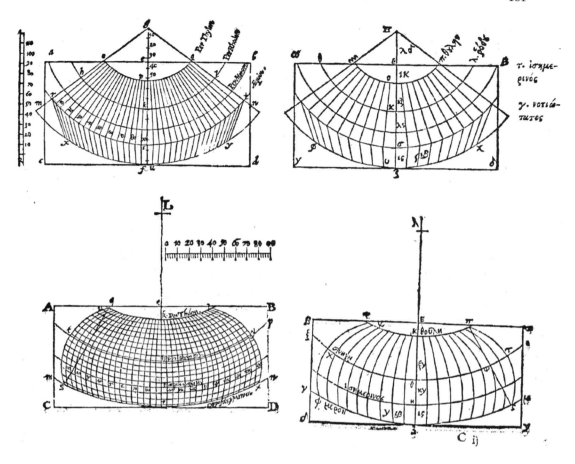

8.8 Ptolemy's first (above) and second projections. Longitudinal interval: left, 5 degrees; right, 10 degrees. Northern limit: Thule, 63° north; southern limit: Meroe, 16° 25' south. Gerard Mercator, *Claudii Ptolemaii … geographicae libri octo* (Amsterdam, 1605).

Rhodes as the latitude of his standard parallel.) Other parallels are spaced at their true intervals along this radius, forming arcs concentric with the standard. On the sphere, the circumference of the standard parallel depends on its latitude and on the radius of the earth. On the map the arc of the standard parallel is given its correct length, but it will not be a full circle because its radius on the projection is greater than on the globe. This arc can be divided into 360 equal parts to mark the intersections of the meridians at one-degree intervals. The meridians themselves are straight lines connecting these intersections with the vertex of the cone.

Noteworthy features of the resulting network are that the pole is an arc and not a point, that all the parallels of latitude are arcs and not complete circles, that the parallels are correctly spaced along the meridians, and that the most important point used in constructing the projection, the vertex of the cone, does not appear on the map. Ptolemy also made a virtue of keeping the equator in its correct proportion to the northern limit of his map, which stood at latitude 63 degrees, but this seems to be a consequence of starting with the same parallel as Marinus. If the new standard parallel had been much further north or south the proportions would have been less correct. The real surprise occurs south of the equator where Ptolemy put his projection into reverse, bending the meridians inwards to make his southern-most parallel the same length as its northern equivalent. After the sixteenth century most cartographers rebelled against this unsightly expedient.

Next, and without preamble, the *Geography* switched to a radically different method of mapping the known world. The central meridian was again a straight vertical line and the parallels were again correctly spaced concentric circular arcs. Three parallels, those of Thule (sometimes identified with the largest of the Shetland Isles), the Tropic of Cancer, and the equator, were correctly divided. This gave three points on each meridian, the meridians themselves being interpolated, not now as straight lines but as arcs of circles drawn through each meridional trio of points. The only other problem was how to fix a centre and a radius for whichever line of latitude governed all the others. Here Ptolemy's directions are too intricate to be summarised.[26] Part of his justification for this difficult alternative graticule was that its curving meridians were more reminiscent of a globe.

It is often argued that Ptolemy drew no maps himself but only made lists of coordinate values. If so, a guilty conscience might help to explain why he took the trouble to describe a second choice of projection when so few of his successors would bother to describe their first choice: as a consolation for not getting an actual map, two verbal statements might with luck seem better than one, if not rejected as a case of protesting too much. In the event, these projections were a remarkable triumph. If Ptolemy had never lived, European cartography might well have been dominated by rectangular or circular graticules for a much longer period.

Extending Ptolemy's graticules

Ptolemy's own preference was for his second method, the first being included only in recognition of his readers' mental incapacity – their laziness, as one translator puts it.[27] Among fifteenth- and sixteenth-century editors of the *Geography*, the two systems were to prove about equally popular. In maps expressing post-medieval world-views Ptolemy's influence was less direct. Indeed some cartographers ignored both him and Marinus, reverting (in a mathematical if not a strictly historical sense) to the simplicity of the plain rectangle. Whether such shapes should be accorded the title of plat carrée is an interesting question. A marine chart of *c.*1520 has been given chronological priority in this respect since it has both a latitudinal and longitudinal scale,[28] but projections can exist without graticules, as we have seen, and for any sixteenth-century world map a plat carrée attribution may be worth

8.9 Modified cylindrical projection. Giovanni Magini, *Universi orbis descriptio ad usum navigantium* (Venice, 1596).

considering if (a) there is no evidence pointing to any other projection (b) latitude observations seem likely on historical grounds to have been used in constructing the map (c) there are straight lines crossing at right angles which, even if not numbered or otherwise identified, look as if they might be meridians and parallels. Before adducing any particular formula, however, it would be advisable to measure a few lines, for example the total width of the map as compared with the north-south distance between the tropics. On this basis Francesco Rosselli's world map of 1508 can probably be counted as a plat carrée; also Laurent Fries's map of 1522. On other contemporary maps the latitude and longitude intervals are not similar enough to qualify, and then one must look for the most likely standard parallel. In Giovanni Magini's world map of 1596, for instance (Fig. 8.9), this would appear to be about half way between the equator and the pole.[29]

For devotees of Ptolemy the main problem presented by post-medieval exploration was the much larger area, including the Americas, that now had to be

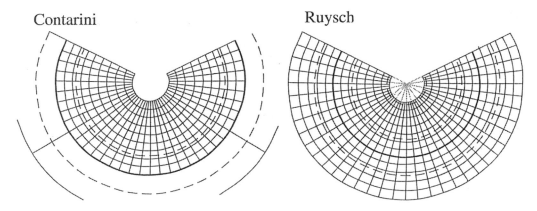

8.10 Graticules at 10-degree intervals for world maps by Giovanni Contarini, 1506, and Johannes Ruysch, 1507, showing the earth's full extent from east to west. The thick line is the equator, the broken lines the tropics. The smallest arc represents the north pole in Contarini, 70 degrees north latitude in Ruysch.

disposed around his central meridian.[30] The simplest course was to extend the Ptolemaic graticule outwards in all four directions, as in the well-known world maps of Giovanni Matteo Contarini and Johannes Ruysch (Fig. 8.10). By abandoning Ptolemy's awkward change of direction beyond the equator, these authors effectively 'purified' the simple conic projection. However, in their hands the result was unattractive to look at and suffered from mapping some of the world's emptiest and least-known latitudes on a larger scale than anywhere else. Subsequent cartographers had to accept that Ptolemy's first projection for the *oikoumene* was just not suited to the entire globe. Its influence survived, however, in the lengthy and successful career of the simple conic as a vehicle for regional mapping.

In the long run Ptolemy's second and more ingenious projection attracted a larger following among post-Ptolemaic geographers. It too was extended laterally to embrace the new discoveries, at the same time undergoing some important internal modifications. The standard parallel could be defined, as usual, by a cone wrapped round the sphere. The slant length of this cone became the radius of the equivalent parallel on the map. Other parallels were evenly spaced concentric arcs, but now each of them was privileged by being individually divided in its correct proportions on the same scale as the central meridian. By connecting these divisions, the non-central meridians could then be drawn as continuous curves. The result has sometimes been called cordiform or heart-shaped, though the aptness of this comparison depends on how much of the earth is included (Fig. 8.11).[31]

The cosmographic heart

In modern projection literature cordiform networks are usually classified according to the choice of standard parallel. If this is anywhere between nought and ninety

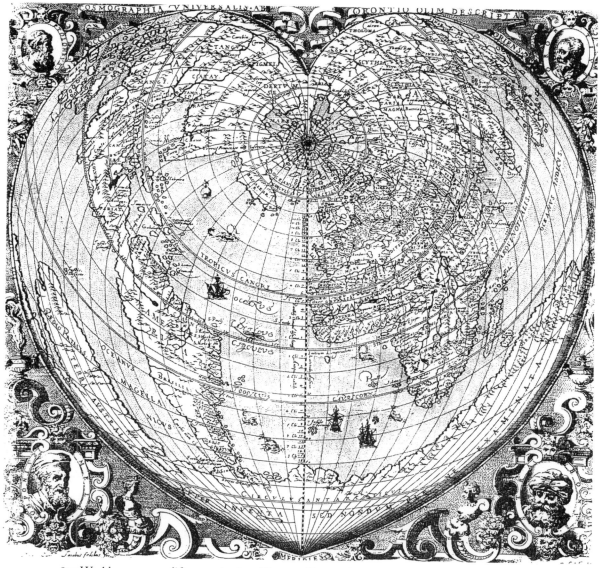

8.11 World map on cordiform projection. Giacomo Franco, *Cosmographia universalis ab Orontio olim descriptio* (Venice?, 1586–7?), derived from Oronce Finé's map of 1534.

degrees, the projection is one of the most famous of all time, later named after the eighteenth-century French cartographer Rigobert Bonne. An early example, with a standard parallel at approximately 47 degrees north, is the world map of Bernard Sylvanus.[32] The higher the standard parallel, the more compactly the polar regions are mapped. As a limiting case the parallel may be conceived as located at the pole itself. This is how later authorities would interpret the second and best known of the three projections invented by Johannes Stabius and described in 1514 by

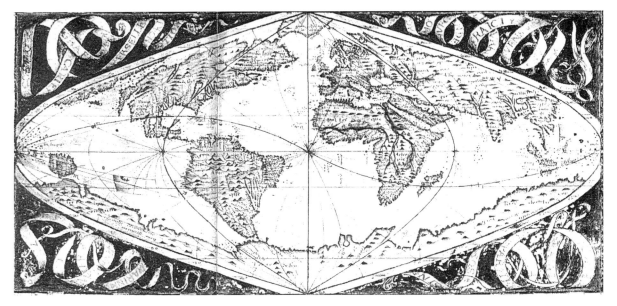

8.12 World map on sinusoidal projection. Jehan Cossin, MS, 1570. See note 34.

Johannes Werner, though on a superficial view the Werner outline looks much the same as Bonne's.[33] When the standard parallel was moved to the equator, on the other hand, this projection became unmistakably different and deserving of its own name – the sinusoidal, in reference to the sine curves that now formed the non-central meridians. This rather ugly design, also recognisable from its straight parallels, was used in the late sixteenth century by Jehan Cossin (Fig. 8.12)[34] and Jodocus Hondius.[35] Later the sinusoidal became a popular projection for both world and continental maps among French cartographers influenced by Nicolas Sanson, from whom, together with the English astronomer John Flamsteed, it is often named. (Flamsteed used it for star maps.) None of them is known to have pointed out that relative areas are correctly shown on this and other Bonne-related projections.[36]

An advantage of Bonne and Werner was to give wide longitudinal exposure in latitudes where new geographical knowledge was accumulating in the earliest years of the sixteenth century.[37] On the other hand their shapes were increasingly distorted eastwards and westwards from the central meridian. The easiest way to improve the general appearance of these projections was to trim off the oddest-looking parts of the world. It was not uncommon for instance to stop short at 80 degrees north and 50 or 60 degrees south. Another remedial measure, adopted by Oronce Finé (1531) and Gerard Mercator (1538), was to project the northern and southern hemispheres sideways as separate 'hearts' that made contact on the equator with the central meridian continuing in the same straight line from one heart to another. On a long view, however, the best future for projections of the Bonne family was in the mapping of single continents and countries. Indeed the whole of

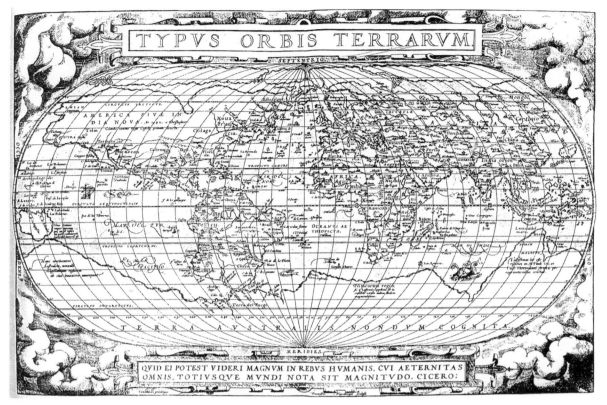

8.13 Oval projection. Abraham Ortelius, *Typus orbis terrarum* (Antwerp, 1570), from Ortelius's atlas, 1570, p. 1, with geographical information from Mercator's world map of 1569. See also Fig. 2.6.

regional cartography might almost be seen as a refuge for projections that had fallen behind in the quest for an ideal world map.

So far we have been following two broad streams of historical development, the first involving circles and radii, the second cones and (by extension) cylinders. These streams may be conceived as merging to produce the kind of oval projection chosen by Francesco Rosselli in the first decade of the sixteenth century.[38] There was now a straight vertical line as central meridian, a straight equator at right angles to it at twice its length, and other lines of latitude parallel with the equator, their lengths decreasing towards the poles. On these terms the sinusoidal would qualify for inclusion – not altogether appropriately, though for many decades it was too uncommon for its status to be of much interest. In other maps, more convincingly oval, for instance those of Battista Agnese (*c.*1540) and Abraham Ortelius (1570: Fig. 8.13), the poles appeared in semi-cylindrical style as parallels of latitude half the length of the equator. The core of the graticule, up to 90 degrees east and west of the centre, was a circle, and the outer meridians were semicircles. In another group of maps, including that of Benedetto Bordone (1528), the meridians were elliptical.[39] Near the equator and central meridian, most oval and cylindrical graticules were much alike. Elsewhere the angles on every oval map diverged considerably from the truth.

For want of documentary evidence it is impossible to state the criteria by which a sixteenth- or seventeenth-century map-user judged the graticules he saw in print. We can only observe (using for instance the wide range of early printed world maps reproduced by Rodney Shirley) what contemporary cartographers managed to get away with.[40] One's impression is that most people's expectations from a renaissance world map were a moderate level of realism: the grossest errors of distance and direction should be avoided but there was no point in expecting anything more. And as Ptolemy had observed, the clearest mark of realism was for the earth's circles to appear as curves. The main difference among curvilinear maps was in the amount of angular distortion. The worst irregularities occurred near the edges of a standard cordiform graticule, in just the areas that had been attracting attention as possible channels for north-west and north-east navigation routes from sixteenth-century Europe to the Pacific.

Such weaknesses may suggest a non-scientific explanation of the vogue for cordiformity, and in this connection some writers have drawn attention to the spiritual significance of heart-shaped symbols for the religious sect known as the Family of Love, which apparently had Ortelius as one of its members.[41] In strictly geographical terms a better compromise between distance and direction was a curvilinear outline divided by a straight equator. Within this basic framework two circles created less distortion than a single oval, and for a public that was growing more cartographically mature the break between the hemispheres was easier to understand in say 1590 than it might have been a hundred years earlier, especially now that the Atlantic formed a convenient boundary between old and new worlds of roughly comparable extent. (Northern and southern hemispherical maps were always less popular than western and eastern.) The hemisphere was also more realistic in the sense of approximating to what people saw when they looked at a globe, and indeed as globes grew increasingly familiar so hemispherical maps became more common. For some reason, anyway, hemispheres seemed to be driving out ovals at the end of the sixteenth century.

Projection theory: interest and indifference

By contrast, rectangular world maps were handicapped by displaying four fictitious corners and by exaggerating distances remote from the equator to an extent that could eventually reach infinity.[42] This made rectangles something of a poor relation in the family of sixteenth-century graticules, accounting for only about one in seven of Shirley's world-map facsimiles up to the fateful year 1569. Where the perpendicular style did justify itself was in facilitating the use of a protractor to set out compass directions on marine charts: as Mercator remarked, sailors did not want to

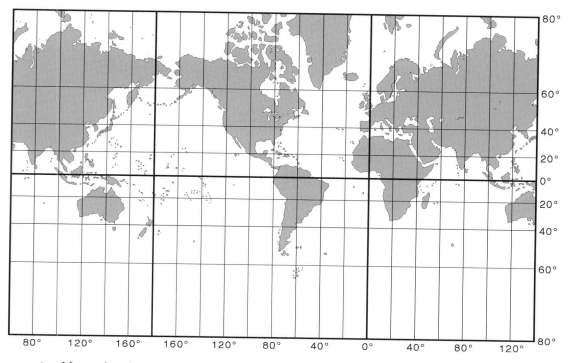

8.14 Mercator's projection to 80 degrees north and south, with modern coastlines. See also Figs 4.15, 15.1.

see meridians and parallels 'so strangely distorted that they cannot be recognised'.[43] Most early sea charts were in manuscript and therefore outside Shirley's terms of reference. Among printed maps it was only a small minority that made oceans as prominent as continents, and these are often identifiable from titles like 'Carta marina'[44] and from subtitles referring as if apologetically to hydrographic or navigational uses.[45] Maps carrying this description were invariably rectangular.

The disadvantage of the plat carrée for navigation was that unless a ship's course followed one of the four cardinal compass directions, a map-bearing taken to its destination would not actually follow the right course. In 1569 Mercator triumphantly overcame this difficulty with his 'increasing latitudes' (Fig. 8.14). In his own words[46]

> If you wish to sail from one port to another, here is a chart, and a straight line on it, and if you follow carefully this line you will certainly arrive at your port of destination. But the length of the line may not be correct. You may get there sooner or you may not get there as soon as you expected, but you will certainly get there.

Mercator had modified the plat carrée by progressively increasing north-south distances away from the equator so that at any point on the map the scale was the same in every direction (although different from the scales at other points), with the further result that loxodromes or lines of constant bearing were now straight. On the other hand, areas towards the poles were greatly exaggerated by comparison with those at lower latitudes, and the poles themselves were forever excluded from the map by virtue of being infinitely large in all directions.

The idea of projections evolving one from another may serve better as an expository device than as a serious statement of historical fact. This caution will perhaps be more effective if, before saying more about typology, we pause to summarise the differences between renaissance attitudes and our own. The Ptolemaic revival certainly inspired a keen interest in graticule-construction after many centuries of neglect. But as we have seen, geography was not the only factor determining the appearance of a map. Several sixteenth-century projections were devised for their entertainment value by mathematicians or artists who probably felt little desire to improve the comprehensibility of future geography books. Few map-users would have welcomed Albrecht Dürer's coverage of the world on twelve pentagons, for example; even the prospect of building the pentagons into a dodecahedron seems unlikely to have aroused much enthusiasm.[47] At the theoretical level, what mattered most was how many different shapes the earth could be made to assume, and here it was the globe itself, the whole of God's home for mankind, that caught the map-maker's imagination: individual continents or countries lacked the quasi-mystical attraction of the terrestrial sphere, which is one reason why countries have a mercifully simpler projection-history than planets.

Given the literary backwardness of cartography in general it is hardly remarkable that so few pre-nineteenth-century experts wrote books on the art of flattening a sphere. But we may still wonder why Mercator, in a long account of mathematical geography, should apparently regard a globe as the only method of representing the earth's surface that needed attention. For all his essay said to the contrary, there might just as well have been no such thing as a map; and yet Mercator was in the very act of introducing the world's most famous atlas. When graticules were not worth so much as a casual reference one could scarcely expect them to be classified, analysed or (for that matter) named.[48] Perhaps unsurprisingly, Mercator's choice of projections in his own atlas was not particularly consistent.[49]

Another obstacle to classification was that early geographers felt less inclined than their present-day successors to regard each projection as a self-sufficient and unchanging entity, and more ready to introduce minor variations for the sake of immediate convenience, forcing even the most diligent modern commentators to take refuge in such vague descriptions as 'a sort of conic version of the trapezoidal projection' or 'apparently a rudimentary prototype of the equidistant conic with two standard parallels'.[50] As late as 1677 John Adams's map of England is said to have combined the stereographic projection for one parallel of latitude with something more like Bonne's for the other parallels.[51] In the spirit of a newly self-conscious scientific culture Adams was happy to take credit for this achievement. Other map-designers were less assertive. Their modesty was not always rewarded, for when

8.15 Anticipation of
Mercator's
projection (south
above) by Erhard
Etzlaub, Europe
and adjoining
regions, 1511.

projections did start acquiring names these often turned out to commemorate the wrong person, as happened with Guillaume Postel, Nicolas Sanson, John Flamsteed, Rigobert Bonne, Jean Louis Lagrange and perhaps even Gerard Mercator.

More about increasing latitudes

In some respects the history of Mercator's projection was normal enough. Like many others it took time to establish itself. Thirty years went by before 'waxing' latitudes began to appear in other cartographers' world maps and their first major successes came even later, in Robert Dudley's *Arcano del mare* (1646), *Le Neptune françois* (1693) and a number of Dutch sea atlases.[52] However, with improvements in maritime surveying the navigational superiority of a conformal chart to the plat carrée became increasingly significant, not only on a world scale but in mapping comparatively small areas: Cook's choice of Mercator for New Zealand was an important step forward in this respect. On land the corresponding advantage was most apparent in high-latitude regional mapping (where the plat carrée's north-south and east-west scales differed most), as was attested by several late eighteenth-century examples made for the Hudson's Bay Company.[53] After this the only real surprise came with the Victorian era, when Mercator won overwhelming popularity for a kind of land-based political world map to which it was almost completely unsuited.[54] Some historians associate this strange development with an imperialistic desire among European cartographers to make their own continent look larger than it was.[55]

An equally pertinent aspect of the Mercator story is how map-makers and map-users thought of the graticule itself, as distinct from the geographical image it transmitted. Subconsciously, at least, Mercator's was probably seen as the first major projection in which the resemblance between plane and sphere had been orchestrated to obey the strictest possible kind of mathematical rule. A simple conical representation within a single hemisphere is rule-bound in a sense that falls well short of strictness, namely that each of its parallels, like those of the globe, is shorter than all parallels drawn at lower latitudes. Mercator went beyond this relatively modest level of precision. His plan was for numerical variables on the map to change not only in the same direction as the corresponding variables on the earth but in the same arithmetical proportions, and not just as an unintended by-product of a particular surveying or plotting process but by the deliberate choice of the author.

The last proviso has important implications for the historian. In any sphere of action it is hard to tell which choices are deliberate without some insight into the chooser's mind. This is why credit is seldom given to Erhard Etzlaub for anticipating Mercator's projection in the year before Mercator was born (Fig. 8.15).[56] Etzlaub left no verbal account of his map. Mercator did. (It came in a panel covering part of North America.)[57] Such explicitness was itself unusual: despite the precedent set by Ptolemy, it would be a long time before post-Mercatorian inventors of new projections would feel moved to write about their work in an accompanying text. But in 1597 William Barlow described at length how a Mercator graticule could be drawn, a lesson driven home two years later in more strictly mathematical terms by Edward Wright.[58] It looked odder still when non-academic cartographers began to move in the same direction. By the end of the century two world maps had acknowleged a debt to Mercator, though admittedly without saying exactly what had been learned from him. In 1610 he was even the subject of map-historical debate, with Wright arguing in the margin of his own map that although the new system had been 'credited by some' to Mercator its essence was already to be found in Ptolemy's *Geography*.[59] By the 1670s 'according to Mercator's projection' had become a common formula in the sub-titling of world maps. The motive for all this publicity is not very clear: perhaps it was a well-deserved respect for Mercator's theoretical contribution; or perhaps his map was thought to look so strange that somebody needed to take the blame for it.

Until late in the eighteenth century Mercator's only rivals in the theory of projections were men of modest ambitions and almost negligible influence.[60] But with the flowering of German mathematics there came a breakthrough. The achievement of Johann Heinrich Lambert in 1772 can perhaps best be seen as a revolution in projection typology.[61] Previous literature had been mainly descriptive, with an emphasis on methods of construction.[62] The new idea was to create a matrix of taxonomic pigeonholes by combining the notions of class (as in 'conical'), aspect (as in 'transverse') and property (as in 'equivalent'). Lambert has been credited with 'literally' devising an infinite number of new projections,[63] but in fact he left ample scope for later mathematicians to extend and refine the idea of a 'property' – though as it happened none of the resulting networks came into regular use until after 1850.[64]

Projecting regional geography

It was among regional and topographical maps that projection theory had most effect in the eighteenth and early nineteenth centuries. At sub-continental scales renaissance cartography showed none of the fecundity, complexity and mathematical elegance that dominated the mapping of the world as a whole. On this humbler level the approach of nearly all early cartographers was unassumingly pragmatic:

> If you would make a map of one of the four quarters of the world, or of some large empire or kingdom [wrote William Alingham], then you must project your hemisphere very large, and cut off such a piece or part of it as is capable to contain the quarter or part of the world you would project.[65]

In fact it was not a hemispherical projection but a plat carrée that best lent itself to this kind of primitive surgery, possessing as it did the unique characteristic that every extract looked the same as every other. However there was still the disadvantage that plat carrée scales were seriously distorted in all but the lowest latitudes. A better expedient, in a small country as well as in Marinus's *oikoumene*, was a modified cylindrical projection in which scales along the meridians matched the scale of a parallel running through the middle of the map.[66] In 1466 Nicolaus Germanus replaced Marinus's cylinder by a correct division of the northernmost and southernmost parallels so that the meridians now converged to form a trapezium (Fig. 8.16). Next, Mercator reduced Germanus's errors by moving the two standard parallels from the upper and lower edges towards the middle of the mapped area. Ptolemy's maps had been captioned with a note distinguishing longitudinal and latitudinal scales, a helpful gesture remembered in due course by Mercator but not by many other cartographers.

With the decline of the plat carrée it became less desirable to cut pieces out of a larger graticule. Instead the cartographer had to choose a standard parallel that

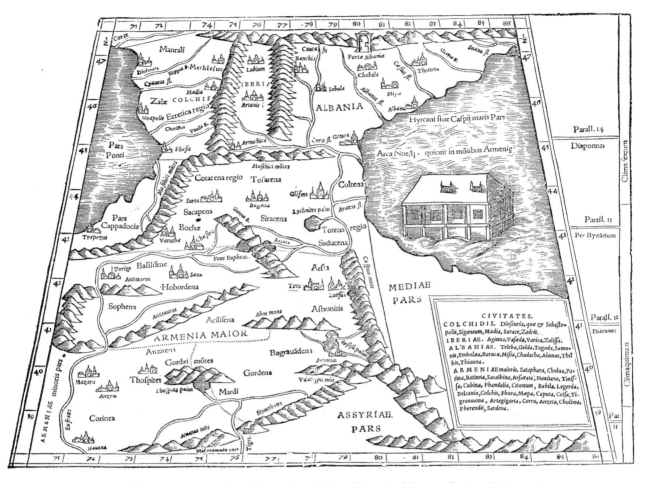

8.16 Trapezoidal projection. *Tabula Asiae III*, from Sebastian Münster, Ptolemy's *Geography* (Basle, 1540), with Noah's ark. See also Figs 2.7, 15.6.

suited the latitudinal situation of his map, and for trapezoidal or conical networks he also had to make sure that his central meridian was properly central: as John Green succinctly put it, 'particular maps require a particular projection'.[67] The consequences of disregarding this advice appeared when Nicolas Sanson and Nicolas de Fer drew a seriously mis-shapen British Isles with all its convergent meridians sloping to the right (Fig. 8.17).[68] (Such a case should be distinguished from the comparatively harmless tilting of a complete rectangular network, as in Henry Hondius's map of Hesse.)[69] As for the design of purpose-built regional projections, well into the eighteenth century it could be dogmatically stated that 'in maps of kingdoms only, the … meridians and parallels are expressed by straight lines.'[70] In the same vein Germanus's trapezoidal projection was still being recommended by the *Encyclopaedia Britannica* as late as 1823, though conical

8.17 Externally placed central meridian. Nicolas Sanson, British Isles (Paris, 1734), presumably abstracted from a map of Europe.

projections were actually now quite popular, having received favourable publicity when Russia was mapped on this system by Joseph Nicolas Delisle in 1745 with two standard parallels.[71] The next logical step for a regional map was to divide all parallels correctly, as in Bonne and the sinusoidal. It is not known how far the correct representation of area (in the sense of square-mileage) was a motive for taking this further step. The mid-nineteenth-century situation is summarised by John P. Snyder's census of continental maps in five major atlases dating from 1820 to 1859. The scores are: Bonne 15, sinusoidal 7, equatorial stereographic 3, Mercator 3, simple conic 3, globular 1, polar stereographic 1, trapezoidal 1.[72]

Projecting national surveys

For topographical surveys of whole countries the problem of projection assumed a distinctive form. In this genre the scale was too large for errors to escape notice, and they were made more evident by numerous intersections of geographical detail with a regular system of sheet lines. The first attempt at a solution was in the Cassini map of France, plotted as a transverse plat carrée with the meridian of Paris substituting for the equator. (For a limited area the resulting graticule looks something like Bonne's.)[73] In a small country the distorting effect of distance from the north-south 'equator' would be kept within tolerable limits. Though carefully thought out, this was the kind of projection that could have emerged more or less by accident as a by-product of the plotting process.

The Cassini projection was also chosen for the Ordnance Survey of Britain, perhaps out of admiration for Cassini and his colleagues or perhaps to benefit from the simplicity of the requisite calculations.[74] In England however matters were complicated by an early decision (subsequently rescinded) to publish the national survey as a series of county maps. Later, south-east England was projected on the Greenwich meridian, south-central and south-west England on the meridian of 3 degrees west. Then in the mid-nineteenth century, with the revision of the Survey's one-inch map, a meridian through Delamere Forest in Cheshire was adopted for the entire country.[75] In a single Cassini projection for the whole of Britain distances would be in error by up to 1 in 530 and angles by up to 3 minutes 14 seconds.[76]

The first rival to Cassini's projection for use in a national survey was that of Bonne, chosen in 1817 for the French official *Carte d'état major* at the recommendation of an officer who happened to be Bonne's son, though it also had the merit of involving less distortion than Cassini, which was probably seen as a more important advantage than its correct representation of areal differences.[77] It is typical of cartographic literature that no writer has compared the maximum errors of these two projections on the margins of any particular national survey area in simple quantitative terms; what can be said with confidence is that in countries much larger than France there would be strong arguments against using either of them. A possible alternative was to introduce an element of independence into the separate sheets of a multi-sheet map by increasing the number of standard parallels and central meridians, thus rationalising the Ordnance Survey's original policy for southern England. This was the basis of the polyconic projection devised in *c.*1820 by Ferdinand Hassler, first director of the United States Coast Survey.[78] But since no national map series appears to have been published on Hassler's principles during the period of the present study, the subject of polyconics need not be pursued.

It may however be worth mentioning that the ultimate preference for national surveys was Lambert's Transverse Mercator. Unlike Cassini and Bonne, this was a conformal projection well suited to the characteristically twentieth-century purpose of directing artillery fire at invisible targets.[79] As in Cassini, its scale increased away from the central meridian, but with Mercator a single number would suffice as a 'scale factor' for correcting measurements taken across the map in any direction from a given point, whereas Cassini needed a different scale factor for every compass bearing. This is a modern solution, not adopted by the British Ordnance Survey until 1935, but it just creeps into the present discussion by having been chosen for C.F. Gauss's map of Hanover in 1821.

CHAPTER 9

One fair card or map

Aⁿ FTER SURVEYING AND protraction, we move now to the circumstances in which these processes were put to use. On 'developmental' grounds the history of map-making might be expected to proceed from the easy to the difficult, that is to say from small areas to large. Accepting this hypothesis, if only as an aid to exposition, we may begin somewhere in the renaissance period with the simple idea of a 'locality'. For present purposes this was an area in which it had become feasible to measure and map everything of importance with the greatest precision currently obtainable and with a good chance of profiting from the sale of the results. The customer for a local map in this sense was often a single client, either individual or corporate, whose order had been placed in advance with the cartographer to meet a particular need.

Varieties of local map

Some of the earliest subjects for local map-making were estates, fortifications, and towns. In each case there was a practical reason for seeking a high level of accuracy. Perhaps the most important function of an estate map (Fig. 9.1) was to set out the limits of a property, though there were differences in the amount of supplementary map-detail considered necessary to help identify the course of a mearing line across the ground: early Russian estate surveyors, for example, seem to have been particularly dedicated to the recording of single-point boundary markers.[1] Another function was to help determine the value of a piece of land, whether for sale or rent. Accessibility and productivity were of special significance in the valuation process, between them justifying a fairly comprehensive portrait of the landscape, but the factor that made most demand on the surveyor's expertise was surface area. Here he faced an awkward problem of definition. The map is flat, the earth uneven. Should not the estate surveyor measure along the slope to determine how much ground was bringing profit to the lessee or purchaser? The answer might depend on whether the land was under grass, which blanketed the whole surface, or tree crops that grew vertically upwards. One enthusiastic eighteenth-century surveyor calculated how many more carrots could be grown per acre on a 20-degree slope than on a level surface.[2] On a map, of course, lines are always measured across flat paper with a horizontal scale, and whatever else he did the surveyor had to provide the data from which the length of these lines could be plotted. In the final map the textbook said

A Map of A Farm lying at Larkfield in the Occupation of Abraham Walter &c.

PART of Bradborn Park

	The Contents of the Land	Ac: R: P:
A:	The House, yard, & Orchard	0 — 2 — 10
B:	The Green Field	6 — 0 — 20
C:	The Pond Field	4 — 0 — 15
D:	The Great Quickset	6 — 0 — 04
E:	The Little Quickset	3 — 1 — 08
F:	The Hoath Field	5 — 3 — 37
	Summe of Acres —	26 — 0 — 12

Explan͡ation of Notes in y͡e Map

X: Is Mr Thomas Goldings Plott of ground with his House & Barn standing thereon; (Y) is Edmund Walter's Plott of ground & his House & Shop ~ Z) is George Wray's Plott of ground with his House Maulthouse & Barn ~ ⌂) The Court-house standing on Larkfield-green, Being the place where the Court is held & Constables chosen for y͡e Hundred of Larkfield ——— Boundary Lines, Signifying that these Fences are not maintained by this Land ~ ╪) Stands in the Table of Trees and the like (╪) may be found in y͡e Plott of the House and Orchard: And Shews that in the Fence between the Orchard & yard & George Wrays Orchard there is 6: Timber Elms, 7: Pollard Elms, and two Ashes &c:

Table of Trees				
Tim Elm	Poll. Elm	Yong Elm	Poll Oak	Ash
6	7	—	—	2
4	24	—	—	2
6	15	3	—	—
16	18	14	—	1
7	10	2	—	—
7	2	12	—	—
4	9	3	1	1
3	2	2	—	3
—	2	—	2	24
1	2	12	3	9
7	10	—	—	—
1	5	—	—	—
—	5	—	1	3
72	111	48	7	45

9.1 Estate map with reference table, original scale 1:3960. Abraham Walter, East Malling, Kent, MS, 1681. F. Hull, *Catalogue of estate maps, 1590–1840 in the Kent County Archives Office* (Maidstone, 1973), Fig. 9.

that complaints about arithmetical error should be anticipated by shading those areas where a slope had been corrected. Considering how many writers repeated this advice it is a pity that no historian has told us whether it was ever taken.[3]

A single estate map could be completed – and paid for – fairly quickly, but with the progress of agrarian change its author would doubtless find additional opportunities for similar employment in the same district. In such circumstances a recognised surveying industry could be expected to flourish and maintain itself. A competent practitioner could hope to remain at work within a limited area, becoming well known to his local clientele (and to future historians) and also qualifying as a member of a nationally recognised fraternity with its own methods, its own training system, its own conventions and its own social status. His mapping activities could be combined if necessary with other locally based occupations such as schoolteacher, valuer, land agent, quantity surveyor, architect, builder and engineer. There might be room for academic debate about the effectiveness of chaining, traversing, plane tabling and theodolite triangulation in any given situation, but on broad technical issues like this there was generally a strong

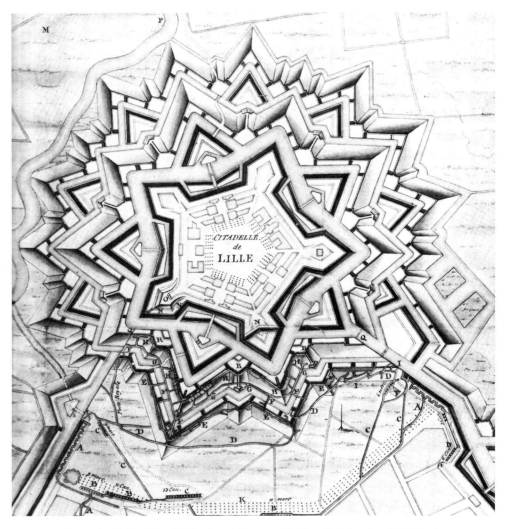

9.2 Fort plan. Lille, French Flanders, MS, 1708. The Royal Collection, RCIN 726013.
© 2008 Her Majesty Queen Elizabeth II.

measure of agreement among local surveyors, at least within any one country. Seen from a distance, this was a world in stable equilibrium, where technology harmonised with economics and supply was balanced by demand.[4]

Renaissance artillery forts were 'cartogenic' in a way that medieval castles could never be: they were larger, more elaborate in plan, composed of broad earth banks rather than narrow stone walls, and less dependent for their efficiency on features built in a vertical plane (Fig. 9.2). The fort therefore stood in greater need of planiform display, not least as a means towards estimating the cost of its construction and maintenance. Mapping and design went hand in hand, and both tasks were thought to lie outside the competence of a local surveyor. They required

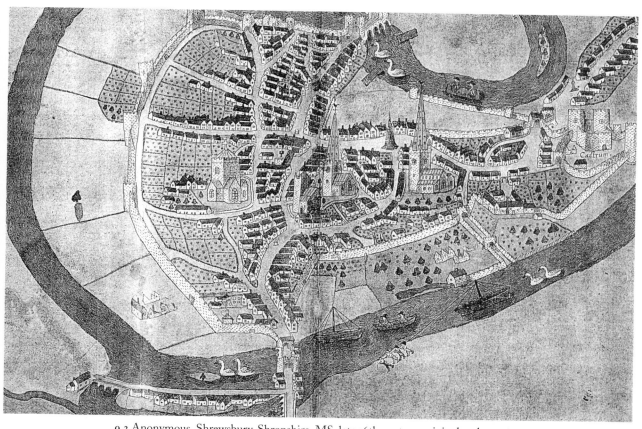

9.3 Anonymous, Shrewsbury, Shropshire, MS, late 16th century, original scale *c*.1:5800.
British Library, Royal MS 18. Diii, ff.89v–90.

expert knowledge, soldierly discipline, and a close contact between employer (in this case an army commander) and employed.[5] Many sixteenth-century European military plans were made by immigrants from Italy, where modern fortification techniques had first developed.[6] Even when the engineers were being drawn from an indigenous population, as in late seventeenth-century England, they still formed a distinct group that might not have been expected to take much part in other kinds of map-making. For modern landscape historians, the weakness of this cartographic genre lay between the fort itself and the margin of the map, an aureole of unfortified town or countryside that one suspects to have been filled in by some cartographers from their own imagination, conceivably to prevent a potential enemy from discovering the exact location of the fort.[7] This difficulty might sometimes be partly met by setting fort and town plans more snugly within a frame that was circular rather than rectangular.[8]

The mapping of complete towns and cities has a more complex history.[9] Many urban cartographers drew on a legacy of pictorial impressionism that helped to make planimetric truth seem less urgently desirable (Fig. 9.3). John Norden thought

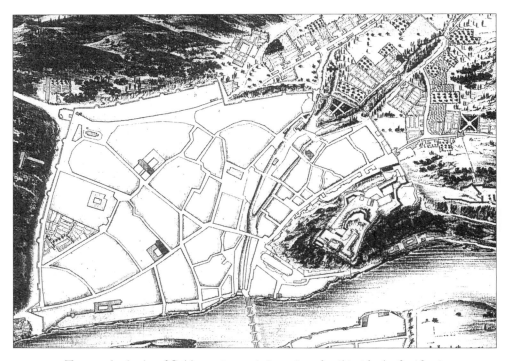

9.4 Town and suburbs of Coblenz, 1834, omitting minor detail inside the fortifications.
Michael Swift, *Historical maps of Europe* (London, 2000), p. 126.

that towns should be mapped 'briefly' as well as 'expertly' and not long afterwards
John Speed was demonstrating in a series of one-day urban surveys that up to a
point these two aims were capable of being combined.[10] Other authorities, such as
Egnatio Danti, laid more stress on the need for measurement in town plans.[11]
Perhaps the most compelling argument in favour of exact distances and bearings
was that walled towns became functionally similar to forts as soon as municipal
defences began acquiring a penumbra of banks, ditches, counterscarps, ravelins,
hornworks and the like. Some cartographers emphasised this point by the drastic
expedient of omitting much of the detail inside the urban perimeter, perhaps just
the minor buildings (Fig. 9.4), perhaps the street-edges as well.[12] At any rate, it was
in the surveying of defensible towns that the concept of an exact scale seems first to
have been imported from chart to map.[13]

On the same theme, Ralph Agas in 1596 advised surveyors to give 'just measure'
of streets 'for paving thereof'.[14] His argument gained weight when London needed
rebuilding after the great fire of 1666. Among the maps inspired by this crisis the
most remarkable was that of John Ogilby and William Morgan (1676), their scale
of 100 feet to an inch having been suggested by an eminent man of science, Robert
Hooke. Such spatial extravagance encouraged these authors to adopt a number of
major innovations: streets shown in their true widths, buildings drawn in plan

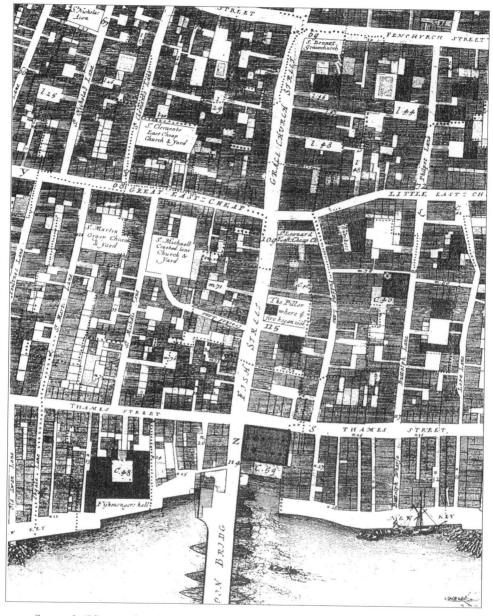

9.5 Streets, buildings and yards, original scale 1:1200. From John Ogilby and William Morgan, *A large and accurate map of the city of London* (London, 1676).

rather than pictorially and individual houses distinguished from their neighbours even when they were joined together (Fig. 9.5).[15] With Ogilby and Morgan the English town plan had come of age. The best eighteenth-century examples were not substantially different.

Local surveys were seldom valued as contributions to the scientific knowledge of a whole community. On the contrary, some landowners seemed anxious to restrict the circulation of their estate maps, in order to maintain an advantage over their tenants or perhaps over neighbouring proprietors, and the motives for keeping fort plans secret were obviously even stronger. It is not therefore surprising that maps of these types should make no reference to latitude and longitude. Town plans were a different case in this respect, because burghers might feel pleasantly flattered if anyone took the trouble to establish their true position on the globe. The simplest and least conspicuous way of recording coordinates would be a marginal citation of one latitude and one longitude for the town as a whole, adapting the policy of minimal disclosure followed for small islands by many cartographers from Ptolemy onwards. In fact such precision was seldom considered worthwhile.[16] Equally rare, and even more unexpected, was the appearance of detailed latitude and longitude scales on a single town plan, as in Johannes Mejer's atlas of Schleswig-Holstein.[17]

A context for regional mapping

Local operations like those just described were almost inevitably discontinuous in territorial coverage. Collectively they might range over a wide area, as with a survey of royal forests or defensive frontier posts, but they could not be readily combined into a single regional map. Since all the land of a civilised country was likely to be owned by someone or other, an optimistic historian might expect private surveys of different estates to be more easily aggregated, but in practice most patterns of land ownership were too broken and irregular for any such synthesis to emerge, and a region embracing many properties, together with a normal quota of forts and towns, would almost certainly have to be mapped in its own right as a separate operation. Considered in the abstract, such larger enterprises might seem technically comparable with a local survey. Thus Cyprian Lucar argued in 1590 that the same instrument, namely the plane table, could be used for a field, fort, camp, town, city, lordship, shire or 'country' but the value of this advice was a matter for debate in which even a single expert might sometimes give conflicting opinions.[18] The fact is that few textbook writers were sufficiently interested in regional as distinct from local surveying to care much either way.

Since errors tended to accumulate with distance, a regional survey had less chance of being accurate than a local survey. But it was economic and managerial circumstances that did most to distinguish different scales of mapping. To cover a large territory might well be beyond the means of an individual freelance operator, who at a certain radius from his home would begin to incur additional travel and living costs, to say nothing of the cultural frictions experienced by intruders in a strange environment. It was also difficult to organise surveyors into groups. One feature common to the economics of map-making and agriculture as industrial processes was a high ratio of land to labour, which meant that surveyors working for a large employer would be widely dispersed and correspondingly hard to supervise.

This rule was not without exceptions, some of them remarkably productive at least for a short time. In England John Ogilby's road surveys could certainly be so described. More durable associations, later, were the firms of John Rocque[19] and John Cary.[20] Yet there were never enough practices of more than family size to make regional surveying an identifiable profession. One barrier to growth, at all periods, was imposed by an elementary mathematical fact: within any given territory, the larger the area of each assignment, the fewer surveys would be needed to cover the whole. This set a limit to the number of firms engaged in regional mapping, which in turn made it unprofitable to publish separate manuals of instruction on the survey of counties or provinces – or to include more than a few paragraphs about this subject in a book that might devote hundreds of pages to the local practitioner. It is consequently difficult to say just how regional cartographers went to work, a problem complicated by the diversity of their education and experience – some being estate surveyors, some engineers, some clergymen and some gentleman amateurs whose main intellectual preoccupations might be either scientific or antiquarian.

In this connection it is remarkable how few individuals devoted more than a small share of their working life to any one kind of regional survey. Rarely indeed were all the provinces of a country mapped by the same surveyor. Financial insecurity must have been a strong deterrent; likewise sheer fatigue and failure of will. Norden wrote with feeling that 'to observe singularly and precisely, will require the whole time of a man's ripe years, to effect the whole description of England'.[21] One wonders how many more counties he would have done himself if government support had not been withdrawn after his first eight. Contemporary French experience was even less fortunate, with Nicolas de Nicolay being ordered to make maps of each province but failing to get beyond two or three.[22] Eighteenth-century surveyors did little better, as witness the uncompleted English county mapping programmes of John Warburton, John Rocque and Thomas Jefferys.[23] A more successful enterprise of this kind would probably be found to depend on some kind of hidden government subsidy, as in early nineteenth-century Ireland where William Larkin's seven county maps were partly compiled from the road surveys he made for the post office.[24] In all this, Christopher Saxton's county atlas of England and Wales stands alone, even accepting that he too had received a measure of state support. Like the author of the fourteenth-century Gough map, Saxton has been much celebrated by English historians, in neither case with quite the degree of astonishment that their work deserves.

Regional and local surveys also differed in the character of their clientele, a difference that naturally affected the content of the resulting product. The larger the area shown on a map, the more numerous were its potential purchasers and the

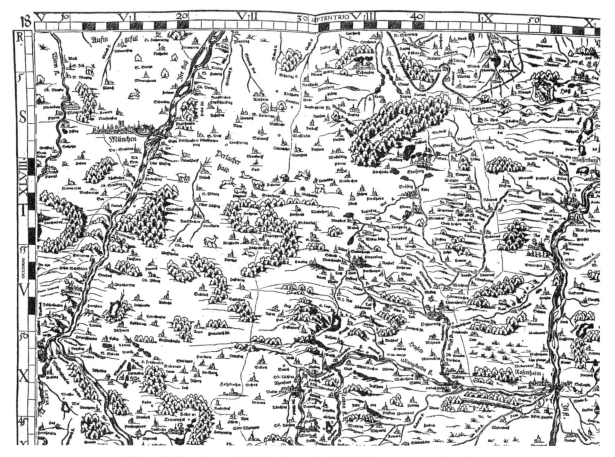

9.6 From Philipp Apian, *Bairische Landtaflen XXIIII*, sheet 18, original scale 1:144,000 (Ingolstadt, 1568).

more profit could be gained by printing and publishing it. Wider coverage necessitated a smaller scale. This was not just a question of physical convenience, because an unwieldy map could after all be divided into sheets of manageable size and presented as an atlas. More important was the need to safeguard the promoters' investment, a task not eased by the average reader's lack of curiosity about minor topographical details in remote places. Something could be done in the new era of print and publication by enticing potential customers with catalogues, prospectuses and newspaper advertisements – important sources of factual information and sometimes error for map historians from the early eighteenth century onwards.[25] But the real problem for the regional cartographer was in choosing how much information to omit.

In the last resort, the economic health of even a single regional map might depend on an infusion of public money. The importance of government patronage can be illustrated in a negative sense by dwelling briefly on varieties of printed map that had little obvious value to a national or local administration. Several categories come to mind: (1) radial maps, showing the country up to a certain distance from a major city, in England typically from ten to thirty miles; (2) maps of natural regions such as the Fens, the lakes of Killarney or the Rhine Valley; (3) maps of historic or ethnic provinces with no current political standing, such as eighteenth-century Germany, Italy or Wales; (4) news maps illustrating battles, sieges, natural disasters, shipwrecks, and crimes; (5) thematic maps of interest mainly to physical or social scientists. Until well into the nineteenth century maps in each of the foregoing classes were relatively uncommon, most of them being compilations rather than original surveys. The steadiest markets were probably for road guides and navigation charts, whose subjects can perhaps be regarded as a special kind of natural region.

By and large, however, the most successful regional maps were those that the governors of some political or administrative territory had a reason for consulting, to coordinate military defence, to identify political power bases, to organise administrative jurisdiction, to initiate plans for public works or to estimate tax-paying capacity. As a rule, cartographers were most attracted to the largest official subdivision of a kingdom or empire that could serve as a practicable unit of surveying: typical examples were the English shire, the French or Dutch province, the Swiss canton, the German duchy and in due course the American county. Apart from their significance in the operations of government, these areas often inspired a degree of regional patriotism, providing the would-be cartographer with a convenient framework in his quest for both topographical data and financial support as well as offering a suitable unit for historical or statistical inquiry to map-users of more scholarly disposition. For the local surveyor seeking to widen his horizons, such territories also presented fewer technical difficulties than a whole kingdom. Many famous sixteenth-century maps fell into this category, including Sebastian Cabot's Gascony and Guyenne, Martin Waldseemüller's Rhineland, Gerard Mercator's Lorraine and Philipp Apian's Bavaria (Fig. 9.6).[26]

The early regional cartographer's themes

In England the county has been described as the largest region for which maps could accommodate the essential facts of human occupation.[27] The most suitable scale for this level of treatment was open to debate, with published maps growing larger as each generation of surveyors tried to outdo its predecessors. Under Queen Elizabeth I English county maps varied from about 1:300,000 to 1:150,000.[28] These scales left room for coasts, rivers, hills (to be further considered in chapter 12), forests and major settlements, as well as the boundaries of the next smallest territorial division below the county, which in most of England was the hundred or wapentake. There was also ample scope for the grading of settlements and other

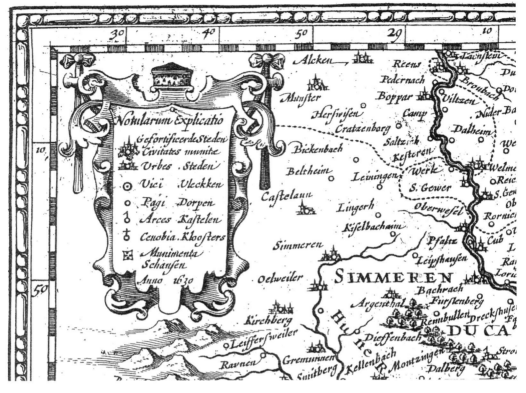

9.7 Topographical map with key to symbols. From Joannes Janssonius, *Nova descriptio Palatinatus Rheni* (Amsterdam, 1630).

sites. Saxton seems to have distinguished cities, towns, parish villages, chapelries or hamlets, and castles or manor houses.[29] Norden's additions included beacons, bishops' sees, forts, hospitals, houses of different grades, lodges, market towns, mills, monastic foundations and 'places ruinated and decayed'.[30] Similar distinctions were drawn in regional maps on the European mainland (Fig. 9.7). These often carried explanatory legends, in which the smallest category was usually *vicus* (hamlet), though some classifications extended even further down the settlement hierarchy to the *viculus*, the *sedes nobilis* and in at least one German example the 'Entselshof' or single dwelling.[31] In fact it was in lands of strong German influence that cartographers set most store by a thematic diversity that could encompass vineyards, mines, water mills, wind mills, iron forges, glassworks, spas, universities, sites of battles and various kinds of antiquity.

This leads to a point worth making about any cartographic record based on field work, which is that its author would involuntarily encounter many striking phenomena outside the land surveyor's normal purview, and might with luck have room for some of these items in his field book and on his map at virtually no additional cost in time or effort. Note the following oddities from the cartobibliography of a single English county. Christopher Saxton, 1575: 'ye breache' beside the Thames between Plumstead and Erith; Philip Symonson, 1596: a gate giving access to Waterdowne Forest; John Norden, 1605: 'Old Winchelsey whose Ruyns lurk unseen under the sea waves'; John Speed, c.1610: an unexplained tent symbol west of Sevenoaks and elsewhere; Robert Morden, 1695: 'The Ruines of the [unnamed] Priory'; John Andrews and collaborators, 1769: an invisible line of Roman road joining Canterbury with Richborough; William Mudge and the Board of Ordnance, 1801: 'Pembury Tree', shown by a normal if unusually large tree-symbol but unexpectedly found worthy of an individual name; Christopher and John Greenwood, 1829: a 'Land Mark' (not further particularised) in the Isle of Thanet. No one would pick up a map of Kent in anticipation of finding these places, but cartography would be poorer without them.

The great mystery about sixteenth-century and early seventeenth-century maps, common to many cartographers and many European countries, is the absence of even a rudimentary communications network. The exceptions were bridges and – very infrequently – 'passes' through woods or bogs.[32] One almost gets the impression that earlier maps had been superior in this respect, though admittedly there are too few survivals of medieval date for such a comparison to be carried very far. Of course roads did still exist on the ground, though perhaps not yet so well differentiated into major and minor as they were later, and every surveyor must have spent a good deal of time walking or riding along them. Cartographically they first became common in the Low Countries, perhaps because larger scales were necessitated in that area by a higher population density and a more complex drainage pattern, and also because there were very few hills to compete for the cartographer's space. This unusual concern for road coverage is well shown in a group of Dutch regional maps added to Mercator's atlas in the 1630s (Fig. 9.8).[33] For Europe as a whole, however, it can only be assumed that in the renaissance period stocktaking had become a more important function of land-based maps than way-finding. On a longer view, it seems likely that as roads attracted more attention from topographical surveyors, so too were they mapped (space permitting) at successively smaller scales.

How to make a regional map

Whatever our explanation for the neglect of roads, cartographers guilty of it are unlikely to have depended very much on traverse surveying. Their preferred technique was more probably triangulation. Points suitable for instrumental sightings can certainly be identified on many of their maps. Parish church-towers were everywhere a dominant feature, and towns were often represented by similarly vertical building-profiles alongside a small circle enclosing a dot. Relief too was usually shown as a series of individual summits, some of them localised enough to

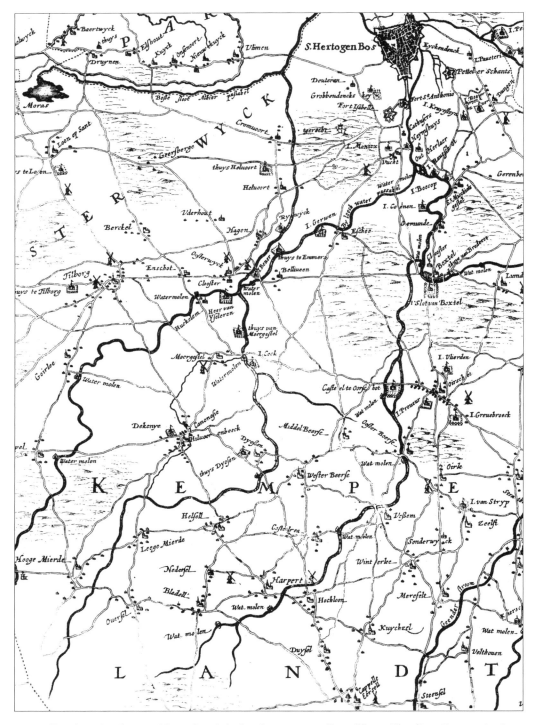

9.8 Dutch regional map with roads, original scale *c*.1:215,000. From Henry Hondius, *Shertogenbosch* (Amsterdam, [1634]).

have been used as survey stations. In Wales instructions were given for Saxton to be 'conducted unto any tower, castle, high place or hill to view that country' and it has been suggested that in England he chose to triangulate the hill-top beacons erected as part of a national defence system.[34]

The word 'triangulation', though unavoidable in any account of post-medieval cartography, may sometimes carry pretentious overtones not necessarily appropriate for the realities of Saxton's lifetime, and in this connection we should not forget the contemporary surveyor who absolved his fellows from the need to measure and calculate as punctiliously as if weighing out saffron.[35] In such circumstances the compass could be expected to play a part in measuring angles, and it may be significant (though not conclusive) that many maps of the period are oriented towards magnetic rather than true north.[36] Whether a county needed dignifying with latitudes and longitudes was a matter of opinion and on this point a regional map-maker could choose between two contrary cartographic traditions. In geographical science meridians and parallels had been popularised by the Ptolemaic revival and later by Mercator's atlas, until by Ben Jonson's time English theatre audiences were expected to know that latitude was something they might find on a map.[37] But in the survey of estates and fortifications, as we have seen, field astronomy was usually felt to be irrelevant, its technicalities ignored by even the most conscientious textbook writers. On the middle ground of regional map-making the experts showed little unanimity. John Norden took a negative view of geographical coordinates.[38] John Speed had some difficulty making up his mind.[39] Robert Norton assumed that a single latitude and longitude determination could provide a graticule for a whole map without attempting to explain the difficulties involved in this procedure.[40] John Love and several of his imitators wrote vaguely of 'taking' the latitude at three or four places in a county, with no mention of how the results might be assimilated to the same surveyor's terrestrial distances.[41] For these writers, coordinates were not so much a control system as a cosmetic after-thought, a status hinted at in John Wing's complaint that 'latitude is erroneously stated in several great towns'.[42] In a maritime territory latitude and longitude might seem more deserving of notice, though there appears to be no proof that county maps with coastlines ever played much part in practical navigation.

From about 1540 onwards most regional maps carried a numerical scale-statement, a tacit claim that at least one distance had been determined by measurement across the ground. Yet for a long time the pundits remained surprisingly casual on the question of base lines. William Bourne glossed over the subject in his account of triangulation.[43] Nathanael Carpenter advised readers to reckon distance 'experimentally from your own knowledge or some certain relation of travellers',[44] omitting to add that the relations of travellers are notoriously uncertain and that distances cannot count as 'knowledge' unless obtained by some reliable method. Even more incautiously, John Gregory in 1649 professed himself content with 'common reputation' as a substitute for linear measurement.[45] For Venterus Mandey it was sufficient to inquire about the itinerary mileage or perhaps just the travel time between two selected places.[46] Even John Green, most uncompromising of English cartographic experts, was prepared to rely on vague generalities when the subject of scaling a triangulation came up.[47] The same applied to a would-be scientific account of mid-eighteenth-century trigonometrical operations at the Cape of Good Hope.[48]

If such laxity seemed tolerable in the triangulator's all-important base line it is unlikely that any great refinement was applied to distances between purely local landmarks. And how did the surveyor deal with linear ground-features, which are not always easily reducible to finite collections of points? Given a list of all the villages within each territory, administrative boundaries could be roughly inter-polated on a map of village-sites, a procedure implicitly anticipated in Ptolemy's tables and explicitly recommended by Arthur Hopton in 1611.[49] Lists of this kind could have been a necessary aid to tax-collectors at any time since the invention of writing. Interpolative 'threading' was certainly consistent with the rather unspecific curvature of the territorial divisions shown on many early regional maps, where political boundaries crossing land often look almost as non-committal as those drawn through water (Fig. 9.9).[50] Only in unusual circumstances would inter-polation be rejected as prohibitively venturesome: it was strange indeed to deal with an ethnic divide by simply writing the words 'Wendisch' and 'Deutsch' one above the other in more or less randomly chosen locations along the presumed boundary, a dozen times each, without attempting to separate them by a line.[51] (Perhaps this was a map that its author had forgotten to finish.) As for rivers and shorelines the most likely assumption is that these too were drawn freehand, though by the late seventeenth century county surveyors were being advised to run a traverse along the sea coast.[52] A good example of double planimetric standards was Apian's survey of Bavaria where, despite the existence of a triangulation, distances were either 'eye-sketched' or deduced from travel-times, sometimes resulting in positional errors of more than a kilometre.[53]

Successors to Saxton and Speed

It seems reasonable to ask how contemporaries apprehended the difference in intellectual status between triangulated points and sketched lines within a single map, a conceptual gulf that may well have been wider in the sixteenth century than at any time before or since. To bridge the gap, or even to accept its existence, would surely need a high degree of mental sophistication, but no such quality is displayed in contemporary surveying literature: perhaps the subject just lay undigested somewhere in the collective unconscious. When the issue of accuracy did arise it was approached obliquely rather than head-on. In Pembrokeshire George Owen characterised one distance as 'by the measure of Mr Saxton's map if he hath truly

9.9 Interpolation as a probable method of boundary mapping, Saintonge, France, original
scale *c.*1:360,000. From Gerard Mercator, Henricus Hondius and Joannes Janssonius, *Atlas
or a geographicke description of the world*, ii (Amsterdam, 1636), p. 319. See also Fig. 13.2.

calculated the same', which at any rate made no bones about including 'true
calculation' among a map-maker's duties. Owen also criticised the outlines of
Saxton's Pembrokeshire even where geometrical precision was no more than a
minority interest: Saxton should not have made St Anne's Head at the mouth of
Milford Haven look curved when it was really pointed. Particularisation could be
just as culpable as generalisation: 'there are no such small creeks to be seen within
the bay as Mr Saxton in his maps hath noted down'.[54] Such views hardly fit the
assumption made by some modern historians that the early map-user was more
easily satisfied than his modern counterpart (Fig. 9.10).

Nor was Owen the only topographical writer to be aware of Saxton's limitations.
A map of Kent by Philip Symonson (1596) won praise from another county
historian for having 'not only the towns and hundreds, with the hills and houses of

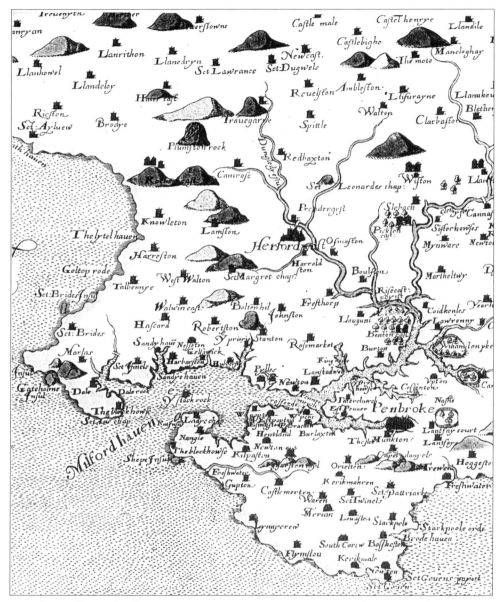

9.10 From Christopher Saxton, *Penbrok comitat'*, original scale 1:193,807 (London, 1578). St Anne's Head is the headland nearest 'Sct Ans chap'. The creeks disputed by George Owen lie between the villages of Nangle and Kilpaston.

men of worth … more truly seated: but also the sea coasts, rivers, creeks, waterings and rills … more exactly shadowed and traced, than heretofore, in this, or any other of our land (that I know) hath been performed.'[55] However, it was nearly a century before such claims became a matter of routine. Two broad historical trends are relevant here, one general and one particular. First there was a change of mental

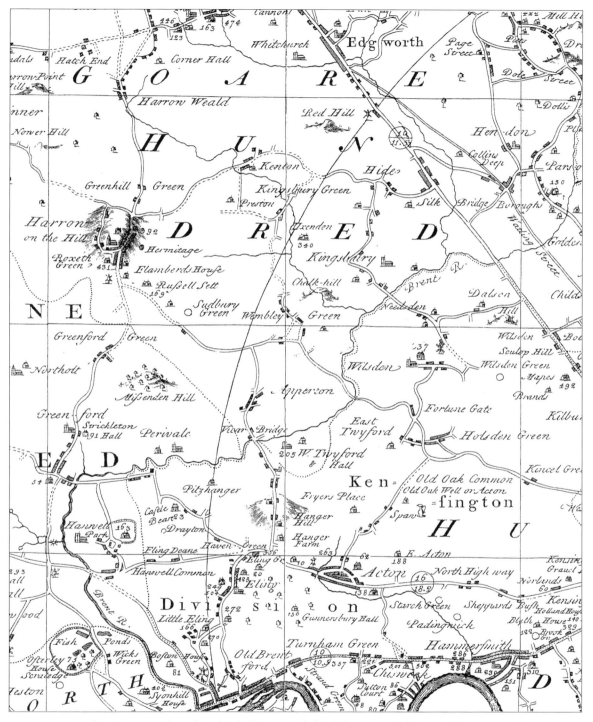

9.11 County map with roads individually surveyed. Circled: measured distances between market towns in miles and furlongs compared with previously estimated mileages. From John Warburton, map of Middlesex, scale 1:63,360 (London, 1749).

9.12 Easthampstead parish, Berkshire. Left, from John Rocque's county map, 1752. Right, from an estate map by Josiah Ballard, 1757. Paul Laxton, 'The geodetic and topographical evaluation of English county maps, 1740–1840', *Cartographic Journal*, xiii, 1 (1976), p. 45.

climate accompanying the institutionalisation of scientific research. Geography was one of many studies and activities to be affected by this trend. Meanwhile economic prosperity was facilitating peacetime travel to a degree previously unknown, a development which for Englishmen received its most spectacular cartographic recognition in the surveys of John Ogilby.[56] If Ogilby could measure all the roads of a whole kingdom, surely it was not beyond the power of man to tackle those of a single county, where the demands of a larger scale were mitigated by a reduction in the total mileage to be perambulated.

John Oliver's map of Hertfordshire acknowledged Ogilby's example (and

implicitly disparaged Saxton and Norden) by including an account of the 'measured miles, furlongs, and poles, on all the roads … a work never performed in any county before'.[57] Four years later in Cornwall Joel Gascoyne claimed to have drawn not just roads but coasts 'in all their meanders' and settlements all 'in their true places and distances'.[58] In 1724 Richard Budgen enlarged on the notion of delineating roads 'exactly': it was to be done 'according to their several angles, turnings, bearings and situations', as in his own new map of Sussex.[59] Contemporary maps go some way towards justifying such claims (Fig. 9.11), but what actually happened on the ground remains a matter of speculation. It might be argued that since road networks were generally denser than village networks, especially in areas of dispersed rural settlement, the effect of introducing roads would be to enlarge the scale of county maps, and that this in turn would eventually encourage the adoption of more accurate survey methods. Meanwhile a complete regional road system could be surveyed as a set of interconnecting traverses, most of them closed rather than open, and even without triangulation this could form a practicable basis for a county map, of low to medium grade according to whether or not wheel and compass had been preferred to chain and theodolite.

As it turned out, high cartographic aspirations in late seventeenth-century England were soon to reveal themselves as somewhat premature. It was not until after 1750 that the character of county surveying began to change fast enough to justify the idea of a revolution.[60] Several causes were now involved. The most consequential was the effect on the landscape of accelerating industrial and agricultural development. This reinforced the trend towards larger scales, though not so large as to threaten the concept of a county map. In eighteenth-century England one inch to one mile was clearly the most convenient option. It left scope for some features at least to be checked by a vigilant amateur: new roads and canals, for example, could be verified from independent engineering surveys, parks and farmhouses from a landlord's estate maps (Fig. 9.12). Meanwhile a new institutional source of criticism was the Society of Arts, which after 1759 gave premiums for 'accurate' county surveys on which its judgement was often supported by certificates of merit from knowledgeable local residents.[61] These possibilities naturally put the county surveyor on his mettle. In other European countries the landscape changed more slowly, but similar influences were at work.

From what is known about the history of surveying instruments we should expect a higher standard of precision in the eighteenth than in the seventeenth century, and there are various signs that this kind of improvement was being achieved. One sign of progress was an increasing preference for geographical over magnetic north to define the upper margin of a topographical map-sheet. Then there was the issue of latitude and longitude. Many eighteenth-century county maps had reportedly been 'regulated' or 'corrected' by 'astronomical observations'. Some cartographers, like Thomas Milne (Hampshire and Norfolk) and Andrew Armstrong (Durham), even stated where and by whom these observations had been made, and within what margin of accuracy to the nearest second.[62] Inside the relatively narrow limits of an English county there was a case for not attempting more than one high-grade astronomical operation, even if this left the cartographer with the problem of how to space his meridians and parallels.

Another claim to scientific status was the marginal triangulation diagram, which, rather surprisingly, seems to have been pioneered in England by the unsuccessful Restoration atlas-maker Moses Pitt.[63] At the end of the following century no less than eight English county maps had included the same kind of information, recording station names, sight-lines, and sometimes numerical angles and distances.[64] Here, by implication, was a foretaste of the 'orders' – primary, secondary and tertiary – that came to be distinguished in later control surveys. Equally clear was Thomas Jefferys's reference to the 'great' angles he had measured in Bedfordshire; excluding angles, evidently, that were to be regarded as less than great.[65] The term 'secondary' would have been appropriate for at least some of the observations by which William Green 'angled to all the surrounding country' from a 'primary' station at Warton Cragg in Lancashire.[66] The same applies to Peter Burdett's note on his 'series of great triangles reduced to an horizontal plane' in Derbyshire: 'by which the most eminent places of this county were projected, and from whence all the inferior parts were drawn in a like manner, the whole taken with instruments graduated with great care'. An optimistic surveyor might take bearings to fifty or sixty non-primary objects from a single station, some of them involving an element of indeterminacy, like the extremities of an oval-shaped island.

Harder to find in the historical record are any documents that close the gap between a surveyor's theodolite angles and the coasts, rivers, roads and boundaries of the landscape. Only one set of field notes from an early nineteenth-century county survey is known to the present writer: these relate to the work of William Edgeworth and his assistant William Hampton in the Irish counties of Longford and Roscommon.[67] They are not necessarily comprehensive, having perhaps been originally augmented by free-standing sketch maps that no longer exist. This source deserves a comprehensive study: at present we can only note that, in one small part of Longford, Hampton's field book locates twenty-four sites of angular measurement (none of them principal triangulation points) on or near roads and rivers totalling some seven miles in length, an average of about 480 yards between stations. In that distance, equivalent to just under half an inch on the finished map, a narrow river might exhibit ten or a dozen definite changes of direction. However, with careful sketching the results could still be creditable enough.

Mapping the nation state

We must now move upwards in the territorial hierarchy, this time to the plane on which monarchs are crowned, ambassadors appointed and wars waged. From the fourteenth century onwards, territories of national extent were mapped more often than before and, to modern eyes, more effectively. But did mapping of this extent imply surveying? The question arises at two levels. First, is there documentary evidence of national cartographers taking the field? Secondly, do the extant maps have any appearance of being based on 'strict dimensuration'? Consider for instance the extraordinary detail given for England in the Gough map of *c*.1360. Here it is only by guesswork that historians have invoked a rudimentary trilateration of distance estimates,[68] a regrettable state of uncertainty concerning a map which, despite its faults, remained the principal cartographic source for England over a period of at least a hundred and fifty years.[69]

By the early sixteenth century the rise of modern European cartography was coinciding with a realignment of political forces in which sovereign states increased their wealth and power, governments grew more efficient, and frontiers were extended or at any rate more carefully delimited. In 1524 Nicholas Kratzer proposed to construct a new map of England and since he is known to have made inquiries about measuring instruments at the same time his scheme presumably entailed some kind of survey.[70] John Leland was nursing similar hopes in 1546, though his intended production schedule – a mere twelve months – can hardly have left much opportunity for field work.[71] In 1561 John Rudd increased Leland's estimate from one year to two, but even this implied a somewhat unlikely rate of progress – six times faster than Philipp Apian's in Bavaria.[72] Meanwhile William Cuningham had been more modestly content with explaining how readers could make their own national map.[73] All these projectors must surely have envisaged something more exact and informative than the Gough map and its derivatives, otherwise there would have been no point in putting themselves forward.

In practice a country's cartographic prospects could be expected to vary inversely with its surface area. Failure was certainly no surprise in a kingdom as large as France, where Jean-Baptiste Colbert tried in vain to get a national map finished by Nicolas Sanson.[74] Conversely, sheer smallness might be thought to account for the cartographic precocity of mini-states like sixteenth-century Venice, Florence, Milan and the Papacy, which did indeed produce an impressive array of *ad hoc* regional maps, though without yet being surveyed by modern methods from end to end.[75] More telling evidence comes from countries where comprehensive survey programmes are known to have been left unfinished. In some respects the most remarkable case was Spain, where a major project by Pedro de Esquivel was abandoned in *c.*1590 and soon forgotten.[76] Smaller national surveys with ostensibly better chances of success were those of Ireland (Robert Lythe, 1567–71)[77] and Scotland (Timothy Pont, 1583–1601),[78] neither of them brought to a conclusion. In 1589–92 Tycho Brahe's proposed map of Denmark, another territory of seemingly manageable extent, got no further than the tiny island of Hven.[79]

Whatever graphic evidence has been preserved or lost, one technical question remains difficult to ignore. A map may derive from a single survey in which any two points are connected by an unbroken chain of first-hand observations, however long and circuitous that chain may be. Alternatively, it may have been created in the drawing office by joining two or more independent surveys. All modern cartographers would regard a compilation as inferior, other things equal, to the 'one fair card or map' of 'an whole province or region' aspired to by ambitious Elizabethan surveyors.[80] Surely a serious historian should place any important and interesting map in one or other of these two categories. Sadly, most authors have felt unable to oblige. In many cases, one must admit, there can be no easy way of doing so. In fact the only countries with claims to much attention in this respect are Spain and Portugal, where the maps drawn from Esquivel's data have been associated by recent scholars with a surviving list of coordinates for some three thousand

9.13 Iberian peninsula with sheet lines of larger-scale sectional maps, from Escorial atlas, MS, *c.*1580–90. Geoffrey Parker, 'Maps and ministers: the Spanish Habsburgs' in David Buisseret (ed.), *Monarchs, ministers and maps* (Chicago, 1992), p. 131.

locations accompanied by 'bearings from two separate observation stations', together with rough estimates of distances.[81] It seems generally agreed that these constitute a record of field observations and not just a gazetteer fabricated *ex post facto* from a draft of the map itself.

As for internal evidence, the only obvious examples are suggestive rather than conclusive. Perhaps the clearest case is that of Saxton's atlas, which covered first the whole kingdom of England and Wales and then the individual counties. Differences in scale and orientation within a set of county maps are not necessarily relevant here: they may have arisen in the course of re-drawing material derived from a single survey. A more instructive test is whether, after any necessary re-scaling and rotation, the common boundaries of adjacent counties fit tightly together. In Saxton's case they do not: his Rivers Waveney and Little Ouse, for example, are noticeably more sinuous along the borders of Norfolk than in the neighbouring county of Suffolk which he mapped on the same scale. This makes it seem probable that he surveyed each county separately and that his national maps were created afterwards in an act of synthesis.[82]

In William Petty's published atlas of Ireland (1685) the format is similar but all the county maps are all on the same scale and there is no difficulty about joining them.[83] We might infer that Petty surveyed the whole island in a single campaign, but it remains possible that disagreements along boundaries have been removed by judicious 'assimilation' – fudging, as critics would call it – at some intermediate stage in their passage from surveyor to engraver. A different arrangement again is presented by the Escorial atlas recording Esquivel's surveys of Spain. Here there is no problem of boundary-matching because each sheet forms a rectangle filled to the edge with geographical detail and the whole assemblage matches a surviving small-scale gridded outline of the Iberian peninsula that was clearly meant for an index to the sectional sheets (Fig. 9.13).[84] In this case the hypothesis of compilation is even less attractive, and if the index is set aside (as indexes usually are) there is probably no more than one Spanish map involved here.

What seems to emerge from this discussion is that until the eighteenth century there were very few empires, kingdoms or republics whose territories had been comprehensively surveyed and mapped in a single operation. How then could progress be achieved? The textbook remedy was to begin with a national triangulation of the kind advocated for Ireland by William King as early as 1709.[85] But where were the star performers eligible for such a demanding role? Might the work of regional surveyors be somehow coordinated to provide a substitute? In Britain any county triangulation was likely to include a few prominent landmarks in adjoining counties, and this made it seem feasible for a net to be spread across administrative boundaries 'to facilitate the continuation and connection of … county surveys', as Peter Burdett put it in 1762.[86] By the end of the century such links had been established for Lancashire, Cheshire, Derbyshire, Staffordshire, Leicestershire and Warwickshire with outlying single stations in Lincolnshire and Shropshire. In Ireland surveyors of a later generation did the same for Counties Mayo, Roscommon and Longford. But if directed by more or less independent county representatives such an enterprise would be bound to lack consistency; and under a single private organisation it would have taken an intolerably long time. Even if the triangulation could be managed there was still the question of an astronomical control, described in 1756 as 'not an affair in the reach of a private purse.'[87] In a national setting, then, the choice was between a government triangulation and no triangulation at all. The subject became topical in the 1780s when the English section of a geodetic link between the observatories at Greenwich and Paris was organised by the board of ordnance and the Royal Society under the direction of William Roy. The outcome can hardly be seen as an act of high policy. In fact according to legend the cross-channel triangles grew into an all-British control survey only when someone told the master-general of the ordnance about a

9.14 From Christopher Greenwood, *Map of the County of Somerset, from actual survey made in the years 1820 & 1821* (London, 1822), original scale 1:63,360.

theodolite that had unexpectedly been left unsold – the cartographic equivalent of acquiring the British Empire in a fit of absent-mindedness.[88]

If private surveyors were incapable of triangulating a whole country, they might at least be expected to benefit from the government's efforts to do the job for them. A hint to this effect was the deliberately restrictive title 'Trigonometrical Survey' for what later became the Ordnance Survey of Great Britain and Ireland, almost suggesting that everyday ground detail formed no part of the Survey's business. There were signs that the implied division of labour between control and detail surveys might actually come about when Joseph Lindley used seven of Roy's stations in an otherwise independent county map of Surrey.[89] Of course, a common triangulation would not save the participant regional maps from differing among themselves in scale, content, local accuracy and general appearance. But with the progress of (non-cartographic) scale economies in British business life, might not a uniform series of detail surveys be within the capacity of a single firm? If so, the most likely candidate was Christopher Greenwood, who claimed in 1818 to be directing 'the largest private establishment of surveyors, that were ever united in this kingdom'.[90] By 1834, Greenwood had published thirty-five English and Welsh counties, mostly at one inch to the mile. Fifteen more, and he would have made his own national map (Fig.9.14).[91] It was a remarkable career, but not remarkable enough to deter the government from issuing an independent detail survey on the same scale.

More about planimetrics

Greenwood's achievement provides an occasion for re-introducing the subject of accuracy tests. We have seen that for local and regional sketch maps W.R. Tobler's bidimensional regression programme generally gives percentage values of between 60 and 90. How do such sketches compare with maps thought to have been based at least partly on measurement in the field? In this respect each category distinguished in the foregoing paragraphs presents its own problems. Early fort plans are hard to evaluate, their features usually having been so much altered on the ground that no points remain identifiable in a modern map. Rural estate maps are often similarly deficient in recognisable landmarks but when they can be tested the scores often exceed 99 per cent, even as early as the mid seventeenth century. By contrast, scores for town plans are sometimes unexpectedly low, perhaps because street-widths were exaggerated, perhaps because the edges of such maps were often less carefully surveyed than the middle. William Smith's Bristol (1568) is only 80 per cent accurate; Rocque's plan of Bristol, 96.3 per cent. Early regional cartography may be fairly represented by Saxton's English county maps, on scales averaging about 1:230,000. A typical Saxton result would be about 99.3 per cent. In a statistical sense, this left little room for improvement, but even so the values achieved by later county cartographers can be seen creeping up. Thus in Somerset Saxton scores 99.37, Day and Masters (1782) 99.95, and Greenwood (1822) 99.96. William Edgeworth's score in the district of County Longford mentioned above is 99.87.

Perhaps the Tobler value of a measured survey should be compared with the readings of a clinical thermometer in the diagnostic significance attributable to small numerical differences. But on any reckoning the above-mentioned Somerset figures seem uncomfortably close together, especially when we reflect that as scores approach 100 an increasing proportion of their shortfall must be due to errors introduced by the modern researcher. Fortunately Tobler offers an alternative. It will be remembered that his programme achieves the best possible fit between the early and standard maps by rescaling and rotating the former. There remains an array of local errors that can be calculated for each point in the sample and expressed in modern units of horizontal ground distance. These can then be averaged to give a single index for the whole map. One advantage of such an index is to allow for the influence of scale. Thus the same error measured on the paper might represent twenty feet in an estate survey and tens of miles in a small atlas map. In Greenwood's Somerset the mean displacement is 0.23 inches on the map which amounts to 1250 feet on the ground. For an Elizabethan county map the corresponding figure might lie between 3000 and 10,000 feet. Such values seem more 'realistic' than the corresponding percentages, but unfortunately little research has yet been done along these lines.

CHAPTER 10

Going in the dark

ONE WAY TO INTERPRET map history would be as a series of divergences whereby each technical breakthrough, instead of obliterating everything else in sight, allows older methods to persist within a narrowed range of opportunity. Thus when the instrumental admeasurement of large areas became feasible in the sixteenth century, the freehand sketch map, far from disappearing altogether, survived in the role it continues to occupy today, as an unambitious and probably ephemeral account of topological relationships. Much the same happened two hundred years later: although in the world's wealthiest societies regional and national surveys had now reached a high level of precision and comprehensiveness, the technology of Saxton and Norden remained appropriate wherever the need for accuracy was less acute. The English military engineer William Roy saw nothing reprehensible in this state of affairs. His own map of Scotland was, he admitted, no more than a sketch; but even as late as 1785 that did not make it anything less than 'magnificent'.[1]

Besides military cartographers like Roy, the chief European exponents of modern sketch-surveying were pioneer map-makers at work outside their own continent. Soldiers and explorers had enough in common to share a chapter in a wide-ranging book. One attitude that united them was an indifference to the notions of territoriality underlying our earlier categories of local, regional and national. In war, and in preparations for war, map-users were sometimes impelled to violate an international frontier. For the geographical discoverer who came in peace from an alien culture such frontiers, if they existed, were probably unimportant and possibly unknown. In the right circumstances both parties might be expected to move forward until the earth was either literally or metaphorically conquered. Exploratory and military surveying were by no means identical, however, and some of their differences, particularly in scale and subject-matter, will soon become apparent in what follows.

Cartography and war

We begin with extensive surveys made by specialist engineers for the use of military commanders. These first became prominent as a seventeenth-century phenomenon associated with the decline of siege warfare and the development of more fluid strategies employing lightly mounted horse and foot.[2] The regions most affected were the frontiers of powerful and mutually hostile kingdoms, especially those ruled by persons of combative temperament like Louis XIV and Frederick the Great. This branch of surveying was distinguished above all by its professional homogeneity. Common practices were disseminated not just through family ties, as in civil life, but through nationally circulating textbooks and national military colleges as well as by the association of practitioners within the framework of a standing army.[3] Soldiers of allied forces fought side by side and learned from their common enemy by translating his instructional literature. Men trained in one country would often enlist under a foreign flag, with results illustrated by some of the better-known military surveyors practising in Revolutionary North America, among them Samuel Holland and Bernard Romans (both Dutch), Joseph Des Barres (Swiss), Claude Sauthier (French), Robert Erskine (Scottish), John Hills (English), and Thomas Hutchins (American).[4]

Outside the ambit of set-piece campaigning, military surveys were sometimes produced for large and previously little-known colonial territories in which a more diffuse spirit of insurrection might erupt anywhere and at any time. Several such theatres of rebellion existed in Elizabethan Ireland, where it was common for defensible sites to be mapped against a physiographical background at scales of c.1:450,000 and with accuracy levels of 70 to 90 per cent. Given time and talent, these scores might be raised to 97 per cent, as with the still under-valued maps of Ulster by Lord Mountjoy's cartographer Richard Bartlett (Fig. 10.1).[5] A better-known example is Roy's sphere of operations in the Scottish Highlands, where Lieutenant General Henry Hawley was so badly off for maps in the 1740s that he described himself as 'going in the dark' to confront the Jacobite rebellion.[6] On both these occasions the maps came too late to help defeat the enemy: in cartography, as in other branches of the soldier's art, many tacticians spent more time preparing for the last battle than for the next one.

When war came it was often fought in wild and poverty-stricken terrain that had hitherto generated little commercial demand for geographical information, but this was not the only reason for the rise of military cartography: by its nature modern soldiering needed a kind of map for which until the eighteenth century there had been no significant market even in regions blessed by peace and prosperity. These maps have been elaborately classified,[7] but if we restrict ourselves to methods of construction the only categories worth noting are reconnaissance maps and topographical maps, and even these may be seen to some degree as a single type without sharply defined subdivisions. At one end of the continuum was the sketch pure and simple, in which estimation entirely took the place of measurement; at the other end was the typical European government survey of c.1900, open-ended in consumer appeal despite its military origin. Both these extreme cases have already figured to some extent in earlier chapters: henceforth attention will settle somewhere towards the middle of the range.

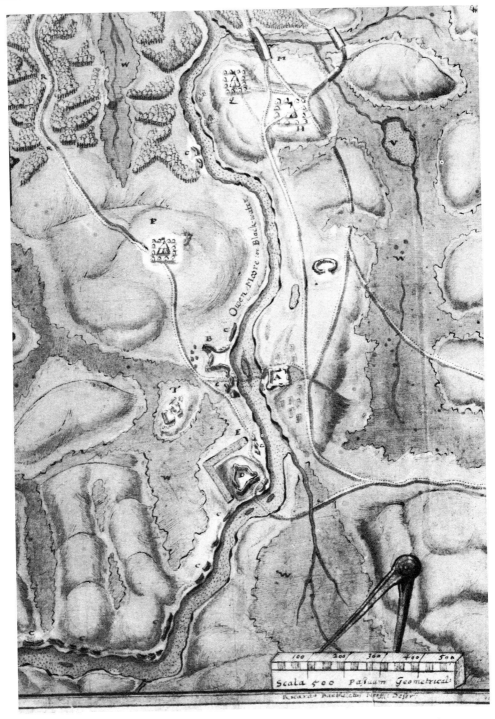

10.1 From Richard Bartlett, campaign map, Blackwater valley, northern Ireland,
original scale *c.*1:12,000, 1602. See note 5.

10.2 From William Roy, Military survey of Scotland, original scale 1:36,000, MS, *c*.1755. British Library, Maps C.9.b.

On the spectrum of accuracy and comprehensiveness a military engineer's position was chiefly governed by the time available to him, which in many cases would be severely limited. In the heat of the American revolution, for instance, mapping a road was reckoned to take no more than three times as long as walking the same distance unencumbered by cartographic duties. An earlier writer, more disconcertingly, wrote of military geographers who could 'flip along from one mountain to another'.[8] Roy's survey of the Scottish Highlands in 1747–51 (Fig. 10.2) was a good deal slower, with each field party needing about a year to cover the equivalent of an English county. In the following paragraphs Roy's achievement will be taken as representative, though an equally good example would have been Charles Vallancey's survey of Ireland in 1776–90.[9]

The soldier-cartographer: in the field

The military surveyor was concerned not with matters of universal importance but with whatever might affect the movement, quartering and provisioning of his comrades. Chiefly this meant roads, hills, rivers, houses, inns, corn mills and forges. Then there were obstructions to manoeuvrability – woods, bog, parkland, and agricultural enclosures as distinct from 'champion' or open ground. Nor was class prejudice an officer's only reason for recording gentlemen's pleasure grounds and gardens: to troops on the march, a resident landed proprietor could offer various kinds of assistance and even hospitality beyond the resources of a small farmer or cottager. Most important of all was the mapping of 'command', or the possibility of looking down rather than up at one's opponents. Military surveying was accordingly dominated by the representation of relief, from the gentlest of inclines to the highest mountain peaks. At the same time there were certain mainstays of civil mapping that held little meaning for the soldier. Administrative and property boundaries would be an unwanted distraction on the field of battle, and antiquities had no claim to notice unless they were fortifications that might conceivably be reoccupied by troops – which no doubt was Roy's excuse for including Roman sites in his map of Scotland.[10] On the other hand all prominent landscape features deserved to have their names shown on a military map, recognisably and pronounceably rather than correctly by any academic standard, and in this respect hills and sometimes roads were at least as deserving of identification as settlements.

The need for haste affected Roy's military survey in a number of ways. In some theatres of war an officer might choose to trace or enlarge a published civilian map and convert it to his own use by pencilling in additional detail.[11] It is unlikely, however, that much time would have been spent in Scotland on the collecting, collating and combining of previous maps. What counted in the Highlands was field work. More than one team of surveyors had to be recruited to get the ground covered in reasonable time, despite the risks of inconsistency and inaccuracy attendant on a division of effort. Altogether six parties were employed, each comprising an engineer officer and half a dozen soldiers. Two men in each contingent were detailed to carry the chain. (That some survey lines were paced might seem a reasonable conjecture from the maps, but there is apparently no documentary evidence for it.)[12] Two others observed fore and back stations with a 'plain' non-

10.3 Military surveyors at work near Loch Rannoch, Scotland, by Paul Sandby, *c.*1750.

telescopic theodolite of seven inches diameter. Both chain and theodolite appear in a famous picture by Paul Sandby, who also executed or supervised the drawing of the maps (Fig. 10.3).[13] The reference to back stations probably implies that the angles were compass bearings and that the 'theodolite' was functioning as a circumferentor, an interpretation strengthened by Roy's preference for magnetic north on the final copies.

All this was clearly the programme for an extended network of traverse surveys, but there were also intersections right and left to fix 'innumerable minute situations', together with such cross-lines as were necessary 'for filling up the country'. The legs of the traverse itself were roads, major rivers and 'numerous' minor rivers – each followed to its source, our informant is careful to say, recalling (no doubt unintentionally) Saxton's cavalier treatment of certain headwaters in northern England. So by a process of elimination it appears that sketching rather than offset-measuring was used for settlement and land cover as well as for relief, and this inference is supported by a further report that each surveyor kept one field book for angles, measurements and intersections and another book for delineating 'the stations and the face of the country'.[14] Perhaps the disposition of traverse stations was accurately plotted in the second book as a control for the sketching; or perhaps 'stations' in this phrase just meant that each sketch was separately taken

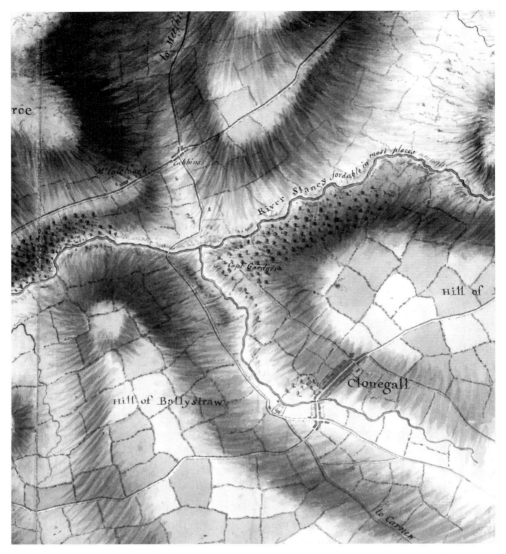

10.4 Schematic fields near Clonegall, Co. Carlow, original scale 1:20,260. From Charles Vallancey, military survey of Ireland, MS, 1776. Royal Irish Academy, MS 12.S.6.

from a single viewpoint incorporated in the survey skeleton. A modern reader may wonder why Roy's surveyors differed from the military engineers of contemporary North America in disregarding the plane table.[15] The soldiers' lack of training was probably a sufficient ground for this decision, though doubtless the Highland climate played its part: plotting the survey data was a suitable duty for 'the winter months at Edinburgh, … where the connection of the summer work of the several surveyors was often the subject of mutual discussion'.[16]

Fences, walls and hedges were important to an eighteenth-century field commander as barriers to vision and the deployment of cavalry. In practice,

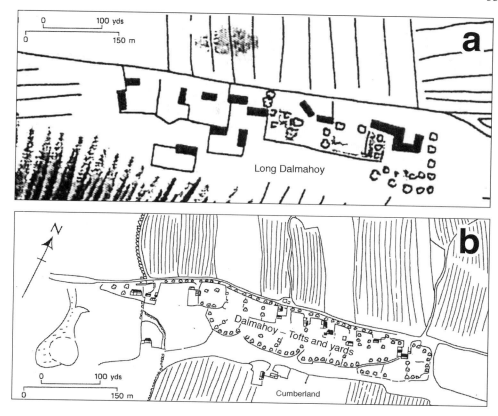

10.5 (a) Part of William Roy's military survey of Scotland, 1747–55, enlarged to the same scale as (b) An estate map of the same area by Thomas Winter, 1749. G. Whittington and A.J.S. Gibson, *The military survey of Scotland, 1747–1755: a critique*, Historical Geography Research Series, 18 (Norwich, 1986), p. 28.

however, there was hardly ever enough time until the nineteenth century for anyone but an estate surveyor to map a network of field boundaries with complete accuracy. Roy, writing of a proposed map-scale not unlike that of his Scottish survey, saw no need to measure 'every turn' though he did insist on 'frequent cuts'. The word 'cut' suggests a straight line driven more or less at random through the fieldscape, successive increments of length presumably being recorded whenever a fence was crossed. Such a mixture of approximation and exactitude seems rather unlikely unless the 'cut' led to something important. More probably the whole field pattern between one road and another would have been non-instrumentally sketched – in some cases, perhaps, more or less invented. As it happened, arable cultivation on Roy's Highland maps was chiefly shown by diagrammatic plough-ridges or furrows with few definite fence-lines. In military surveys of southern Britain and eastern Ireland the enclosures were more obtrusive but equally lacking in conviction (Fig. 10.4), a judgement also applicable to eighteenth-century North American and continental European maps in the same mode.[17]

Roy's methods were only distantly related to those of contemporary civil surveyors whether local or regional. This was not surprising when his labour force consisted of soldiers, doubtless employed because they were cheap and obedient as well as being ready to hand. The difference is also illustrated by his choice of linear measures – chains of 45 or 50 feet instead of two or four perches, and a scale reckoned in yards (1000) to an inch instead of inches to the mile or perches to the inch. His measurements were also less exact than would be expected in a first-class eighteenth-century civil map. There were no trigonometrical points and no latitudes or longitudes. Substantial error has been found in one instance by transferring two adjacent ten-kilometre squares from the Ordnance Survey to Roy's outline.[18] The difference corresponds to a percentage accuracy of 97.28, by no means spectacularly good for its time. Across the Atlantic a different test has been applied to John Montresor's military map of roads between New York and Philadelphia (1777–8), on which the towns were found to be out of position by an average of 3.25 miles.[19]

But planimetric distortion was less important to many map users than the kind of local error or omission that could be detected without taking measurements. Such flaws appear when Roy's military survey is compared with contemporary estate mapping (Fig. 10.5) and again, perhaps more alarmingly, when one version of the survey is checked against another.[20] Whichever draft is in question, boundaries of cultivated land are often suspect, roads sometimes lead nowhere, and there is no attempt to show the exact number of houses in each cluster, or their exact spatial relations one with another. Complete clusters were occasionally duplicated or moved to the wrong side of a road or river. In towns the house plots were often defined conjecturally and the road-exits misplaced.[21] Sometimes whole strips of countryside were left almost empty. Roy himself is unlikely to have overlooked these deficiencies. If war had not broken out with France in 1756, he wrote later, his map would have been completed 'and many of its imperfections no doubt remedied'.[22] This suggests adding and correcting detail within an existing planimetric framework rather than adjusting the entire map to a more accurate triangulation – though, as we have seen, Roy did apparently consider such an adjustment to be feasible.

The soldier-cartographer: in the office

Whatever its accuracy, as a graphic composition the survey of Scotland was true to the principles of military map making. In using a map like Roy's, as in creating it, there was no time to waste. If nothing else, the reader could at least be saved a mental switch from vertical to horizontal symbolism by a policy of planiform representation for hills, buildings and water that was presumably derived from a style long familiar in large-scale fortification plans. Colour appeared abundantly but always in a non-aesthetic role: hills in brown or grey, water in blue, woods in green, cultivated land in buff, roads in ochre, buildings and stone walls in red. Script was used not for long disquisitions, as on so many published maps of the same period, but for indispensable placenames and descriptive terms. There was no decoration for its own sake, and virtually no symbolism of a kind that might need explaining in a marginal key. This last decision may have reflected a characteristically British attitude: continental surveyors were more adventurous in their use of signs, for instance in separating different grades of road and different capacities of harbour or anchorage.[23] Finally the most striking feature of Roy's map, as of most good military maps, was the boldness and clarity with which his draughtsmen drew the hills.

Another thought-provoking aspect of the surveys under review was the relation of scale to content. The normal course of cartographic history has been for maps of any given type to become concurrently larger and more correct. Here was a genre in which a larger scale brought no corresponding planimetric improvement and perhaps even a deterioration. The most common preference for a published English county map was 1:63,360. In Scotland Roy chose 1:36,000; in Ireland Vallancey, initially at any rate, preferred 1:20,160. In eighteenth-century Austria, 1:28,800 was the 'normal military scale'.[24] These were ratios that might suit a manuscript draft intended for engraving at a smaller size but most military surveys were not originally meant for such a purpose. Judged on its own merits their scale was too large for the quality and quantity of information given. Dependence on landscape sketching could have been at least partly responsible for this effect: much of what the sketcher saw was too small to appear on most kinds of map, but having drawn it he might be reluctant to see it edited out and correspondingly glad to choose a format where this would be unnecessary. On a rational view, there should have been a different scale for each level of military organisation – which in modern terms would mean a platoon scale, a company scale and so upwards through the hierarchy of battalion, brigade, division and army corps; but it must have been rare, to say the least, for the subject to be discussed in terms as doctrinaire as these.

A large scale encourages a large sheet-size; and without the discipline imposed by the printing press, a composite manuscript map produced by several cartographers could easily assume an asymmetrical form for which no individual could be held responsible. Thus the original protraction for the Scottish Highlands occupied eighty-four rolls 'of irregular shape and size', which in the unlikely event of their being put together would have measured about 40 by 28 feet.[25] Vallancey's Irish surveys were embodied in a baffling mixture of strips, rectangles, and various other outlines unamenable to geometrical description.[26] This may have discouraged some faint-hearts from consulting the maps at all.

As we have seen, Roy knew his survey to be less than perfect. Its justification lay in maintaining a seemingly more-or-less uniform standard over a large area, an achievement which, to judge from past experience, would not be matched for many

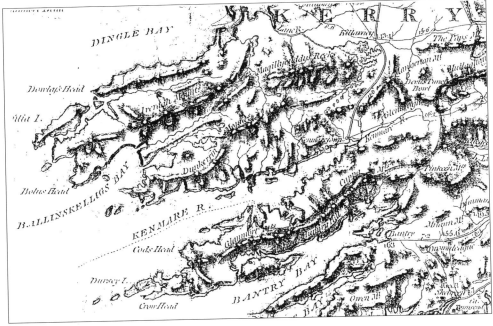

10.6 From Alexander Taylor, *A new map of Ireland having the great features of the country described in a manner highly expressive* … , scale *c.*1:695,000 (London, 1793).

years by civilian surveyors working at any higher level of precision. Moreover, regardless of its military value, Roy's work had a future as a refreshingly underivative national map. When he undertook to continue it outside the Highlands, in 1752, the threat of rebellion had receded. He decided to ignore the Scottish islands, which at least theoretically might still be seen as vulnerable to a foreign invasion, and instead extended the survey through the lowlands, Southern Uplands and borders, all well-travelled and comparatively well-inhabited regions that had long ago ceased to threaten the nation's security. He also took the trouble, in 1787, to have the whole map reduced to manageable size at a scale of about six miles to the inch. How far military maps were intended to remain secret at this period is a pertinent but difficult question. On the whole, little was done to restrict their circulation.[27] At first, it is true, Roy's survey remained almost unknown to Britain's wider carto-graphic community,[28] though it eventually formed the basis for a commercially published national outline if not for any separate county maps.[29] The same happened with Vallancey's military coverage of Ireland as reworked by Alexander Taylor (Fig. 10.6).[30]

Military surveying outside western Europe

Within the framework of military mapping, Roy's style was adaptable to a wide range of scales and convertible, if required, by skilful engravers from manuscript to print. It could be used for mapping past, present and future troop dispositions, or simply as a topographical record. It was certainly compatible with a higher level of planimetric accuracy than was attained in Scotland, except perhaps in the case of agricultural enclosures. This capacity for development had a special appeal in North America, mainly because in that continent there were not yet many civil surveyors with the time and training to tackle extensive tracts of country. A major American preoccupation was the laying out of state borders, many of which were defined in terms of latitude and longitude, and also the boundaries of large land allotments like the 20,000-acre townships adopted for Canada in 1763.[31] In this case astronomical observations clearly had a part to play, and some of Major Samuel Holland's latitudes and longitudes were approved by the Royal Society of London.[32] Such researches would presumably have helped buttress the survey of Canada at 1:24,000 that Holland was proposing in 1765. Here we see military engineers beginning to rival the land surveyor's regional map. Meanwhile the cartographic expectations of the high command were continuing to rise. Such was the opinion of Thomas Jefferson, anyway: in 1786 he credited a new South American survey with 'a precision which qualifies it even to direct military operations in that country'.[33]

Most of the countries so far considered in this chapter were relatively small. In an area the size of Russia different tactics were called for. Ivan Kirilov's instructions to his government's surveyors in 1721 are particularly deserving of attention. In each territory latitudes were to be determined astronomically for every city and for points where main roads crossed the territorial boundary. Distances along the less important roads were to be estimated by the local inhabitants. Each main road was to be measured by the surveyor, and the differences between measured and estimated lengths on these highways would provide a basis for correcting the estimates of the minor roads.[34] A century later there was less dependence on estimation. According to official Russian instructions of 1826–7 the policy for areas where no triangulation had been provided was now to make plane table surveys of all main roads, rivers and administrative boundaries, together with adjacent areas up to one verst (0.67 statute miles) distant from the principal features, more remote points being fixed by intersection wherever convenient. These strip-like plane table surveys served as a control whose interstices could be filled with compass bearings.

Mapping on the frontiers of knowledge

Leaving soldiers on military service, our next subject is what might be called the wilderness map. The cartography of exploration is most conveniently defined by the history of the areas it depicted. Common to these places was their omission from any previous map available to the intending traveller, or at least from any that he felt prepared to trust. Common to the explorers was a determination to travel light. However rich and attractive the country of their choice, they did not expect to put down roots there, but rather to keep on the move, though preferably for not more than a year or two, registering an unbroken stream of new impressions. Before returning they hoped to have done more than 'go and look behind the ranges' (in the evocative words of Rudyard Kipling)[35] but the precise nature of these further aspirations would naturally vary with temperament and circumstance. An explorer might be searching for famous people, perhaps just one famous person: map-making would certainly be easier if there was only a single object to be mapped. In Ethiopia Francesco Alvares's quarry was Prester John. In Venezuala Walter Raleigh had hopes of tracking down a famous chieftain known as Eldorado. In New Mexico it was a cluster of seven legendary cities that caught the imagination of Francesco Coronado. In West Africa René Caillié actually found what he wanted, which was the semi-mythical but sadly disappointing town of Timbuktu.

Goals of wider geographical scope, though often equally hard to attain, included putative inland waterways and arms of the sea. Many explorers hoped to navigate their way across a continent. Even if a river led nowhere they could at least look for its headwaters, especially if these lay among little-known lakes and mountains like the sources of the Nile. Less commonly a delta or estuary might be the prize, as with Mungo Park's fateful downstream journey along the Niger. In the late nineteenth century some explorers hoped to plant a flag in the furthermost Arctic or Antarctic, but at any period the earth's geographical poles would be dubiously relevant to our present narrative (they were already on the map, after all), though they did further illustrate the enthusiasm aroused by single locations as distinct from large tracts of country. By comparison more diffuse objectives, such as enlarging the hinterland of the Canadian fur trade, would seem to have been considered less inspirational.

A surprisingly large number of explorers have chosen to go back for more. For the map historian this impulse may create difficulties of classification, the true discoverer becoming hard to distinguish from the surveyor-cartographer. The overlap between the two is perhaps best demonstrated by Samuel de Champlain, who between 1603 and 1616 made six long journeys in eastern Canada, some wholly exploratory and others covering ground already known to earlier European travellers whose information Champlain blended with his own surveys and with the work of other cartographers.[36] In this respect he was approaching inland lakes and rivers with the same combination of talents that James Cook would later bring to the islands of the Pacific. In general, however, links between exploration and map-making have been weaker on land than at sea. For one thing, a traveller by road or footpath probably knew less than the master of an ocean-going ship about how to fix his exact geographical position and was therefore less likely to leave posterity with any account of having done so. Also, many cross-country journeys could be described verbally, by reference to routes, settlements, territorial divisions and estimates of mileage, in enough detail to guide subsequent travellers through the same area without cartographic assistance. Some of history's most famous expeditions were not shown on any contemporary map that now survives but are still capable of being plotted with assurance by the compilers of modern historical atlases.[37] The travels of Alexander the Great, Ibn Battuta and Marco Polo come immediately to mind, and indeed such 'secondary' mappings are more the rule than the exception at least until the early sixteenth century. A typical case was Hernando de Soto's tortuous route through what are now the southern United States in 1538–43 which though not recorded cartographically at the time has been laid down on a present-day map with the help of a 'U.S. government analysis'.[38]

Mythology aside, the most attractive subject for overland exploration and cartography, one might think, would be a country of temperate climate, fertile soils and easily discovered mineral wealth somewhere on the frontier of existing European colonial settlement. But mapping such a region did not necessarily involve the kind of set-piece journey that is remembered as an historical landmark. Maps could be built up gradually from the knowledge gained by generations of short-range travellers, a sequence of events that we have already postulated (one can hardly expect documentary proof of it) in early sixteenth-century Europe. In wild and difficult country, however, a higher degree of organisation was required. It is true that brave men of substantial private means, like Alexander von Humboldt in Latin America, could still make single-handed geographical discoveries well into the nineteenth century. But from c.1600 onwards an increasing burden began to fall on the kind of expedition that was sponsored by some kind of corporate body. By and large, the more financial support a traveller could draw on, the more mapping he was expected to do. An individual might be uninterested and untrained in making surveys; a large expedition, carefully planned, would be more likely to include the right people with the right equipment. Over the whole time-span of post-medieval exploration a crucial factor here was the general level of map-consciousness in the home country. Many kinds of map could contribute to the gradual formation of a European cartographic culture, but the most important influence (and not so gradual) was surely the progress of marine chart-making in the years following Columbus's voyages. This subject will be considered in the next chapter: the point now is that what could be done so well by sea ought eventually to provide a model for parallel developments on land.

Techniques of exploratory surveying

The outstanding feature of exploratory as opposed to military surveying since the mid-eighteenth century has been the part played by astronomical observation. To some extent this was a response to the sheer length of the typical explorer's journey: the larger the area to be surveyed, as we have seen, the greater the relative accuracy of celestial compared with terrestrial measurement. On this broadest of scales there was also an analogy between mapping and charting: desert, prairie, savannah and tundra all resembled the ocean in their seemingly interminable emptiness. Forest and jungle, it is true, could have benefited from being emptier than they were, and this in its way was no less of a handicap. Common to many such environments was a shortage of natural as well as artificial landmarks that might serve as stations for taking angles. In any case the instruments needed for precise triangulation were too delicate and cumbersome to be moved with safety over long distances on inferior roads.

Another problem endemic to exploration, as to military reconnaissance, was shortage of time. Even where viewpoints could be seen from his route, a traveller could ill afford to reconnoitre a suitable network of triangles, or to set out and chain an accurate base. In fact most kinds of linear measurement were too slow to be practicable on a long journey. The normal standard for reckoning distance was the hour or day of travel time, converted by some more or less arbitrary factor into linear units. Angular measurements were easier to handle, always provided there was something to measure. Bearings could be taken with a compass or sextant, as at sea, though among little-known rocks and mountains there was an extra source of error in the possibility of local magnetic attraction. From compass directions and estimates of distance a traverse could be laid out along the line of march, but the results were almost bound to be seriously inaccurate: hence the need to establish astronomical controls at all suitable points where the expedition could afford to wait for an improvement in atmospheric visibility. One way of processing this kind of survey was to plot the results directly on to a ready-made graticule.[39] So in the end it was latitude, and later longitude, that chiefly distinguished exploratory from military surveying.

We have already defined the explorer as heading for places not reliably shown in any of the maps accessible to him. When he reached his destination this proviso might cease to apply, a possibility foreseen as early as the sixteenth century when Arthur Pet and Charles Jackman were told to take note of any locally-produced maps that might be waiting for them in China.[40] Native maps would be just as much part of an explorer's discoveries as the country they represented, imposing a similar responsibility to utilise what he had found in some constructive way. He would probably feel more admiration for indigenous cartographers on their home ground than if he had been studying their work in the safety of London or Paris. Compared with the colleagues he had left behind, he himself was now so poorly furnished with instruments and artist's materials that he might almost be described in a cartographic sense as 'going native' or, to put it more politely, drawing on a common ancestral heritage too old to be documented by historians. This situation was illustrated when the American explorers Lewis and Clark traced a map in charcoal on bark while negotiating with Indian chiefs and elders.[41] Not that every native cartographer was aiming at the truth: when a Hausa map of 1824 erroneously made the River Niger flow eastwards rather than southwards it was probably a means of encouraging British explorers to take themselves off.[42]

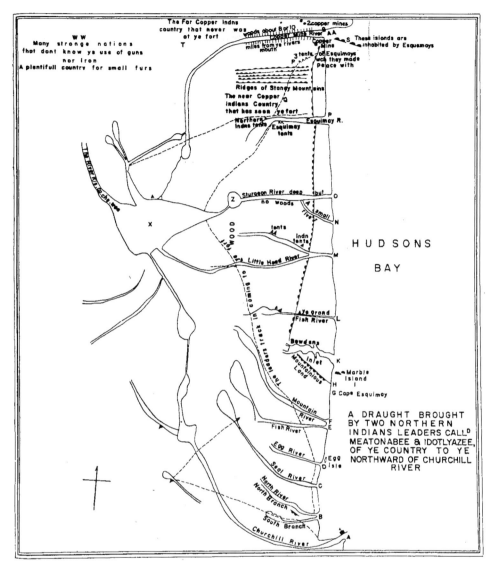

10.7 Native American map showing the west coast of Hudson Bay, *c.*1760, copied and augmented by Moses Norton. Redrawn in Rainer Vollmar, *Indianische Karten Nordamerikas: Beiträge zur historischen Kartographie vom 16. bis zum 19. Jahrhundert* (Berlin, 1981), pp. 68–9.

The best way to show respect for aboriginal cartography was to incorporate its contents into one's own work. This process could take several forms. A native American map might guide the stranger in his choice of a route that he would then plot for himself by using European methods. In doing so, he might prefer to translate its contents from graphic symbolism into a scripted document that gave more scope for approximation (Fig. 10.7).[43] He might just copy his informants' drawing, he might combine it with others of the same kind, he might stretch or compress it to connect with a map produced by his own countrymen. He might add his own kind of compass indicator or other marginal information. (It was a nice touch, if not very useful, to name each cardinal point in a phonetic rendering of a native language that had no indigenous written expression.)[44] All of which implied that American and European cartographies were not as irreconcilable as has sometimes been suggested: after all, nothing like this would have happened with Melanesian stick charts in the Pacific.[45] Here however we are anticipating a later chapter on the subject of compilation.

In the field, original maps were usually drawn in pencil or pen and ink, for the most part without any colour except perhaps a grey wash. If other colours were available the first choice would probably have been blue for water, but at small scales a single line was sufficient for rivers as well as roads, the latter pecked or dotted, the former continuous as far as the author's knowledge allowed. When so many rivers had to be left unfinished an arrow would be necessary to indicate their direction of flow. Explorers had little time for the *minutiae* of relief. Until a surprisingly late date, their hills were usually shown by the quickest and easiest method, namely miniature profile drawings, with little attempt to distinguish relative altitudes or to trace the ground-plan of an upland mass.[46] If mountains actually blocked the traveller's path it might be different; then individual peaks and passes became a matter of concern, and there could even be some reference to numerical altitudes. Where surveyors travelled mainly by water, it was common for the depths and breadths of rivers to be estimated with some care and for breaks in navigation to be marked with self-explanatory cross-ticks. Watersheds in general were also a subject of concern, even where they were not particularly mountainous. In the Canadian Shield, for example, such features were described as a 'height of land' and sketched as a row of hill profiles or, if cartographers realised that these might be taken too literally, as either a single line or a ribbon of shading.[47]

For the rest, abundant writing had long been a characteristic feature of frontier cartography. In a sparsely furnished landscape there was plenty of room for script, and the vagueness of the explorer's knowledge was often better suited to words than to drawings. Since the time of Ortelius, coastlines have sometimes been graphically asserted and verbally half-denied on the same map (Fig. 10.8).[48] An unrepentantly modern example was Dura Brata, marked by Aaron Arrowsmith as two islands in the Caspian Sea with the accompanying note that 'Capt. Woodrofe says he never could find them'.[49] Inscriptions were also often more appropriate than drawings in a thematic sense. Without a ready-made network of towns, prominent buildings and industrial sites, there was no limit to the range of information satisfying Francis Galton's requirement that wilderness maps should 'afford a quick guide for future travellers': famous battlefields, routes of raiding parties or reconnaissance expeditions, herds of buffalo, the estimated populations of villages, the presence of salt in the soil, the first sight of mountains, the last patch of timber. Few of these

10.8 Graphic statement with verbal qualification: 'New Guinea, recently discovered, of which it is uncertain whether it is an island or part of the southern continent'. From Abraham Ortelius, *Typus orbis terrarum*, 1587, in *The theatre of the whole world* (London, 1606), p.1.

features lent themselves easily to non-verbal symbolism. In addition, the names of indigenous nations and territories might be written *in situ*, though probably without any attempt at tracing political boundaries. The scale of eighteenth-century wilderness maps was often large enough to accommodate a level of verbosity that would later be considered odd (Fig. 10.9). A random example: 'In 1746 the governor of Buenos Aires by order of the king of Spain caused the supposed communication of Julian Bay with Campana River into the South Sea to be examined which whereupon was declared to be imaginary'.[50]

Most wilderness cartographers tacitly assumed that a map known to be less than accurate was by nature ineligible for overmuch precision of draughtsmanship. Native maps were especially relevant from this point of view: to be archivally compatible with their European counterparts they had to be reproduced on paper or parchment and furnished with written names and titles – though tactful copyists would instinctively refrain from adding any kind of artistic motif.[51] In the same spirit, explorers' maps were generally lettered in a workaday running hand that made no concessions to the cartographic medium.

Whoever was responsible for their original form, wilderness maps were seen as raw material rather than finished product. In the nineteenth century, admittedly, when specialist geographical agencies had begun to produce their own literature, an explorer's draft might be re-designed for publication without prejudice to its geographical content, probably in the unornamented style that was now fashionable for maps of almost every kind (Fig. 10.10). Of wider historical import is the fact

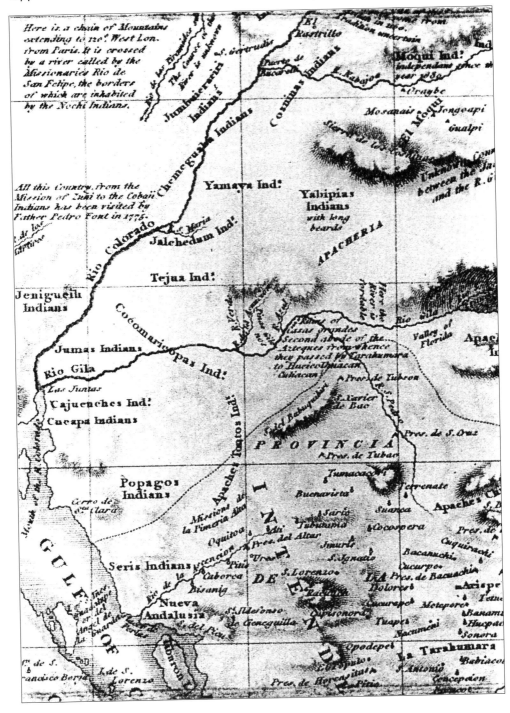

10.9 Inscriptions on an explorer's map. From Alexander von Humboldt, *General chart of the kingdom of New Spain* (London, 1811).

that, even within a broad cartographic synthesis embracing heartland as well as frontier, exploratory traditions would often remain dominant over large tracts of a printed general map. This was especially evident when non-European countries were published in folded sheets at scales of around 1:2 million or 1:3 million. Then the most sophisticated London engravers might sometimes have to write inscriptions like 'Not a tree to be seen' or 'Ill defined and little known, even to the inhabitants themselves'. As late as 1847 a general world atlas could still define linear distances by means of travel-times in countries like Palestine or Persia.[52]

To summarise: physiographically it is rivers and lakes that dominate a typical explorer's map; cartographically it is the graticule, so often considered a pedantic affectation in maps of well-known countries but here deserving the greatest respect as an essential element in the process of data collection.

Three examples

There have been few comparative historical studies of wilderness cartography as here defined: even in R.A. Skelton's classic *Explorers' maps*, almost every chapter is largely concerned with the mapping of seacoasts by seamen.[53] In these circumstances further comment on interior mapping must be confined to a brief selection of particular cases. Anthony Jenkinson in 1559 seems to be the earliest Englishman to have drawn a map from well-authenticated inland travels outside his home country. Oddly enough the result was never mentioned in his journal, though a version of it, complete with date (1562) and author's name, eventually appeared in Abraham Ortelius's world atlas (Fig. 10.11).[54] Ortelius showed the whole duchy of Muscovy with scales of latitude and distance and a variety of elegant 'compartments' and vignettes that helped to fill a double-page rectangular frame. The west and north of European Russia had evidently been obtained from previous maps or descriptions, as also were the rivers flowing from Bokhara and Tashkent. This means that Jenkinson cannot be held responsible for the drainage shown in the vicinity of Moscow, which led undiscriminatingly through interlinked channels to the White Sea, Caspian Sea, Black Sea, Gulf of Finland and Gulf of Riga. It may however have been his own idea to duplicate the River Oxus, misled by the existence of both a Russian and a Chinese name for it.[55] Almost all Jenkinson's geography looks strange to modern eyes, but at least he was right about the elbow of the River Don.

Jenkinson's eight-month journey took him from Moscow to Bokhara, including boat trips down the Volga and through the northern waters of the Caspian. His chosen unit of distance was the longest available to him – the league, usually quoted in multiples of five or ten and never subdivided. (In the non-exploratory section of the journey he had adopted the Russian verst, no doubt deriving the distances themselves from local information.) At Morum east of Moscow he 'took the sun',

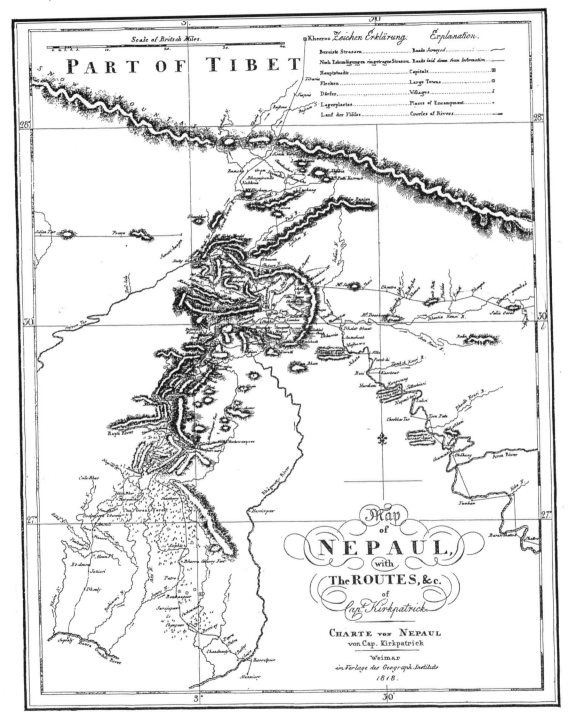

10.10 Nepal: explorations of William Kirkpatrick, 1793 (Weimar, 1818).

10.11 Part of Anthony Jenkinson's map of Russia, 1562, in Abraham Ortelius, *The theatre of the whole world* (London, 1606), p. 104.

and altogether the journal gave latitudes for a dozen or so different places, each to the nearest minute, possibly observed with an astrolabe.[56] Jenkinson's results were sometimes several degrees in error, and there was no reference to longitude in either map or journal, but this was as good a result as could be expected in the time available. One contemporary even called it 'a most exact survey'.[57]

In the English-speaking world President Thomas Jefferson takes credit for promoting the first well-documented scientific journey of exploration to a continental interior. After pondering the matter for many years he found his opportunity in the purchase of Louisiana by the United States from France in 1803. The expedition, numbering forty-five men, was led by two army officers, Meriwether Lewis and William Clark, its main purpose being to find the best route from St Louis via the Missouri River to the recently discovered mouth of the Columbia on the Pacific coast. The leaders were directed to collect information 'by enquiry' and spent much time on the study of native maps, but their own route had to be surveyed at first hand as exactly as possible. Advice as to methods and instruments was sought from the American Philosophical Society and a number of individual experts, while previous knowledge and opinion were synthesised in a map compiled especially for the expedition by Nicholas King.[58]

The detail for Lewis and Clark's own map was obtained by compass traverse, with distances along the rivers from the log line and overland by calculation from travel times. This method reflected Jefferson's erroneous opinion that a half-day land portage would be sufficient to carry the explorers across the continental divide. In fact their mileages were by no means uniformly reliable even on the waterways, the log being much affected by complex currents and by river debris.[59] The hazards of time-and-distance reckoning on foot were more predictable: of a map made by one team-member, Robert Frazer, it was later remarked: 'Where the party moved along fast, Frazer's map is foreshortened, and where the party struggled along slowly ... it is strangely elongated'.[60] So the traverse clearly needed strengthening by observations for the latitudes and longitudes of tributary-mouths, rapids, islands and other landmarks.[61] The principal instruments were the sextant, an artificial horizon (more necessary on an irregular land surface than at sea) and a chronometer. Longitudes were to be found from lunar distances, using tables computed by an American university professor of mathematics, but their values would not be finalised until after the expedition was over. When it came to the point the lunars were less satisfactory than estimates of overland distance, possibly because successive readings varied too much about the mean.[62] William Clark's map of the north-west, with longitudes by dead reckoning, was finally published in 1814, and remained the standard authority on its subject for more than forty years (Fig. 10.12).[63]

In the history of exploration Francis Galton (1822–1911) is important not so much for the ground he covered as for the vigour and clarity of his writings. In some ways he seemed to be reverting to a humbler level of professionalism than that of Lewis and Clark. Inheriting private means, he could afford to travel for the fun of it, and in his case – not a common one, it has to be said – fun included the making of maps. By this time there was a more specialised learned body to act as adviser: Galton's plan to explore Damaraland in south-west Africa was approved by the Royal Geographical Society in London, although he had to teach himself most of what he needed to know about surveying and in general, to borrow a phrase he made famous, about the 'art of travel'. Eventually he could take pride in his competence with the sextant, which on one memorable occasion he applied to an

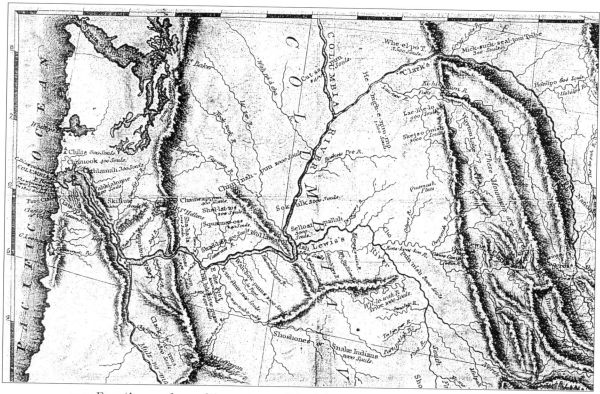

10.12 From 'A map of part of the continent of North America … shewing Lewis and Clark's rout over the Rocky Mountains in 1805'. Brian M. Ambroziak and Jeffrey R. Ambroziak, *Infinite perspectives: two thousand years of three-dimensional mapmaking* (New York, 1999), p. 48.

anatomical purpose by measuring angles ('from a modest distance') across the body of a Hottentot woman.[64] Like most contemporary African explorers Galton was especially interested in lakes and rivers, and he began with the intention of finding a new route to Lake Ngami, which had recently been discovered by David Livingstone. Then he changed his mind and in 1850–2 made one north-south and one west-east journey, mainly by ox-cart, across the territory later to be known as Namibia, covering a total of 1700 miles.

Galton differed from earlier explorers in making considerable use of triangulation, using either azimuth compass or sextant, though sometimes he could find no suitable landmarks. He also observed fifty-three latitudes and, by the method of lunar distance, six longitudes. The margins of his map were divided at ten-minute intervals (its convergent meridians not identified as forming any particular projection) and his figures for individual stations were given to the nearest minute. In awarding Galton a gold medal the RGS made a point of praising these observations for their accuracy, though how they could be checked in London is not altogether clear: presumably accuracy in this case meant consistency. As usual in

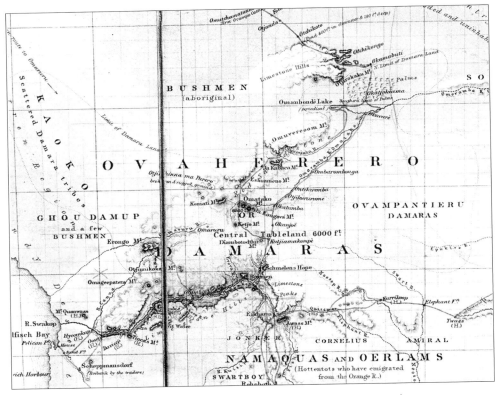

10.13 From Francis Galton's map of Damaraland (London, 1853).

such surveys, distances were clocked rather than measured, one hour's travel being equated with 2.5 miles.

By not trying to cross a whole continent Galton gave himself time for a more extended style of mapping than many of his predecessors had found possible (Fig. 10.13).[65] Admittedly his scale was small (about 1:4,300,000), and his cartographic style remained in many ways that of an explorer. His route was drawn prominently in red, and at a distance from it he sometimes openly depended on informants, as when reporting 'a grassy treeless plain said to extend to the sea'. Even when recording his own experience he often preferred inscriptions to graphics. On the other hand much of his message was geographical in the twentieth-century sense of the word. Deserts were shown 'chorochromatically' in a yellowish tint and fertile regions in shades of green that grew deeper as the land improved. There were many references to vegetation, and some to rock types and altitudes. There was even a 'southern limit of palms' though this disappointingly proved to be a parallel of latitude. Placenames were classified as Damaran, Hottentot or European. Perhaps most striking of all, the map was colour-printed by lithography. In this as in other aspects of his career, Galton prefigured a new age.

CHAPTER 11

All marked with lines

CHARTS IN THE MOST common cartographic sense are maps entirely devoted to representing a body of water and the features perceptible from points on or just above its surface. This specification is sometimes more inclusive than one might expect: Mount St Helens, clearly visible to a look-out on George Vancouver's ship *Discovery*, stands eighty-five miles inland from the Pacific coast. Not far away the same definition proved surprisingly narrow when Vancouver missed the mouths of the Columbia and Fraser Rivers. The truth is that an explorer can do his job without seeing anything but water and sky. The course he follows from A to B is a record of the intervening space, a 'track chart' in which the ship's route appears as a line. For no land to be visible from the track would constitute a negative but meaningful geographical proposition, whereas a blank sheet without the line might be no more than a helpless gesture of agnosticism. Among the facts important to future navigators are variation of the compass, depth of water, character of seabed, tidal range, times of high and low tide, strength and course of currents, perhaps even the appearance of the sea surface itself.[1] Until the nineteenth century, however, chart makers were slow to map more than those obstacles that a ship was likely to encounter through physical contact.

The world of the portolan

The earliest extant artifacts that belong undeniably in the present chapter are a product of Mediterranean Europe from the fourteenth century onwards. They were the famous portolan charts or, to use a permissible solecism, portolans. There are some 180 maps and atlases of this type surviving from before 1500 and many portolan features continued to appear on charts of sixteenth- to early eighteenth-century date.[2] Of all major map-types, portolans are the least well documented in contemporary non-cartographic records, suggesting that only a small and closely-knit community was involved with their production and use. In these circumstances our knowledge of the maps themselves must depend on internal evidence and particularly on their treatment of identifiable geographical features, which thanks to an abundance of placenames are gratifyingly numerous. A typical full-size portolan showed well-captioned coastlines for the Mediterranean and Black Seas, together with the Atlantic and North Sea littorals from north-west Africa to Denmark, probably at a scale of about 1:6 million. Small islands were often picked

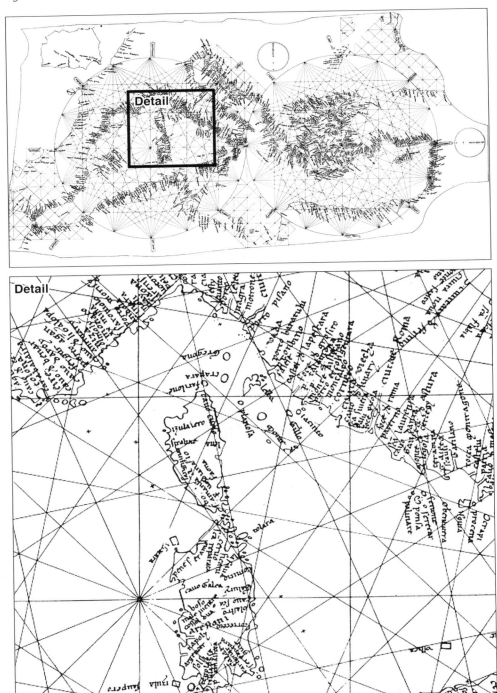

11.1 The oldest extant portolan chart. Carte Pisane, late 13th century. A modern reconstruction.

out in colour, but there was never much interior detail, and none with anything like the refinement and accuracy that distinguished the coasts (Fig. 11.1). Some charts were richly decorated, but these are usually regarded as presentation copies. Charts meant for use at sea were sometimes mounted between wooden boards that could open like a book.

Much historical research has been inspired by portolans. One almost unanimous conclusion, at first sight hardly a surprise, has been that the larger specimens were probably made by piecing together operations of more localised extent.[3] From our point of view this process counts as compilation rather than surveying and would belong to a later chapter if it were possible to say much about it. The origin of the smaller charts remains uncertain. (And even they were not so very small.) Circumstantial evidence on the subject is hard to interpret. Portolans seem to make a sudden appearance, with even the earliest examples managing to accommodate the whole of the Mediterranean coast. Other parts of the portolan domain had been incorporated by the early fourteenth century, the Black Sea first and then the Atlantic margins of North Africa and non-Scandinavian Europe.[4] The whole of this area was united in a mesh of criss-crossing straight lines that radiated from a number of regularly spaced points, often being made more easily distinguishable by variations in their colour: these were what Gemma Frisius called 'nautical lines for steering ships' and what maritime historians call rhumb lines (Figs 2.8, 11.2, 11.8).[5] In fact, to be 'all marked with lines' was a characteristic feature of European sea charts for several centuries.[6]

Beneath the portolans' appearance of consistency were hidden a number of significant variations. The Atlantic coasts showed peculiarities in scale, projection and accuracy that suggest a different survey method from the Mediterranean, as well as a different period of origin.[7] It has been speculated for instance that position-fixing may have been done astronomically on the oceanic margins from an early stage, a theory supported by those charts that featured latitudinal scales referring explicitly to the Atlantic but not to the Mediterranean.[8] It has also been suggested that different miles were used inside and outside the Mediterranean, causing portolan scale-ratios to be smaller (by about 18 per cent) in the Atlantic than elsewhere.[9] In a broad view these are complications that many historians have preferred to ignore. The important point, we can agree, is that lengths and breadths greater than those of any contemporary European state were being mapped with unprecedented accuracy.

The Mediterranean and Black Seas were no more frequented by medieval shipping than the coastal waters of north-west Europe, and no more subject to the kind of governmental power that was capable of organising a major survey. The distinguishing character of both seas was physical rather than political or economic. Both were enclosed areas of manageable north-south extent. Shoals and sandbanks were not a serious danger and (in contrast to the Baltic Sea) soundings played little

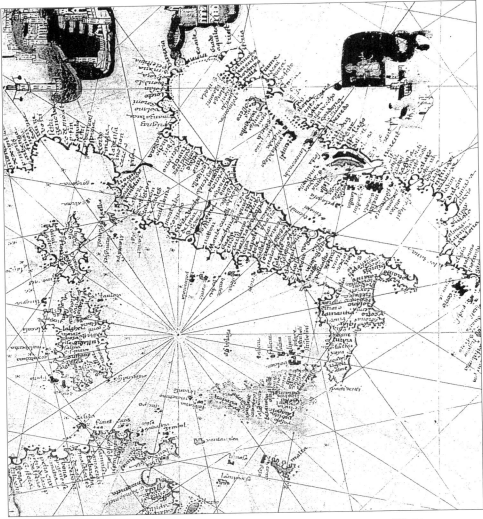

11.2 Detail, portolan chart. Matteo Prunes, MS, 1559. Library of Congress, Washington.

part in position-fixing.[10] Winds and visibility generally favoured the navigator, and observation was made easier by the prevalence of high land near the coast. Maps were not a necessity in these congenial waters, and navigation appears to have flourished without their help for many centuries. The fact remains that charts did come to seem desirable – eventually – and we must therefore seek some altered circumstance that brought about this change of attitude.

Under scrutiny a typical portolan coastline appears as a schematic succession of short straight lines and small semicircles. Coasts were sometimes broken off at river mouths as if from uncertainty as to how much of an estuary should be included (Fig. 11.2). Seen from a greater distance the portolan outlines have been described as 'uncannily' accurate,[11] so much more so than other contemporary maps that we can surely dismiss the possibility of their having been surveyed from the landward side. They were certainly too good to be interpreted as any kind of 'mental' construct. In this respect they compared favourably with Ptolemy's *Geography*. Another striking difference from Ptolemy is in respect of latitude and longitude. As a system of reference, geographical coordinates were well known in the middle ages,[12] and a modern map-reader might feel tempted to regard some of the lines in a portolan rhumb network as parallels and meridians, though without numerical annotations such ideas must remain open to doubt. Given the chart-makers' desire for accuracy (and given their readiness to spend time drawing geometrical patterns), one would expect to find degrees and minutes spelt out if any had been supplied by astronomical observation. It is not only modern map historians who have remarked upon this point. In a 1630 sea atlas by Joao Teixeira one sheet called itself 'Chart of the Mediterranean Sea by latitudes', another (with a different shape) 'Levant by sailing routes'.[13]

Latitudes and loxodromes

So what techniques were left if celestial observation is ruled out? The Mediterranean was large enough for a cartographer's choice of projection to make a measurable difference to the shape of his map, and the portolans were accurate enough to permit at least some inferences about their mathematical basis. In the absence of a graticule it seems unlikely that they were designed with any particular projection in mind. The point is rather that, as we have already seen, some systems of surveying tended to produce certain 'accidental' projections when the results were laid down by whatever seemed the least laborious method. For instance a survey based entirely on latitudes and longitudes with results given in degrees would appear as a plat carrée. Perhaps we can begin by ruling out any projection not likely to have been generated accidentally.

At first sight, if none of the lines on a chart are meridians or parallels they can do nothing to identify its projection. In theory this deficiency might be made good with the help of scale-statements specifying linear distances. The portolan charts were the earliest European maps to include such statements, typically in the form of a divided bar or ribbon and usually without specifying any units (Fig. 11.3). Portolan measures of distance have accordingly attracted much historical attention, some of it directed to finding a pre-medieval origin for them – not that this would necessarily prove anything about the antiquity of the charts themselves.[14] Though often roughly drawn, the scales were meant to be treated with respect: some charts show signs of having been tightly laced across a wooden board to stop the vellum from shrinking.[15] Sometimes there were two similar scale bars, one vertical and one horizontal, implying (if they were representations of linear distance) that the scale

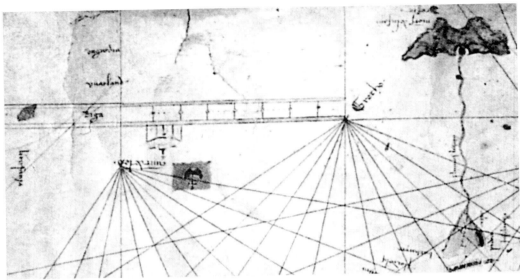

11.3 Portolan scale-line. Graziosa Benincasa, Western Europe, Mediterranean and Black Sea, 1469. British Library, Add. MS 31315, ff. 4v–5.

was the same in both directions, presumably throughout the map.[16] This ought to eliminate any projection with a square graticule. Equal perpendicular scales are compatible with a zenithal equidistant projection, which would be created 'accidentally' by the kind of survey described in an earlier chapter as radial. But a radial maritime survey is so unlikely – where would be the point of origin? – that it seems better to interpret these scale lines in a different way. More probably they expressed an assumption that the earth's curvature could be ignored, which unfortunately deprives them of diagnostic significance for our present purpose. So it may be more advisable to identify projections by comparing charted geographical features with their counterparts on a modern map whose projection is known.

It has long been thought that Mediterranean portolans conformed to a projection which today is familiar to all geographers but which in the thirteenth and fourteenth centuries was almost certainly unknown.[17] That projection is Mercator's, in which the scale increases outwards from a selected great circle according to a definite formula. The historian's problem is whether this formula fits the portolan charts. It must be confessed that in a narrow latitudinal zone like the Mediterranean the gap between distances on the rival projections may be no more than about ten per cent, but several recent writers have voted for Mercator after taking careful measurements (Fig. 11.4).[18]

The Mercator formula ensured that any straight line on this projection must trace a constant azimuth on the globe, crossing every meridian at the same angle to form a loxodrome. If the straight lines on the charts were loxodromes, then the fundamental portolan survey process seems likely to have entailed following a line

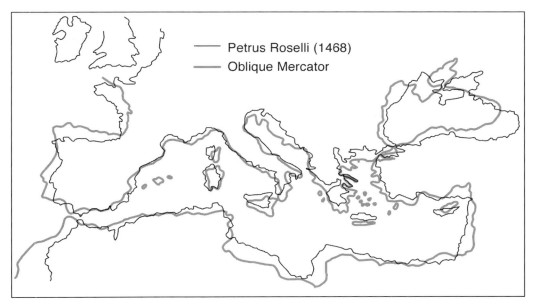

11.4 Mediterranean and Black Seas, from a portolan chart and on a modern Mercator's projection. W.R. Tobler, 'Medieval distortions: the projections of ancient maps', *Annals of the Association of American Geographers*, lvi (1966), p. 358.

of invariant direction from one landmark to another. (There is no need to assume that the alignments thus surveyed were identical with those drawn across the finished map.) On any other hypothesis it would surely be an improbable coincidence for loxodromes to be charted as straight.

With rare exceptions, a loxodrome was not the same as the side of a spherical triangle, so the lines on a portolan chart could not be the result of triangulation or trilateration. This meant that the only other widely-used survey method available to the portolan surveyor was the traverse, whose shape would necessarily depend on the determination of distances as well as bearings. Traverses were more reliable if closed than open: hence, on this theory, the greater accuracy of the portolan charts in the basins of the Mediterranean and Black Seas than on the open coasts of the Atlantic. But even within a closed traverse there may have been unsupported branch lines that served the same purpose as an offset measurement in land surveying. This possibility gains support from an exception that 'proves the rule': some charts made the east side of Italy unnaturally straight, perhaps because a traverse leg had been mapped by mistake as if it were a coast. Another advantage of the traverse theory is in allowing portolan charts to have been based on non-graphic sailing directions, such directions being themselves equivalent to a land surveyor's field book.[19]

Traversing involved the determination of numerous linear distances. The only measures specified in this case, and that infrequently, were miles, a unit that varied notoriously from one country to another. Historians have put much scholarly effort into identifying one or more portolan miles, partly by means of cartometric analysis, but without throwing much light on contemporary methods of marine surveying.[20] It might be argued that since fourteenth-century seamen had no means of taking measurements across the water, their leagues and miles could only have been calculated from astronomical data, but sixteenth-century seamen could quickly judge to the nearest league how far they lay from a visible landmark (at more than ten leagues it would probably be out of sight) and there is no reason why their medieval predecessors should have been any less skilful.[21] As on land, the reliability of such estimates for any given route was doubtless proportional to the amount of traffic that had been using it. Furthermore many Mediterranean ships were galleys, which could travel more easily in a straight line than sailing vessels, a fact verifiable by observing the ship's wake, though we may wonder how many mariners ever measured their progress by counting oar strokes.[22]

Chart and compass

A traverse also depended on bearings. North was given by the pole star, south by the noonday sun, and in the Mediterranean Sea winds bringing distinctive weather might be known to blow from a certain approximate direction. The surveyor's problem was to measure the angle between these directions and the course of his ship, and here the main point at issue is the role of the mariner's compass. The connection of magnet and chart was certainly noted at an early, albeit not quite medieval, date. In the East Indies a Malay ship's captain 'carried the compass with the magnet after our manner', and also consulted a chart marked with lines.[23] On a sweeping view of history, chart and compass were introduced at about the same period, and the first European references to the magnetic needle described it as an instrument habitually used by sailors.[24] Compass bearings made it safer to navigate over long distances with no sight of land, and the laying-off of such courses was obviously easier with a chart. Again, if the compass was mainly a seaman's tool its genetic association with sea charts may be strengthened by the comparative torpidity of European land-based map-making in the thirteenth and fourteenth centuries. If chart makers had used any other method than the compass we should have to explain why the same method was not also applied to topographical mapping in countries like Spain and Italy.

No doubt the magnetic hypothesis for portolan charts can be advocated with varying degrees of tenacity: compass bearings could after all have been combined with other kinds of azimuth in any proportion the historian likes to choose. A 'strong' interpretation faces several difficulties. In one sense the compass made its historical debut too early. Its reported use in c.1180 preceded the first mention of portolan charts by nearly a hundred years.[25] In its mature form, however, the compass came too late, for it was not until the early fourteenth century that angles are known to have been easily read from a needle rotating on a graduated card.[26] However, failure to record an innovation is hardly a proof of its non-existence.

Another clue to the use of the compass is the relation of the charts themselves to magnetic north. The earliest examples had no definable orientation. Their rhumb lines were uncaptioned, and within a single map the placenames and symbols might be aligned in many different directions. The 'top' of a directionally non-committal map could perhaps be defined as the side furthest from the reader when he unrolled the parchment by pulling with his right hand on the 'neck' or tab that projected from many such sheets. (The choice could depend on which end of the Mediterranean seemed more important at the time – east during the Crusades, west in the age of Columbus.) Not until 1375 did rhumb lines begin to radiate from compass roses in which north, south, east and west were identified either by words or by conventions that were universally understood.[27] Of charts without compass roses, the most one can say is that they were usually drawn on more or less rectangular skins of vellum, that two of the rhumb lines can usually be identified as more or less parallel to the sheet edges, and that it may be clear without taking measurements that these lines run more or less from north to south and west to east.

On closer inspection, the matter is not so simple. In many charts the Nile delta appeared due 'east' of Gibraltar, whereas it actually lay further south by about five degrees of latitude, placing the 'top' of the charts not towards the geographical pole but somewhere closer to medieval magnetic north. The degree of rotation differs from one map historian to another, but that is only to be expected from the imperfections of contemporary instruments and the variability of modern research methods. The general consensus favours a twist of about nine degrees, encouragingly similar to the magnetic declination of c.1300 as reconstructed by geophysicists from the lava flows of Mount Etna.[28] But the uncertainty of these figures would make it hard to test any particular theory by reference to temporal and spatial variations in terrestrial magnetism. One can only say that there is apparently no sign of any systematic increase or decrease in this phenomenon between the earlier and later charts, a fact consistent with the hypothesis that an otherwise unrecorded archetypical portolan survey was made at an early date, perhaps around 1270, and afterwards repeatedly copied.[29] Then there is the possibility of synchronic variation from one part of the Mediterranean to another, which might be a means of verifying a composite origin for the charts.

The main issue here, however, is whether the 'portolan twist' can be explained without reference to terrestrial magnetism. It has been suggested, for instance, that given a more or less rectangular area with its long axis running west to east, any trilateral survey must necessarily appear twisted when laid down on a plane surface.[30] But which end of the map will be turned upwards? The answer seems to depend on whether the draughtsman has been working from west to east, or vice versa, or from the middle outwards. Perhaps plotting was more likely to proceed from left to right than vice versa, thus pushing the meridians of a north-pointing map in an anticlockwise direction. All the same, chart-makers without a compass would be expected to find some way of checking geographical north at the 'far' end of such a survey, and in general the portolan twist must be allowed to carry some weight as evidence in favour of compass surveying.

From sea to ocean

Outside the Mediterranean and Black Seas the higher cost of mounting expeditions put commercial profit ahead of scientific progress as a motive for geographical research. At the same time the research itself had suddenly become more difficult. Europeans faced the world's oceans without the 'pre-cartographic' knowledge of physical phenomena that had been so useful to them on the portolan sea lanes. Coastlines were no longer disposed in basins of convenient size, so ships' courses did not readily assume a triangular or polygonal form and traverses could not be checked against themselves. Winds, currents and degrees of visibility were unpredictable. Variations in magnetic north were wider and apparently less regular. The scope for combining terrestrial and maritime surveys was limited by the danger of landing among unfriendly inhabitants: even in the heyday of the 'noble savage' a seasoned explorer could advise navigators leaving Europe to anticipate a hostile reception.[31] Mathematically, long distances made life more more difficult for both authors and users of charts as the earth's curvature became too pronounced to be left out of consideration. More importantly, on a trans-oceanic voyage supplies of food and fresh water were bound to be uncertain and there was more risk of losing crew-members through accident or disease. Jacques Cartier in the St Lawrence estuary took pride in not 'losing a day nor an hour'.[32] Other explorers too were usually in a hurry, and at sea they could often hope to gratify this impulse: with fair winds an eighteenth-century sailing vessel might cover a degree and a half of latitude in a day.

In short, among the navigators of Columbus's time oceanic charting was a novel experience, a fact with important practical implications for the future availability of historical data. Most aspects of portolan production and use had long since become a matter of routine, not worth spelling out for posterity. Many accounts of early voyages outside the Mediterranean were by contrast carefully preserved, sometimes in literary narratives published not long after the event by sympathetic editors but often as daily records kept by the ship's officers. For this reason alone the historian has the sense of entering a different era after 1492. Unlike land travellers and coastal pilots, the ocean navigator needed maps. He would take as many with him as he could get, passing judgement on their merits and drawing attention to their errors. One captain wrote of examining 'each man's chart … and found them to agree in

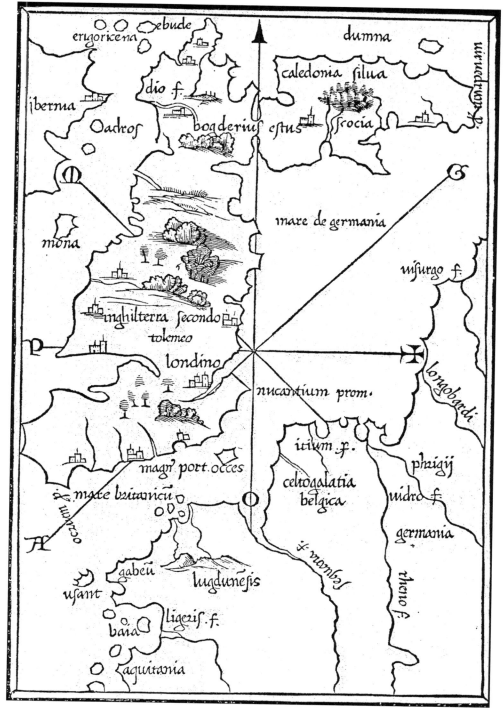

11.5 Woodcut island map, Britain, *Libro di Benedetto Bordone nel qua si ragiona de tutte l'isole del mondo* (Venice, 1528).

height [of the sun], but some contrary'.[33] It is a moot point whether new charts were plotted by ships' officers *en voyage* or laid down retrospectively from their log books. Somehow, at any rate, the world's maritime cartography was revolutionised.

How quickly and how widely the new facts became known was another matter.[34] Repositories of charts were established by governments and trading organisations, but without being made accessible to the general public. A sea captain's employers would want his charts returned to them, not laid open to potential enemies or competitors. According to Humphrey Gilbert, Spanish or Portuguese pilots giving information about politically sensitive discoveries were liable to the death penalty.[35] Nor were the supposedly libertarian English immune: Martin Frobisher was forbidden in 1578 to publish or 'give out to others' any charts made during his quest for the north-west passage.[36] Henry Hudson's mutinous crew escaped hanging because at least they had behaved properly in this respect. Even in a more enlightened age the French explorer Louis Antoine Bougainville told how in Batavia a pilot breaking the rule of confidentiality could be whipped, branded and exiled to a distant island.[37] Similar attitudes were encountered by British naval officers in Japan.[38] A more belligerent manifestation of security-mindedness was the deliberate falsifying of charts in order to lead an enemy astray.[39]

The result of all this secrecy was for marine cartography to fall into national schools, different in outline and content as well as in the language of their placenames. Thus an English author could complain that the Spaniards had 'spread numerous islands and shoals over their charts of the Pacific Ocean which cannot be found by the navigators from England'.[40] Some national differences inevitably persisted when governments began to assume a more active role in commissioning and monitoring hydrographic research, in establishing permanent agencies first for the preservation and then the actual construction of charts on a comprehensive plan, and finally in marketing the results. Early official hydrographic services included those of France (1720), Denmark (1784), Britain (1795) and Spain (1800).[41] Reluctance to publish lasted longer: it was only in 1823 that British charts became available to the people whose taxes paid for them. In this more peaceful century the surveys of different nationalities became more alike in substance if not in style.

Meanwhile private publishers had been contributing to maritime cartography from a relatively early stage in the history of printing. Woodcut maps of islands appeared in books of sailing directions before the middle of the sixteenth century (Fig. 11.5). Later one or two charts in the portolan mode were cut or engraved in Venice.[42] However, it was in northern Europe that chart-printing made most rapid progress and here it was soon to be combined with the practice of publishing maps in book form. The first atlas of charts was Lucas Janszoon Waghenaer's *De Spieghel der Zeevaert*, published at Leyden in 1584 (Fig. 11.6) and surely owing much in form if not content to Ortelius's already famous land atlas. Dutch pre-eminence in this new medium became still more obvious when the first book of printed charts to

11.6 Title page, *The mariners mirrour* (London, 1588), English translation of Lucas Jansz. Waghenaer, *Spieghel der zeevaert* (Amsterdam, 1584).

cover the Mediterranean was produced by William Barents in the North Sea port of Amsterdam. In fact a bound volume was not the most user-friendly vehicle for distributing charts. From the producer's standpoint, however, a book would always be better business than a sheet. Waghenaer's successors in the Netherlands were Willem Blaeu, Jacob Colom, Anthonie Jacobsz, Gerard van Keulen, Pieter Goos and Hendrik Doncker.[43] Subsequently other countries added their own sea atlases to the pile, including the *Neptune françois* in France and the works of John Seller and Greenvile Collins in England.

Maritime surveys of the sixteenth century

How, in the post-medieval era, could seamen find their way across an ocean? It might be thought that bearing and distance alone could establish the position of a ship on the globe or on any chart in which north-south lines ran parallel to each other and perpendicular to west-east lines. In practice reckonings of distance, whether by log or estimation, were seriously affected by winds and currents. For instance, the reason why the north coast of the Yucatan peninsula appeared excessively long on early charts was that their authors' ships had been making such slow progress against the trade winds.[44] Hence the importance of geographical coordinates: given the length of a degree in leagues, latitude calculated from an estimate of linear distance could be set against latitude from the sky. Columbus was reputedly the first explorer to take this opportunity.[45] The log books of his sixteenth-century successors were crammed with astronomical determinations, normally from the sun, sometimes from the pole star, in most cases with hand-held quadrants, astrolabes or cross-staves. This was all very different from the comparative paucity of references to latitude, let alone longitude, in reports written by land travellers at the same period – the main exception, as we have seen, being on journeys through the kind of desert environment in which the land could be as empty as the ocean.[46] All in all it is not surprising that scales of latitude were the most innovative feature to appear on sea charts from the age of discovery.[47] As if to emphasise this break with the past, the new degree-divisions were often left completely unrelated to the network of old-style rhumb lines drawn on the same map.[48] When margins were divided for longitude the same phenomenon was observable: vertical rhumb lines would often meet the border of a map in the middle of a degree-division. This remained a feature of many reputable charts well into the seventeenth century.[49]

At that time it was unusual for a pilot beyond sight of land to know exactly where he was, especially in a longitudinal sense. In the typical renaissance navigator's log book, east-west distances were expressed as leagues and where longitude as such was quantified this seems likely to have been done 'by account' – by some non-astronomical means, in other words. Theoretically the difference of longitude

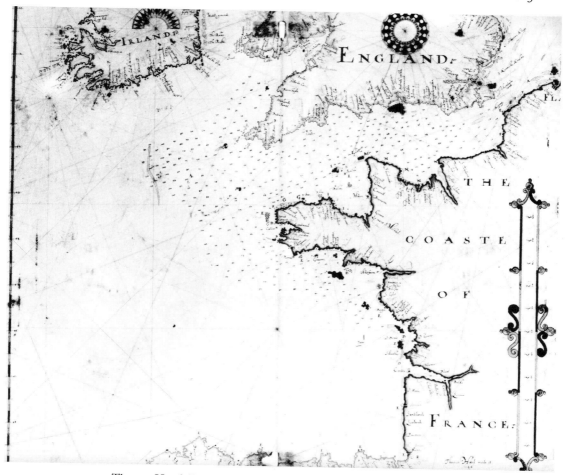

11.7 Thomas Hood, Bay of Biscay and English Channel, MS, 1596. See note 52.

between any two points was obtainable from their latitudes and from the compass bearing of one from the other. But of course latitudes could not be observed in bad weather and bearings were not always reliable. Given the even greater uncertainty of distance-reckoning it was easy to see the need for some independent method of finding longitude. The tone of the following passage from 1648 is as instructive as its content:

> The pilot was a Portuguese … who no doubt understood his business; he daily computed how many leagues we ran, according to his judgement, for in sailing from east to west there is no certain rule. This is a subject has employed many, and does at present, to find the fixt longitude, but I believe to no purpose. Some who slept more than the pilot, would have it we had run more leagues, and said, we were past the Islands of Thieves, now called Marianas: there was much debate, and wagers laid: the pilot was nettled, and swore they should not be seen until next Sunday morning. Everybody looked upon it as a piece of Portuguese positiveness. Trinity Sunday came, at sun-rising, he sent up to the round-top, and said, 'This day before eight of the clock we shall discover the islands'. It was very strange; about half an hour after, he that was at the top-mast cried out, 'Land ahead. Land'. They all stood amazed, and not without cause.[50]

Despite these difficulties over 'positiveness', most world charts carried a scale of longitudes from the mid sixteenth century onwards, though in regional coverage on larger scales it was often considered not worthwhile to venture an opinion on so contentious a subject.

This brings us to the charting of minor detail. In 1588 an English expedition to northern Siberia was directed to map 'the proportion, and biting of the land, as well the lying out of the points, and headlands'[51] but with latitude and longitude still uncertain there were limits to the refinement appropriate for a faraway coastline. Once the navigator's position on the globe was known, the bearing of any coastal feature visible from his ship would be given by the compass and the corresponding distance would probably be estimated. If a navigator could see a small island where the chart said it should be seen, he would be reasonably well content: the island's exact shape could be ignored. And here we must qualify the impression of extreme accuracy given by the portolan charts, much of which comes from contrasting them with non-maritime maps of the same period. This perception may have obscured another fact that has never been given much emphasis by historians: when scales were enlarged, and survey areas reduced, sixteenth-century mapping in the portolan tradition could be surprisingly unrealistic. For instance, compare the squashed appearance of Cornwall (ignoring different orientations of the whole peninsula) in the charts of Thomas Hood[52] with the more modern outline achieved on shore by Christopher Saxton (Fig. 11.7).

Cook and his contemporaries

Even after 1700 many European charts of African, Asian and American coasts continued to resemble portolans in various ways[53] and we may note in this connection that as late as 1739 the most famous English school of portolan-style chart makers was still not quite defunct.[54] No doubt the apparent strength of the medieval tradition owed something to the sheer rapidity with which so many of the world's coastlines were being mapped in the age of Columbus and Vasco da Gama (Fig. 11.8). Behind this facade of consistency, however, the charts of the eighteenth century were changing in almost every respect. One noteworthy development was a new awareness of the technical difficulty of maritime surveying, a point memorably expressed in James Cook's complaint that few seamen of his time were able to draw a chart or sketch.[55] The change of attitude showed itself in the advent of a hydrographic instructional literature as exemplified by John Roberts's *The elements of navigation* (1764), Alexander Dalrymple's *Essays on nautical surveying* (1771) and Murdoch Mackenzie's *Treatise of maritim surveying* (1774). Most practical surveyors, one must admit, remained as uncommunicative at sea as on land,[56] so it is only an assumption that any these books told the truth about what was actually going on.[57]

11.8 Portolan features outside the Mediterranean. Diogo Homem, chart of South Atlantic, 1558. British Library, Add. MS 5415A, no. 7.

The main technical improvements of the period have already been considered in earlier chapters. These changes naturally affected the form of the surveyor's final product. Quality deserved perpetuation in print, and fine cartographic detail could best be rendered by engraving on copper. More accurate surveying encouraged a demand for cartographic assistance from navigators who might previously have done without it, so that in the mid-eighteenth century it had become common for charts to be published of quite small areas like the Downs off east Kent or Waterford Harbour and Kenmare Bay in Ireland. Historians have written surprisingly little about the choice of scale for sea charts, but it seems clear that greater accuracy worked in favour of enlargement: a chart of the western Mediterranean in the early 1600s at something less than 1:4,000,000 might well in the following century be more than four times that size. On the whole there seems to have been plenty of room for variation. That was certainly Alexander Dalrymple's impression of current practice in 1779, though his own preference was for a rigid structure, with larger and smaller scales maintaining a uniform relationship.[58] Half a century later,

Francis Beaufort was moving in the same direction. 'With regard to scale', he told a subordinate in 1831, 'I do not wish to fetter you, nor do I care for any uniformity'. Later he became rather more positive:[59]

> For straight and sandy shores or when a steep but forbidding coast denies all access, or whenever the dreary nature of the soil proves that no inhabitant could probably exist, you will … adopt a scale of one inch to a mile; but when headlands and receding bays show that shelter may be obtained, or where the soundings project to a considerable distance, or where from a teeming population it may be evident that some intercourse must take place with the sea, then the scale should be in no case less than two inches and generally three inches to the mile.

With outlines accurate enough for exact courses to be laid down by ruler and compasses, publication in single large sheets was clearly more practicable than on successive pages of a bound volume.[60] (Nor for that matter did a division into separate rectangular panels offer any particular advantage.) Another consequence of larger scales was that charts could become less dependent for their intelligibility on supplementary written sailing directions.[61] The sea atlas responded to these pressures first by narrowing its territorial range from the world as a whole to the coasts of a single country and later, in the nineteenth century, by disappearing altogether.[62]

A further consequence of greater accuracy was a sharpened distinction between different kinds of chart. Of course the sea offered none of the ready-made categories – towns, forts, counties, kingdoms and so on – that facilitate the classification of land maps. But in 1748 Lewis Morris could still suggest a three-fold division of the marine surveyor's subject matter: first, large areas ranging in size from the South Seas to the English Channel; secondly, the coast of a particular country such as Britain or France; and finally an individual harbour, bay or road.[63] Between the first two types it seems difficult to find a clean break, but the distinction between harbour charts and others is worth pursuing. It was already latent in the work of Lucas Waghenaer, who mapped inlets and estuaries not individually but at a larger scale than the more open coasts on either side of them (Fig. 11.9). As a less misleading alternative, an important harbour might be segregated at a suitable scale as an inset to a larger chart, in the manner of the town plans accompanying English county maps.[64] On the other hand some harbour charts were getting a sheet to themselves before the middle of the sixteenth century, the first English example apparently being Dover in 1540.[65] Other candidates for special treatment were small islands, archipelagoes, lakes and straits (Fig. 11.10). In all this, the clearest ultimate basis for classification can be easily stated: when did the surveyor leave his ship? As early as 1497 Vasco da Gama, approaching the Cape of Good Hope, unloaded an exceptionally large instrument to take observations at St Helena Bay.[66] He is the first hydrographer known to have worked on shore.

The need for land-based measurements became more evident in the seventeenth century. When Greenvile Collins began surveying the harbours of England in 1681

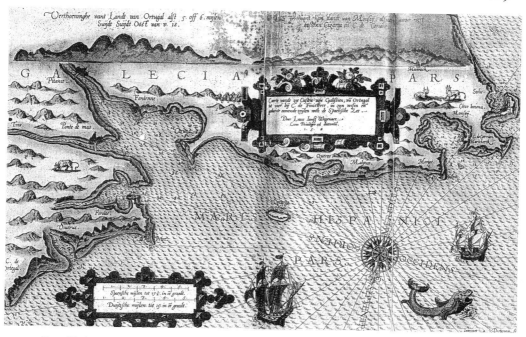

11.9 Cape Finistere, north-west Spain. Lucas Jansz. Waghenaer, *Spieghel der zeevaert* (Amsterdam, 1584).

he was specifically told to do it with a chain, a proviso unexpected enough to catch the attention of a layman like John Evelyn.[67] Another importation from maps to charts was the idea of triangulation and base measurement, recommended to seamen not long afterwards by Edmond Halley. Murdoch Mackenzie in 1749 is the first British surveyor recorded as taking this advice with a 3.75-mile base of frozen lake-surface in the Orkneys. The difference from normal land surveying was that the chart-maker's base had to be near the seacoast. This could make the actual measurement an unfamiliar experience. In Queen Charlotte's Sound, William Wales and William Bayley chose to fire a gun, another expedient originally proposed by Halley.[68] Joseph F.W. Des Barres actually measured through the water with a sounding line laid out on the sea bed, though without managing to explain his technique in comprehensible language.[69] A still less probable method was for rockets to be detonated at a known altitude and an angle of slope taken by theodolite to the point of explosion.[70] All of which makes it strange to find so few references to a ship's mast functioning as the linear standard in a tacheometric survey as suggested by Dalrymple.[71]

From the base, bearings would be taken to as many landmarks as could provide well-conditioned triangles, including steeples and other prominent buildings together with rocks, cliffs and promontories. The advantage of surveying on land was that the observer could move quickly to new viewpoints, and his instruments could be securely planted on their stations. Plane tabling, impossible on shipboard,

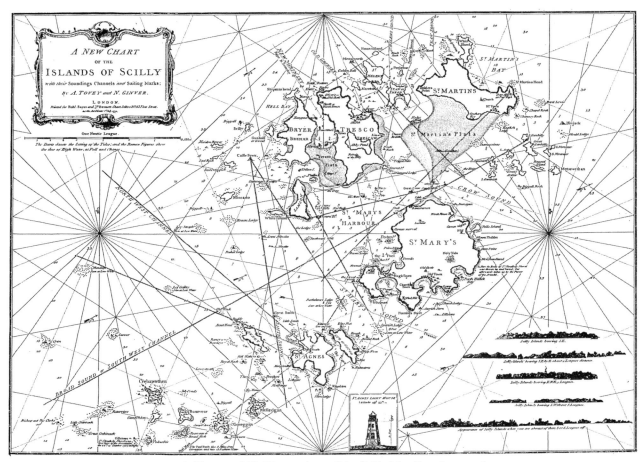

11.10 A. Tovey and N. Ginver, chart of the Scilly Isles (London, 1779).

could also be used for filling in coastal detail: one remembers the lessons on this subject that Cook requested and received from the military surveyor Samuel Holland.[72] The trigonometrical details of such a land-based coastal control system were sometimes published in the margin of the resulting map, as had been done with several county surveys; an example was Robert Laurie and James Whittle's chart of west Pembrokeshire in 1812.[73] Networks of this kind would be extended as far as possible out to sea by intersection, using buoys and other temporary marks on sandbanks or even in open water. The rule with these more detailed surveys, then, was to spend as little time as possible afloat, though of course not all requisite features were observable from the land: soundings had to be taken over the seabed, at positions fixed by bearings to a convenient coast, and a boat would probably be needed to traverse the boundaries of an individual shoal. Depths were generally the last features to be surveyed. A writer of 1775 refers to 'such of Mr Des Barres' draughts as wait only for the soundings'.[74]

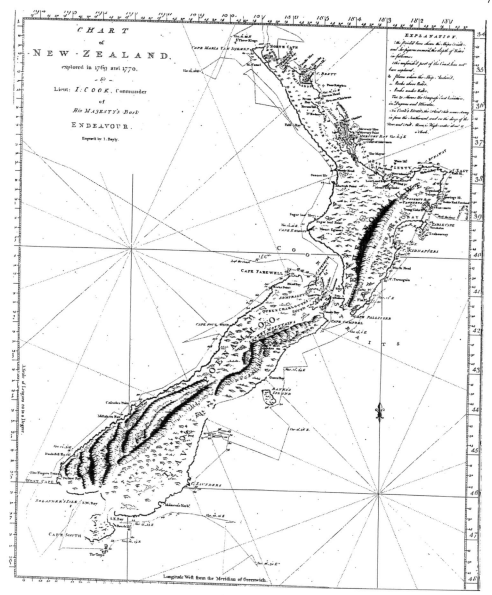

11.11 James Cook, chart of New Zealand (London, 1772).

Might it not have been easier for eighteenth-century maritime surveyors simply to work outwards from the coastline as already shown on a good land map? Perhaps they would have denied that there was any such thing as a good land map: most of them seem to have proceeded on that assumption. Only in the case of a trigonometrical skeleton were chart-makers ready to depend on land surveyors, and then it was best if the land surveyors proved their competence by working for a government department. An example from 1801 was the suggestion by Britain's first naval hydrographer, Alexander Dalrymple, that a new maritime survey of coastal waters should be based on points supplied by General William Roy.[75] Later the favour was reciprocated when Ordnance maps began to draw their off-shore information from Admiralty charts.[76]

Opposed to triangulation was the kind of 'running survey' brilliantly conducted by Cook along the coasts of New Zealand, Australia and Pacific North America. This was in effect a series of well-controlled traverses. The survey vessel would follow the straightest possible course from one 'ship station' to another, measuring the distance by the log line – or perhaps, if two ships were working together, by vertical angles to a masthead.[77] Landmarks would be fixed by horizontal angles, intersecting either on the coast or (with the help of the station pointer) on the ship itself. At intervals, if possible once a day, latitude would be taken from the sun's altitude and longitude either from lunar distances or by chronometer.[78] A problem in running surveys was how far to follow each indentation backward from the coast. One policy was to explore every important-looking inlet in a small boat, its width being estimated visually, its length chronometrically, only the direction being instrumentally measured, and that by the easiest available method. Before turning to rejoin the ship a latitude would if possible be taken from the sun.[79] By these methods Cook might hope to cover a hundred miles of coastline in a week. The results would clearly never be infallible, especially when weather conditions precluded close observation of a coast that it was impracticable to revisit. Among the features in New Zealand that Cook had to leave doubtful were the heads of Tasman Bay and Palliser Bay, the isthmus (in fact non-existent) crossing the Fouveaux Strait, and the channel (also non-existent) separating Banks 'Island' from the mainland (Fig. 11.11).[80] Navigators more pressed for time than Cook were naturally less successful. La Pérouse reported in 1786:

> We sailed too fast along the coast to have seen everything and we are answerable only for the features we determined with precision and the trend of the coast, excluding the contours of the bays and the various opening[s] of which we did not sight the end ... [I]t is quite possible that we have omitted shoals in the space between the coast and our frigates, which we have no reason to suspect.[81]

The chart-maker's text

As already suggested, one effect of the larger scales necessary for more precise measurement was to give the draughtsman more space and so to encourage a widening of his thematic range. Much of the additional information – on landing places, for instance, or sources of fresh water and timber – could be expressed by verbal description. More important map-historically are the chart-maker's graphic devices. Early sixteenth-century sailing directions often featured profiles of the land as seen from an approaching ship. Waghenaer incorporated these into charts by

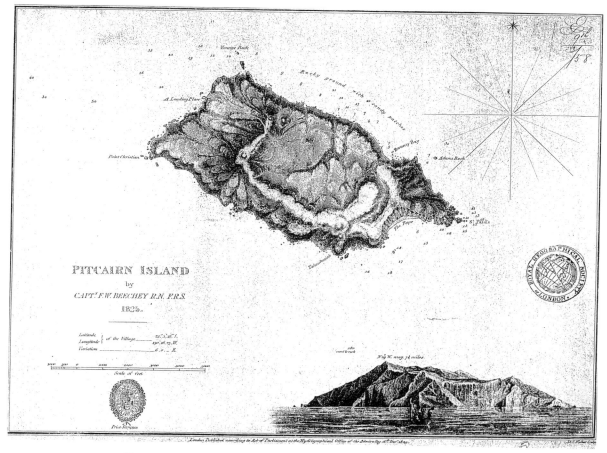

11.12 Chart with coastal view. F.W. Beechey, *Pitcairn Island … 1825* (London, 1829).

placing them directly above the appropriate stretch of coastline. Profiles, being harder to generalise than plans, were most effective on charts of limited extent. At smaller scales they had little to offer, especially on a coast without strong relief. Objects easily recognisable on the map, such as river mouths, would sometimes be impossible to find in the profile, and where two sets of names were given they often failed to correspond. Beginning as no more than silhouettes, these drawings were improved in the late seventeenth century by the addition of identifiable houses and churches on or below the skyline, though chart-makers were warned against including woods and hedges – 'because such things may be cut or felled down, and so your mark is lost'.[82]

By Cook's time the purpose of these marginalia seemed to be changing. Now, as if in half-acknowledgement of defeat, it had become common for published charts to feature only a single profile. As a consolation for becoming less numerous, coastal views developed into impressive specimens of landscape art, highly informative

11.13 From William Heather, chart of Spithead, the Solent and the Isle of Wight (London, 1797). *Inset:* explanatory key transferred from another part of the same chart.

about on-shore topography if not very useful for position-finding (Fig. 11.12).[83] The ultimate remedy would be to map visible relief in the same way as on land: once slope shading had superseded profile symbols (a development to be traced in the next chapter), relief representation on charts was generally of high quality. After all, as a navigational aid hills could do more for the sailor, in default of other visual information, than for the landsman. This argument was taken a stage further when the British naval hydrographer asked for heights on charts to be determined trigonometrically, so that future users could determine the distance from ship to shore by taking an angle of elevation.[84]

As well as hills, Waghenaer had introduced the charting of localised guide-points such as church towers. In the following century the significance of these features was emphasised by straight 'leading lines' drawn between pairs of landmarks observable from vessels entering an estuary or harbour (Figs 7.6, 11.13). By this means the navigator was reminded to take action – either changing or maintaining course – when one chosen object lay directly behind another. In their heyday, the popularity of leading lines was doubtless related to the gradual decline of the portolan-style rhumb. The latter had survived the transition from manuscript to print, different directions now being distinguished in black and white as heavy lines, fine lines and dotted lines instead of by different colours. The printed convention was hardly an improvement aesthetically, and may have discouraged later cartographers from prolonging the network from sea across land in the time-honoured portolan style. In their original Mediterranean habitat, rhumbs could still make a good showing as late as the 1800s.[85] But with accurate protractors they had gradually become superfluous, and indeed positively misleading unless the projection was Mercator's: it was better for directions to be reckoned from a grid of meridians and parallels.

On a water surface rhumbs were in danger of conflicting with other kinds of information. Some of this, such as crosses for localised rocks, was as old as the Catalan Atlas.[86] Stipple for shoals and anchors for anchorages also had a long history. Such conventions were greatly extended by Murdoch Mackenzie. As well as recognising various degrees of exposure for rock, sand, shingle and mud, he had symbols for the safest anchorage, 'where a vessel may stop a tide'; the safest channel; 'an eddy within which there is very little stream of tide'; overfalls, or rough breaking seas; whirlpools; and the direction of movement for flood tides. Also recorded by Mackenzie were beacons or perches, figures for minimum depth in fathoms, initial letters for the character of the sea bottom, and roman numerals for times of high water on the full and change days of the moon.

The first separate printed key to official British charts is dated 1835.[87] In its replacement of some thirty years later there were more than a hundred *in-situ* verbal abbreviations, including 'P.D.' for 'position doubtful' and 'E.D.' for 'existence doubtful', and more than sixty graphic conventions for different kinds of point, line and area.[88] As far as possible these resembled the symbols on contemporary land maps, but self-evidence was perhaps more difficult to achieve at sea: it could certainly present problems with a category like 'edible seaweed'. Colour was not formally admitted to the code, but sometimes appeared on manuscript charts to distinguish different kinds of foreshore.

Soundings in fathoms had been first printed on maps by Waghenaer. They were shown as spot depths in arabic numerals 'which is always to be understood at half flood or ebb', though on this last point Waghenaer was unable to speak for posterity. The sextant and station pointer made possible a great proliferation of soundings, not just along established channels of navigation but wherever they could be taken, so that many modern charts would feature numerous parallel chains of closely-spaced numbers marking the passage of the survey ship, sometimes with more than twenty readings per square inch of map. More will be said about soundings in the next chapter.

The systematic mapping of tidal information over large areas is generally thought to have been pioneered by Edmond Halley. Arrows or 'darts' for the direction of tidal currents were more or less self-explanatory, but there was a new element of convention in the use of roman figures to indicate the number of hours between the moon's northing or southing and the time of high water.[89] A good convention was to make the arrows long or short in proportion to the duration of the current.[90] By the mid-nineteenth century tidal streams at flood and ebb were being distinguished by feathered and plain arrows respectively, and 'black balls' showed the hours at which different speeds were recorded.[91] The philosophy behind these measures was to saturate the chart with information, making it, in the words of Matthew Flinders, 'as full a journal of the voyage as can be conveyed in this form'.[92] So even without portolan-style rhumbs a nineteenth- or twentieth-century chart could look distinctly crowded, with none of the unity that topographical portraiture derives from the visible presence of the landscape.

CHAPTER 12

The true features of the ground

HILLS HAVE COMMANDED attention from all map-making cultures in all historical periods and at all scales. Being visible from a long distance they were a useful aid to self-location, both for the traveller feeling lost and for the surveyor choosing a point of optimum visibility. At closer quarters relief, including that of sea, lake and river beds, formed a serious barrier to movement, and this could give it an important political role in the delimitation of territorial boundaries. Altitudes and gradients also affected the design of roads, canals, dams, irrigation schemes and other public works, besides influencing the disposition of troops and artillery on a battlefield. In agriculture and estate management height and slope gave valuable clues to land quality, as well as being a necessary element in the precise calculation of surface area. Finally, in modern times mountains have been found worth recording as objects of natural beauty.

It was not unusual in early maps for hills to appear as compact symbols not much larger than churches or trees – hardly adequate recognition that the relief of the earth's surface belonged to a distinct semiotic world containing elevational equivalents for every horizontal feature on the map with a few additional characteristics peculiar to itself. The third dimension, like the other two, could be envisaged in either numerical or purely graphic terms. In a quantitative sense distance was measurable by either lines or angles in both horizontal and vertical planes of representation. Horizontally, absolute measurements were multiples of the quantity specified in a scale bar. Vertically they were recorded in standard units above or below a datum plane. In both cases, there was a non-reciprocal relation between yards and degrees: from all the lengths in a horizontal triangulation one could determine the angles, and from all the altitudes on a land surface one could determine the gradients, but neither of these calculations was reversible.

The concept of the datum imparts a mathematical quality to every location on the surface of an exact map. With altitude, as with latitude and longitude, each point has only one correct value – except in the unusual case of a vertical or overhanging slope. What makes these characteristics interesting, as we shall see, is that with the same exceptions they vary not discretely, like the property of having or not having a parish church in a map of villages, for instance, but as a continuum. One more aspect of location must be noted before we pass from abstract to concrete. Some attributes of a place are mutually irreconcilable in the sense that the same point cannot simultaneously be part of a building and part of a corn field. In the landscape, altitude is not like that: it precludes nothing and underlies every-thing. On a sheet of paper there is no such happy state of integration: whatever is being represented, ink excludes ink. Outside the world's deserts, therefore, any general or topographic map that showed relief would be the scene of opposing and perhaps incompatible demands for the same space.

In all this there is one overriding difference between horizontal and vertical. If maps are by definition flat, then local relief must be inherently resistant to cartographic portrayal, like the larger relief of the earth's curvature. For this reason any normal surveying process was designed to ignore – we might almost say abolish – vertical distances, from which it follows that relief representation could never be a process of simple 'mimesis', another reason why hills make trouble for the cartographer.[1]

Hill features: a sideways view

Relief maps began with observers trying to reproduce what they saw, which was either a profile or a shadow. Profiles or silhouettes were so easy to copy, and to understand, that they must have been drawn independently by many different artists at many different periods (Fig. 12.1).[2] As landscape features the profiles visible from any point had a real existence, and could be recorded in linear form on a map of appropriate scale and accuracy, though without a record of precise altitudes this might be a hazardous operation.[3] Some relief forms presented no profile, for instance parts of a concave slope. And where profiles did exist, they were not uniquely characteristic of any topographical feature, but only relative to a particular line of vision, so that the same site might display an indefinite number of them.

Since maps by their nature made no provision for an observer's viewpoint, the silhouette, like the notion of relief in general, was in one sense a concept antipathetic to cartography. In these circumstances the simplest expedient for the map-maker was to choose a supposedly representative profile and draw that. Unfortunately natural forms had no 'privileged' aspects equivalent to the front, side and rear elevations of a rectangular building. Ideally, the right view was the one that did most to distinguish a hill from other hills. Apart from the work involved in identifying the requisite image, this notion assumed that relief could be analysed into separate entities, whose individual forms would dictate where the observer should change his angle of vision. It is hard to imagine any such method of selection being systematically applied across a real landscape, and few students of early maps would hope to identify the place from which even a single hill was actually sketched.[4] At times a cartographer might try to match an unusually distinctive shape, like that of Table Mountain near Capetown,[5] but such verisimilitude is not to be taken for granted: for instance, one looks in vain on pre-nineteenth-century maps of Ireland for the famous cliffed frontage of Ben Bulben

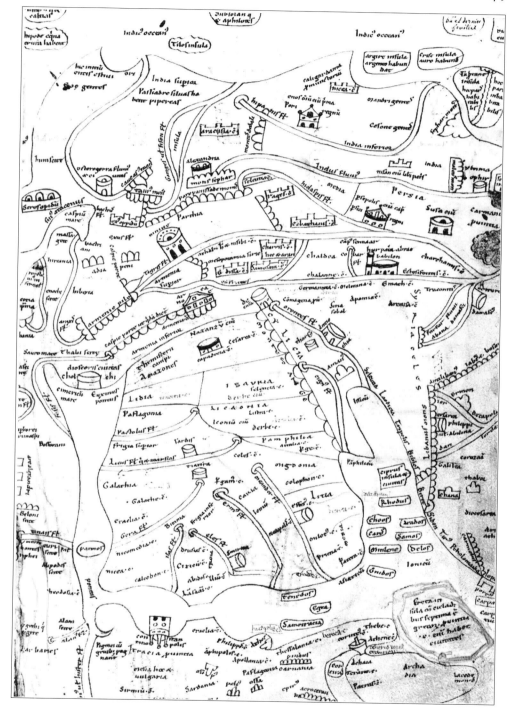

12.1 Early hill profiles. Asia, 12th century, to illustrate the writings of St Jerome. British Library, Add. MS 10049, f. 64r.

in County Sligo. In general, the most likely subject for realistic treatment would be a well defined upland, small enough to fall within one field of vision and surrounded by plains on every side, somewhere like Ayer's Rock in Australia – a rather unusual type of landform as it happens.

There was one case, however, in which the subjectivity of the profile could be disregarded, and that was where slopes lay symmetrically disposed about a vertical axis. A conical or hemispherical hill avoided the sketcher's dilemma by presenting the same appearance from every direction, but since in the real world such hills were unusual, the simplest kind of profile symbolism involved reinventing the landscape to suit the limitations of the artist's medium. Here, as elsewhere in cartography, falsehood could be made acceptable by reclassifying it as convention. When this happened, the individuality of early profile symbols often owed less to their subject-matter than to a vertical equivalent of 'personal curvature'. It then became easy for a map to exaggerate the natural characteristics of the average hill: in early Swiss cartography some profile symbols look more like nuclear explosions than any kind of land form.[6]

Other artists came nearer to flattening nature out. In extreme cases all claims to realism might be abandoned, as in a recommendation of 1698 that hill signs should be 'in shape like a bell'.[7] It is easy to see why such symbols have sometimes been read not as pictures but rather as an unusual kind of word.[8] Many early cartographers seem to have accepted this quasi-verbal interpretation: in copying a predecessor's map they seldom followed the relief features exactly line for line, or even hill for hill. Deliberate miscopying of a predecessor's profile symbols was strikingly illustrated when Jodocus Hondius re-drew John Smith's Virginia in 1618 and again in the English county maps copied by Joan Blaeu from John Speed's *Theatre* (Fig. 12.2).[9] In the work of eighteenth-century cartographers like Emanuel Bowen, Thomas Kitchin and Thomas Conder, identical small hill symbols were often evenly scattered over an entire territory, almost as if simply denoting land as opposed to water (Fig. 12.3). Bowen and Kitchin's *Royal English atlas* (1762), for example, makes Hampshire as hilly as Cumberland. So far as is known, the first serious response elicited by this proclivity was disapproval of Thomas Jefferys's failure to take the Appalachian ridges literally enough when he plagiarised a well-known map of British North America by Lewis Evans.[10]

Refining the profile symbol

Conical hill signs could be varied in a number of ways, realistic or schematic, to make their meanings more definite. Profile lines, like their referents, were either sharp or rounded, high or low, steep or gentle, concave or convex, long or short, complex (that is with some slopes reversed) or simple, multiple or unitary, separate or overlapping, and bounded or unbounded on their lower sides. These were all

12.2 Hill profiles freely interpreted, original scale *c*.1:140,000. *Above*, from John Speed, *The countye of Monmouth* (London, 1610). *Below*, as copied by Joannes Blaeu, *Monumethensis comitatus* (Amsterdam, 1645).

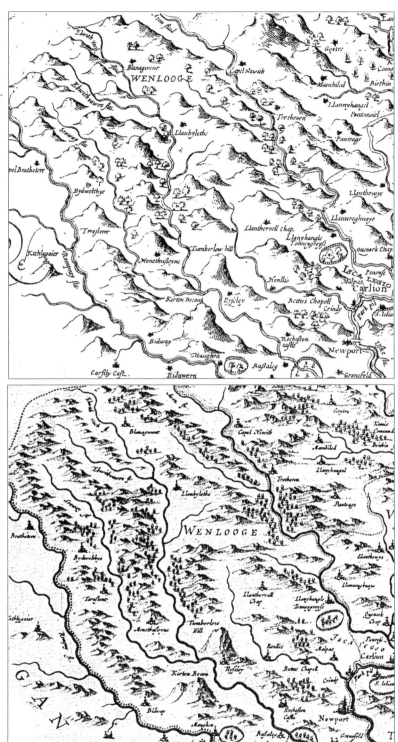

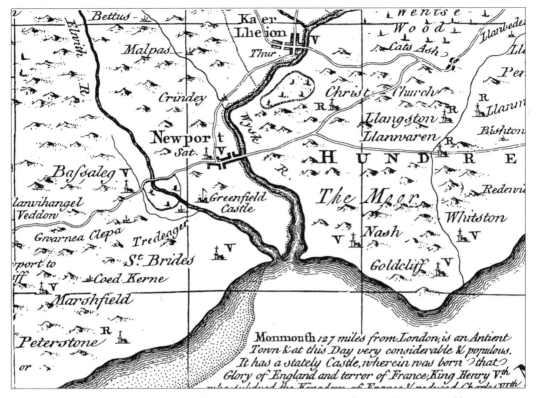

12.3 Standardised conventional hillocks, original scale 1:126,720. Emanuel Bowen and Thomas Kitchin, Monmouthshire in *The royal English atlas* (London, *c.*1763). In fact 'The Moor' and large areas around Nash, Marshfield and Peterstone are almost completely flat.

concessions to visual truth. More thoroughly conventional was the kind of enlarged hill symbol whose size indicated not physical bulk but some less blatant claim to notice. Even within a single map not all hills had to be equally abstract. One might be veridical, others stylised – a semiotic difference perhaps more familiar in the representation of buildings. Apart from physical magnitude, some persuasive reasons for individual treatment were oddity of shape (already noted), familiarity from historical literature, presence of volcanic activity, proximity to the source of a major river, or the possession of a well-known name. If a hill carried a prominent tower, beacon or boundary-marker, it might be the only relief-feature for miles around to catch a cartographer's attention: perhaps this was why in Warwickshire William Smith took notice of Windmill Hill near Alcester while ignoring a nearby escarpment that had no conspicuous buildings on it.[11] A hill might also be made larger because an egocentrically-minded cartographer had chosen to climb it (Fig. 12.4).

Size on the map was not the only weapon in the cartographer's armoury. Another way to individualise profiles was by embellishing their facades with additional detail such as stream courses, natural terraces or rock outcrops. But these

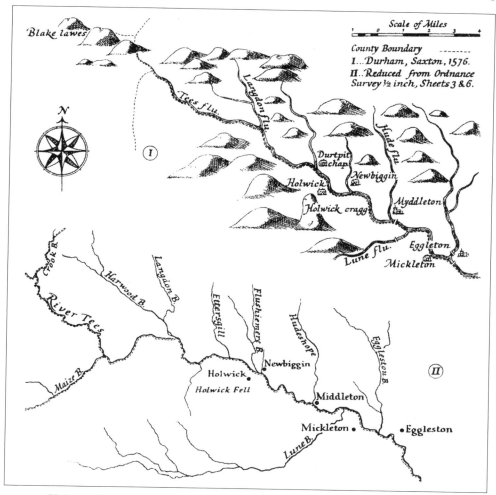

12.4 Holwick Crag (Holwick Fell), Upper Tees valley, north Yorkshire, a likely viewpoint in Christopher Saxton's survey of Co. Durham, 1576. *Above*, Saxton. *Below*, Ordnance Survey. Gordon Manley, 'Saxton's survey of northern England', *Geographical Journal*, lxxxiii (1934), p. 312.

too were often just a method of suggesting three-dimensionality rather than a record of any particular observation. There is however a feature that many hill symbols have in common. Since higher slopes generally occupy less space than lower, profiles differ from other methods of relief representation in helping to define the top and bottom of a map. This rule was sometimes broken, notably by certain early marine charts in which, to facilitate map-reading at sea, the bases of the hills ran parallel to the nearest coastline rather than to the lower edge of the map-sheet.[12]

Like profile-lines, shadows had a real existence on the earth's surface (Fig. 12.5), but since they depended on the changing position of the sun their shapes were just as variable. Apart from its evanescence a shadow was not particularly easy to draw. In practice artists often simulated light and shade by hatching composed of single

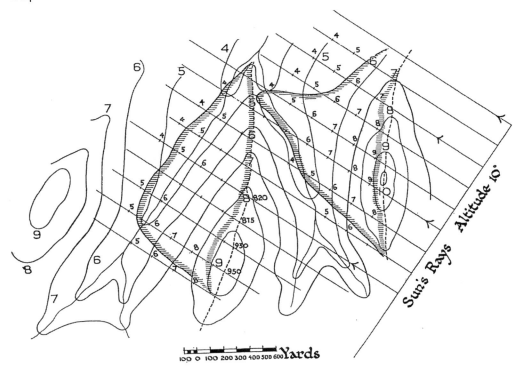

12.5 Modern contour map of sunlight and shadow. Frank Debenham, *Exercises in cartography*
(London, 1937), p. 64.

pen-strokes – a method for which engravers seldom found any alternative (Figs 2.7, 12.6).[13] A widely adopted practice was to illuminate hills from the north-west, not in fact a very common source of sunlight in the terrestrial hemisphere where most of the world's maps happen to have been produced. This custom is easier to explain in the drawing office than in the open air: indoors, the 'sun' shines from the upper left, light being partially excluded on the other two sides of the map-rectangle (or so we are told) by the shoulder and arm of a right-handed cartographer. The importance of this famous convention can be easily exaggerated. In many maps light was assumed to come from several different directions, chosen to give an effect of solidity rather than to depict the land surface at any given time of day.[14] Indeed there might be so much shading on a hill symbol that where its upper edge was faintly drawn the word 'profile' could lose nearly all its significance.[15]

   Shaded or not, profiles undeniably had their limitations as a mode of portraiture. If deployed too enthusiastically they could drive almost everything else off the map (Fig. 12.7).[16] They were unsuited to very mild relief and to certain extensive or asymmetrical landforms like plateaux and escarpments; on Philip Symonson's otherwise excellent late Elizabethan map of Kent, for instance, the scarp of the North Downs had to be misrepresented as a row of sugarloaves. Even where an

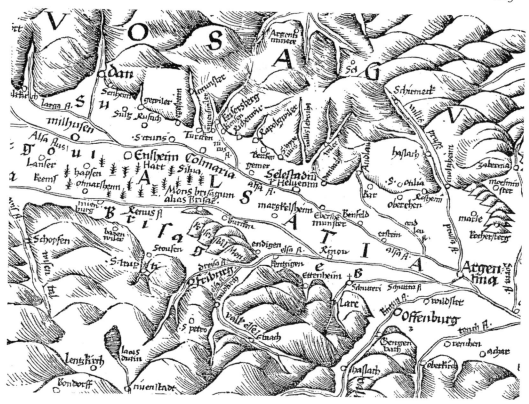

12.6 Hill profiles with illumination from upper left, original scale 1:507,000. Martin Waldseemüller, upper Rhine, woodcut (Strasbourg, 1513).

individual hill was recognisable, readers could be left uncertain whether the apex of the symbol was meant to coincide planimetrically with the summit on the ground, though a cartographer might sometimes hint at an answer to this question by his handling of a genuine hill-top settlement. A profile could also look ineffective at very large scales:[17] here some early map-makers fell back on what was for cartography the even more primitive medium of verbal language, not always used with maximum economy.[18] Profiles had the further disadvantage of obliterating or obscuring other detail. Lastly, they could make it seem impossible to travel downhill in a northerly direction, and perhaps it was no accident that the increasing representation of roads on European regional maps coincided with a search for alternative methods of relief portrayal. In some ways it is surprising that profiles lasted so long. Perhaps it was because in the sixteenth and seventeenth centuries Europe's leading cartographers were the Dutch, who avoided the need to show hills by inhabiting a country that was largely free of them.

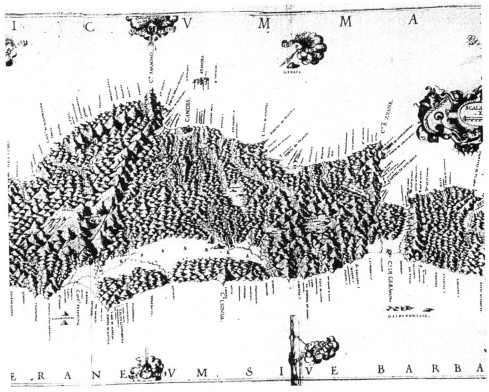

12.7 Closely spaced hill profiles. From Francesco Basilicata, Crete, 1612, MS. See note 16.

Shading without profiles

The treatment that most cartographers eventually adopted for relief was to forget about profile lines and shade the whole extent of each slope in rough proportion to its gradient. This practice had some affinity with shadowing the side of a profile symbol illuminated by an imaginary light, but in the simplest kind of shading the light would shine vertically downwards: less of it was reflected from the steeper slopes, which would accordingly be shown darker on the map. It was a revolutionary idea. Except for dead flat surfaces, every point in the landscape now had some mappable relief. Slope shading therefore obliged the surveyor to take a comprehensive view instead of just picking out the literal and metaphorical high spots. In manuscript maps the shading could be a continuous wash (Fig. 12.8), for which in the long run the most popular colour seems to have been grey. The engraver could achieve a similar effect with bunches of closely spaced parallel lines which, if they ran down the steepest gradient, as eventually became the rule, could be conceived as mapping imaginary stream courses in a region of abnormally high rainfall

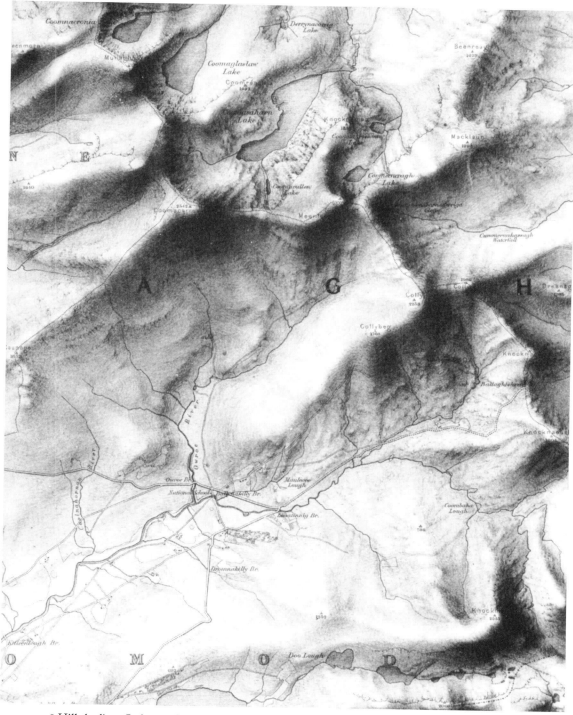

12.8 Hill shading, Ordnance Survey of Ireland, one inch to one mile, sheet 183, MS on printed base, 1872. J.H. Andrews, *A paper landscape* (Oxford, 1975), pl. XVII.

12.9 Printed hill features, Ordnance Survey of Ireland, one inch to one mile, sheet 183, hachured edition, 1882.

(Fig. 12.9).[19] This analogy may explain why such lines in practice came to be drawn 'vertically' rather than obliquely or horizontally, though as we shall see horizontal shading was eventually popularised by the advent of instrumental contours. But having introduced the word 'oblique' we must immediately recognise its ambiguity as between the angle from which a landscape is viewed and the angle from which it seems to be lit. In discussions of fully developed slope-shading the latter sense is more likely, as will soon become clear.

Vertical shade-lines eventually came to be known as 'hachures', though this word was not imported from French into English until the method itself was becoming unfashionable and so never had time to be thoroughly anglicised as 'hatchure'. In manuscript maps the distinction between hachures and area shading was less clear-cut, brushstrokes and penstrokes occasionally being found together in a single artistic composition.[20] Perhaps this explains why the terminology of hill mapping has never been properly systematised. 'Hill shading' or 'slope shading' may include hachures as a special case, but these phrases are sometimes reserved for shading in which no individual lines are perceptible. A neat if arbitrary solution would be for 'slope shading' to include hachures and 'hill shading' to exclude them.

The merit of slope-shading was to accommodate considerable variations in relief. One of its advantages was that individual hachure lines could be drawn to taper downhill, a guide to the direction of slope that was not available with a continuous colour wash. On the other hand there was also a danger of hachures being mistaken for a different kind of line: some engravers responded ill-advisedly by curling them into crinkles; others, more enterprising, experimented with aquatint, mezzotint and other stippled or granular effects to produce an entirely non-linear shade.[21] On very steep slopes, black hachures were liable to overshadow other detail, but in that kind of country there was probably not much other detail anyway. This last generalisation became less plausible when mining and manufacturing settlement began to spread along narrow valleys like those of South Wales, but by that time hachuring was too well established to be readily given up.

In manuscript cartography, on the other hand, wash shading could be transparent enough for lines, point symbols and script to overlie or underlie it without disastrous results. As estate maps, cadastral surveys, town plans and engineering designs became more numerous, profiles lost credibility by their failure to give satisfaction at very large scales. At the other end of the spectrum, as in an atlas map, hachuring has been given a bad name by the much-quoted simile of the 'hairy caterpillar'.[22] The truth is that hachures were potentially as effective as any other method of simplifying relief, but their constituent lines, like shading in general, were undoubtedly among the most difficult features that a cartographer ever had to draw. It was particularly hard to maintain consistency over an area too large to be covered by a single artist. In relief maps of Ireland derived from the Ordnance Survey's hachuring, for example, the south-western mountains never seem quite bold enough. Hachures suffered the additional handicap that fine grooves in a copper printing plate would start to wear out after a comparatively small number of impressions had been taken off.

From elevation to plan

The distinction between large-scale plans and small-scale maps remains important when renaissance hill portraiture is considered in more strictly developmental terms. Cadastral surveys are often overlooked in discussions of relief mapping simply because they so seldom attempted to show physical features in a third dimension. Reference must nevertheless be made to surveys in which, for architectural or archaeological reasons, all the buildings were represented by ground plans. In such cases the same technique might by analogy be applied to hills, especially with gradients too steep to be ignored: a precocious example is the thoroughly modern-style slope shading in Leonardo Bufalini's mid sixteenth-century plan of Rome.[23]

At smaller scales, the historical change from profile lines to planiform shading can be rationalised in various ways. It may for instance be conceived as a process of aggregating profiles seen from different angles. This finds some warrant in the road strip maps of John Ogilby's *Britannia* (1675), where two nearby summits were often projected 'upwards' in opposite directions, with either a hill or a valley indicated (one can hardly say 'simulated') between them. Occasionally a few of Ogilby's shadow lines seem to have escaped from underneath a profile and to be functioning as an independent record of relief – as hachures, in other words.[24] On the whole, however, his dual profiles bear little resemblance to hill shading as later understood (Fig. 12.10) and it was not surprising that Ogilby should revert to upright symbols when he turned his attention from individual roads to complete counties.

More worthy of comment are certain primitive hachure-like lines that seemed to run diagonally down a slope instead of following the steepest path. These can turn up almost anywhere, at first usually in small amounts and often for no particular reason. There are well-developed if inconspicuous examples in Egnatio Danti's map of Perugia dating from 1584.[25] Others by Johannes Andreas Rauhen appear in Joan Blaeu's atlas of 1663.[26] This technique was also illustrated by Ogilby, and by a large number of early eighteenth-century cartographers working to a similar scale.[27] Such proto-hachuring is not easily described in words. Sometimes it looked like the profiling of numerous parallel cross-sections taken through the same hill in the manner of a sliced loaf. Alternatively the lines might represent an oblique view of an upland mass, rather as if hachures were evolving by degrees from profiles as the artist's imaginary angle of vision was gradually increased from horizontal towards vertical.[28] Such interpretations are sometimes hard to reconcile with the reality of cartographic practice, where it was common for relief features to be seen

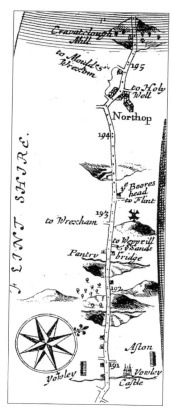

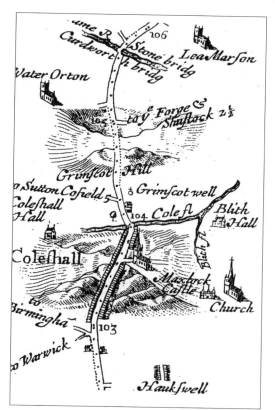

12.10 Hills in John Ogilby, *Britannia* (London, 1675), original scale 1:63,360. *Above left:* profiles, upright and inverted, Northop, Flintshire (pp. 46–7). *Above right:* profiles with added shadow lines, Coleshill, Warwickshire (pp. 42–3). *Below:* 'proto-hachures', east of Bridgwater, Somerset (pp. 62–3). See also Fig. 10.1.

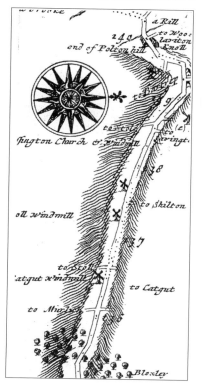

from many different directions on the same map. But at least it seems probable that hachures did somehow develop out of shadow lines.

Even within a given range of scales the exact chronology of the shift from profile to planiform remains obscure. One problem for the historian is that for a time hachures may have seemed more suited to specialists than to the general reader. Thus Speed sometimes used them for natural features in his manuscript town plans – the front of Lincoln Edge is a good example – but on printed versions of the same maps they are restricted to man-made slopes.[29] In English cartography as a whole the change has been located by one authority at about 1680 – at almost the same time, it so happens, as Thomas Burnet was complaining about the misrepresentation of mountains on earlier maps.[30] However, it was in military campaigning that ignorance of what lay behind a hill became not just disagreeable but dangerous; so the most likely hypothesis at present is that profiles first began to lose credibility in the mapping of fortifications and their approaches, and that this change had its origin in France, the most extensively and scientifically fortified country of the seventeenth-century world. Perhaps the progress of slope-shading was encouraged by the contemporary vogue for three-dimensional relief modelling among French military experts.[31] At any rate, from now on it was army officers who dominated the history of hill mapping: to pick up another linguistic clue, it was they who started characterising natural relief by phrases like 'the true features of the ground' as if there were no other features in existence.[32]

In seventeenth-century France, military hachuring was sometimes to be seen in print.[33] Early English survivals are mainly manuscript, typical examples including maps of the lower Medway (1725) and of Dover and its harbour (1737) in the topographical collection of King George III.[34] This style of hill work reached its culmination in William Roy's survey of post-Jacobite Scotland. Meanwhile there had been several occasions when the hachure concept could have penetrated the world of estate cartography along with surveyors leaving military service. A good example was a survey of lands near Guildford in Surrey made for the Duchess of Marlborough in 1728 by John Peter Desmaretz, a draughtsman in the Ordnance office at the Tower of London.[35] One variety of estate map where there may have been a particular sympathy between officers and other kinds of gentleman was the depiction of ornamental parks, where the practice of early eighteenth-century English landscape architects was as far as possible to eliminate walls, hedges and other visible boundaries from the earth's surface. In mapping this kind of open terrain, profile lines would have been an unwanted distraction, whereas vertical wash shading looked exactly right, as can be seen from Thomas Hogben's survey of Leeds Castle, Kent, in 1748.[36] As it happened, hachuring on printed maps was first made familiar to English readers by a specialist in parkland depiction, John Rocque.[37]

The sheer difficulty of hachuring no doubt explains the reluctance of many eighteenth-century cartographers to face its challenge wholeheartedly. In some maps there is a feeling of disharmony between the downward and the horizontal view, as if their authors were bewildered by living in a period of transition. One response was to be less generous with all forms of relief than had been customary in the heyday of the hill profile. Compare the sparsely furnished county maps of Richard Budgen (Sussex, 1724) and Benjamin Donne (Devon, 1765) with the hummocky textures achieved many years earlier by Saxton and Speed in the same counties. An equally cautious policy, in its way, was the restriction of hachures to a few areas where by modern standards profiles were considered exceptionally unsuitable, among them a 'high plain 70 leagues long' between Lakes Huron and Michigan that was the only hachured feature accompanying innumerable profiles in several eighteenth-century maps of North America.[38] Other cartographers were less timid, if also less consistent: hachures and profiles were being blatantly intermixed in an ambitious topographical map of Germany as late as 1789.[39] And it was profiles only, at least on occasion, for a number of British cartographers until the end of the century and beyond, not only part-timers like Peter Fannin and Edward Hasted but also such leaders of the profession as Aaron Arrowsmith.[40]

Amid all the complexities of this long transition period, three modest generalisations seem worth considering. First: hachuring was more likely to be influenced by, or replaced by, the profile concept in little-known regions where a cartographer felt (literally) unsure of his ground. Secondly: the smaller the horizontal scale, the more durable the tradition of profiling. Thirdly: where different methods of relief portrayal were combined, slope shading was more likely on lower, profiles on higher ground.[41]

Meanwhile, the most spectacular single victory for hachures had come with Cassini's published large-scale national survey of France, though, as it happened, the relief features in this case were not particularly good – 'whimsical and sometimes amazingly barren' in one critic's opinion.[42] Improvement thenceforth was achieved not so much by changes of principle as by taking more care to make the map reflect the slope. It was in this spirit that hachures were recommended by the *commission de topographie* in 1802. Pending the advent of quantitative methods to relief representation, the last important issue to arise was where the sketcher's imaginary light should be located. The French commission advocated oblique illumination from the north-west at a 50-degree angle to the earth's surface;[43] but for the definitive *Carte d'état major*, begun in 1824, a vertical system was preferred – perhaps for the sake of its greater simplicity.[44]

Hill sketching in the field

However we explain the change from profile to planiform, allowance must be made for different methods of field work and different horizontal scales. Where a large region had been triangulated on a small scale, with no room for the tracing of minor undulations, it might seem appropriate for a surveyor to treat relief as merely a collection of summits. Each of these could be fixed by intersection or perhaps just placed by eye within its appropriate triangle on his line diagram, no doubt using the same kind of compact profile symbol that would appear in due course upon the finished drawing or engraving. Maps of a slightly larger scale would probably be surveyed by traverse with distances booked in a narrow column and most of the surveyor's attention fixed on detail adjacent to the main lines. One author advised traverse surveyors to 'note down also when you ascend a hill, and when you come to the top thereof, and when you descend the same, and when you come to the bottom thereof'.[45] No doubt this was Ogilby's method in his *Britannia*: one of its consequences was that roads always seemed with curious perversity to be crossing a hill at its highest point.

With smaller areas and larger scales, observations were less likely to be booked as a series of numbers and more likely to be plotted in the field, either by plane-tabling or as some kind of sketch. An attempt could then be made to treat breaks of slope as linear features and survey them in the same way as roads or rivers. This was how Cassini's surveyors went to work, to judge from what is known of their hand-drawn 'minutes'.[46] It was also the most important precept in Basil Jackson's *Course of military surveying* (1838), though Jackson added a recommendation to start at the highest hill and work downwards, omitting small irregularities altogether rather than exaggerating their size. Shading could consist of either vertical or horizontal lines as 'best suited to the nature of the ground'.[47] Such advice if taken literally would have reduced every landscape to a combination of ridges, gulleys and escarpments, but no doubt a skilful topographer could accommodate features other than linear slopes by drawing them *in situ* with pencil or pen. In sketching on a small scale it was natural for hills to be shown obliquely because that was how they were so often seen. The difficulty now was that an observer's view of the landscape would depend on where he took his stations, and at each station his interest would be partly absorbed by non-physiographic detail. It was a problem accentuated by the inherent nature of the earth's relief, extended continuously across the horizon in ever-varying forms that could seldom be extrapolated with the same confidence as a man-made landscape or even a river system.

The only safe way of integrating different viewpoints was to draw the rest of the map first and add the hills later. This insight ought to have marked a recognisable threshold in the history of survey methods, because there was now a possibility that map-making would involve two visits to the ground, separated by a spell of office work. Under such a regime, hill portraiture would surely benefit by engaging more of the surveyor's attention or by the employment of specialist sketchers. At any rate such two-stage surveys appear to have become more common in the early nineteenth century, notably in the making of English Ordnance Survey maps.[48] These developments certainly coincided with a considerable improvement in the quality of British hachuring, but in one sense they came rather late in the day. A century earlier, it would have made a tidy historical generalisation if the advent of sketching on ready-made outline maps had coincided with the adoption of a vertical sight-line for shading and hachures, but in the absence of any sequential record of real-life survey procedures at the period in question it is impossible to comment further on this point.

What does survive is a considerable literature, again largely of nineteenth-century date, about how hills should be added to an existing map. Although most of this writing was meant to be educational, it is often extremely hard to understand. Army officers were taught relief-depiction by instructors chosen for their eminence in other forms of artistic achievement, and by observation of studio models – ranging from spheres and pyramids to human bodies – that bore little resemblance to any natural landscape. This at least makes clear that some kind of oblique lighting was envisaged, because vertical representation could surely have been made almost automatic by estimating the steepest gradient at each point or small patch of ground and then applying a proportionate shade. Although in the age of romanticism many writers emphasised the aesthetic side of relief portrayal, oblique shading was also justifiable for narrow semiotic reasons. A strictly vertical wash could distinguish flats from slopes, but not uphill from downhill, whereas if light was known to come from the north-west, a conical form with additional shading on its south-east side immediately became identifiable as a mound rather than a hollow. This argument was perhaps somewhat academic, because in the average topographical map the rivers, roads and settlements would usually serve to distinguish vertically-shaded ridges from valleys. The other advantage of obliquity was to reduce the amount of heavy shading in areas of strong relief, though on what Jackson called 'tame ground' it was still desirable that every slope should carry some shading, however light. In the end he seemed to settle for a mixture of methods. On the highest, steepest and most rugged slopes he even found room for 'freedom' and 'spirit', no doubt because in this environment minor features had only negative military significance.[49]

Relief by numbers

Sketching without measurement lasted longer in the mapping of relief than in any other branch of professional cartography, and its history overlaps with that of numerical methods by a wide margin. Systematic quantification began with the navigator. The fathom of six feet was essentially a seaman's measure of depth which perhaps began as successive gatherings of a lead line, each held with arms outstretched (this is what the word 'fathom' originally meant) as it was drawn out of the water. Weighted sounding lines had been familiar to Herodotus, and by the seventeenth century they were being graduated with colour-coded leather tags and

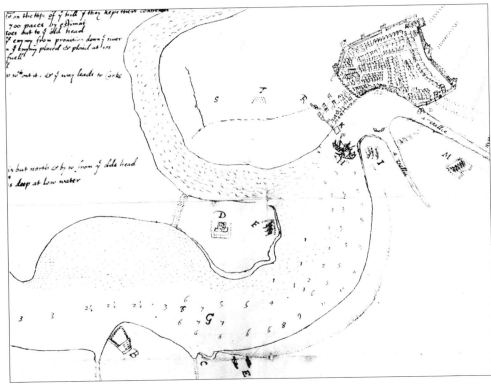

12.11 Early soundings, Kinsale harbour, Ireland, 1601, original scale c.1:10,000. G.A. Hayes-McCoy,
Ulster and other Irish maps c.1600 (Dublin, 1964), xiii.

made in various lengths to suit different depths.[50] Since it was impossible to sketch
the form of the seabed by direct examination, soundings were for many centuries
the only guide to underwater relief (Fig. 12.11). Given the antiquity of Europe's
oldest sea charts, the spot-depth was late in achieving cartographic recognition, but
when it did appear in the 1560s it soon made up for lost time.[51]

On land the most conceptually primitive system of relief depiction by numbers
was that of 'command', which involved the ordering of hill summits to match their
relative altitudes above or below a specified datum.[52] Historically if not logically this
also arrived rather late, appearing in military surveys towards the end of the
eighteenth century. On surviving maps altitude more often took the form of spot
heights expressed in standard units and associated with various kinds of engineering
work. An exceptionally detailed instance was the Elizabethan plan of Dover
attributed to Thomas Digges, which in a marginal key gave heights for fourteen
points while otherwise showing relief by hill profiles characteristic of the period.[53]
Among non-specialist cartographers the land spot height remained a curiosity –
even a freak, to judge from Christopher Packe's map of altitudes in east Kent.[54]
Only when mountain scenery became fashionable with tourists was there anything

like a regular trend in this direction, inaugurated by the altitude figures in J.B. Ramond's *Carte physique et minéralogique du Mont Blanc* of 1797.[55]

Ramond's source of heights was the barometer, which was also chosen by Thomas Jefferson for his survey of Virginia.[56] Among nineteenth-century European surveyors, however, more accurate if more complicated methods of height-determination soon came to be preferred. One medieval technique made familiar in the earliest surveying literature depended on the application of elementary mathematics to the measurement of vertical angles.[57] Given the horizontal length and the gradient between two points, their relative altitude could be found from a table of tangents. If the chain was carried instead along a hillside of uniform slope, the sine of the angle could replace the tangent. Similarly the height of an inaccessible mountain could be found from angles taken to the summit in both dimensions from either end of a measured horizontal base. Even with the best instruments refraction (greater with tilted than with horizontal sight-lines) could cause altitudinal errors of several feet. Hence William Roy's predilection in 1785 for the admittedly more laborious process of levelling.[58] By 1838 this technique was accurate enough for sea surfaces to be compared across the width of south-west England from Portishead to Axmouth.[59] In the following year levelling superseded vertical angles when the British Ordnance Survey began to establish a control network of altitudes as the vertical counterpart to its primary triangulation.

Whatever the surveyor's method, spot heights and spot depths created a problem that became harder to ignore as accurate vertical measurements proliferated. This was the question of a datum plane. Some writers, including Thomas Jefferson as late as 1816, were content for each upland to be measured from a nearby valley floor but the leaders of the surveying profession were less easily satisfied.[60] Roy in 1766 wanted the base for his proposed triangulation of Britain to be 'reduced to the level of the sea', even from altitudes as low as those of Wiltshire or Cambridgeshire.[61] In 1802 the French *commission de topographie* took the same view.[62] But how could a tidal sea possess a single level? In a marine chart, low water would be the most appropriate base level because it showed the minimum depth of water available to shipping. For the engineer, concerned with road maintenance, flood control and kindred matters, it was high tide that set limits to his theatre of operations. For general maps the obvious compromise would be mean sea level.

The first contour lines

The long-term influence of contouring among map-users left its mark on the English language when the symbols took lexical priority over what they denoted: for present-day cartographers a 'contour' is something one sees not on the ground but on a map and it would be intolerably pedantic to insist that the correct expression is 'contour line'. The lines in question were those joining points on the

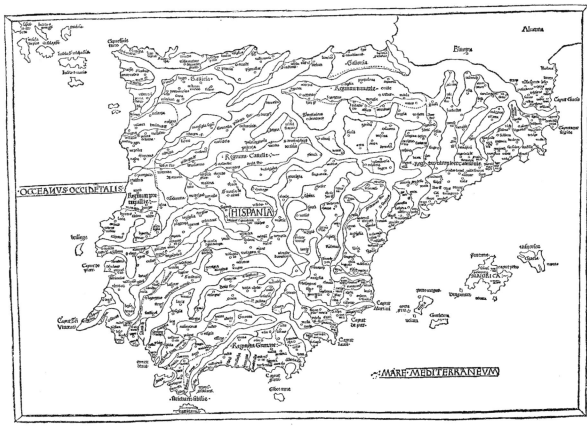

12.12 Contour-like boundaries of upland masses, Iberian peninsula. Claudius Ptolemy, ed. Dominus Nicolaus Germanus, *Geography* (Ulm, 1482).

earth's surface at the same vertical distance above or below a single datum which would preferably be made known to the reader. One of their virtues was to give altitudinal information not only for these points but also, admittedly with less precision, for the whole area between one line and the next. Quantitatively, then, contours recorded both height and gradient. Qualitatively they could show almost every landform known to verbal language, together with an infinity of others yet unnamed. In fact contours were perhaps the cleverest innovation in the entire history of cartography, which makes it remarkable that they must apparently be credited to a number of different inventors. Some ancestral images, we must admit, are of somewhat dubious legitimacy. In an early printed edition of Ptolemy's *Geography* each upland massif was surrounded by a planiform girdle within which the individual hill profiles were drawn by hand (Fig. 12.12).[63] On at least one larger-scale sixteenth-century map a contour-like line can be seen running not round the base of a slope but about half way up it.[64] Later, in the historical twilight of profile symbolism, such lines seem to have settled down along the edge of land liable to

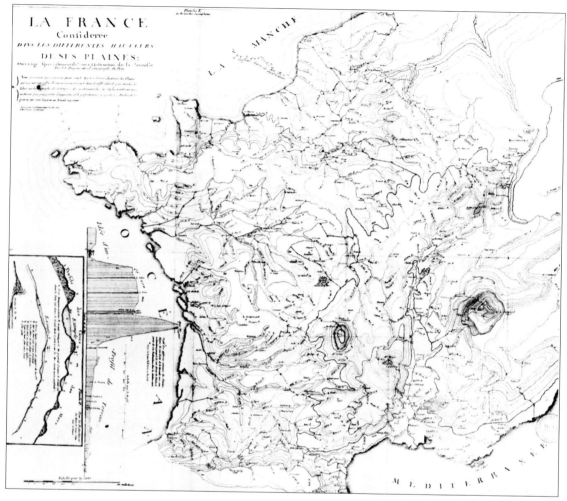

12.13 Jean-Louis Dupain-Triel Jr, *La France considerée dans les differentes hauteurs de ces plaines* (Paris, 1791).

flood.[65] This kind of binary symbolism could usually be left to explain itself: verbal glosses, like 'All within this yellow circle is high ground' (1567), are rare enough to be worth collecting.[66]

For modern students contouring is hardly recognisable as such unless there is, somewhere on the map, a slope with more than one line on it. In this more restricted sense development occurred as a series of apparently separate episodes, beginning in the Netherlands with Pieter Bruinns (1584) and ending with the French commission of 1802, which recommended contours for large-scale maps wherever suitable data were forthcoming – without suggesting that such data should be made available for any but small tracts of country.[67] In this long and not very eventful gestation period the nearest approach to a climax came with Jean-

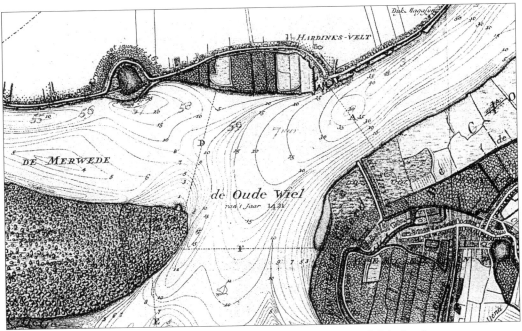

12.14 The first printed contour map. Sub-surface contours in feet. Original scale 1:10,000.
From Nicolaas Samuel Cruquius, *De Rivier de Merwede* (The Hague, 1729–31).

Louis Dupain-Triel's contour map of France (Fig. 12.13),[68] but several other developments are worth recording. Like spot heights and for the same reason, true contours began under water (Fig. 12.14). On land they were being drawn soon after 1774 – an event now claimed as a 'first' on a roadside monument in Scotland – though not actually published until four years later.[69] Contours resembled spot heights in being pioneered not by professional land surveyors or general map publishers but either for use in engineering works or to satisfy a purely scientific desire to get at the truth. Charles Hutton's were intended to throw light on gravitational anomalies by measuring the mass of a single mountain. John Churchman's, in 1804, were simply an addition to geographical knowledge.[70]

Contour surveys were so laborious – according to an English Ordnance Survey officer nearly three times as costly as hill sketching[71] – that they remained unusual for a considerable period. As with hill shading in the seventeenth century, it was the military tacticians who first saw the need for change, and for several decades almost all the world's land contour maps were made by the French army, though it was the United Kingdom Ordnance Survey, in a period of declining military influence as it happened, that first sought to make contours a standard component of a national large-scale map.

Like spot heights and spot depths, the first contours were instrumentally derived. For charts they could be inserted by eye on the draughtsman's table among

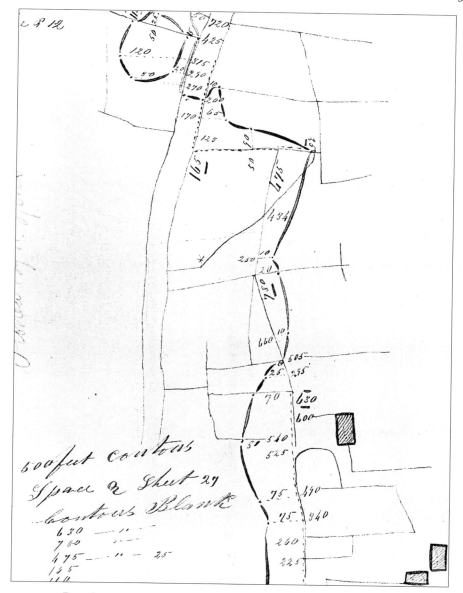

12.15 Page from contour-surveyor's fieldbook, Ordnance Survey of Ireland, 1856. Ordnance Survey office, Dublin.

a mass of individual soundings that would have deserved mapping for their own sake even if contours had never been thought of. Above sea level the weaknesses inherent in such mechanical interpolation would be visible to any observant dry-shod map-user taking a walk, which meant that land contours, in the strictest sense of the term, could be surveyed only by placing a level and staff at the required height, carrying these instruments horizontally round the hillside, and recording each position of the staff as part of a planimetric survey (Fig. 12.15). The stations had to lie sufficiently close together for the connecting lines to be sketched without serious risk of error. When contours were still a novelty the staff positions might be thought worth showing on the finished map as dots.[72]

The contourer's ultimate dependence on estimation between fixed points had one corollary worth noting with reference to earlier periods. By widening the horizontal gap between consecutive stations the survey could become progressively more approximate until in the end there would be no stations at all, reducing the contour to an unnumbered 'form line' derived entirely from sketching. At this point Jackson's views on horizontal shading become relevant. To what extent form lines had been developing independently may be difficult to say, but on many eighteenth-century maps the hachures give an effect of terracing as if derived from a form-line sketch.[73] There was indeed some affinity between a contour and a break of slope, although of course such breaks are not necessarily horizontal. Non-instrumental form lines were certainly plentiful in the early nineteenth century;[74] they played a major part in the field work of the Ordnance Survey, but were always 'rotated' into hachures before appearing in print.

Contours: towards maturity

Without some supplementary information, contours were not always easy to read. So could relief maps be made more comprehensible by in some way synthesising pictures and numbers? In particular, could slope shading become mathematically precise without sacrificing its graphic appeal? The answer was affirmative, as long as enough contours were drawn first. In 1799 a German army officer, Major Johann Georg Lehmann, published a scale of exact thicknesses for hachure lines, but this seems to have been thought 'too pedantic' for general use, even in the land of its birth (Fig. 12.16).[75] In Colonel J.E. Van Gorkum's alternative system, lines were drawn at right angles to the contours, only for these refinements to be described by a British expert as 'not suited to our national character'. They certainly added nothing to the original contours, and in retrospect it seems strange that they were ever taken seriously by readers of any nationality. The furthest the Ordnance Survey would go with its continental colleagues was sometimes to eliminate sketches by translating contours directly into hachured form as an indoor operation, though it always felt rather guilty about doing this.[76]

Then there was the possibility of characterising the contour symbol itself. Its usual style was a fine black line, dotted or dashed by a familiar convention to indicate the absence of any physical barrier. The representation of individual levelling points by larger dots was given up when the surveyor's instinct to cover his tracks reasserted itself. Dotted or not, the lines were sometimes hard to separate from other detail. Contours could best be made more comprehensible by increasing

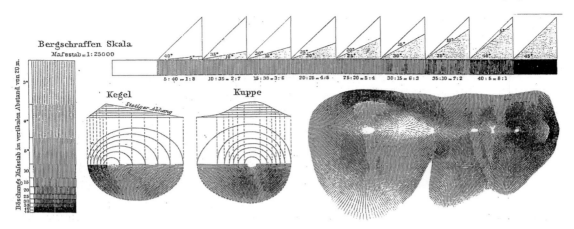

12.16 Quantitative hachures. Emil von Sydow, *Methodischer Schul-Atlas* (Gotha, 1888), pl. 5.

their visual impact and reducing the number of assumptions involved in reading them. Particular adjustments of this kind, like many innovations, were mainly devised by individual cartographers and then left to languish without publicity for several generations. One was uniform vertical spacing. The interval between contours would depend on the cartographer's objectives – in engineering plans it might be as narrow as one foot – but for topographical maps there was eventually to be an empirical formula: optimum vertical interval in feet equals 50 divided by the map's horizontal scale in inches to the mile.[77] Other improvements included the thickening of every tenth contour for easier reading; the dotting of supplementary contours inserted locally at less than the standard interval; and a change of line-colour to distinguish different kinds of underlying surface such as vegetation, bare rock or ice. Most striking visually was the differential colouring of the spaces between the contours as a speed-reader's guide to altitude: this kind of 'hypsometric tint' was tried from 1815 onwards, but did not become widely familiar until near the end of the century.[78]

Meanwhile the British Ordnance Survey had been proceeding more cautiously. Its 'scale of shade' was devised in *c*.1830 not for the public domain (in fact no specimen of any such reference-aid is known to survive) but as an intra-departmental contribution to the art of hachuring. The Survey's engravers were advised 'to increase the breadth instead of the number of lines in etching hills where it may be necessary to have greater darkness', the purpose of the greater darkness being to indicate higher altitudes as distinct from steeper slopes.[79] ('Etching' in this case was the use of acid to widen the engraved hachure-lines on a printing plate when a darker shade was required.) The scale was applied with considerable discretion. It had to be. For one thing no instrumental contours were yet available on the maps due to be hachured. As a substitute guide for the hill draughtsman, pseudo-contours were interpolated between spot heights, thus creating a distinctive black and grey hypsometric style of unpublished map – with 'steps' rather than layers – that later caused some mystification outside the department.[80] More importantly, the Survey was already committed to the idea of oblique illumination, which meant that it was never possible to carry the same intensity of shading all the way round a hill. Despite their subtlety of conception, Ordnance Survey hachures could be alarmingly obtrusive, in mountainous country enshrouding all other kinds of detail with almost impenetrable gloom.[81]

In the end the most legible solution for relief was to put both slope shading and contours on the same map in a different colour from the rest of the landscape, thus defying the French experts of 1802 who advised against multiple symbolism. In point of fact such hybrids had never been unusual. On eighteenth-century maps an eye-catching profile was often perched on top of a hachured hill. Later, spot heights were freely used on contour maps to give the exact heights of summits, cols and other points of special interest. Hachures were retained alongside contours for certain minor landforms, such as man-made embankments, that were too small to exceed the contour interval. It seems a pity that the triangular symbols now used for trigonometrical stations are more likely to have been inspired by surveyors' diagrams than by hill profiles: otherwise we could claim that every phase in the history of relief representation has left its legacy on the modern map.

CHAPTER 13

In no case arbitrary sounds

A SURVEYOR HAS MORE TO do than sketch or measure what is visible. He must also investigate placenames and territorial boundaries, as well as determining the precise functions of man-made landscape features whose purpose is not immediately apparent. Most words used on maps are either descriptions or proper nouns. Some descriptions refer to an object such as 'church' or 'mill', others to some characteristic of an object like 'deciduous' or 'arable'. Such terms could all be replaced by symbols and explained in a key outside the body of the map.

Names differ from descriptive terms in denoting not types but singularities. Although logically this is a clear distinction, in the natural languages of the world it can sometimes be obscured by historical accident. A passenger buying a ticket to London expects to be deposited in one particular city: he is not inviting the travel agent to choose at random from all the locations on the planet that happen to be called 'London'. A more probable reason for confusing names with descriptions is that one category has often evolved rather slowly from the other, so that any surveyor collecting names in a large territory can reckon to find some words indecisively poised above this semantic threshold. A clear proof that the line has been crossed is the addition of a descriptive word to what was once a synonym of itself: thus etymologically 'Isle of Sheppey' means 'island of the island of sheep' and 'River Avon' means 'river river'. In this respect a contradiction can serve the same diagnostic purpose as a tautology: once the words 'Nameless Island' had been written with initial capitals on a map it was not unreasonable for a geographer to insist on regarding them as a name.[1] In many such cases the loss of connotative meaning could speak for itself, but some expressions, like 'Devil's Punchbowl' or 'Wayland's Smithy', might seem to retain a descriptive function without actually doing so. The Philadelphia placename 'Point No Point' can hardly be considered an effective remedy for this state of affairs. Sometimes it was length or clumsiness, rather than factual content, that blocked the grammatical passage of a noun-phrase from common to proper. Neither of these characteristics had the effect of disqualifying 'The Man and His Man' in Cornwall, which is undoubtedly a name.[2] On the other hand John Davis's 'Furious Overfall' at the entrance to Hudson Bay would always have the look of a description, as most non-cartographic writers and some map-makers seemed to acknowledge by putting an indefinite article in front of it.[3]

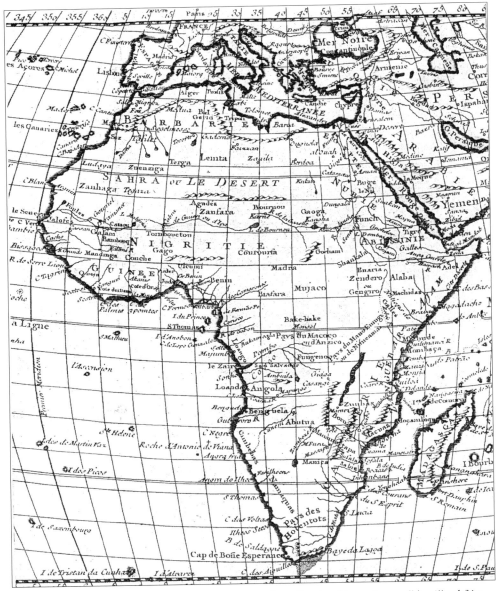

13.1 Regional names without correlative boundaries. Jean-Baptiste Bourguignon d'Anville, Africa, (Paris, 1727).

Names have often been complementary to boundaries. In that case they could not be precisely meaningful unless accompanied on the map by some kind of areal definition, a fact successfully obscured on small-scale maps where many two-dimensional features have been collapsed into points or lines. In this sense certain names might be termed incompletely significant, referring to a region with no precise limits such as 'The Cotswolds' or 'The Dukeries'. Such denominations could

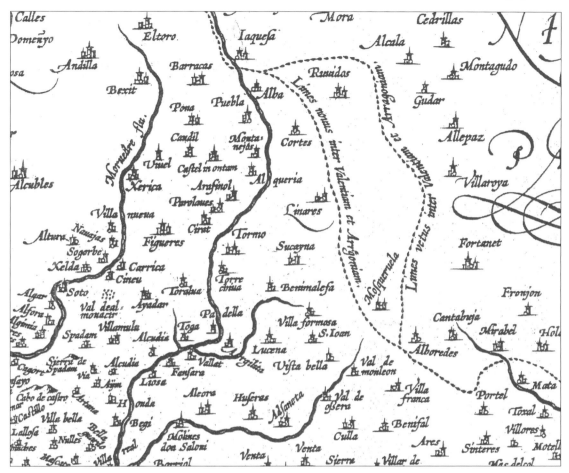

13.2 Old and new boundaries between Valencia and Aragon, Spain. Abraham Ortelius, *The theatre of the whole world* (London, 1606), p. 19.

expand or contract their scope for various reasons, one of which was the average map-maker's dislike of empty space: hence the ease with which an unbounded Ethiopia spread southwards and westwards across sixteenth-century maps of Africa, eventually embracing part of the Atlantic Ocean.[4] Cartography has never quite managed to cope with this situation. A common practice was simply to insert the name and nothing else, leaving its extension to be interpreted as the land or water underneath the appropriate letters of the alphabet together with a circumjacent area of indeterminate size (Fig. 13.1).

Where the recording of a boundary seemed both practicable and desirable, the cartographer had several procedures at his command. Sometimes delimitation could take place on the draughtsman's table, following the line of streams, fences, walls or roads that had already been mapped in their own right. Otherwise there were two possibilities. One was for significations to be pointed out by local residents, who led the surveyor across the landscape as if in re-enactment of a traditional bound-beating ceremony. Where informants disagreed, the conflicting versions might both have to be surveyed and the argument settled later. Perhaps this is why Ortelius showed an old as well as a new boundary between Valencia and Aragon (Fig. 13.2). Alternatively, the boundary was extracted from documentary evidence, either as a verbal description or as a map (the latter probably with a smaller scale than that assigned to the current cartographer) and then identified for survey in the landscape. The best-known examples of this process occurred in treaties between sovereign states. The simplest case, in theory if not in practice, would be to create, on the ground, a real-life version of the parallel or meridian so insouciantly described in words and numbers by the politicians. A surveyor might also be required to squeeze a broad frontier mountain range or belt of wetland into a single line, replacing a geographical no man's land by a genuine boundary. A not dissimilar situation could arise in the mapping of rivers. Should the name of a stream be carried upwards to the most 'important' of its sources? And did this mean that the riverine surveyor must measure widths or even depths before deciding what to write?[5]

Words versus numbers

Names like descriptions were seldom as indispensable as they seemed. In theory they could be superseded by numbers in a geographical coordinate system, though an exception would have to be made for 'Greenwich' or whatever place was used to give the coordinates a point of origin. A string of numbers could also represent a digitised territorial boundary. In the absence of geographical coordinates an arbitrary numerical reference grid could be devised for each map, though alphanumeric systems (used from the sixteenth century onwards) were clearly not meant as name-surrogates since the letters of the alphabet unlike numbers are incapable of subdivision (Fig. 13.3).[6] A unique grid reference, however it originates, is in some ways preferable to a name: numbers may occupy less space than words, they are internationally understood, they have no frivolous or otherwise distracting associations, and being infinitely partible and multipliable they can describe every location in the universe without taxing anyone's inventiveness.

But despite its advantages numerical referencing has not yet replaced verbal nomenclature. This is because to most people the neutrality and unemotiveness of numbers make them dull and forgettable, whereas names with their wealth of historical allusion and poetic resonance evoke stronger feelings than anything else in cartography:[7] no one with any sensitivity could argue that in a well-ordered society the only function of names on maps is to help the reader find his way from place to place. Words have also proved especially inviting to the cartographic fault-finder or busybody who could back 'constructive' suggestions with the argument that names can be changed on his advice without damaging the rest of the map

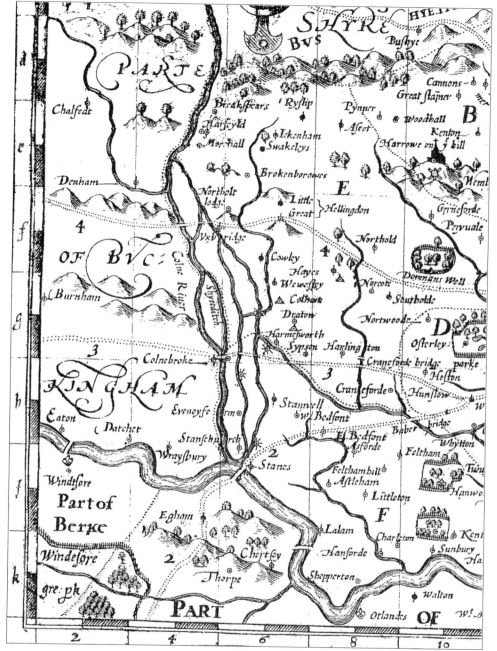

13.3 Early alphanumeric grid. From John Norden, Middlesex, *Speculum Britanniae* (London, 1593). See also Fig. 9.6.

(Fig. 13.4). Some early cartographers, like Laurence Nowell and Humphrey Lhuyd in Elizabethan Britain, were serious students of placename-science, and probably acquired their interest in map-making through this channel.[8] Scholarship like theirs

13.4 Contemporary MS additions and corrections to placenames. From Abraham Ortelius, *Eryn: Hiberniae Britannicae insulae, nova descriptio: Irlandt* (Antwerp, 1573). National Library of Ireland.

has often been combined with a cheerful tolerance of the layman's toponymic foibles. But nobody's foibles should be carried too far. For all their subjectivity names must in the end be held in common by producer and client, which may not

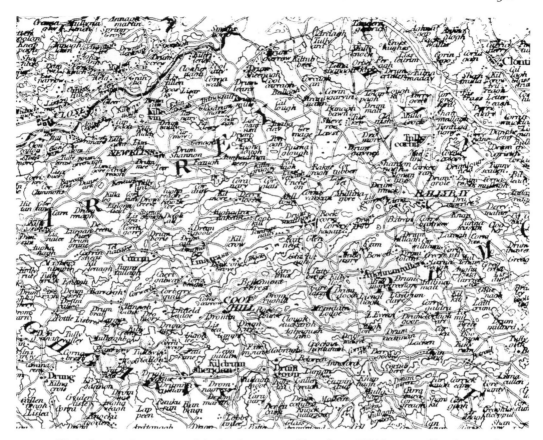

13.5 High-density placenames, original scale 1:253,440. From James Wyld, *Ireland* (London, 1839).

always happen if a surveyor insists on excogitating them out of nowhere. A good rule of cartographic practice is that when two people agree on a name a third person can do both of them a favour by keeping out of the way.

The average map-reader's leaning towards verbal communication has one potentially unfortunate consequence: he happens to have chosen a medium in which words can easily be overdone. A purist might even argue that script is by its very nature anti-cartographic, since the essence of maps is to make patterns that can be apprehended instantaneously rather than being read *seriatim* in a succession of independent visual events. It was certainly regrettable when inscriptions became so numerous as to spoil the aesthetic impact of the cartographer's 'artwork' – a real danger in regions with more current placenames than could be shown at a practicable scale on any single map (Fig. 13.5).[9] Once again the cartographer had to make a choice. The obvious rule was that, other things equal, each place's claim to representation should depend on how many people would benefit from knowing its whereabouts. The answer to this question has depended on historical circumstances.

Names must obviously be found for new towns or new local government areas. More interestingly: although physical features may be unchangeable, people have varied in their desire to talk and write about them. For instance, seas and oceans needed no identification until the advent of long-distance navigation, and many mountain peaks could safely remain anonymous until sportsmen and scientists took up climbing.

The capriciousness of placenames goes far beyond the reality they denote. A town or a headland can be perceived as a single object. The surveyor measures it and it stays measured. He would probably wish to treat nomenclature in the same way. But names are not normally inscribed on the landscape. They inhabit a world of their own that is only partly material, or more truthfully a number of worlds, for no name can be regarded as unique. The implications of this last statement deserve a moment's thought. It is not just that the same place may have more than one name, for instance in Ireland 'Dunleary' and 'Kingstown'; or that a single name may have more than one form, such as 'Dunleary' and 'Dun Laoghaire'; or that a single form may have more than one spelling, such as 'Dun Laoghaire' and 'Dun Laoire'; or that a single spelling may undergo more than one kind of diacritical treatment, such as 'Dun Laoghaire' and 'Dún Laoghaire'. More than that: a name, far from being immutably laid up in heaven, is rather to be identified with the sum of its individual occurrences, whether as patterns of sound waves or as marks on a visible surface. In that sense there are as many 'Dun Laoghaires' as occasions when these two words have ever been written or spoken.

The quest for authority

After these preliminaries we may proceed to the simplest possible historical situation, that of a surveyor at work among monoglots who spoke the same language as himself. He collected the names by asking for them, and here we must at once distinguish between speech and script. Conceivably there might be nothing

Situations.	Descriptive Remarks.
Lies in the S. East of this parish bounded by Moore Park & Glansheskin townlands, By Kilclugh in the parish of Macrooney and by Ballyderown & Moore Park in Kilcrumper parish. Eastern Division, all in the barony of Condons & Clongibbons.	Is the Property of Lord Mountcashel by dea for ever contains a 2 to all of which is flat and dry dea, in an excellent state of Cultivation The houses and roads are in good repair Pays for Co. Cess

13.6 Ordnance Survey name book, County Cork, Ireland, *c*.1840. National Archives of Ireland, OS 88 Cork 277, by permission of the director.

to link these two media: a spoken name need never be committed to writing and a written name need never be read aloud. Yet neither source could be confidently dismissed. Their joint claim was neatly acknowledged by Edmund Gibson in 1722. In revising an early map, he wrote, a cartographic editor could follow either 'the way of writing' or 'the common way of pronouncing among the people'. Only names that failed on both levels should be corrected.[10] The hydrographer Francis Beaufort would evidently have agreed. 'You will of course follow the impression of your ear', he told a junior naval surveyor in 1831, complicating matters immediately afterwards with the recommendation that some knowledgeable person should be asked to write the names being collected.[11]

A survey team of deaf mutes would certainly be at a disadvantage, but left to his own devices the average cartographer, himself confidently literate, would doubtless feel happier with written forms. For one thing, their relative durability facilitated the process of verification in a way that was impossible with speech until the invention of sound recording. An inscription could enter the realm of cartographic discourse through several channels. Local residents might be asked to write down a name. Or the surveyor might begin by obtaining some kind of gazetteer, perhaps from an official source such as a tax assessment list or population census; he would then ask the inhabitants to check the spellings and help pair off the names with whatever visible features he intended to map. If, as sometimes happened, his source of data was itself a map, doubts must begin to threaten the distinction between survey and compilation that has been recurring in the present book. Perhaps it should be stipulated that to count as a surveyor, rather than a compiler, one should seek non-cartographic corroboration *in situ* of the names on any pre-existing map with claims to be treated as an authority.

Not all the spellings collected in the field for any one name were likely to be identical. Hence the advisability of keeping a name-book or similar document, in which each locality has a whole page to itself, with one column for variant forms and another identifying their authorities (Fig. 13.6). The reward for the surveyor's diligence was then to be confronted with another problem: which source should be preferred? Again the obvious guide was democracy: consult as many instances of each name as possible and choose whichever version occurred most often. But should a published record such as a directory or newspaper have a 'block vote' proportional to the number of copies printed? And what about people disenfranchised by illiteracy but still capable of making a phonetic contribution? From here it was a short step to choosing certain informants on 'merit', based on education, wealth, social status, cooperativeness of disposition, length of family association with the locality and so on. It was in this spirit that Cassini preferred a name 'spoken or written by the landlord, priest or government representative for the parish' when mapping the villages of France.[12] In the South Seas Cook's policy was even more autocratic: as a guide to names in the Friendly Islands he chose a single native Polynesian and ignored everyone else.[13]

By now another rule of procedure has come into view: names should not be allowed to change too fast, which itself was a special case of the labour-saving principle that maps should not need to be revised too often. The kind of educated informants chosen by Cassini might occasionally have strange ideas on the subject of language, but at least they would be unlikely to introduce a new spelling out of pure ignorance or carelessness. In any case, belief in correctitude as such was not the only reason for resisting orthographic change. As Christopher Packe pointed out, there could sometimes be an unbridgeable gap among name-users between the 'conveyancer or antiquary' and the 'naturalist'.[14] A map-maker might seek to guard the rights of property by copying from ancient deeds and leases rather than from present-day notice boards or newspapers.[15] Proprietorial considerations would be especially relevant in the 'bespoke' mapping of a particular estate, where the landlord might legitimately (in his own eyes) choose names unrecognised by popular usage. Apart from issues of ownership some map-makers just believed in conserving old placenames as such, if only as a way of giving practical significance to historical researches in which they took a personal satisfaction. Carried to extremes, this attitude has been held to justify a resurrection of the undeniably dead, perhaps after many centuries of interment. Morecambe Bay in Lancashire owes its name to an eighteenth-century writer who rashly identified it with Ptolemy's Morikambe.[16]

Sound and sense

Imagine next a people with a single written and spoken language, inhabiting a country rich in placenames that happened to have left no documentary traces past or present. In general, a surveyor would transcribe the names according to accepted orthographic conventions which even where there were no written authorities

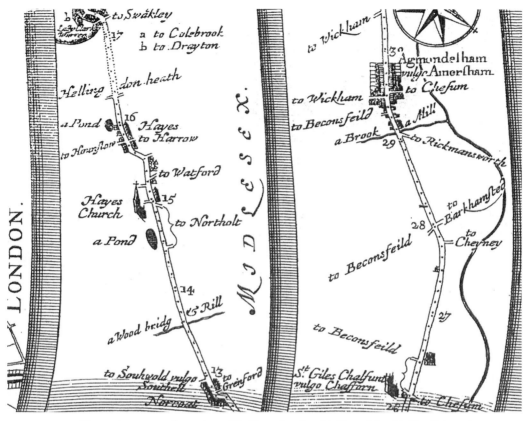

13.7 Alternative names, Amersham, Chalfont, Southwold. From John Ogilby, *Britannia* (London, 1675), pp. 22–3.

would seem to validate the notion of a 'correct' name. There were various difficulties with spoken names, however, especially for a surveyor ignorant of the local language. Respondents could deliberately mislead the interrogator, a not improbable reaction to an official survey threatening higher rents or taxes. He might erroneously give placename status to what was meant only as a description, or even to some totally inappropriate speech-act such as a greeting or a comment on the weather.[17] Or he might simply mishear the sounds addressed to him, like the sixteenth-century French cartographer whose English county names included 'Daoncher' and 'Hamecher'.[18] A problem here was the failure of some scribal conventions to accommodate certain long-term linguistic changes, as illustrated by the shortening in English speech of 'ton' to 't'n' or ''n'. Other examples may be studied in John Ogilby's *Britannia* (1675), where the popular alternative spellings (identified as 'vulgo') usually contained fewer phonemes than the supposed originals (Fig. 13.7).

Ogilby's recording of aliases was a natural response to the vagaries of pronunciation. In estate cartography, as we have seen, it might seem especially important to defend the rights of property by preserving 'divers of the ancient names' as well as 'the most known names in these days'.[19] Cartographers who preferred to be decisive could easily find themselves on the 'wrong' side of the line between sound and script, as when Wapping in east London was mapped cockney-style without its final 'g'.[20] But there were some obvious arguments against diversity: lack of space, reluctance to shirk authorial responsibility, and the need to discourage readers from suggesting alternatives of their own. Certainly the British Ordnance Survey seems in general to have set its face against the publication of variant forms.[21] However, a modern reader taught to respect this rule will find aliases surprisingly common in early maps of almost every kind.

The intimate if illogical relationship between speech and text has been a factor working to keep names within their original language even when their constituent words have long since been dropped from the dictionary. In principle, it is true, the association of names with languages might sometimes be open to debate. Certain names in fiction, particularly science fiction, have been genuinely devoid of linguistic context; which means that where nationalistically-minded placename enthusiasts disagree it might be just possible for a neutral observer to appeal outside the circle of familiar languages. But most names originated as descriptive phrases belonging firmly inside the circle, and many cartographers would have agreed with Isaac Taylor's dictum that the words of a map should be 'in no case arbitrary sounds'.[22] This point is easily forgotten when 'corruption' has gone too far for original meanings to be recognisable. Hence the distinction between transparent names like Newcastle or Redhill and opaque names like Coventry or Leeds. The line between these categories can sometimes be a little fuzzy. A name might be transparent to a linguistic scholar but not to a layman, or in one semiotic context but not another, making 'Adelaide' transparent as a placename and opaque as a personal name. But for practical purposes it is a serviceable distinction.

As a class, transparent names were at first sight more deserving of attention. Their affinity with particular languages was agreeably unambiguous and should have made them less liable to miscopying than opaque names. They might also be genuinely helpful in a geographical sense, telling the map-reader what he could expect to see on the ground and justifying the inclusion of a column for descriptive comment in the surveyor's name book. The advantages of transparency might sometimes turn out to be spurious, it is true: Newport in Shropshire is not new, nor is it a port. But at least such 'delusory' names could provide a further historical bonus by way of recompense. And most people preferred using words that they could understand: this is why opaque names were often reconstructed to make them transparent, with the Cornish 'maen' (a rock) assuming human form on English maps in names like 'Old Man' and 'Deadman'. Sometimes the results of this process can surely be described without equivocation as incorrect. No one could argue that Ortelius saw 'St Kilda' as an arbitrary triad of syllables that just happened to look like a personal name: he must really have believed in a non-existent saint called

Kilda.[23] Altogether it was understandable when a twentieth-century surveyors' rulebook appeared to challenge Isaac Taylor by insisting that 'no attempt shall be made to follow supposed meanings'.[24]

Opaque as well as transparent names can often be assigned to particular languages. 'Florence' and 'Venice' are as English as 'London' or 'Liverpool', and there need be little hesitation about classifying 'Giovanni', 'Ian', 'Ivan', 'Jean', 'Johanne', 'John', 'Juan' and 'Sean'. So opacity is not synonymous with anarchy: each language has its own way of handling different sounds, regardless of whether those sounds mean anything. Take the common Irish element 'baile' denoting an inhabited locality, as represented in modern names like Ballyfermot and Ballymun. Both the sound and the spelling of this prefix have been variously anglicised, but over the centuries 'Bally' became the most common form on maps of Ireland, no doubt because by analogy with other words ('rally', 'sally' etc.) it matched the way most English-speakers pronounced the syllables in question, whereas 'Baly' for example (common in sixteenth-century maps) had the disadvantage of evoking the sound of the English word 'scaly'. 'Bally' was accordingly approved by the Ordnance Survey, and to a modern eye 'Baly' looks wrong. Why not use the original Irish form? Because for most readers that looks wrong as well, its sound to all appearances being unnecessarily signalled twice, first by the medial 'ai' and then by the terminal 'e'. (In fact 'baile' in Irish speech sounds nothing like 'bail' or 'bale'.) The pragmatics here are simple enough: spellings that encourage mispronunciation should if possible be avoided – unless the cartographer was willing to accompany his map with an explanatory text.[25] This problem did not arise in languages that had never been committed to writing. For an eighteenth-century explorer in the Pacific there was no alternative to crude phonetics.

Names and languages

There is of course an important sense in which all anglicised forms of 'baile' are wrong, and true correctness attainable only by restoring the original Irish-language word. Attempts to reform toponymy on this principle are at least as old as the seventeenth century. Hiob Ludolf was an expert on Ethiopian languages whose map of Abyssinia in 1683 made improvements to corrupt forms of place-names adopted by the Portuguese.[26] In eighteenth-century Wales, William Morris complained that previous cartographers were ignorant of his country's ancient language. He wanted Welsh names to reappear 'in their purity'.[27] At this point the issue of transliteration arises. It is arguable that in the recording of Russian, Arabic or Persian nomenclature purity as Morris saw it could be achieved only by retaining the original characters in which those languages were written. It has long been agreed however that for purely practical reasons any foreigner with a printing press and a supply of type must be allowed to keep his own alphabet.

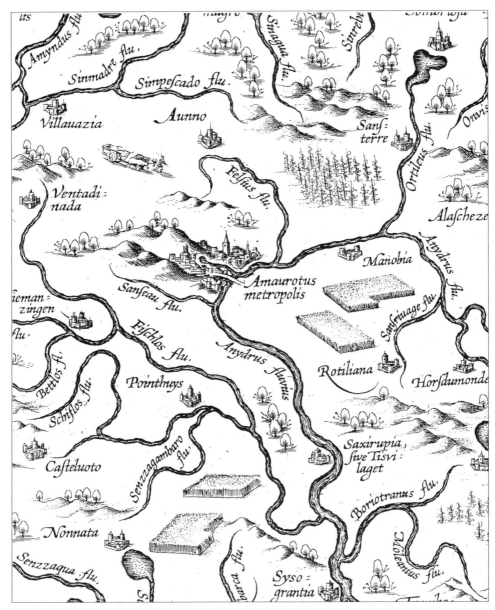

13.8 Imaginary names inspired by European languages. From Abraham Ortelius, *Utopiae typus* (Antwerp, *c*.1596).

This was not a problem peculiar to cartography. It arose as soon as people of one country began writing for their own consumption about a country with a different script from theirs. The obvious treatment for discordant alphabets was a process of transposition in which each letter in one language would be replaced by the corresponding letter in another, and when that happened the transliterator might

13.9 Bilingual and trilingual placenames in mid-Wales. From Humphrey Lhuyd, *Cambriae typus*, 1568 in Abraham Ortelius, *The theatre of the whole world* (London, 1606), p. 13.

claim to be preserving the original orthography. But not all alphabets contained the same number of characters, and in some languages the symbols corresponded not to letters but to syllables, as in Amharic, or even to complete words, as in Chinese. In these circumstances it was natural to fall back on phoneticism, as would happen if the foreign alphabets had never been encountered. (Cartographers could show devotion to duty in this predicament by separating the syllables of an outlandish new name with hyphens, as with Aaron Arrowsmith's 'Slee-ki-eet-ack-coo' among the headwaters of the Columbia River.)[28] The reform of the anglicisation process began during the era of Samuel Johnson's dictionary. John Green in 1747 used a conventional English alphabet and a systematically romanised Chinese syllabary for the spelling of non-European names. In 1788 Sir William Jones recommended treating consonants as in English and vowels as in Italian, a rule later adopted by both the Royal Geographical Society and the British Admiralty for 'barbaric and savage languages'.[29]

Which brings us from the language of names to the language of maps in general. Here the practicability and legitimacy of translation were admitted without demur on all sides. John Speed presented his new map of Poland as 'done into English', Nicolas Sanson's maps were 'rendered into English' by Richard Blome, and so on. Even a map with no names could still be classified linguistically by words like 'nord', 'equateur' and 'designé par'. Of course there might still be a mixture of languages. A cartographer could even change loyalties between one name and another on the same sheet; Ortelius did this in mapping Utopia, presumably as a gesture of internationalism (Fig. 13.8).[30] Where space and commitment were available the same place might be named in two or three different languages, as in Humphrey Lhuyd's map of Wales (Fig. 13.9) and William Brassier's map of Lake Champlain (1762).[31] Otherwise there was only one logical solution: names should be in the same language as the rest of the inscriptions on a map.

What language should that be? Once more it seems tempting to invoke the democratic spirit, this time as a licence for each cartographer to achieve maximum comprehensibility by adopting whichever medium appealed to the largest number of potential users. Indeed he would probably do this anyway for commercial reasons. In European history the main complication has been the prestige of the classical languages. Greek was less 'map-friendly' than Latin because of its accents and other extraneous signs.[32] Latin also had its own political role as an appropriate vehicle for the names of islands, traditionally regarded as belonging to the pope.[33] More generally, Latin transcended boundaries of both time and space and until the seventeenth century could claim to be the vernacular of international science and scholarship.[34] Modern-language aliases in sixteenth-century map-titles were often prefixed by the qualification 'vulgarly'. But should vulgarity be avoided in the body of the map by labelling Paris 'Lusitatia' and Strasbourg 'Argentina'? As early as 1571 Georg Braun raised this issue by challenging Ortelius about the acceptability of Latin.[35] Perhaps it was best to face the inevitable and produce different maps in different languages, with each cartographer sailing under his own colours. That was the opinion of William Dampier in 1693: 'I write for my countrymen and have therefore for the most part, used such names as are familiar to our English seamen, and those of our colonies abroad.' He then spoiled the effect with the feeble escape-clause: 'yet without neglecting others that occurred.'[36]

The cartographer as name-giver

We must now discuss the conferring of names that were entirely new. The verb 'to name' is ambiguous in this respect: compare the tabloid newspaper headline 'We name the guilty men' with the royal speech-act 'I name this ship *Queen Mary*'. Naming in the second sense, for which 'dubbing' might be an appropriate term, is clearly not what happened when the names of a particular language, or a particular historical epoch, were more or less unintentionally changed in the course of transmission from one user to another. The difference is that dubbing was contingent on individual willpower: cartographers could sometimes avoid such acts of creativity, and in that case would generally have been wise to do so. Once more the principle at stake was ease of comprehension. The case against coining a new name was that other names might already exist, and that two names with one referent were an obvious recipe for confusion. More than two were even worse: the Falkland Islands

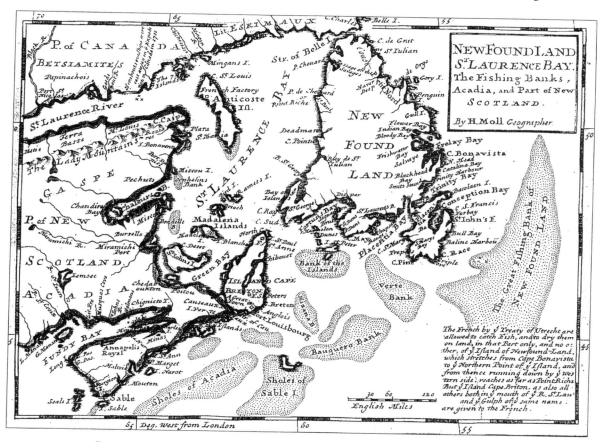

13.10 Pointe Riche, southern limit of French fishing rights in western Newfoundland, here moved from its 'French' position at Cape Ray to the more northerly location favoured by British opinion. From a late (*c*.1732) edition of Herman Moll's *Newfoundland*.

have been variously known as Alencam Islands, Davis's Southern Island, Falkland Islands, Hawkins's Maidenland, Isles Malouines, Maiden Islands, Malvinas, New Islands, Nouvelle Islands, Sanson Islands, Sebald's Islands and 'Seven Islands of Virgin'.[37] Such differences are often associated with different language-groups, but not always: English has always been the language of the *Irish Times*, but when reporting Anglo-Argentinian hostilities in 1982 this newspaper consistently referred to the Falklands as the Malvinas.

The worst consequence of over-naming was that a single feature might be mistaken for two or more features and then appear more than once on the same map. Here was an advantage of following native toponymy that had nothing to do with political correctness: it enabled the traveller half-lost on a strange coastline to check whether the name on his chart was familiar to the local inhabitants, who might never have heard the names invented by previous explorers. At least an importation did not have to be imported from too far away, a plausible reason for labelling Australia with a Maori name.[38] The principle of toponymic self-government was clearly stated by John Green in 1753[39] and by the time of Cook, La Pérouse and, later, Francis Beaufort it had come to be generally accepted:[40] one unphilosophical writer even credited indigenous peoples with the 'true' names.[41] In fact it remained possible for native practice to vary from tribe to tribe and from time to time. Such variability made trouble for early cartographers in French Canada, where it was mentioned in 1689 as an argument for suppressing Indian names altogether.[42] Somehow this difficulty was lost sight of by well-intentioned European geographers in the centuries that followed.

In general, the search for native names grew more intense as modern cartography developed – thanks to higher professional standards rather than through premature feelings of collective post-imperialist guilt. One foretaste of a more sensitive future appears in Samuel Hearne's account of northern Canada. His 'No Name Lake' (1795) was neither accident nor joke, but simply an acknowledgement of how unusual it was not to find any precedent in native usage.[43] It was for physiographic names that the explorer felt most confidence in his own toponymic authority. Even towards the middle of the nineteenth century a South American cataract could be named after King William IV[44] and an African lake (previously Ukerewe) after Queen Victoria – royalty surviving as the last stronghold of the dedicatory principle in cartography as later of the hereditary principle in other spheres.

The opposite danger to the proliferation of aliases was for different features to be mistakenly conflated under a single name. A troublesome example was Pointe Riche in Newfoundland, which French and British geographers of the 1760s placed in different positions more than two hundred miles apart on a coastline whose other features had been mapped in much the same way by both parties (Fig. 13.10).[45] Such is the power of logophilia, however, that a deficiency of names has always been less common than a surplus.

These problems naturally did little to facilitate the progress of geographical discovery. When he came upon an uninhabited island, how could an explorer be sure that none of his predecessors had given it a name? It would be inconvenient to put off a decision until he got home, because here and now there were charts to be updated and a journal to be kept: for the sake of intelligibility alone a place had to be called something. The same sense of urgency was likely to determine who chose a name. Ideally, this privilege should perhaps have been reserved to whichever agency – monarch, parliament, commercial organisation, learned society – had paid for the explorer's journey, but in practice it was easier to make some choice or other on the spot. No doubt the captain would be willing enough to gratify a patron, who would in any case look less egotistical if 'his' island had been named by someone other than himself.

For these reasons a discoverer was generally accorded the power of naming his discovery. Early explorers like Columbus were happy to take this opportunity, and by the late sixteenth century the privilege was being formally conceded by their employers. Thus in 1580 Arthur Pet and Charles Jackman were authorised by the Company of English Merchants to give 'apt names at your pleasure' to any new harbours or headlands they encountered on their way to China. Eight years later, in closely similar instructions to James Bassendine and others from the English ambassador in Russia, the incautiously chosen word 'pleasure' was replaced by 'discretion'.[46]

On the whole, naming was indeed a pleasure, which it was rare for an expedition-leader or map-maker to renounce voluntarily. Captain S. Volkerson took this self-abnegatory stance on the west coast of Australia in 1658;[47] and nearly a century later John Mitchell offered no substitutes for certain new but long and allegedly barbarous, unauthoritative and useless names that he had refused to include in his map of North America. More enthusiastic explorers have generally accepted without prompting that the right to dub must depend on the absence of previous names. When James Hall baptised a headland in honour of the Danish king in 1605 it was on the understanding that 'none other before him hath named it'.[48] Eminent contemporaries who followed the same line included Walter Raleigh and Henry Hudson.[49] Explorers who broke the rule might find their work undone by successors with more demanding standards. Luke Foxe in 1635 claimed to have 'restored all the names of capes, headlands, and islands, formerly given by Captain Davis, Mr Hudson, and Sir Thos Button (which since have been infringed upon) unto their first appellations, both in my book and map.'[50] This was probably an implied criticism of Captain Jen Munk, who in 1614 had taken the liberty of renaming Hudson Strait and Hudson Sea.[51] The more famous the explorer, the more respect his coinages deserved, even from the agents of a rival government. This was Bellingshausen's suitably deferential argument for not replacing Cook's uninspired choice of 'Sandwich Islands'.[52]

Passing from ideals to reality, we must now distinguish cases of obedience and disobedience to what might be called the 'when-in-Rome' principle. There could be no criticism of dubbing as applied to previously unknown and uninhabited countries or to coastal regions whose interiors would probably remain inaccessible for many years. Nor were intruders always to blame for not making contact with the natives, a circumstance that explained why Cook learnt fewer indigenous names in New Zealand than in other more hospitable Pacific islands. In such cases it might be appropriate for non-native names to be withdrawn after further consideration, as when King George's Sound gave way to the supposedly native American form Nootka Sound.[53] Perhaps the same policy should have been extended from the physical world to the creations of colonial settlers, though in the first instance these could surely be expected to carry immigrant names. Then there were previously anonymous natural features of more interest to newcomers than to aboriginals. An example of special interest to cartographers would be a previously nameless hill adopted as a point for triangulation or intersection, Mount Everest being perhaps the best-known example.[54] Shoals on charts fell into the same category. More noteworthy were differences in scale of perception between the mappers and the mapped: Cook could find no native denomination for the island that he decided to call New Caledonia, concluding that 'probably, it is too large for them to know by one name'.[55] Later, no doubt, a local population would be encouraged to invent whatever names their ancestors had neglected to provide, but it is hard to imagine this happening before the twentieth century.

A different situation arose when an existing name was discarded as in some way unsatisfactory. It is hard not to sympathise with a classical Roman geographer's complaint that the names of peoples and towns in Africa were 'absolutely unpronounceable except by the natives'.[56] He was not the last writer to make this confession. Even at the peak of the Enlightenment some North American native words (including, presumably, names) were 'not to be pronounced by Europeans'.[57] Apart from being too hard for the cartographer's fellow countrymen to speak or spell, a name might be too long for the space available on the map, or too easily confused with some other name that was just as well established.

More controversially, the motives for revision could sometimes be political. Thus a name in the English language naturally gave the impression that Englishmen had been the first Europeans to see the feature concerned, which in turn would seem to invalidate claims to sovereignty by any rival government.[58] So when the king of Portugal asked for charts 'with new placenames' in 1503 he doubtless expected these names to be recognisably Portuguese.[59] In the same spirit Louis de Freycinet chose Napoleonic versions, including 'Golfe Bonaparte' and 'Golfe Josephine', for places in south Australia recently mapped and named by the English.[60] Even a native name could support a pre-emptive strike in a battle for onomastic supremacy among colonial powers, or perhaps among individual explorers. It was not unknown, however, for geographical common sense to outweigh political principle. Mapping Australia in 1744, Emanuel Bowen preferred to follow the Dutch names, 'that if hereafter any discoveries should ever be attempted all the places mentioned may be readily found in the Dutch charts which must be procured for such a voyage.'[61] He later followed the same policy with French names in Louisiana.[62]

New names from old

The invention of new names attracted a growing body of precedent that map-makers were generally expected to treat with consideration: most of them did take their responsibilities seriously though some discoverers, such as Cook and Vancouver, clearly enjoyed using as many different name-types as they could think of. Names were generally understood to deserve a measure of dignity, however, and teases or jokes like the beautifully sarcastic 'Painter's wife's island' had little chance

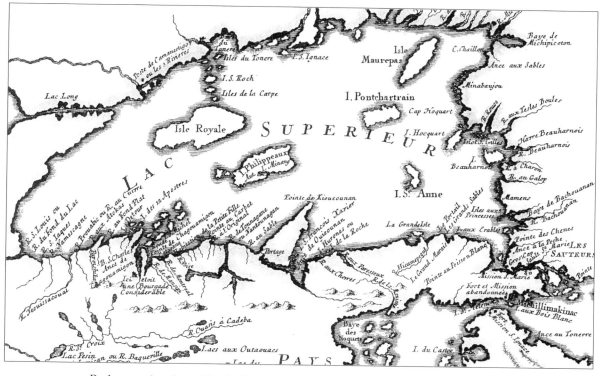

13.11 Real names, imaginary islands. From Jacques Nicolas Bellin, *Partie occidentale de la Nouvelle France ou du Canada* (Paris, 1755). The names Maurepas, Philippeaux and Pontchartrain were those of French government ministers.

of long-term survival.[63] This is not to say that derogatory names were always inappropriate, though it might be seen as a danger signal if a cartographer felt obliged to add an apology, as at 'Lousie Hall so called by the scholars [of Oxford]' in Ogilby's *Britannia*.[64]

The easiest course for the innovator was to fall back on a name that already existed elsewhere. As a palliative for homesickness the transfer of placenames was probably more common among settlers than explorers, though in Australia Matthew Flinders upset this rule by naming some twenty features near Spencer Gulf after places in his native county of Lincolnshire.[65] Not every such choice was dictated by simple nostalgia. A name might recall some genuine physical resemblance to a distant prototype, as with Cook's Portland or Vancouver's New Dungeness. Nor was importation necessarily a one-way process even among Europeans, as can be seen from 'Welsh Potosi' in John Evans's *A new map of Aberystwyth* (1824).[66]

Many imports came from not from places but from persons or other animate beings. These could usually be read as dedications. Columbus was an early exponent: he knew he was renaming rather than naming,[67] but evidently felt justified by his choice of material, which started with God and then descended in succession to the Holy Virgin and to the king, queen and prince of the state employing him. Such sequences could if necessary be continued downwards to the lowest level of approval and even beyond. One can imagine unpopular individuals providing names for natural hazards such as a bog, rock, sandbank or swallow-hole. The Devil has certainly played this part on a number of occasions.

The deferential society was not without its pitfalls. On any given expedition or in any given country, the status of a discoverer's patrons should clearly be proportionate to the size or significance of the features that carried their name. Would King George III really want to share the royal appellation with anywhere as wild and barren as South Georgia?[68] When many features clamoured for attention it could be difficult to avoid afterthoughts: an explorer's initial encounter in an archipelago might not turn out to be the most deserving in any non-chronological sense.[69] On the other hand it was a pity for desperate expedients like 'First I[sland]' and 'Second I[sland]' to be given immortality on a published map.[70] Even when the ships were safely home such problems could not always be forgotten. In mapping the Clarence archipelago John Ross is said to have added six imaginary islands to the three real ones 'so that the Clarences and Fitzclarences might have one apiece' (Fig. 13.11).[71] Not surprisingly, royal or aristocratic nomenclature has often aroused a certain amount of resentment, but at least none of it is likely to rival Committee Bay in Canada for the title of world's most boring placename.[72]

After royalty and nobility, the most likely models for derivative naming were the explorers themselves. A modest commander might await suggestions from his subordinates, who could generally be relied upon to strike the right note. The same relationship could happily be turned around by christening a feature after a junior crew-member who drew attention to it, as with Collins Cape honouring Henry Hudson's boatswain.[73] Matthew Flinders sought the best of both worlds by naming Flinders Island after an officer who happened to be his own brother.[74] Discoverers' names have sometimes been conferred retrospectively as the products of historical research. 'Torres Strait' was the creation of Alexander Dalrymple, who in the 1760s discovered a forgotten report written more than a century and a half earlier by the first European navigator to pass between Australia and New Guinea.[75] This impulse to 'keep it in the family' could also be extended to an armchair geographer, as at Cape Plancius and Hakluyt's Headland, but sadly there are no conspicuous Mercators, Orteliuses or Sansons on the modern map. In the last resort a dedication might be purely personal and private: 'this foreland', a British officer in Alaska told his superiors, 'I beg leave to call Nancy's, a favourite female acquaintance of your humble servant'.[76] By comparison, for an ex-student to name a feature after his professor may seem positively public-spirited.[77] In fact there is no limit to the dedicatory process as long as the number of individuals and communities in the world exceeds the number of mapworthy geographical features.

Some dedications were generic rather than individual. Examples were the Spanish language names chosen by Vancouver to acknowledge his cordial relationship with the Spaniards he met in Pacific North America.[78] Others observed the rule of etiquette that prevents persons of high rank from being addressed directly. An example is Virginia or Maidenland, named by John Hawkins for Queen Elizabeth I in 1593. Further down the social scale such reticence was less appropriate: there is something excessively modest about a name like 'Lieutenant's Lake'.[79]

Some impersonal name-sources

Not all names could be copies of other names. But in cartography, unlike fiction, totally original coinages are less common than might be expected. One large and generally unproblematic category was the perceptual name, based on sensory knowledge of the feature concerned. This after all is how toponymy seems most likely to have evolved in pre-literate societies. Examples are Broad Sound, White Rock, Long Island or, in metaphorical vein, Hen and Chickens. In Drake's New Albion the whiteness of chalk cliffs was coupled with an ancient word for Britain, providing 'affinity, even in name also, with our own country'.[80] More scientifically respectable allusions could be geological (Sandy Hook), zoological (Galapagos), botanical (Grass Cove), or criminological (Isle of Thieves).

Not unnaturally, cartographers were sometimes inspired by their own discipline to create positional names like Antipodes Islands and Eight Degree Channel. Some of these included a not altogether desirable element of egocentricity or indexicality: the Southern Ocean was defined with reference to one person looking in one direction at one time – the year 1513, when Vasco Nunez de Balboa stood on the Isthmus of Panama well within the northern hemisphere. This brings us to such long-sighted specimens as the South African Algoa Bay on the way to Goa in India, or Lachine on the St Lawrence River supposedly leading to China, the latter perhaps a case of an ironical name.[81] Then there was the descriptive naming of shapes too large to be visible anywhere but on a map, such as Footprint Lake, Heart Lake, Horn of Africa and (many centuries older) Nile Delta.[82]

A commemorative name marked an event contemporary with the discoverer's visit. Saints' days appealed especially to Catholic explorers from Spain, Portugal and France, though the festivals celebrated in Easter Island and Christmas Island could be inspiring to Christians of almost every denomination. Secular events were more easily forgotten. Posterity was unlikely to share an explorer's pleasure in providentially finding sacks of meal at 'Mealhaven'.[83] Naming places after ships is pardonably sentimental – and helped to discourage jealousy among seamen vying for personal recognition – but it cannot be said to 'travel well' in a historical sense. Feelings could be memorialised as well as facts (Fig. 13.12). Examples included Point Deceit,[84] Cape Comfort and, most famous of all, Cape of Good Hope.

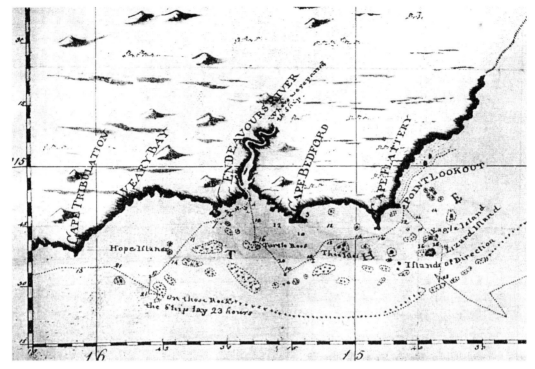

13.12 Placenames with attitude. James Cook, coast of New South Wales, 1770.
British Library, Add. MS 7085.34.

Travellers of the nineteenth century probably paid more heed than any earlier generation to the preference of responsible opinion for native forms as expressed by newspaper and journal editors, civil servants, senior military and naval officers and academic pundits. In a climate of cultural nationalism, name-givers grew more concerned with propriety, gravitas and fear of ridicule. Even an individual's right to name his own discoveries would eventually be disputed by right-thinking commentators.[85] On the whole, however, it is hard to ignore the fact that names can give harmless pleasure by their differences of origin and meaning. Apart from achieving 'accuracy' in the same sense as the roads and rivers on a map, that is by conforming to non-cartographic experience, they should be brief, easy to pronounce, clearly distinguishable from other names, and not liable to arouse unintended emotions. Of course the world abounds with exceptions to all these rules and in the last resort a good name has to be defined as one that no map reader wishes to complain about in public.

CHAPTER 14

Shadowed and counterfeited

MOST MAPS HAVE BEEN produced by copying other maps. As soon as the world's first cartographer laid down his pen (or whatever he was drawing with) the way was clear for the world's first copy. In principle, copying is easily distinguished from the processes of compiling and editing described in later chapters. As a matter of definition, the compiler can increase the geographical content of each map he works on; the editor can decrease that content; the copyist can do neither. In practice, needless to say, the distinction may be a good deal harder to enforce. The long history of map-copying has never been written. Of course imitation is not a process peculiar to cartography, but whatever his or her subject the copyist's methods have evidently been thought too familiar and uncomplicated to justify publishing any systematic account of them.

For the student of reprographic processes this problem is often exacerbated by a shortage of evidence. Maps withdrawn from a repository for copying sometimes failed to make their way back: if the new version was seen as an improvement, the old one might be intentionally destroyed. There is an analogy here with the 'foul papers' of a literary artist: whoever first used the emotive word 'foul' in such circumstances did some disservice to the cause of history. Another hazard for posterity's map scholars has been that some forms of reproduction actually involved the sacrifice of the original, for instance by cutting through the lines of a paper map placed upside down on a printing platform.[1] In the absence of these lost sources, any analysis along the following lines is bound to be imperfect.

Another obstacle to the study of copying is its association with the morally objectionable idea of plagiarism. When Samuel Pepys found an English chart-maker passing off 'the very same plats with the Dutch' it is impossible not to detect some undertones of disapproval.[2] But for an existing map to be 'shadowed and counterfeited'[3] was seriously reprehensible only when it involved deliberate deception or when it deprived the original author of his legitimate reward. It is the injustice of such encroachments that has created the legal concept of copyright, recognised in some form or other by almost all countries with a printing industry. This is a monopoly power over the reproducing, publishing and selling of one's own creation, either in perpetuity or for a specified period. It became a serious issue only when technical progress made it easy to lose large sums of money as a result of unfair competition. Copyright could be held to exist in all printed matter, or it could take the form of a 'privilege' received from a government for some particular work. The first such privileges were granted to book publishers in fifteenth-century Venice.[4] From the next century onwards the practice was extended to maps, the privilege often covering ten years and sometimes more, though seldom above thirty.[5]

In England the position was regularised by a statute of 1734 protecting for fourteen years (later increased to twenty-eight) any maps that bore their publisher's name and their date of publication.[6] Little seems to have written about the legal side of map reproduction either before or after the passage of this act.[7] The only case widely known among historians occurred in 1785, and included a thought-provoking comment from Lord Mansfield: 'Whoever has it in his intention to publish a chart may take advantage of all prior publications. There is no monopoly of the subject … but upon any question of this nature the jury will decide whether it is a servile imitation or not'.[8] On the whole, west European law seems to have succeeded in preventing root-and-branch imitations of works already published in the same country, but did not deter anyone from reproducing the geographical content (Mansfield's 'the subject') of an earlier map.

Methods of copying

Copying methods can be arranged in order of technical advancement and potential accuracy. The crudest was to redraw a map some time after inspecting the original, like a candidate in an old-fashioned 'unseen' school geography test. This must have been how the sixteenth-century German Emperor Maximilian could 'jot down an impromptu map' (in the words of a modern historian) of any place within his domains.[9] A kindred spirit in the following century was Sir John Scot, praised by Joan Blaeu for having 'such an excellent memory that, without any chart or book, he drew the shapes of areas, the places, boundaries … the cities, the rivers, and many other things of this nature',[10] though it seems doubtful whether Blaeu actually published any maps that he knew to have originated in this hit-or-miss fashion. Another feat of memory was a map of the upper Mississippi reconstructed in 1674 by Louis Jolliet after his original had fallen in the river.[11] Before cameras became available an extant map might be copied 'by eye', perhaps as a task for a secret agent whose memory was not as good as Scot's and Jolliet's. Such inexactitude would almost certainly produce considerable errors in distance and bearing as well as omissions: the larger the map, the greater the probable error, both absolutely and proportionately.

One way to limit the copyist's fallibility was in effect to replace a large map with a series of small ones, dividing its surface by visible lines into small bounded figures, and then making a separate freehand copy of each figure on an identical network of lines. This could be most easily done by means of a square grid. The method of squares was described by Richard Eden in 1572[12] and could still be recommended more than a century later, in comparison with a growing number of alternative techniques, as 'the best and exactest way that can be'.[13] Such grids are to be seen pencilled across many early manuscript maps. By varying the size of the squares copying could be combined with either enlarging or reducing. Within each square the draughtsman was dependent on his unassisted vision, though dividers could be used for transferring the positions of points where geographical detail cut the grid lines. Proportional dividers, an instrument known to Leonardo da Vinci,[14] were helpful when a map was being redrawn on a different scale.

Other copying methods required one map to be directly superimposed upon another. 'Pricking through' was a phrase that explained itself: with the original laid on a blank sheet, a sharp point was pushed through both surfaces to transfer significant locations from one to another in the form of pin-holes – especially towns, mountains, or angles and intersections strung along the course of a linear feature such as a road or river. The second map was then finished by connecting the pin-pricks with freehand drawing. Alternatively, powder could be shaken through the upper pin-holes on to the lower sheet, whose surface would thus remain undamaged – a surprisingly effective procedure, as long as perfect accuracy is not required. With points used as a substitute for lines, these perforations anticipated the modern technique of digitising. Another method was to go over the upper outline with a hard pencil or drawing point, impressing its detail into the map below as a pattern of grooves without actually piercing the surface of either sheet. Both holes and indentations often remain visible on early maps.

In a 'transfer', the original (upper) and new (lower) map were separated by a third sheet dusted with blacklead and turned face downwards. Tracing now took place as before, but in this case pressure could be reduced and the new image came out as black lines rather than colourless depressions. A not dissimilar process was to trace the original in lamp-black and 'rub down' the resulting image on to the new surface. As we shall see, the word 'transfer' was also used more specifically for the reproduction of an engraved image on a lithographic stone. More familiar to modern readers is the tracing made directly on to transparent paper, another operation likely to leave tell-tale marks on the original map. In 1838 one generally knowledgeable writer referred with an air of unfamiliarity to 'what is called tracing paper',[15] but long before this product became available commercially a home-made version of it could be produced by treating ordinary paper with linseed oil, a process described by the fourteenth-century artist Cennino Cennini.[16] Maps on tracing paper were brittle, dimensionally unstable, and not always easy to read. An alternative was the 'glass for tracing plans from the light' mentioned by James Cook in 1768.[17] Here good quality drawing paper could take the place of a transparent overlay. The clean paper was placed above the original map, and both rested on a glass surface mounted above a strong light – in pre-electric days, probably daylight from the office window.

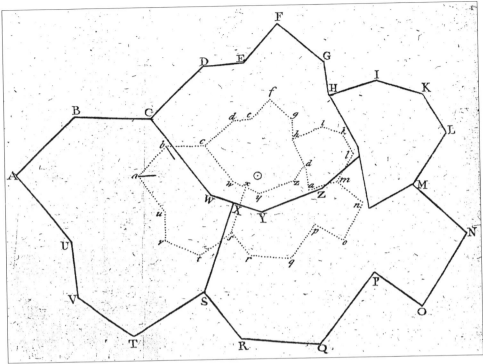

14.1 Map-enlargement by the radial method. Robert Gibson (ed. Patrick Lynch), *A treatise of practical surveying; demonstrated from its first principles* (6th ed., Dublin, 1810), pl. 12.

Copying could also be performed instrumentally. One appliance sometimes recommended was the 'reducing scale', apparently a brainchild of the early seventeenth-century surveyor Aaron Rathborne.[18] This entailed superimposing the old and new sheets, pinning a pivoted ruler through some convenient point near the middle, and then rotating the ruler so that each feature could be given its correct bearing and correct rescaled distance from the central 'origin' (Fig. 14.1). It would certainly be easier if the copy was plotted on tracing paper, but even then such a device might well rank among those described by a later surveyor as serving more to amuse than to edify.[19] More elaborate was the pantagraph, invented by Christopher Scheiner in 1603 but still a novelty sixty-five years later and not widely known until it had been improved in the middle eighteenth century.[20] This used a system of beams and joints with a tracing point for the original and a pencil for the new drawing (Fig. 14.2).[21] Another such device, the eidograph, was an arrangement of wheels, arms and tracing points introduced in 1821 and said to be more accurate than the pantagraph for reductions by up to two-thirds.[22] Its optical counterpart, the camera lucida, had been invented fourteen years earlier.[23] There was one danger in any method that included tracing or pricking through: the map might at some stage become accidentally reversed. This possibility has been held to explain the transposition of east and west in a map of Iceland by Gerard Mercator's son Arnold.[24]

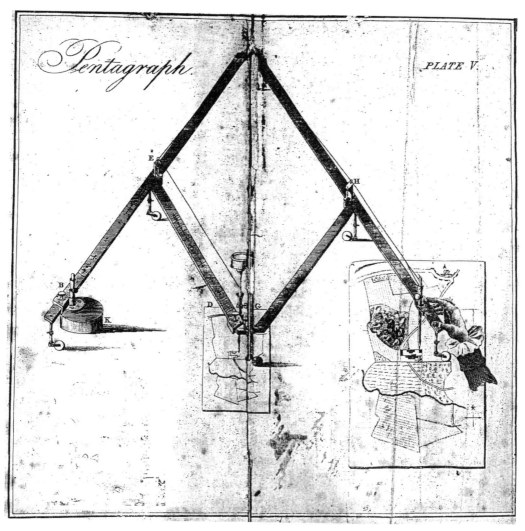

14.2 Map-copying by pantagraph. A. Nesbit, *A complete treatise on practical surveying, in seven parts* (3rd ed., York, 1824), p. 237.

The sovereignty of print

A copy might in Francis Drake's time be described as correct 'to a hair',[25] but all the methods reviewed above were liable to error. As often happens, Ptolemy's is the first recorded cautionary voice: 'continually transferring a map from earlier exemplars to subsequent ones tends to bring about grave distortions in the transcriptions through gradual changes.'[26] Tracing was also a slow process that there was no way of accelerating: each successive copy took as long to prepare as the last. The technique that avoided this endless repetition was printing, which apart from its economic benefits had the advantage that successive impressions from an engraved plate were identical in appearance with the plate image as well as with each other – except that the image had to be cut in reverse, which could be made easier by tracing the original map on the back of itself.[27] Once engraved, the copy would enjoy a degree of dimensional stability not to be expected from paper or parchment.

When a blank plate had been made potentially legible by waxing or varnishing, an image could be recreated on it in reverse by any of the copying methods mentioned above, except of course for direct tracing on a transparent medium.[28] As we have seen, none of these methods was infallible, and a map could still suffer some mishap on its journey from manuscript draft to printing surface: hence Robert Plot's anxiety lest the degrees of latitude in his map of Oxfordshire had 'moved in the [en]graving'.[29] At least one eminent eighteenth-century cartographer hesitated to take the risk, and made a point of drawing his own outline on the plate before allowing the engraver to start work.[30]

The earliest printed maps date from the fifteenth century. At first, in Italy, engraving was done on copper. This metal could be cheaply purchased in large sheets that were durable and yet soft enough to be easily incised and if necessary re-incised. In cartography, engraving has generally meant cutting into the plate with a sharp knife known as a burin or, for fine uniform lines, a needle or drypoint. With recurrent images some labour-saving device might be substituted, such as a punch or stamp for a dot symbol, or a toothed wheel for a broken line.[31] In certain methods of pictorial engraving the plate was 'bitten in' by acid to produce a variety of shades and tints: the best known applications of this technique were aquatint, mezzotint and etching. Occasionally a stone tool would serve to engrave an effect of continuous shading.[32] These more complex processes were little used in map production, probably because cartography could achieve most of its desired effects without them.[33]

Engraving of all kinds was more difficult than drawing or painting on paper. It was generally learned in youth by apprenticeship, thus militating against the acquisition of other skills including the art of the surveyor. Practitioners might move into various kinds of profitable business activity, map-publishing among them, but seldom into different branches of cartographic technology. Many engravers spent part of their time on subjects other than cartography, such as portraits, heraldic devices, landscape views, flower pictures, anatomical diagrams and so on. But there was also a tendency for map engraving itself to be subdivided. A cartouche or title panel, for instance, would often be cut by a different hand from the adjoining map. Cartouches were unknown in the British Ordnance Survey, but it employed different engravers for outline, ornament, writing and hachures. (Ornament meant distinctive areal 'characteristics' for wood, pasture, marsh etc.) The abilities required for printing were less arcane. The plate was inked, and the surplus ink wiped from the flat portions of its surface. Sheet and plate were then passed like a sandwich through the rollers of a press, transferring to the paper the

14.3 Woodcut map with woodcut names. From *Tabula nova Heremi Helvetiorum* in Martin Waldseemüller (ed.), Ptolemy, *Geography* (Strasbourg, 1513). See also Fig. 12.6.

14.4 Woodcut map with some letterpress names: the Balkans. From Sebastian Münster,
Nova Graecia in *Cosmographia* (Basel, *c*.1550).

ink that had remained in the metal grooves. Before the nineteenth century, multi-colour printing was achieved (though not very often) by superimposing on the same sheet successive images from two or more plates each charged with a different ink.[34]

From the 1820s onwards, as maps were bought and sold in larger editions, copper was sometimes replaced by the more durable medium of steel, allegedly with the result that 'finer lines could be etched closer together on the surface, allowing for more subtle variations in shading, definition and hachuring',[35] though some experts claim to find steel and copper engravings indistinguishable.[36] Historically, as we shall see, this problem was overtaken by the advent of cheaper reprographic methods.

An alternative material for printing-platforms, especially popular in sixteenth-century Germany, was wood. Like copper, this could be used as a medium for line engraving; but normally the term 'woodcut' meant that map detail, instead of being incised into the printer's block, was preserved in relief by removing superfluous wood, so that in printing it was only the inked ridges that made contact with the paper. Woodcuts were of coarser appearance than impressions from metal plates, and the printing surface was easily damaged (Fig. 14.3). On the other hand in a relief system the crudity of woodcut lettering could be avoided by combining the line-work of the main block with the letterpress reproduction from metal type of names and other script inside or outside the map frame (Fig. 14.4). This had the advantage that the same map could be printed in alternative editions with names and inscriptions in different languages. Words could be treated in three ways. In the first, type was set up in a separate forme and united on the paper with line-work by double printing. Secondly, type could be fitted into rectangular holes cut in the map-block. Finally a mould could be taken of each complete word as set up in conventional type and a cast of molten metal poured into it, solidifying as a thin stereotype that could then be cemented on to the block. Much care and thought have been given to the identification of these processes as revealed on surviving sixteenth-century maps, but the essence of the woodcut technique remains the same.[37] Its importance in the long term was to facilitate the use of maps as book illustrations.

Meanwhile copper plates were gaining ground for sheet and atlas maps and remained in regular use until the nineteenth century. An important extension of the technique was the electrotype process of 1839. Previously, duplication had been effected by re-engraving a map *de novo* on a fresh sheet of copper, perhaps as a substitute for a worn plate, perhaps to speed up the printing of a map in heavy demand.[38] Henceforth, however, the printing surface could be copied by a combination of physics and chemistry. This involved depositing a layer of copper electrolytically over an engraved plate to produce a matrix on which the incised lines stood out in relief. If the same operation was then applied to the matrix, the result would be a facsimile of the original printing surface.[39] Ridges on a matrix were more easily removed than incisions on a plate, and portions of the matrix that

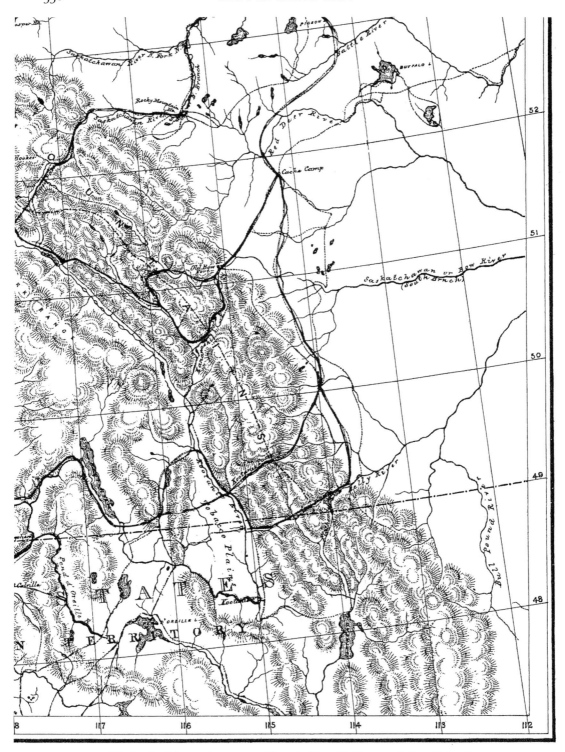

14.5 Printing by lithography: western Canada. From H. Parsons, *British Columbia*,
lithographed by W. Oldham, 1862.

were scraped in this way would appear on the duplicate plate as clean blank spaces in which new detail could be inserted by the engraver. A contemporaneous application of electrolysis was cerography, in which the map was engraved in a coating of wax on a metal surface, and then printed from an electrotype of the incised wax. The advantage, as with woodcuts, was that letterpress names could be mounted in the printing platform. This was an American invention and remained most popular in the United States.[40]

A slightly earlier nineteenth-century innovation, more widely adopted than electrotyping, was lithography, in which a reversed image of the map was drawn in specially prepared ink on a flat limestone slab, another medium on which erasure was comparatively easy. Printer's ink applied to this surface would then adhere to the drawing but not to the empty spaces, thus allowing the image to be impressed on to a sheet of paper brought into contact with the stone. This technique was quite different from engraving on stone, as practised in the official survey of early nineteenth-century Bavaria.[41] Although only half as expensive as copper engraving,[42] lithography was less suited to fine detail (Fig. 14.5), but many nineteenth-century printers got the best of both worlds by engraving a map on copper, transferring the image to stone and then printing from that. The original plate, preserved from the wear and tear of everyday use, could then print a second transfer when the first one wore out, an operation that could if necessary be repeated a number of times.[43] Many nineteenth-century maps were of hybrid origin, engraved material reprinted from transfers being combined with additional detail drawn directly on the stone. Not all such mixtures were harmonious.

Even after copper-plate printing was well established a small number of manuscript copies (up to ten, according to a nineteenth-century source)[44] could be produced for less than the cost of engraving a single plate. Hand copying therefore kept its place where demand was limited, as with confidential estate maps or military surveys.

The propagation of error

'How unsafe it is to copy anyone', Abraham Ortelius was told by a candid friend soon after publishing a large collection of map copies in his *Theatrum*.[45] Indeed, accurate reproduction of any sort has always been a task of some difficulty. Could Newton or Einstein have transcribed a 20,000-word text without a single mistake? On a map the difficulties were accentuated. Copyists of prose were guided by their understanding of its sense: if words were inadvertently omitted a reader had a fair chance of noticing that all was not well, even when the ideas in transmission were unfamiliar to him. There were many map features for which this kind of reassurance would clearly be impossible. Also, the correction of prose texts is made easier by our need to absorb them sequentially: at any given moment the literary proof-reader

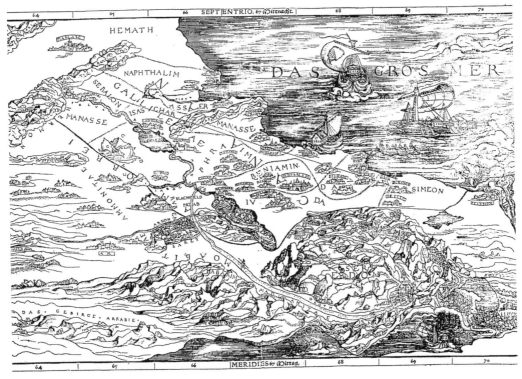

14.6 Engraver's reversal of a Biblical map-image, the Exodus. *Above*: Christopher Froschauer (Zürich, 1525), reversed. *Below*: W. Vorsterman (Antwerp, 1528), corrected.

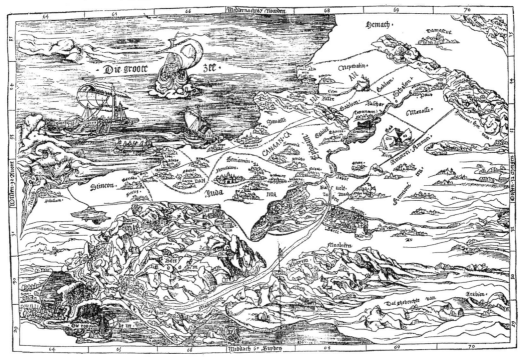

knows how much has been done and how much remains. A gridded overlay could be used to impose a serial character on the process of checking a map, but graphics still seemed to tax the corrector's self-discipline more severely than prose, not least because the grid would usually embrace a number of inscriptions running from one square into another. Error often spread beyond the body of a map: it was not unknown for the numbers or even the words in a cartographer's scale statement to be miscopied, with obviously calamitous results.[46] Altogether, Thomas Jefferson's experience in mapping Virginia was probably not unusual: in his first proof there were 172 errors, and that from an engraver reputed to be the best in London.[47]

Let us now review the main sources of error in copying. No method that involved estimating distances or directions could be totally reliable. In particular, many freehand draughtsmen were prone to increase or decrease their scale unwittingly as they worked their way across the page. Topological errors were less common. Sometimes a narrow isthmus might be overlooked to make an island out of a peninsula, as when Scotland was separated from England[48] or the Peloponnesus from the rest of mainland Greece,[49] though accidental islands of this kind should be distinguished from those created by extrapolating from original but imperfect survey data. Most copyists would be tolerably careful about such fundamental issues. A more common if less spectacular case was where three roads or rivers were wrongly made to converge on two points instead of one, or vice versa. It was in fact surprisingly difficult to trace a complex outline with no errors at all; this can be shown by comparing the ink lines on an original map with the indentations left by the copyist and noting how often the two tracks fail to coincide. Another problem was that paper would sometimes shift in the course of tracing, even when upper and lower sheets had been pinned together. The result might be a kind of fault zone or shatter-belt running across the copy, its gaps and overlaps concealed by a half-unconscious process of 'distribution'.

In tracing and eye-copying alike, omissions were by no means unusual. If a draughtsman left a gap in what ought to have been identifiable as a single line, his mistake would probably be noticed before the map left the office. Harder to detect were missing placenames, towns, branch roads, river tributaries, small lakes and islands, short lengths of boundary common to not more than two territories, and patches of colour or shading. 'Dyslexic' eyeslips, usually associated with writing, could also occur in line-work: on a river, for instance, directions were sometimes transposed in copying a double meander loop.[50] Such errors could occur non-dyslexically when an engraver forgot that the image on his plate should be the reverse of what the reader was meant to see (Fig. 14.6). Inversions of this kind were especially characteristic of engraved marginal ornament, where the only harm they could do was aesthetic.[51] The consequences were more serious if magnetic north was printed on the wrong side of true north.[52] Transpositions in the body of a map might be easy enough to detect, as for instance when the name 'Mespil House' was

printed in reverse on James McGlashan's map of Dublin in 1846, but some readers
may have been misled when the St Lawrence appeared east rather than west of
Newfoundland on a map by André Thevet[53] or when Johann Christoph made a
similar mistake in copying a plan of Cambridge.[54]

Even printers were not invulnerable to errors of copying. Plates might become
accidentally scratched, the scratch reappearing in print as a faint and meaningless
line that could conceivably be mistaken for an otherwise unrecorded river or
boundary (Fig. 14.7). Fractures on the edge of a plate might cause part of the
printed detail to become displaced on the paper.[55] In at least one case a plate is
known to have printed an image of the rivets used to repair it.[56] Even without such
traumatic experiences the copper itself would eventually become worn down by
repeated passages through the press, obliterating finely engraved detail such as
minor hill features or waterlines along the coast. Where different colours were
printed on the same sheet from different plates, their images not infrequently failed
to 'register', perhaps causing black roads to run across blue lakes. In addition, some
shrinkage usually took place when paper was dampened as a prelude to copper-plate
printing: with a large sheet-size, especially, this made the scale of an impression
measurably smaller than that of the plate it was printed from. Peter Perez Burdett
claimed to be unaware of this effect when a 'small contraction' in his map of
Derbyshire drew critical comment from the Society of Arts.[57]

The copyist as interpreter

Next for analysis are less straightforward kinds of error involving the use of
conventions. Copyists were expected to use their common sense. They ignored blots
and slips of the pen, as well as differences in the thickness and density of individual
lines or in the dimensions of small symbols that were obviously intended to
maintain a standard size. Nor would they necessarily reproduce every dot in a
stippled sandbank or every shadow-line on a hillside. Everyone knew that in these
cases an individual mark had no separate significance and that what counted was
the overall effect.

But having assumed a degree of independence that no reasonable person would
deny him, the copyist might then be tempted further. Consciously or otherwise, he
would now make an effort to record not what he saw, but what he thought was
represented by what he saw, whether a physical object on the ground or a letter of
the alphabet on another map. This was an important distinction, because beliefs
unlike raw sense-data have the property of being either true or false. Two antitheses
were involved here, conventionalism versus realism and copying versus editing. Take
for instance the schematic point-symbols and line-characters used by cartographers
for buildings, settlements, boundaries or railways. Copyists might alter these, not
for the sake of visual effectiveness (that would make the copyist an editor) but more

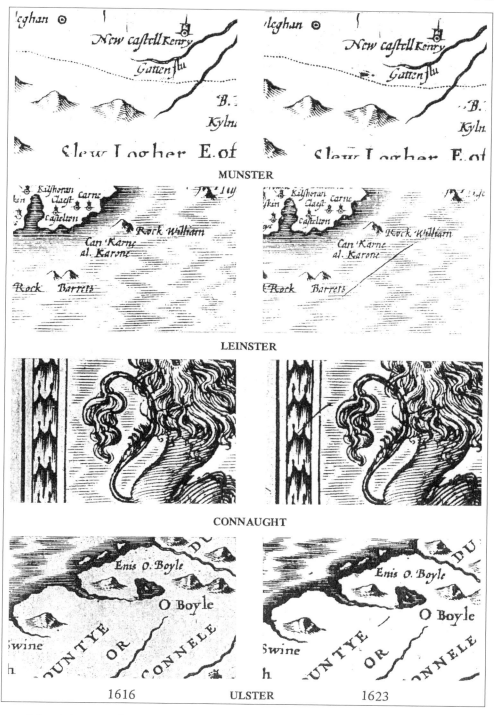

14.7 Printed plate scratches appearing in later editions of maps by John Speed. Andrew Bonar Law, *John Speed: maps of Ireland* ([Dublin], 1979).

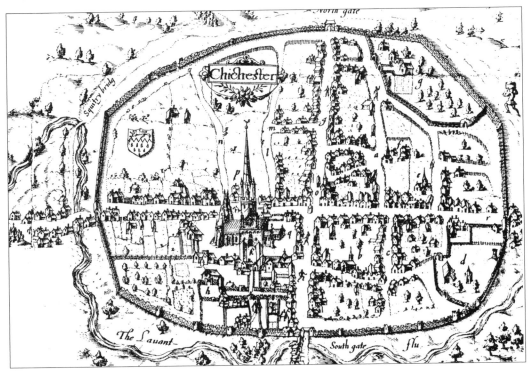

14.8 Chichester, original scale *c.*1:7000, from John Norden, *Sussex* (London, 1595/6).

or less automatically, to match their own personal style. In particular they often found it hard not to modernise antiquated conventional and decorative features especially subject to the influence of fashion, even in estate maps where the testamentary value of a proprietorial record might be increased by stylistic evidence of age. Often they must have found the old forms too unfamiliar to be easily imitated.[58]

More important for later historical geographers was the conventionalisation of quantity. Modern researchers seeking to estimate a pre-census population have sometimes counted the number of houses shown on an early map as the basis for a carefully considered multiplication sum. A time-travelling early copyist might well find this procedure reprehensible. A good example is John Speed's plan of Chichester in 1610, usually regarded as a copy of a plan made by John Norden fifteen years earlier.[59] In the block bordered by East Street, South Street, West Pallant and North Pallant Norden shows some twenty-five houses fronting one or other of the streets (Fig. 14.8). Speed shows about thirty-four (Fig. 14.9) – not quite what Drake's contemporaries would have described as copying 'to a hair'. Nor were Speed and his engraver over-punctilious with the copying of built-up areas from manuscript to print in his own town plans.[60] Clearly in such cases the number of house-symbols was simply not regarded as a matter of substance.

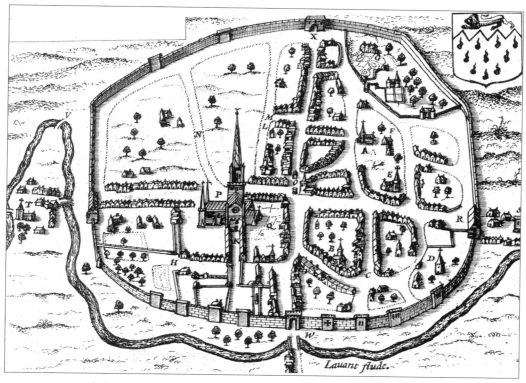

14.9 Chichester, original scale *c*.1:6800, from John Speed, *Sussex* (London, 1610).

The tracing of lines, the copying of point- or area-symbols and the writing of words were usually different processes, perhaps separated by a short but significant lapse of time. For instance coasts, rivers and roads might be drawn first, with towns, area-fillings and placenames inserted later. This interval made it all too easy for a church to be placed at the wrong road-junction and a town on the wrong side of a river, or for woods to be coloured as if they were lakes. Such eye-slips could have serious results, the most flagrant example in the whole of cartographic history being the transposition of 'Europe' and 'Africa' in the Hereford *mappamundi*, perhaps the result of employing a second copyist.[61] Even when there is no positive evidence of copying it may be difficult to blame such errors on the original cartographer: surely no one who knew Wales as well as Humphry Lhuyd would have placed Newport at the mouth of the River Wye instead of the Usk.[62]

Conventions in particular were subject to interpretative errors that did not arise when outlines had been followed mechanically. A line symbolising a territorial boundary could easily be mistaken for a river: this explains some but by no means all of the erroneous hydrographic connections to which early cartographers were especially prone, a phenomenon well illustrated – to pick an example at random – in maps of Corsica published between the sixteenth and early eighteenth centuries.[63] Other fictitious rivers began life as uncoloured profile symbols, an error occasionally seen in John Speed's county maps. Spectacular misunderstandings occurred when marginal line-work was treated as a boundary between land and water. A famous case was the suspiciously rectangular island which appeared as 'Antillia' on several fifteenth-century maps of the Atlantic and which probably originated as a box enclosing a placename.[64] A related error was to mistake an initial letter 'O' for a circular settlement symbol, as when a single house-cluster called Shipton Oliffe in Gloucestershire became split into two villages, Shipton and 'Liffe'.[65] Certain conventions were so subtle that a copyist might simply fail to notice them. Mercator's world map of 1538 differentiated conjecture from fact by leaving some countries without a definite coastline, but heavy offshore shading made this feature easy to overlook, and there was at least one copyist who gave both kinds of coast the same emphasis.[66] A similar distinction got lost when Lewis Evans's map of eastern North America was copied by Thomas Jefferys.[67]

Point symbols could also be misinterpreted. Robert Lythe's original survey of the barony of Idrone in Ireland no longer exists, but we may be sure that he did not give every church the kind of massive square tower and tall pointed spire that appears on the only surviving manuscript copy of this map.[68] Again, in Baptist Boazio's *Irelande* (1599), also derived from Lythe, there were large building symbols coinciding with most bishops' seats and almost every placename with an initial 'C' (for 'castle') was given a solid-looking tower and a flag, again in defiance of the original survey.[69] It is in cases like this that copying merges into editing, and editing into compiling.[70]

Script and transcript

Handwriting was especially susceptible to quasi-editorial adaptation. In transcribing a written document a clerk did not expect to risk accusations of forgery by exactly reproducing the appearance of the original right down to the colour of the ink and the sizes of every loop and tail: no honest man copies a word by tracing over it, and indeed a copyist faced with an unfamiliar script might give up altogether and rest content with total anonymity rather than attempt an exact imitation.[71] In this respect cartography resembled ordinary prose. Thus on English manuscript maps earlier than the 1570s the inscriptions were often written in a 'secretary' hand, many of whose characters have become unrecognisable to a modern reader. Yet this style of writing is almost never seen on printed maps of any period, presumably because engravers had chosen to substitute their own kind of roman or italic script.

Among all kinds of inscription placenames have invariably been the most accident-prone (Fig. 14.10) – a fact acknowledged in Cornwall by Joel Gascoyne when he stood over everything else on a county map of 1699 but admitted that his names 'may not all of them be according to their originals'.[72] Among the characters

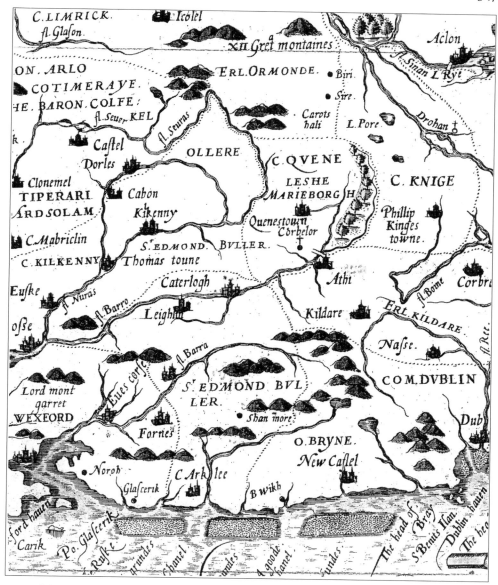

14.10 Miscopied names, south-east Ireland. Jodocus Hondius, *Hyberniae novissima descriptio*, 1591. Examples include Buller (Butler), Fornes (Ferns), Knige (King's), Ollere (Ossory).

most frequently confused were 'o' for italic 'a', 'i' for 'l', 'c' for 'e' or 'o', 'n' for 'u', 'g' for 'y', 'k' for 'h', 'cl' for 'd', 'm' for 'nn', 'b' for 'G', 'C' for 'G' and 'S' for 'ʒ'. Capitals and miniscules were hard to distinguish for 'c', 'm', 'n', 'o', 'p', 's', 'u', 'v' and 'w'. Dyslexic transposition could change 'Stratford' into the not-implausible 'Startford'. Words might be wrongly divided (fission) or run together (fusion), a common occurrence when unfamiliar abbreviations were misinterpreted. Thus in early English maps of

Ireland 'C[astle] Lander' was copied as 'Clander', 'B[ally] Amore' with two mistakes as 'Ramore', 'Clare mon[asterium]' as 'Claremon', and 'F[luvius] Inch' as 'Finch'. The last of these errors was reversed when the River Fleask appeared as 'F[luvius] Lyx'. Then there was the risk, already encountered in Gloucestershire, of a two-word name being mistaken for two names, a danger bravely outfaced in the Irish Ordnance Survey's deliberate choice of forms like 'Newtownmoneenluggagh' and 'Englishgarden'.

A name could lead a copyist astray by reminding him too vividly of some other name. In John and Charles Walker's map of Leinster (*c.*1845) the town of Castleblayney appeared as 'Carrickmacross'. Since Carrickmacross was a real town no more than an inch away on the same map, it needed only the first two letters to trigger the necessary association in the mind of an inattentive engraver. Such confusions were especially probable where one or both of the offending words occurred in ordinary language as well as in toponymy, a coincidence that caused John Rocque to engrave 'Old Town' (a common placename elsewhere in Ireland, though usually written as one word) in County Dublin instead of what seems more likely to have begun as the descriptive caption 'Old tower'. Contrariwise, the first three letters of Saxton's 'Thefofog' (modern Tycoch) in Herefordshire were predictably misread by Pieter van den Keere as an English definite article. These associative errors became especially revealing if more than one language was involved, as when the English placename suffix 'wick' was 'naturalised' by a sixteenth-century continental engraver into 'wik'.[73] In a table of linguistic equivalents drawn up by an eminent Dutch cartographer of the same period the Dutch name 'Grauesendt' (Gravesend) was conscientiously anglicised to 'Grauvvesdendt'.[74] This kind of anomaly can sometimes help in assigning nationalities to the people concerned in creating or copying an imperfectly documented map.[75]

Some principles of cartographic genealogy

It is not just the fact of copying that interests a map historian, but the exact sequence of events involved, and here we must briefly introduce the idea of family relationships. A family in this sense is a group of maps that are all descended from the same archetype. It is sometimes possible to express such genetic relationships diagrammatically as a tree or stemma, in which 'being copied from' is equivalent to 'being the child of'.[76] Most maps possessed some characteristics which were not indispensable to the nature of a map and which did not correspond with any kind of reality on the ground. Examples were the exact designs of north indicators, scales and systems of conventional symbols. If different sets of these features were alike, it might be for either of two reasons. First, however separate the two maps' individual life-histories their authors might have been members of a single 'school' distinguished by common forms of decoration or convention. There were some

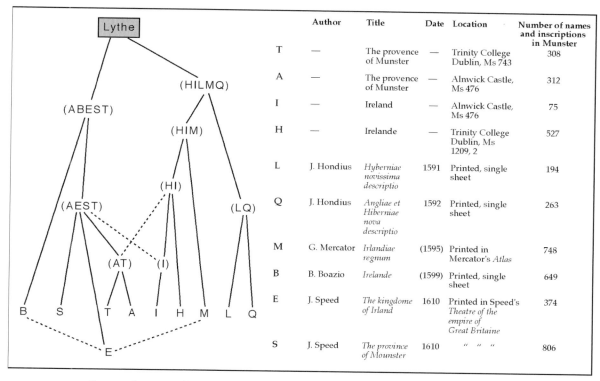

	Author	Title	Date	Location	Number of names and inscriptions in Munster
T	—	The provence of Munster	—	Trinity College Dublin, Ms 743	308
A	—	The provence of Munster	—	Alnwick Castle, Ms 476	312
I	—	Ireland	—	Alnwick Castle, Ms 476	75
H	—	Irelande	—	Trinity College Dublin, Ms 1209, 2	527
L	J. Hondius	*Hyberniae novissima descriptio*	1591	Printed, single sheet	194
Q	J. Hondius	*Angliae et Hiberniae nova descriptio*	1592	Printed, single sheet	263
M	G. Mercator	*Irlandiae regnum*	(1595)	Printed in Mercator's *Atlas*	748
B	B. Boazio	*Irelande*	(1599)	Printed, single sheet	649
E	J. Speed	*The kingdome of Irland*	1610	Printed in Speed's *Theatre of the empire of Great Britaine*	374
S	J. Speed	*The province of Mounster*	1610	" " "	806

14.11 Suggested stemma for maps derived from the Irish surveys of Robert Lythe (1567–71), based on placenames in the province of Munster. Brackets indicate hypothetical ancestors, defined by the reference-letters of the maps believed to be descended from them. Convergent lines of descent are shown by broken lines: see below, chapter 15.

things however that could not be derived either directly from the landscape or from any peer-group's reservoir of cartographic custom. Imagine a large town of irregular shape and indefinite external boundaries being mapped by two cartographers on different scales and in different styles. Imagine also that on both maps all four margins intersect the straggling outer suburbs of the town in exactly the same places. Such a coincidence could hardly be due to the authors having mapped the same subject from within the same cartographic tradition. More likely in this case is our second possibility: that one map had been copied from the other or that both had been copied from a common exemplar, or at any rate that both maps were members of the same family in the sense defined above.

With maps, as with human beings, family resemblances can often be detected almost at a glance without the help of conscious itemisation and analysis. It is easy to see for instance that certain early sixteenth-century maps of England and Wales must be closely related to the famous Gough map of nearly two hundred years earlier.[77] But more careful study will probably be needed to establish the exact relationships within a kin-group of this kind. The resulting stemma may contain both terminals and intermediaries, and some of the latter may form nodes from which two or more lines of descent diverge. Suppose now that a stemma is to be drawn for three maps of the same family, A, B and C. If in some particular detail A and B agree with each other and disagree with C, then it is likely that C occupies a terminal rather than a nodal position on the family tree and the stemmas that can be classed as improbable are those in which no line of descent, whether straight or angled, can be followed between A and B without passing through C.

Besides extant maps, the stemma will often embrace inferential or hypothetical maps whose existence is deduced either from the form and content of other maps in the same family or from independent historical evidence (Fig. 14.11). To establish the exact nature of a stemmatic relationship may require a close scrutiny of individual variations among which simple copying errors are the most important. In this process shared errors, compared with other recurring map features, may have the evidential advantage of not being derivable from either reality or cartographic custom, and thanks to the range and depth of human fallibility they are also helpfully numerous. In fact there are so many ways of going wrong that we may be tempted to consider it unlikely for two independent copyists to make the same mistake. This means that an error repeated is probably an error copied. It is also often a fair working assumption that once an error had slipped through, no later copyist would be knowledgeable enough to correct it. The easiest errors to study are those involving words, because letters of the alphabet, unlike other kinds of shape, exist as discrete categories amenable to statistical analysis.

In practice, it was clearly possible for a map to disappear after having had only one copy taken from it, and for this to happen several times in succession within the same family, leaving the historian with a number of hypothetical maps threaded along a single branch or a single trunk of his stemma like beads on a string. There might be historical evidence for 'uniparous' hypotheticals, and even without such evidence the map itself may arouse suspicions of copying. For instance where Mercator named Ballymore in Ireland as 'Palemone' it might seem unlikely that he or any other individual would have combined two or more bad mistakes in a single fixation, so this may look more like a coupling of successive errors by different copyists.[78] It will probably be impossible, however, to say how many non-surviving copies succeeded each other on a single unbranched line of descent. It therefore seems best to avoid an infinity of alternative stemmas by assuming that the only hypothetical ancestors admissible in a family are those standing where two or more lines diverge.

We have already laid down some of the conditions, for a group of three maps, in which a number of theoretically possible stemmas can be dismissed as improbable. The further significance of such cases depends on the exact nature of the agreements within the group. Let us continue to assume that A and B agree against C. If A and B seem erroneous and C correct, or if A and B seem more erroneous than C, then it follows: (i) that C is not descended from either A or B and (ii) that either A is descended from B, or vice versa, or that A and B share a common ancestor which is not C and which is not shared by C. In this way a large number of theoretically possible stemmas can be eliminated.

Ideally such arguments, pursued from name to name across the whole map, should in the end leave only one possibility unrefuted. If this does not happen, we should choose whichever of the surviving stemmas includes the fewest hypothetical intermediaries. If every conceivable stemma has been eliminated this is for one of two reasons: either some variations must have been duplicated, or corrected, by chance coincidence; or else some lines of descent must have converged instead of diverging. In assigning probabilities to these options, different patterns of variation can be weighed in proportion to the frequency of their occurrence. But in any such assessment some errors must be recognised as intrinsically more probable than others. For instance mistaking an italic 'a' for an 'o' was easier than copying 'o' as 'oo'. Sometimes such *a priori* probabilities may be hard to evaluate. Certain spelling variations that would now count as erroneous were wholly permissible before the eighteenth century, for instance in English the choice of 'y' or 'i' and the presence or absence of a terminal 'e'. Then legibility may have depended on the physical state of a map that is no longer available for inspection. Thus for a copyist to omit the initial letter or letters of a word might seem antecedently improbable; but if on the original map the ink had flaked away from those letters, or if they had been obscured by a band of heavy colour, this error could easily have been repeated by a succession of independent draughtsmen.

The copyist's contribution: a summary

Enough about the copyist's capacity for harm: what good did he do? On the face of it, none. Indeed, having for the moment excluded the compiling and editing processes, we have made his ineffectiveness a matter of definition. Of course each copyist had his own skills: a sharp eye, good sensori-motor coordination, and the ability to persist with an uninviting task for hours at a stretch. But these gifts had little connection with the subject-matter on which they were brought to bear. An artist might copy only one map in his entire career and yet still do it as well as anyone else could. If he copied two maps from two different sources, nobody would blame him – in his capacity as copyist – for any contradictions between them, just as no modern letterpress printer in a free country would lose professional credibility by accepting the work of authors with different opinions on controversial subjects. In short, it was unnecessary for a map draughtsman to have any knowledge of geography or even of cartography – which means that the tendency of copyists to sign their work may bring less benefit to the map historian than might be hoped. However, these people can still take credit for a major quantitative service to mankind: by multiplying a single image, they extended its usefulness temporally, spatially and thematically. They could prolong the life of a map when the original was in decay, not just by making more copies but by making individual copies more durable, as for instance when a paper map was redrawn on parchment or on a metal printing plate. Furthermore, they could allow it to be coloured, augmented or annotated while leaving the original inviolate. No wonder so many map copyists seem to have decided that if a thing was worth doing, it was worth doing badly.

Not badly in every possible sense, however. There was one way in which a copyist could improve upon his model without exceeding our definition of his role. He could increase its marketability by making it look better. This was probably not very difficult. In the first draft of a map, the author had to concentrate on getting the lines to fit together, and until this had been done it was pointless even for the most talented artist to worry too much about refinements of presentation that could safely be postponed until the making of a fair copy.[79] We are not discussing extraneous decorative features here: they must be classified as editing. Our concern is with the copyist who gave lines a more uniform width and density, standardised colours and fitted them more exactly into the appropriate spaces, regularised the size and form of script and improved its legibility. In all these respects the influence of printing was especially important. The worst engraving was never as bad as the worst penmanship: it took so much work and skill to put a line on to copper that time had been flagrantly wasted unless the result looked right, on first acquaintance anyway. This difference appears in the fact that among eighteenth-century readers one way of praising a manuscript map was to say that it resembled a copper plate.[80] At this level, copying was a scarce talent, not necessarily combined with any other. In fact, reversing the sense of the previous paragraph, we may say that the more beautiful the map, the more likely it is to have been redrawn without substantive modification from a single source.

CHAPTER 15

Cunningly compiled and made

SOME MAPS DERIVE NOT from the author's own experience but from data originating with previous researchers whether as verbal texts or graphic representations. In this chapter compilation means making a map from two or more such independent sources. (We ignore the sense in which a surveyor could be said to compile a map from the pages of his own field book.) As history unfolds compilation might be expected to become first more and then less widespread relative to other cartographic activities. After all, man's earliest map-making endeavours must have been based on direct observation, presumably of small areas. But as maps accumulated, so more material became available to the would-be compiler. In 1780 the English topographer Richard Gough quoted an estimate that among 16,000 maps hitherto published, the number of 'originals' was not more than 1,700.[1] Eventually, however, with advancing technology and expanding budgets the demand for the compiler's services would diminish. Once a whole country had been accurately surveyed at the largest practicable scale, all future maps (outside the thematic category) could be based on that one survey and its revisions: in the terminology of this book, compiling would have given way to editing.

The logic of conjunction

The course of cartographic development has run less smoothly than the foregoing simplistic model would suggest. Many of the earliest known maps depicted very large areas, up to and including the earth as a whole, that could not possibly have been visited in a single campaign. However, it was only during the European middle ages that compiled maps began to survive in sufficient quantity for cartographic 'families' to be recognised and their genetic relationships unravelled. Given a choice, some map-makers would prefer to place the different outlines before the public and let each reader form his own judgement. Thus a renaissance geographer might copy the Ptolemaic *oikoumene* and a medieval *mappamundi* on different pages of the same atlas, as if conceding equal merit to both genres.[2] Something of this attitude was inherited by Abraham Ortelius, for example, despite his professed disapproval of cartographic bet-hedging.[3] Even in the eighteenth century, a map-maker could reportedly offer to print California as either island or peninsula according to the wishes of each individual customer.[4] There are also claims that a single feature might deliberately and without notice have been placed

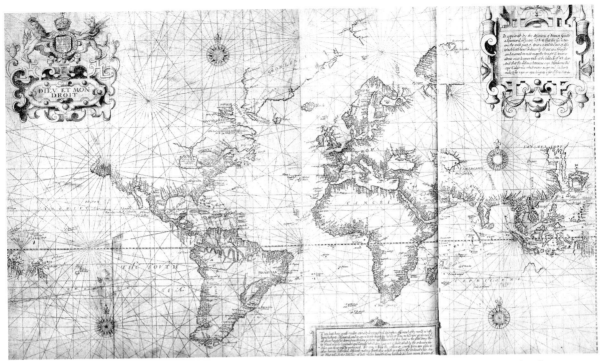

15.1 First-hand observations as a cartographer's only source. Edward Wright, world map, in Richard Hakluyt, *Principal navigations, voyages, trafiques and discoveries of the English nation* (London, 1599).

in two different locations on the same map for no other reason than to avoid taking a decision.[5] These are unusual examples, it is true: more often the cartographic record consists of straightforwardly declarative maps within which the influence of different archetypes can in the right circumstances be distinguished by visual comparison.

At this stage a line must be drawn between the two cartographic philosophies of perfectionism and pragmatism. Some compilers were unperturbed by empty space. They would set themselves a standard of accuracy and reject everything that fell below it, or so at least they liked to imply. The most obvious measure of self-discipline was to depend only on those maps that were known to be first-hand records of particular surveying expeditions (Fig. 15.1). The alternative was to do as well as possible with the least unsatisfactory data discovered for each region. Even then there might still be areas for which no geographer had offered any hypothesis, however speculative: this is probably how Ptolemy would have explained his failure to map the whole world. Some material accepted by a compiler of pragmatic disposition might be weak to the point of outright flimsiness. Consider Richard Hakluyt's rash advice that Ortelius should 'place before our eyes the Strait of the Three Brethren in its correct position, since there is always hope that at some time it may be discovered, and by marking it on the map the error of those cosmographers will be refuted who deny that such a strait exists.'[6] In the same spirit, Pieter van den Keere offered blatant speculation 'lest those less knowledgeable might think something was missing here, until something more certain emerges'.[7] Such 'tactical' mapping of unverified detail clearly had its dangers. So did the process of combining one source with another, or what John Norden called 'the tedious conferring of so many disagreeing plots together, which cannot be truly reconciled by greatest care'.[8] In short, the divide between cartographic 'hawks' and 'doves' would be drawn in different places by different map-makers. Ptolemy was more scientific in this sense than Crates of Mallos, Mercator than Ptolemy, Delisle than Mercator, d'Anville than Delisle, and so on throughout the period of 'enlightenment'.[9]

Perhaps the strongest impulse to compilation came from the very different cartographic histories of coasts and interiors. The portolan charts of Europe and north Africa featured long stretches of authentic coastline strikingly different from anything to be found on medieval *mappaemundi*. Behind these coasts the chart-makers left empty spaces that cried out to be filled – necessarily from some other source. The problem worsened as Columbus's successors outside Europe showed how far and how fast a portolan-type coastline could be carried across a previously unknown expanse of global surface. The surveying of interiors was much slower (even on the loosest definition of the word 'surveying'), both because of physical impediments to travel and observation and because many countries had to be conquered before their geography became fully intelligible. Meanwhile the territorial consolidation of Europe's own sovereign states required each metropolitan political unit to be mapped all the way to its frontiers: if such areas were too large for a single survey, this too was a situation that could be dealt with only by piecing different sources together.

The compiler's credentials

A compiler had to know what raw materials existed and where to find them. He could buy maps through dealers; he could beg, borrow or steal them from private and public collections; he could employ agents in foreign countries; or he could cultivate the friendship of other cartographers and local experts, a process well documented in the extensive correspondence left to posterity by Ortelius and Mercator.[10] He might dedicate a published map to a person of influence who could be expected to respond by laying open some otherwise inaccessible collection. It is even said that Mercator agreed to survey the whole duchy of Lorraine purely as a cover for getting inside the duke's library at St Dié.[11] The compiler also needed a knowledge of geographical and topographical literature as an aid in assessing the credentials and reputations of his informants. It was not sarcasm to describe a sixteenth-century world map as 'cunningly compiled and made'.[12] Faced with an artifact of this complexity historians have to match the cunning of the author. In some ways their task is harder, for how can a present-day student judge what level of inclusiveness to expect from an early map-collector in a world without catalogues, bibliographies, academic journals, learned societies or international conferences? Almost the only advantage that such a milieu could boast over modern times was the dominance of the Latin language. In this respect political and linguistic frontiers were becoming a more serious obstacle to post-medieval scholarly communication. John Green would freely criticise his fellow cartographers in eighteenth-century France but he did not blame them for knowing less about English sources than he did.[13]

What were the compiler's more personal qualifications? In the words of a Victorian authority:

> A geographer is so many-sided that it is not easy to give a comprehensive definition of the term. He should have been trained by years of land or sea surveying, or both, and by experience in the field in delineating the surface of a country. He should have a profound knowledge of all the work of exploration and discovery previous to his own time. He should have the critical faculty highly developed, and the power of comparing and combining the work of others, of judging the respective value of their labours and of eliminating errors. He must possess the topographical instinct; for, like the poet, the geographer is born – he is not made.[14]

This was the voice of utopia. In the real world, compilers could come from anywhere, but particularly from other branches of cartography. Field work, interestingly enough, was never a major reservoir of compiling talent. To the most conscientious surveyors, no doubt, compilation would probably seem on a par with copying or (not to put too fine a point on it) plagiarism. Perhaps the most likely scenario for a surveyor was to become so preoccupied with one particular country that he would continue to study maps of it after his own opportunities for direct observation had ceased – a sequence of events more likely with an explorer of distant lands than a maker of estate maps or other local surveys nearer home.

A more promising recruit to the business of compilation was the specialist in history, archaeology or natural science who needed a base on which to plot his own thematic researches, or the literary geographer in quest of ready-made illustrations for a forthcoming book, like Thomas Jefferson in 1786 with his *Notes on Virginia*.[15] Such men could easily be driven into grass-roots cartographic research by the inadequacy of existing sources. But perhaps the most likely apprenticeship for a compiler was experience in the processing of other people's maps whether as publisher, draughtsman or engraver – or possibly even as colourist, not yet a child's employment in the seventeenth century as it was in the nineteenth but rather an 'honourable and lucrative profession'.[16] These were all activities that might strengthen a practitioner's critical powers as well as familiarising him with possible source materials.

15.2 Northern hemisphere with corrected length for the Mediterranean Sea. Guillaume Delisle, *Hemisphere septentrional pour voir plus distinctement les Terres Arctiques* (Paris, 1714).

It was in eighteenth-century France that compilation reached its zenith, first with Guillaume Delisle and later with Jean Baptiste Bourguignon d'Anville. Neither put any of his talents into surveying, travelling, engraving, instrument-making or any map-related activity apart from turning primary into secondary statements. Their special merits lay in the quantity, variety and up-to-dateness of their sources, their gift for interpretative analysis, and their unwillingness to be intimidated by the reputations of earlier cartographers ancient or modern.[17] At the time Delisle must have impressed knowledgeable readers by earning praise from his ultra-critical English contemporary John Green.[18] In retrospect his boldest achievement was to reduce the length of the Mediterranean by a well-deserved fifteen degrees (Fig. 15.2). D'Anville scored by writing commentaries on his own work and by amassing what has been described as 'the largest map collection ever formed by a single individual', which he secured for posterity by presenting it to the

French ministry of foreign affairs.[19] Its scope may be briefly illustrated from one of Europe's smaller and more peripheral kingdoms, not at this period a country of profound theoretical or practical interest to geographical science. D'Anville collected approximately 120 separate national and regional maps of Ireland, some of which he found time to study with close attention, correcting William Berry's names in the Aran Islands and also moving the whole archipelago several miles to the north-east.[20] There was surely no other non-Irish cartographer before the nineteenth century who would have attempted this kind of correction to someone else's map of Ireland – very few Irish cartographers, if it comes to that.

Literary and documentary sources

In mapping remote countries, a compiler might find no cartographic data-base to work from, only the literature of description and travel. In that case his task would range from difficult to well-nigh impossible. It certainly presented hazards not to be found in even the worst of graphic documents. A minor example of literary influence was to map Latin placenames in an inappropriate grammatical case after extracting them without alteration from a prose text.[21] More serious was Matthew Paris's verbally inspired amalgamation of two quite different rivers, the Warwickshire and Hampshire Avons.[22] In a broad view, these are entertaining trivialities. The main point to note is that, on the whole, features definable by short phrases or individual words were easy enough to portray in what could be conceived as purely generic terms. A compiler's shapes might be made more determinate by helpful supplementary matter in his source-text: for instance, a land mass might have been compared to some everyday object such as a bird's egg or the leaf of a plane tree, or perhaps to a triangle or some other familiar geometrical form – analogies that must themselves have originated with maps subsequently lost to view.[23] In general however the problem with a source composed of abstractions was that the cartographer would be compelled by his chosen medium to particularise, the exact opposite of what verbal language teaches us all to do and an almost certain cause of error. This is where 'personal curvature' might come into play. At all events it took a brave compiler to resist the temptation of the 'wiggly line' and to highlight the difference between knowledge and guesswork by making his unexplored coastlines straighter than convention would normally permit (Fig. 15.3).[24]

Assembling maps from verbal data would generally be considered an expedient of 'first resort', appropriate only where geographical science had not yet advanced very far from vagueness towards precision. But the same practice was also observable at the other end of the production line when a surveyor found unexpected gaps in his knowledge that it was too late to make good by field work. Thomas Jefferson's map of Virginia was already in mid-edition when he sought to bring future copies up to date by asking a correspondent where the Ohio River met the western

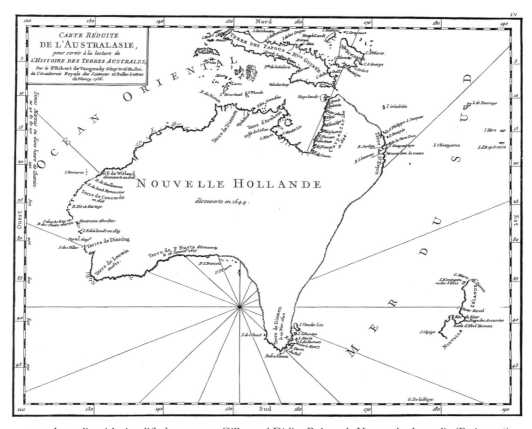

15.3 Australia with simplified east coast. Gilles and Didier Robert de Vaugondy, Australia (Paris, 1756).

boundary of Pennsylvania: with letter-writing as the medium of communication the answer, like the question, would probably have come in words rather than graphics.[25] More localised were a proof-reader's requests from the same period that a town should be drawn larger and a river narrower – in both cases without saying by how much.[26] Even in the British Ordnance Survey an examiner could ask for a Fenland dyke simply to be 'made wider'.[27]

The main difference among non-cartographic sources was between words and numbers. Geographical essays would often specify mileages for the lengths and breadths of countries, and occasionally gave the latitudes and longitudes of their extremities or their principal towns. Similar and seemingly more authentic information occurred in the narratives of particular journeys and at this point a further distinction needs to be drawn. Verbal accounts were usually judged inferior to maps, but good numerical data might be ranked higher than almost any kind of graphic display, especially if they were latitude and (eventually) longitude observations recorded in ships' journals.[28] Other ways of reckoning distance were bedevilled by the problem of units and here d'Anville set new standards with his extraordinarily pertinacious investigations into pre-metric measures all over Europe and Asia. Of course numerical values were not necessarily self-consistent, and where they differed a compiler might feel justified in striking an average. A well-documented instance of this procedure is the longitude assigned by James Rennell to Cape Comorin at the southern tip of the Indian peninsula in 1792.[29]

Maps from maps

Next comes the making of new maps from old. How did the compiler choose from among the latter? There might be documentary evidence that a possible source-map was based on first-hand knowledge. At least it might be attributed with certainty to a cartographer of repute – or perhaps to someone working for an employer of repute such as a national government.[30] It might even have earned a certificate of merit or a prize.[31] Sometimes a reputation could be checked against more specific biographical data. When Francis Drake mapped the southernmost islands of South America he was taxed by at least one critic with not having spent enough time in this region to know much about it.[32] Personal qualities apart, the preoccupations of a compiler and his proposed source might differ too widely for any fruitful relationship to be possible: hence the recommendation of some chart-makers that charts should be made independently of land maps.[33] As for internal evidence of authenticity, a map might conform to reality in a general way but still not be entirely plausible – the lines unnaturally straight, the bends too angular, the point-features too uniformly spaced. By contrast well-articulated geographical outlines, especially those bearing names, were unlikely to be completely fictitious. Otherwise what probably influenced a compiler most was sheer quantity of detail: the larger the scale of a possible source-map, the greater its opportunity for resembling nature, and the more likely it was to have originated with a true survey. But perhaps the most plausible hope of success was for a map of *terra incognita* to overlap with reliable portrayals of more familiar subject matter, inviting estimates of its value to be extrapolated, however precariously, from one region to another.

Another principle of selection was illustrated in the first century AD when Pliny found that each of the authors he had studied 'gave the most careful description of the particular region in which he was personally writing'.[34] A possible cartographic corollary was that a small area had more chance of being accurately shown on a map wholly devoted to itself than on a map depicting some larger area of which it formed part. This was because the author's attention had been more tightly focused on his subject. In other words, where a map of Norfolk contradicted a map of England, the map of Norfolk was more likely to be correct. A ground-breaking application of this rule was the European Samuel Purchas's considered preference for an Asian map of an Asian subject – 'a true China, the Chinese themselves being our guides'.[35] Later the same idea was given a new twist when Nicolas Sanson

15.4 China, with erroneously restored outline of Korea. From Nicolas Sanson, *L'Asie* (Paris, 1683).

contradicted his main source for China by making Korea an island instead of a peninsula, partly no doubt because Chinese geographers were less to be trusted outside the limits of their own country (Fig. 15.4).[36]

With two rival sources of broadly similar content this last principle could justify a copyist in choosing to work from separately published sheets rather than from the appropriate pages of a heavily edited atlas. The compiler faced with a single map could pre-empt the atlas-maker's responsibility for reducing and generalising it instead of allowing a stranger to perform these valuable but hazardous operations on his behalf. Another, more doubtful, mark of reliability in source materials was cartographic style. By the sixteenth century, maps were carrying various artificial or fabricated badges of professionalism such as forms of lettering, graticule, scale statement, north indicator and other marginalia, all capable of identifying an author as one who seemed at home in the world of cartography if not necessarily in the real world. Here, however, there was a counter-argument that grew stronger with the passage of historical time: a map of blatantly amateurish appearance would probably not have been undertaken in the first place unless its maker could claim to be drawing on first-hand experience.

Every kind of copying brings its own mistakes and early cartographers were well aware of this. Surveyors, especially, could recommend their own works as well protected against secondary error, however inadequate in other respects. Otherwise it was reasonable to follow a majority of witnesses, but only if they were genuinely independent, so a good compiler would try to avoid maps that were themselves copies or compilations. Admittedly a given assertion might become more persuasive by passing under the eyes and hands of several draughtsmen without being rejected as incredible by any of them. But in general this argument was probably too flattering to copyists, few of whom knew much about what they wrote or drew. The compiler could therefore profit from determining whether or not one map was genealogically dependent on another. It cannot be claimed that any early map compiler ever put his sources into a stemmatic diagram. But for a cartographer who did contemplate taking this drastic step, the effect of previous transcription errors would best be kept in check if (other things being equal) each family was represented among his own authorities by whichever extant map stood nearest to the apex of its genealogical tree.

The clearest example of a copy was a printed sheet, assuming its engraver or lithographer to have been someone other than the original surveyor. It is true that such a map had probably passed some kind of test by finding favour with a publisher, but another copyist would still have introduced another series of mistakes. This no doubt is why early seventeenth-century Dutch chart-users professed to feel more confidence in manuscripts than engravings.[37] A different kind of objection to printed authorities was that using a map familiar to one's critics might be condemned as discreditably facile: among cartographers, as among historians, one sometimes senses a certain intellectual snobbery in the preference for manuscript sources.

Techniques of cartographic synthesis

We have watched the dedicated compiler attempting to reconstruct the methods of his predecessors. What about his own methods? His work had to be in some way better than anything else available. It might cover a larger area. It might give more detail for the same area. It might give the same amount of detail but with fewer mistakes. In the first case, source-maps had to be joined together along their edges, in the second they had to be superimposed. In the third, each map needed purging of errors, redundancies and inessentials. Butting maps together was usually felt to be respectable enough, at least until the idea of a primary control network had taken root. A revealing example occurred in Jamaica under Charles II when each of nine surveyors was told to make 'an exact description of his own division which may afterwards be reduced into a larger [*sic*] scale'.[38] Apparently no one invoked the danger of combinatorial error.

In real life the three processes distinguished above were not so easily separable. At first sight the simplest case was that of an archipelago in which individual islands could be replaced without involving any other land mass, as when Saxton exchanged one Isle of Man for another in successive maps of England and its adjoining seas that otherwise remained essentially the same.[39] But an exact position and alignment still had to be chosen for each new island, and this problem was actually no different from that of joining separate maps along their edges. In effecting such a junction let us suppose for the sake of simplicity that there were only two identifiable points common to both specimens. The compiler might begin by overlapping the maps to make point A on one of them coincide with its equivalent (A') on the other. However the sheets were now rotated, point B would almost certainly appear at two different positions. The obvious remedy was to assume that the maps had been drawn at different scales and to enlarge or reduce one of them until the two Bs coincided along with the two As. Positions would now have been established for all the other points, but there would doubtless be some awkward gaps or intersections affecting linear features that were required to run smoothly from one contributory map to another. The mis-match might arise from a difference of projection, curable in theory by identifying the projection of one map and redrawing the other map to conform with it. If after rescaling and re-projection the two maps still did not fit, it could only be because one or both of them was in error – and of course being in error is exactly what tends to distinguish an early map. A matter of special concern here is that when measurements were plotted their internal disharmonies often got driven towards the edge of the survey area, which on our initial assumption was where points like A and B would have to be located.[40]

The same problems arose where maps were being superimposed rather than simply joined together, with the further difficulty that the number of settlements and other point-features occupying two different positions was now much larger. A radical solution would be to omit one from each pair of duplicated sites and to adjust networks of linear features by erasing or prolonging them as plausibly as possible. No doubt matters were often resolved as crudely as this, but it may still be

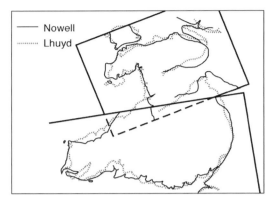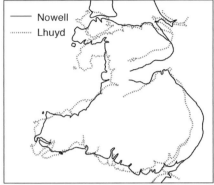

15.5 Outlines of Wales on nearly the same scale by Laurence Nowell (1564) and Humphrey Lhuyd (1573), possibly derived from a lost two-part map whose northern and southern sections they combined in different ways. F.J. North, *Humphrey Lhuyd's maps of England and of Wales* (Cardiff, 1937), pp. 42–3.

worth considering alternative solutions. For instance one might start, as before, with a partial assimilation of the two map-scales, an effect most easily achieved by matching whichever pair of points lay furthest apart. Then, after superimposition, each of the other points could be placed half-way between its two rival locations. What about a place that appeared on only one of the source maps? Here was a chance to adapt a simple method described elsewhere for transferring lost sites from an inaccurate early map to a modern base.[41] This would locate the point at issue on both maps, in positions that could then be averaged as in the previous case. Unfortunately there is no evidence that this or any comparable geometrical procedure was ever followed by an early cartographer, only the slightest of occasional hints.[42]

Some hazards of compilation

When maps were joined along their edges the result could show its origin by several kinds of internal variation whether spatial, temporal, thematic or stylistic. The first category included belts of compression or extension where the fit had been less than perfect. In particular one constituent region might be twisted out of alignment with another (Fig. 15.5); the classic case was the sharp eastward bend of Caledonia (Scotland) on Ptolemy's map of the British Isles, perhaps explicable as a junction of two separate maps after one of them had inadvertently been turned through ninety degrees (Fig. 15.6).[43] The same could happen less conspicuously when one map had been oriented to true north and the other to magnetic north.[44] In William Gerard De Brahm's survey of northern Biscayne Bay, Florida, in 1770 the mainland was plotted on one meridian and the islands on another, the erroneous line being helpfully captioned by the cartographer himself as 'wrong'.[45] Not all such faulty connections are as easily understood. In his map of north-west Europe (1573)

15.6 Supposed faulty junction of England and Scotland in Claudius Ptolemy, *Geography* (Strasbourg, 1513). See also Fig. 11.5.

Christian Sgrooten seems to have tried not very successfully to fit Mercator's British Isles into an independent geometrical framework that included Start Point, the Lizard, Lundy and various other capes and islands, causing these features to appear at first glance as wrongly placed in relation to the rest of Britain and Ireland (Fig. 15.7).[46]

Another cause of trouble was for different authorities to have worked on different scales without the compiler becoming aware of the fact. James Rennell advised map-users that 'it should be a rule observed in all plans, to note how the scale was obtained.'[47] This was more easily said than done, but at least we can look for evidence of distances being derived from more than one source. The scale of Juan de la Cosa's world map (1500), for example, was forty per cent larger in America than in the old world.[48] Spatial unconformities of this kind might be expected to reveal themselves when a square 'distortion grid' is first drawn on a

15.7 North-west Europe with anomalously located coastal landmarks. Christian Sgrooten, 'Generalis descriptio totius Germaniae Inferiori' [1573]. See note 46.

modern map and then transferred by interpolation to its predecessor, the grid lines changing length or direction wherever two independent maps have been joined. In practice, the initial error may have been so extensively rearranged by smoothing out gaps and overlaps that its original form and location are now beyond recovery (Fig. 15.8). In the course of such smoothing, distortions may be pushed into areas where previously they had not existed. Perhaps this explains an otherwise puzzling reference to John Foster's map of New England, that 'being in some places defective, it made the other less exact'.[49]

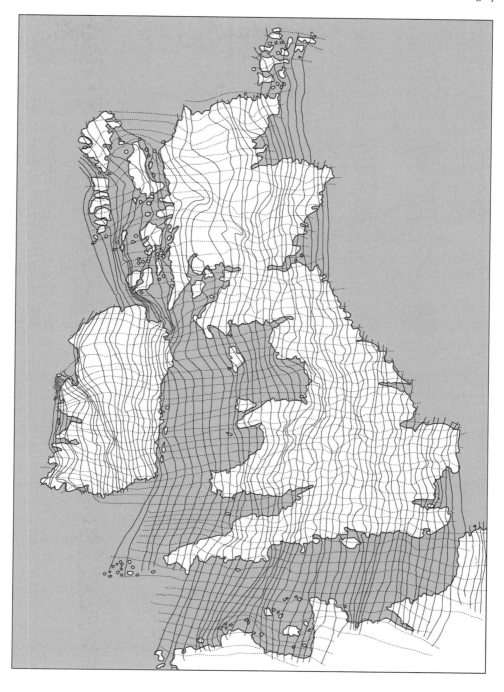

15.8 Modern rectangular grid transferred to Gerard Mercator, *Angliae Scotiae & Hibernie nova descriptio* (Duisburg, 1564) in Walter Reinhard, *Zur Entwickelung des Kartenbildes der Britischen Inseln bis auf Merkators Karte vom Jahre 1564* (1909, reprinted Amsterdam, 1967).

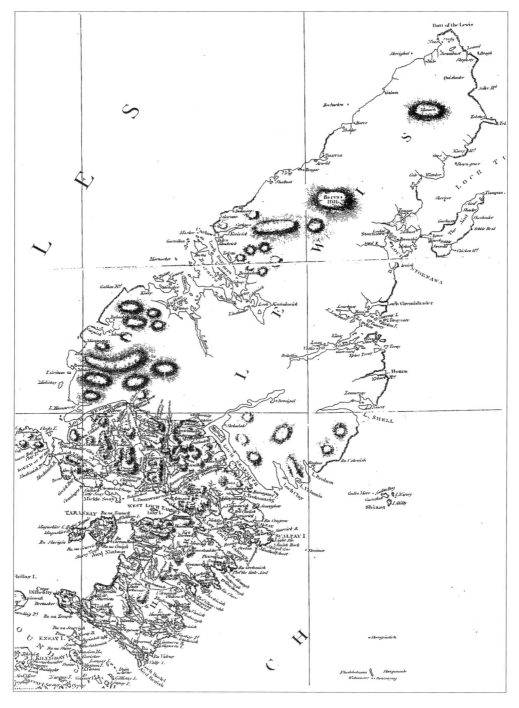

15.9 From Aaron Arrowsmith, *Map of Scotland constructed from original materials obtained under the authority of the parliamentary commissioners for making roads and building bridges in the Highlands of Scotland*, scale 1:253,440 (London, edition of 1810).

Temporal inconsistences could be equally embarrassing. Dates were less common in early cartography than scale statements, and in any attempt to fill the gap a historian would look for events to serve as 'termini' defining the earliest and latest times at which the map in question could have been drawn. A *terminus ante quem* could be given by the death of a territorial lord or landowner named on the map as in possession of his estates. A *terminus post quem* might be the erection on a greenfield site of a building shown as complete and apparently in occupation. It was to be hoped that the later terminus would follow hard upon the earlier, defining the zone of temporal uncertainty as narrowly as possible. But what if *terminus ante* preceded *terminus post*? In that case the map must have been made at two different periods separated by some kind of science-fiction time warp. Such intervals would cease to be problematic if they fell within the likely duration of a single survey campaign, which could easily have lasted for several years. Where the hiatus was too long to be explained in this way, the map became open to interpretation as a blend of two or more differently dated originals.

Another sign of diverse origins is a variation in density of map detail that seems unrelated to the geographical facts. In Aaron Arrowsmith's map of Scotland nobody could doubt that Harris, recently surveyed by an able estate surveyor and consequently full of detail, had come from a better source than the startlingly unfurnished district of Lewis in the same island (Fig. 15.9).[50] Equally significant in this respect are differences in thematic structure from one part of a map to another. In Ortelius's map of Cisalpine Gaul the Roman roads are unjustifiably concentrated in the southern third of the map.[51] In Slovenia and other Balkan territories Mercator shows a large tract of country that differs from the rest of the map in having all its hills omitted.[52] A hydrological example from a much later map of North America is Arrowsmith's sudden outpouring of specialised information – rapids, river widths, altitudes, double lines for minor streams – in one small region between James Bay and Lake Superior.[53] Such anomalies seem to hint at some diversity of origins.

Stylistic variation was prominently on display in a well-known twelfth-century map of Britain, Ireland and a large part of continental Europe associated with the Norman historian Giraldus Cambrensis (Fig. 15.10). Here all the rivers of the British Isles are continued in the form of flow lines some distance out to sea as if between imaginary piers or breakwaters, whereas all the other rivers on the map end flush with the coastline. Moreover, all the British and Irish town symbols differ architecturally from their continental counterparts. Presumably the map had been taken from more than one source, a suggestion supported by an appreciable difference of scale between the British Isles and the mainland.[54] With the increasing standardisation of symbolism in modern cartography, such clues were bound to become less common, but one convincing example was Arrowsmith's switch from profile hills to hachured hills between eastern and western Tennessee.[55] A different

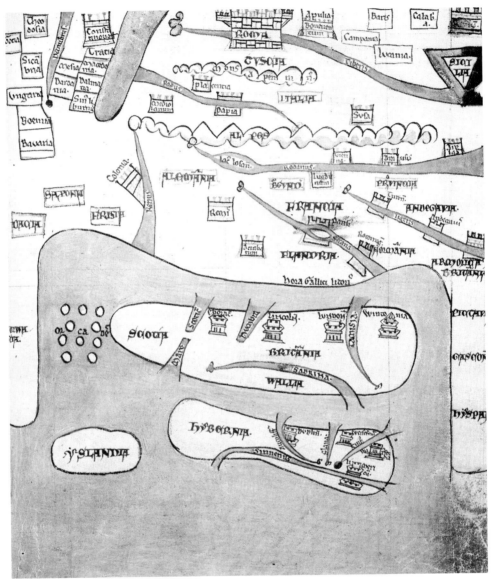

15.10 Giraldus Cambrensis, north-western Europe, *c.*1200. National Library of Ireland, MS 700.

kind of clue that might count as stylistic was the use of more than one language for inscriptions relating to a single country. Thus Jan Jansson's map of north-eastern North America had French names in Nova Scotia ('P. aux Nuneses', 'C. de Sable'), Dutch in southern New England and the Hudson valley ('Hoeck vande Visschers', 'Hend. Christianz Eyland') and English in Virginia ('Trinite Harbar', 'Cape of Feare'), the Atlantic Ocean being designated in a fourth language as 'Mar del Nort'.[56]

From abutment we turn to superimposition. Here too stylistic variations were obviously significant even if they showed no definite regional pattern. A medieval example was the group of names added to one of Matthew Paris's maps in the handwriting of another scribe.[57] From a different era, non-scribal evidence to the same effect became available in Ireland when James Williamson revised a map of County Antrim by James Lendrick. The least attentive reader could hardly fail to notice that Williamson's new houses were in block plan while Lendrick's old ones remained in profile.[58]

At a slightly higher level of complexity, a further brief reference may be made to the idea of 'stemmatics' introduced in chapter 14. Dependence on two or more sources is dealt with in the theory of textual analysis (not very satisfactorily, it must be said) under the heading of convergence or conflation. Suppose a compiler to be studying four related maps of the same country. Suppose also that a number of errors were common to only two maps, and that some of them were shared by A and B, some by B and C, some by C and D, and some by D and A. The most probable family tree or stemma would include a downward-pointing apex, and it should be possible to follow one line of descent between A and B, another between B and C, another between C and D, and another between D and A, in each case without passing through either of the other two maps. The archetype could be any of the four extant maps or a hypothetical fifth map. There might be other clues reducing the number of possibilities to be assessed by the researcher. For instance a map whose individual errors showed it to be terminal could not appear at either the right or left corner of a diamond-shaped stemma. But many cases of convergence would remain insoluble without extraneous knowledge of the dates and archival histories of the maps in question.

A clearer sign of compilation within a single map was the duplication of features which were actually identical but which the cartographer believed to be different because they were differently shown in his sources. An example on the grand scale was the repetition of the River Nile in some early versions of Africa.[59] A more narrowly topographical illustration was Stornoway ('Stornway ca.' and 'Stoy ca.') on Mercator's later maps of Scotland, one of his sources evidently being a map of his own dated 1564.

Assessing the cartographic evidence

When they did resort to compilation, early cartographers seldom had much to say about what they were doing. Such silences were particularly well exemplified in the first modern atlases, where there ought surely to have been room for a modicum of written commentary. Thus by Ortelius's time it had become normal to print a page of letterpress text on the reverse of every atlas map. Here was a chance to set out the author's mode of procedure, but instead such texts seem to have been almost entirely devoted to run-of-the-mill geographical description of a kind obtainable from any competent writer with no special penchant for maps. Some of this material can only be regarded as padding: a historian inquisitive about cartographic techniques in Blaeu's atlas will not necessarily thank the author for explaining how long it would take to walk round the earth if all the seas had dried up.[60]

On the supply side, as we have already noted, compilation was not always seen as a matter for pride. There was certainly no harm in letting the public believe each map to be based on the observations of its nominal creator. With the advent of the atlas, however, any presumption of authorial omniscience was clearly becoming unrealistic. In this respect the atlas medium, disappointingly uninformative in most of its non-cartographic manifestations, at least exerted some influence along the right lines. Since no individual could possibly have measured his way across the world in person, there was less to lose by broaching the subject of sources. From an early stage in the history of map-printing, authorities were often briefly named in editorial accreditations within the body of a map. More will be said about these in chapter 16. Meanwhile the best examples of a renaissance cartographer speaking with his own voice are appeals for readers to contribute material that might be used in future compilations. Such role-reversals seem to have been pioneered by Sebastian Münster in 1528, setting a precedent that even Mercator was not too grand to follow.[61]

The next question is how home-based cartographers compared with the best surveyors and astronomers of the late seventeenth century in giving an account of their work appropriate to the new age of scientific progress and acceptable to the growing class of educated general readers who felt some interest in how cartography was managed. An early move in this direction came when a map by Guillaume Delisle directed attention to explanatory matter in one of the author's own books.[62] By mid-century a large multi-sheet map might sometimes accommodate within its margins a substantial essay – for instance John Mitchell's critical review of North American latitudes and longitudes in 1755, which was at least as long as the average article in a modern academic journal. Then there could be separate publication in the form of a newspaper advertisement, a prospectus, a scientific or scholarly paper or even a book-length memoir. Here the defining moment came in 1744 with the appearance of d'Anville's *Analyse geographique de l'Italie* elucidating a map published a year earlier. Later he did the same for China, Egypt and the Roman empire. This was one of the innovations that set d'Anville above his predecessors.

More strictly cartographic were 'methodological' notes written within the very heart of a map. These included explorers' names and dates alongside a previously unknown coast, a widely practised method of documentation that could also, infrequently, be applied to interior sites. Even less often, such inscriptions have been known to strike a refreshingly technical note. One traveller's journey, in the American mid-west, declares itself to be 'protracted' in Arrowsmith's map of the

United States. Better still, in the same author's South Africa, William Burchell's route was 'here inserted from his own map, and the countries adjoining adjusted to his observations'.[63] It was not unknown for volume and page numbers to appear, footnote fashion, alongside an authority's name in the middle of a map.[64]

Occasionally this kind of documentation inspired its own symbolism. Thus a distinctive boundary such as a broken line might separate areas mapped with different degrees of reliability. Examples range from John Browne's Elizabethan survey of western Ireland ('by view' on one side of the line, 'by report' on the other)[65] to Sebastian Bauman's map of military operations at Yorktown in 1781.[66] In Virginia Captain John Smith drew small anchors to show how far up the estuaries he had pursued his own observations. Hessel Gerritsz's chart of the Pacific (1622) used a colour code to distinguish Dutch and Spanish discoveries.[67] Two of these conventions are particularly worth noting. One, familiar from the early sixteenth century, was to distinguish more and less reliable coasts by more or less conspicuous lines. A special case of this was the littoral symbolism in Isaac Massa's map of northern Russia (1612) with a firm line for Russian and a dotted line for Dutch information.[68] Another was Captain William Edward Parry's map of the Melville Peninsula, which identified by shading the information obtained from an Inuit woman.[69] Equally important was John Green's advice that cartographers should underline the names of towns whose latitudes and longitudes had been determined instrumentally.[70] Such examples could be multiplied many times, but they would still comprise an almost invisibly small proportion of all the maps in existence.

Compiling and revising

The art of synthesis as considered in this chapter should be distinguished from several less fundamental operations that might seem to have more affinity with editing than compiling. First however there is an intermediate type of derivative cartography, in which one source contributes so much more than any others that we may reasonably describe the new map as 'based on' it. Useful in this respect is the eighteenth-century French notion of the *carte citée*, the map that supplies a cartographer with all the coasts, rivers and other physical features in his region that could be regarded as uncontroversial, his own input being mainly classifiable as additions and corrections. An example is John Speed's importation of antiquities and minor administrative territories to the county maps of Christopher Saxton. Here the compiler's reluctance to disturb an original text was betrayed where the boundaries of his new territorial divisions scrupulously avoided colliding with Saxton's hill symbols – respectful behaviour indeed considering how perfunctorily relief features were located by the map-makers of this period. Such deference would also sometimes reveal itself when an author copied decoration as well as topography from a single source-map.[71]

The same asymmetrical pattern of dependency appears in Mercator's regional maps of England. These were taken almost entirely from Saxton, but they also included a very small number of additional names, especially in the north, that drew on sources too obscure to have remained identifiable. The additions were not needed to fill space, because in the same areas Saxton had shown many names that Mercator chose to ignore. It is almost as if this most illustrious of compilers was setting a trap for posterity, defying future readers to charge him with unmitigated reliance on a single predecessor. If so, he seems to have forgotten his own advocacy of preserving 'the production of each individual'.[72] He was also ignoring the advice of his friend Ortelius, who spoke against map-makers 'adding something at their pleasure'. The same point was made in 1717 by Green: 'the map-maker often thinks himself obliged to make alterations from others, that something new may appear in what he publisheth.'[73]

This somewhat lop-sided approach to cartographic synthesis should not be confused with the work of revision. Under the latter heading, the correction of features that were already erroneous when they were first drawn should be distinguished from the replacement of detail made obsolete by the passage of time. Different again was the process of augmentation, which entailed adding matter of a kind not present in the original, as when railways were inserted to make a new 'state' of an early Victorian English county map.[74] Less frequently encountered is the completion of a map by filling gaps within a thematic category already represented on it, a service that William Petty vainly hoped someone else would do for his published atlas of Ireland.[75] Updating, correction, augmentation and completion had a good deal in common and the word 'revision' will here be applied to all four processes, though still excluding changes to marginalia such as titles, dates, dedications or publishers' names and addresses. Nor are we now concerned with the kind of alteration suggested by proof-readers, referees or candid friends before a draft had passed out of its author's control.[76] Note however that the term 'revision' has sometimes signified just this kind of back-stage activity;[77] also that the visible effects of changes to a proof may be indistinguishable from amendments made after publication. In short, revision differed from compilation proper in dealing with one apparently viable map whose identity had been and would continue to be accepted by all parties.

At first sight, the most far-reaching influence on revision was the cartographer's choice of scale. With the smallest and simplest maps there might be little difference between revision and replacement. At larger scales the matter was more complicated. On the ground an individual house, being easier to demolish than an entire town, was more likely to have a short life-span: a map capable of showing single houses would therefore be quicker to fall behind the times. On the other hand, for any given territory a complete revision was more likely to be prohibitively expensive with large scales than with small. In Britain and Ireland this had the

15.11 *Above:* From John Speed's Pembrokeshire, before 1623; *Below:* From John Speed's Pembrokeshire, after 1623.

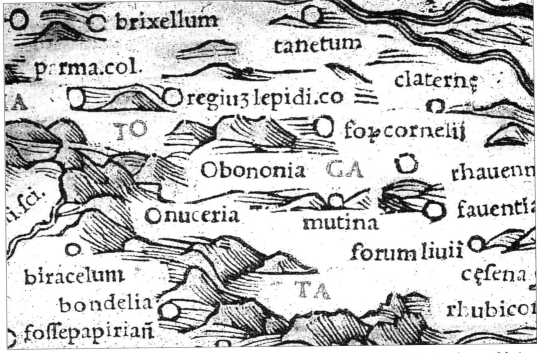

15.12 Northern Italy: two variants from Claudius Ptolemy, *Geography* (Venice, 1511). Misplaced names Mutina and Bononia (*above*), corrected by stick-on labels (*below*). David Woodward, *Maps as prints in the Italian renaissance: makers, distributors and consumers. The Panizzi lectures, 1995* (London, 1996), p. 35.

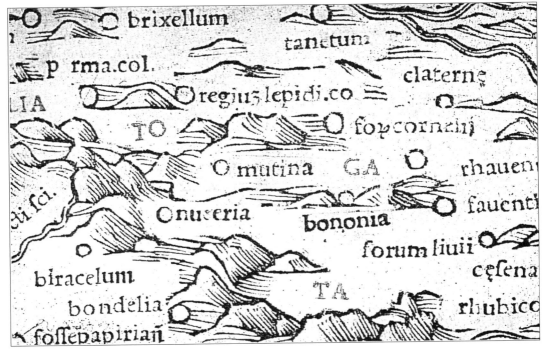

paradoxical result that one-inch Ordnance Survey maps were often more up-to-date than the six-inch maps from which they were ostensibly derived.[78]

Then there were technical considerations, most notably the difference between manuscript and print. In a map drawn by hand on paper, it was hard to remove ink and pigment without doing some physical damage. Additions presented no such problem (Fig. 15.11), though it might be difficult to match the style of the original or even its exact colour, which is how Paul Harvey was able to recognise the name and symbol for Hereford as afterthoughts on the world map now housed in that city.[79] If the worse came to the worst, an individual sheet might be revised by pasting a label over the unwanted detail (Fig. 15.12), a common procedure in the correction of sea charts.[80] On parchment, unlike drawing paper, ink could be more or less neatly erased with a sharp blade. If a manuscript suffered wear and tear it might need completely replacing anyway, and then none of its contents had to be treated as sacrosanct. In correcting printed maps the main distinction was between wood and metal. Names might be changed on a woodblock, but there was no way of adding to its printed geographical features. New words and new lines could easily be engraved on copper, perhaps standing out as unintentionally dark in print if the rest of the plate was already somewhat worn (Fig. 15.13).[81] Just how much could be deleted from a metal surface by hammering, scraping and burnishing depended on the skill and patience of the craftsman and the thickness of the plate. Sometimes the smallest alteration was glaringly obvious, as happened on an anonymous world map of 1600 cited by Rodney Shirley.[82] Elsewhere a map could be virtually rebuilt on its own foundations, for example when Willem Blaeu's world map of 1605 gave way in 1670 to an almost entirely different set of decorations as well as a different geographical outline, both allegedly created on the original plate by Justus Danckerts.[83] In the nineteenth century changes of 'state' began to be facilitated by two new reprographic processes already discussed, electrotyping and lithography. On the whole, however, the use of copper as a printing medium worked in favour of minor revision but against total replacement.

Modes of revision

The process of revision itself could be either 'elective' or 'reactive', complete or partial. An author might set out to do it comprehensively for a whole map or a whole region. Alternatively he might just adopt such suggestions as he happened to receive from well-wishers – or from casual passers-by, one is sometimes almost tempted to add: at any rate 'reactive' changes were more likely to appear random than systematic. No one could have foretold that in Ortelius's map of Spain a perfectly correct name would be erased to gratify a watchful and self-important ecclesiastical dignitary who wanted space to be found for the insertion of his own home town.[84] In general, the likelihood of alteration depended on several factors:

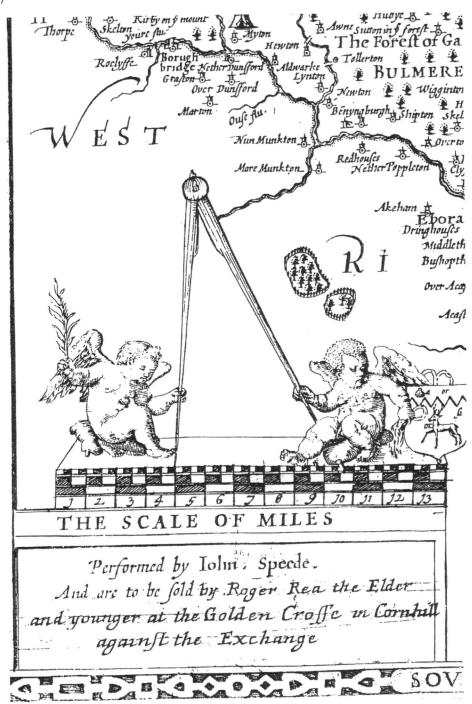

15.13 From John Speed, *The North and East Ridins of Yorkshire*, with original publishers' names and addresses of 1610 (John Sudbury and George Humble) replaced on the same plate in c.1662 by those of Roger Rea senior and junior.

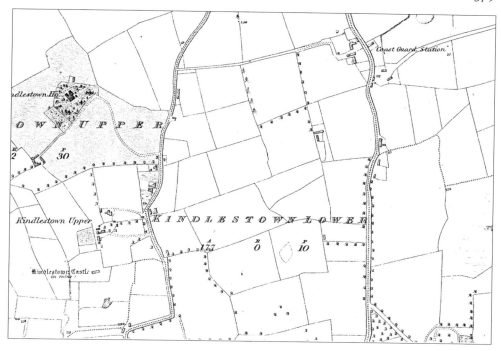

15.14 From six-inch (1:10,560) Ordnance Survey of Ireland, County Wicklow, sheet 8.
Above, as surveyed in 1838. *Below*, as revised in 1885.

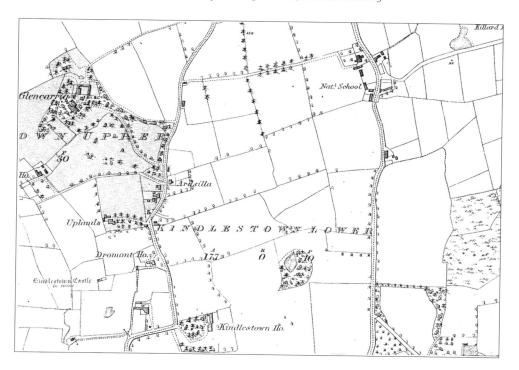

the gravity of the error; the ease of correcting it; the approachability of the culprit; the alertness and censoriousness of the reader. Roughly speaking, the larger the population of educated residents in any country, the more likely were the errors in its maps to be detected. Another 'anecdotal' conclusion is that throughout history placenames have attracted more criticism from non-specialist readers than anything else. The situation was different with maps providing a base for further field work by someone other than the original author, as in an independent survey of geology or vegetation. Serving as a base map in these circumstances was one of the most rigorous tests that any survey could undergo, and field observers at work on an earlier cartographer's outline were seldom backward in reporting defects.[85]

In elective revision the improvements originated with the cartographer himself. On the whole such alterations have been less common than might be expected, and this not wholly for discreditable reasons. During most of human history, the world has changed comparatively slowly. An author might be dead, or his publisher out of business, before enough events had accumulated to make revision seem worthwhile. By that time it would probably be better from a commercial point of view to start a brand new map, a work advertised as original being more easily marketable in large quantities than an overtly later edition.

It was in the management of a state survey office that revision presented its most serious hazards. In theory the procedure was simple enough. The reviser compared the map with the present appearance of the ground, cancelling obsolete detail and inserting new information either directly on to the map surface or as field notes that were subsequently plotted in the manner of an original survey. In both cases the surviving detail played the part of a control network, with the difference that roads, streams and fences were almost certainly less accurate than the points of a triangulation. Here a difficult, indeed an insoluble, problem stood in the reviser's path. He might decide, after preliminary tests, to take his new measurements from selected detail that he thought would have been laid down with particular care. In the last resort he might be forced back on the major chain lines of the first survey or even its trigonometrical skeleton. Where successive measurements contradicted each other, the later values could not automatically be accepted as more reliable. In any case it might prove inexpedient for every small and unimportant error to be rectified: apart from financial constraints, such ruthless honesty could damage the reputation of the surveyors. In the event some pragmatic compromise might be necessary, for instance to alter only those features of a foreshore that could make the old map a danger to navigation, or to correct no errors of less than a certain numerical magnitude. All these problems arose when the first regular six-inch Ordnance maps of the United Kingdom began to be revised from 1845 onwards (Fig. 15.14).[86] Like some other topics discussed in the present book, they were not normally dealt with by early writers offering instruction on surveying and cartography.

CHAPTER 16

Into more finished form

IN 1802 A MAP WAS dedicated to the citizens of Philadelphia by someone identified in its cartouche simply as 'the editor'.[1] To a modern reader, editing would doubtless suggest deliberately changing a map without adding any further information about the area represented. At first sight the most likely candidate for such a task was the person who risked his own money in making the map available, but there were many publishers whom it is hard to imagine contributing directly to the cartographic substance of the product that was nominally theirs: examples familiar to English collectors are John Sudbury and George Humble, John Norton, John Overton and Robert Sayer. On the other hand, in the process of actually constructing maps there may have been some operations that could not be classified as surveying, plotting, drawing, copying, engraving or printing. Sir John Scot had no business connection with any of these activities, and was never given credit on the face of any map, but he has been identified in one modern study as 'virtually editing' a sixteenth-century survey of Scotland for publication by Joan Blaeu.[2] Similarly, without contemporary non-cartographic records we should never know about the 'editorial services' said to have been provided by Thomas Taylor for the Irish county atlas that carries the name of William Petty.[3] Again, it was a later authority, Gerald Crone, who first described the role of John Green as 'knowledgeable and skilled geographical editor' to the eighteenth-century map-publisher Thomas Jefferys.[4]

We might almost define editing as whichever part of the cartographic process is most likely to remain anonymous – so perhaps work seemingly attributable to hypothetical John Scots or Thomas Taylors was really done as a fringe benefit by someone better known as a surveyor or draughtsman. In eighteenth-century America, particularly, 'one or two individuals typically served as a map's compiler, engraver, printer, publisher and retailer'.[5] To which we can add that when there were several participants in the productive process it is often tempting to hold the engraver responsible for the bulk of the editorial work. Sometimes this assumption can be tested against documentary evidence. The engraving of maps for Lysons' *Magna Britannia*, for instance, was divided between two artists: H. Mutlow, who followed his manuscript models slavishly, and Samuel Neele, who made numerous original contributions.[6]

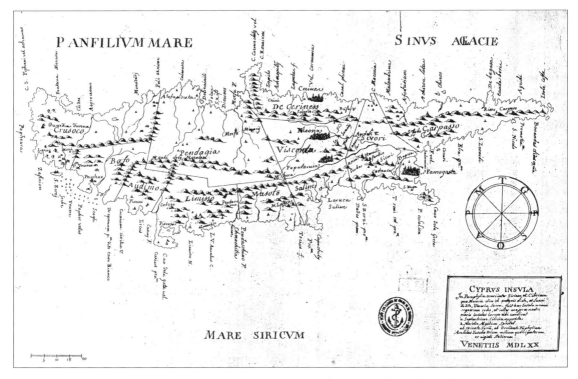

16.1 Anonymous, *Cyprus insula*, in *Pamphylia mari* … (Venice, 1570), with straight territorial boundaries that were omitted in Abraham Ortelius's version of this map.

The editor corrects

An editor may have been actuated by several independent considerations, but these are not always easy to separate, and having identified different kinds of editorial process we should not be disturbed to find them yielding similar results. One reason for changing a map was to make it more correct. Here, as may be clear from the previous paragraph, a certain lack of symmetry is observable in the editor's input. He might eliminate falsehoods from an original, but he could not add new truths to it without becoming a compiler. Destructive or negative editing took place when Abraham Ortelius rejected the unrealistically straight territorial boundaries shown on Giacomo Franco's map of Cyprus (Fig. 16.1) while faithfully reproducing almost all the other information that Franco had to offer.[7] Another example was Antoine Lafreri's omission of dubious directional indicators in his second (1572) edition of Olaus Magnus's map of northern Europe.[8] Also in this category may be Greenvile Collins's latitudes for the Scilly Islands, lost somewhere on the way from manuscript to print in 1693.[9] This was one kind of editorial change that did not necessarily involve copying: matter could be removed from a map by hammering and scraping the copper plate, or by masking the unwanted image in the press with

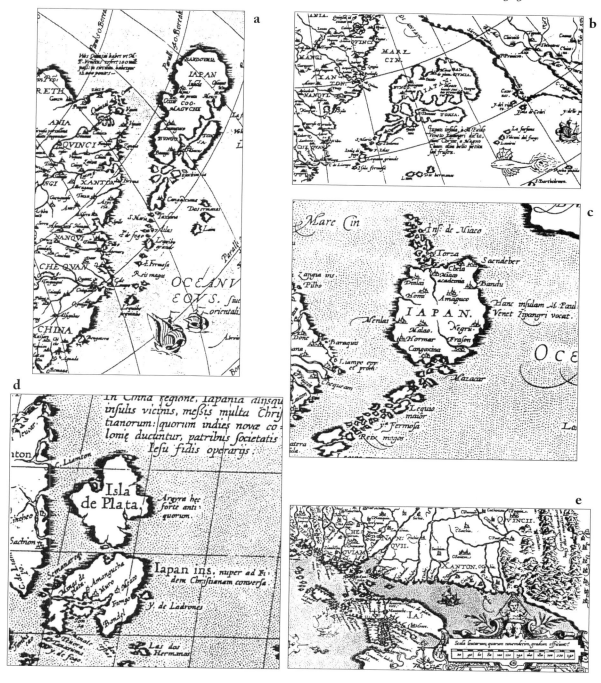

16.2 Abraham Ortelius, variant representations of Japan. (a) *Asiae nova descriptio*, 1570. (b) *Tartariae sive Magni Chami Regni typus*, 1570. (c) *Indiae Orientalis insularum quae adiacentium typus*, 1570. (d) *Maris Pacifici*, 1589. (e) *Chinae, olim Sinarum regionum, nova descriptio*, 1584, north to right. In 1595 all five versions appeared in the same edition of Ortelius's atlas.

a paper shield, or in the last resort by trying to erase some of the dried ink from a printed impression.[10]

The removal of factual disagreements has always been a powerful motive for editorial intervention – not so much within a single design, for except in special circumstances a map cannot disagree with itself, as between different pages of an atlas in which the same territory might successively appear on a world map, a continental map and a regional map. At the least ambitious level of publishing, it is true, this task was often deliberately neglected. Lafreri, for example, acted for the most part less as editor than bookbinder, assembling ready-made maps that he did nothing to harmonise in any geographical sense. He would not have been ashamed of separating North America and Asia by minimum distances that varied from nothing to more than 1500 miles.[11]

When an atlas-maker undertook to re-engrave a set of maps before publishing them, the issue of consistency was harder to avoid. Whatever the extenuating circumstances, Ortelius should not have included five incompatible versions of Japan in the same edition of his *Theatrum* (Fig. 16.2).[12] History had genuinely moved forward when Mercator made some effort to purge his *Atlas* of such anomalies.[13] A further editorial responsibility was that if maps of neighbouring areas were copied and brought to a uniform scale their borders should fit together (allowing for differences of projection), with the same place occupying the same latitude and longitude wherever it appeared.[14] A collection from the post-Mercator era that failed to avoid this pitfall was Jean Le Clerc's *Theatre geographique du royaume de France* (1619), in which half a dozen cartographers and as many engravers left longitudinal disagreements of forty minutes from one map to another. Of course consistency could not always be equated with correctness: after Herman Moll had altered the boundaries in his atlas of England (1724) the counties fitted each other, we are told, 'neatly' but at the same time 'wholly unrealistically'.[15]

Titles and subtitles

So much for the editor's rather limited role in eliminating error. Among his other tasks was the promotion of intelligibility and reader-friendliness. A first-class surveyor might conceivably encrypt the whole of his observations in the form of undifferentiated points and lines plus an explanatory quota of words but the results, however accurate, would earn little respect as vehicles for meaning: points were too small to be easily seen; lines could be interpreted in too many different ways; words offended the spirit of cartography by requiring serial rather than instantaneous perception. Good editors would dispel the resultant confusion by trying to individualise the various elements in the field worker's skeleton. For symbols of restricted currency they sought more widely comprehensible graphic substitutes. Elsewhere, too, they replaced the obscure with the familiar, whether as units of measurement, north indicators, prime meridians, or the language of inscriptions.

If, as the cliché suggests, it is the packaging that sells the product, then the inspection of any complex map should proceed by working inwards from its perimeter. A question that arises in the study of marginalia is how much prior knowledge could be expected from an editor who was not also a compiler. At least he should have known what he was looking at. Before cartography had become widely familiar, distinguishing particular specimens of it was not necessarily a matter of great importance: when a Shakespearean character said 'Give me the map there' nobody answered 'Which map?' Many manuscript maps did fail to identify themselves, but in a collection of any size it was better for each item to carry a name of its own. The same argument could also justify the common practice of varying the decorative content of cartouches and other title-pieces to particularise the successive pages in an atlas: for the browser, there was merit in any feature that helped a map's singularity to 'leap from the page'. Meaningful titles were especially important in maps of small inland territories that lacked the immediate recognisability of a whole kingdom like France or Spain.

Since in normal text-matter the act of reading proceeds from top to bottom of the page, the obvious place for a cartographic title in a European language was above the map it referred to. By the late fifteenth century, in early editions of Ptolemy, this concession to serialism was already well established. Ptolemaic titles were not yet thought to need any special embellishment. However, if purely regional names happened to be written where the reader would expect to see a title there might now be a danger of confusing part with whole – or even exterior with interior if a north-pointing map of England, for instance, was surmounted by the non-titular inscription 'Part of Scotland'. Hence the advantage of each title occupying its own 'compartment'. In cartography as in literature, the wording of titles was a favourite subject for editorial attention. Certainly those given to manuscripts were not always suitable for wider circulation: one need only quote 'A description of the East coast of Scotland drawn out of Wagoner [Waghenaer] and sumqt [*sic*] corrected, but it [*sic*] not fully perfyt & yet hath many errors'.[16] A diligent editor's titles would include the most familiar name and aliases for each major territory and would sometimes enumerate the subdivisions that composed it. Nicolas Sanson was a particular enthusiast for this kind of auxiliary information, accompanying many of his maps with lengthy non-cartographic tables in which brackets were used to show how different orders of territory nested into one another.[17] In relation to an atlas such lists could perhaps be conceived as enormously long titles.

Authors and readers

Apart from naming a map the marginal script-writer would be expected to assign responsibility for it. In early published maps this was often done by means of Latin phrases such as 'apud', 'auctore', 'caelavit', 'composuit', 'delineavit', 'descripsit', 'ex

16.3 Revised titles by Matthias Seutter (d. 1757) and his son-in-law Tobias Conrad Lotter (Augsburg, 1730 et seq.) for a map of New Belgium (New England) first published by Jan Jansson (Amsterdam, *c.*1650).

officina', 'excudit', 'fecit', 'fieri curabat', 'formis', 'incidit', 'invenit', 'procurante', 'sculpsit', 'sumptibus', 'typis' or 'vaeneunt'. Translations of these terms are attempted in several modern reference sources, but they are often too vague to be helpful.[18] As a rule we might expect the workers concerned with any one map to arrange themselves in temporal order as a production line, starting with the surveyor and passing through successive operations to end with the consumer's immediate source of supply. A generalisation worth entertaining is that the later a contributor's position on this scale, the greater his chance of being named on the face of the map.

The commonest inscription was the publisher's imprint. This told potential customers where they could buy the map and discouraged potential plagiarists by identifying the current copyright-holder. Not surprisingly, the replacement of obsolete entrepreneurial names and addresses is perhaps the commonest post-publication plate-change recorded by modern cartobibliographers (Fig. 16.3). Speed's maps of the British Isles, for instance, announced themselves at various times in the seventeenth century as sold by John Sudbury and George Humble, by George Humble alone, by William Humble, by Roger Rea, and by Thomas Bassett with Richard Chiswell, in many cases without undergoing any alteration to their geographical content. Such emendations were sometimes too urgent to wait for an adequately qualified engraver: the correction might then be less well executed than the original map, perhaps even to the extent of near-illegibility.[19]

Next in order of 'visibility' comes the original or primary engraver, the only person able to guarantee his own appearance on a printed map, and the only one whose Latin code-word, 'sculpsit', had the merit of being totally unambiguous. But at this point matters start to get more difficult. Where field workers were distinguishable from compilers, the word 'auctore' was more likely to denote the latter, though without extraneous documentation this hypothesis may be incapable of proof. Surveyors certainly deserved recognition, either as a gesture of goodwill towards a possible future collaborator or as a means of pinning responsibility where it belonged – the latter motive especially powerful in the case of sea charts, where false information could lead to fatal consequences. But none of the above-quoted Latinisms is equivalent to 'surveyed by';[20] nor, if it comes to that, are such common English verbs as 'described' and 'performed'.[21]

All this is in keeping with our idea of the production line. But there were other attributions to which the concept is less easily applicable. These identified a compiler's sources and their function was to give his map a more authoritative appearance than could be achieved by merely claiming to have used the most authentic 'materials', 'observations' or 'authorities'. Thus Martin Waldseemüller's *Universalis cosmographia* of 1507 paid tribute to Ptolemy, Nicolaus Germanus and Claudius Clavus as well as to Marco Polo,[22] while a single map of Hungary in 1528 was ascribed to Lazarus Secretarius, Georg Tanstetter, Johann Cuspinianus and Peter Apian.[23] However, it was Ortelius who most deserved credit for rescuing earlier cartographers from oblivion. More than half the maps in the English edition of his *Theatre* carried marginal references to non-engravers, a few more of whom were noticed in the adjacent text.[24] This lead was only half-heartedly followed by Mercator and Blaeu, it has to be said, though by the end of the seventeenth century a good many acknowledgements had accumulated: in any large European map collection of that period the prize for most posthumous credits would probably be shared by Ptolemy and Mercator with Sanson as runner-up. Authors were especially worth identifying if the reader was likely to have heard good reports of them. In some quarters name-dropping became an end in itself, with cartographers citing sources that they had never consulted, as when Robert Walton's *Africa* admitted an imaginary debt to Joan Blaeu in 1659.[25] Circumstantial references would be more convincing: 'dressée sous les yeux de M. de Buffon' must certainly have carried weight in mid eighteenth-century France.[26] Where debts were more specific – like those of Charles Labelye off the coast of Kent – the benefactor did not have to be so well known (Fig. 16.4).[27]

A map-maker could also expect to be praised by his publisher, directly or indirectly, as well as to be given an opportunity for self-praise. A short and simple step towards accreditation was to treat a cartographer's home country as part of his name: Ortelius's *Theatre* did well to say that Salzburg had been mapped by Marco Secznagel 'Salisburgense' and Hungary by Joannes Sambucus 'Pannonius'. Qualifications were often more fully spelt out, as when Newfoundland was surveyed by Captain John Mason, 'an industrious gentleman who spent seven years in the country'.[28] An unusual stroke of one-upmanship was the official pass authorising a surveyor to visit the battle-front at Boston in 1775, a specimen of ephemera elegantly reproduced in Henry Pelham's famous plan of the city and environs.[29] Another technique that occasionally spread from free-standing advertisements on to the face of a map was to anticipate the modern book publisher's dust-jacket by quoting favourable reviews. Matteo Ricci took this opportunity in his world map of 1602;[30] John Hills's plan of Philadelphia (1796) carried a letter of thanks from the mayor; and at the end of our period John B.Bachelder's survey of Gettysburg did the same with facsimiles of handwritten praise from senior military officers.[31]

Elsewhere marginal recommendations were more impersonal. It was good tactics to slip in at least one expression that struck what could pass for a technical note. The words 'From an actual survey and admeasurement' made an impressive introduction (despite being untrue) to Emanuel Bowen's map of South Wales (1729): pedantic readers might have been hurt by the unnecessary reference to actuality, though this was probably justified by the gradual devaluation of the word 'survey'.[32] 'Improved from several surveys' was perhaps a little too economical, even if most readers probably thought they saw what it meant.[33] In a similar vein, the superfluous-looking phrase 'drawn on the spot' might serve to distinguish a plane-table survey from a protraction based on numerical measurements, as well as giving casual readers a vague sensation of reassurance. From the mid eighteenth century onwards the most widely used English vogue-word for strengthening the titles of medium-scale regional maps was 'trigonometrical', its popularity perhaps due in part to John Rocque's unusually explicit blurb for his plan of London in 1747.[34]

At smaller scales, a title might lay claim to one incontestably reliable source of information, such as a recent voyage of discovery.[35] Or the net could be cast more widely. Aaron Arrowsmith's authorities for South America included both 'scarce

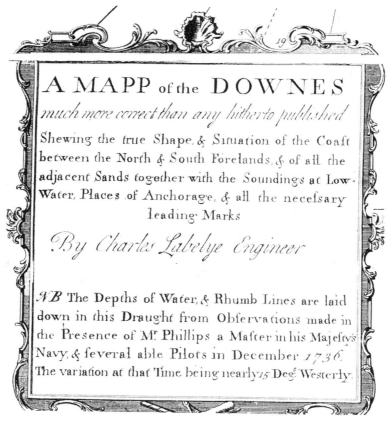

A MAPP of the DOWNES

much more correct than any hitherto published

Shewing the true Shape, & Situation of the Coast between the North & South Forelands, & of all the adjacent Sands together with the Soundings at Low-Water, Places of Anchorage, & all the necefsary leading Marks

By Charles Labelye Engineer

NB The Depths of Water, & Rhumb Lines are laid down in this Draught from Obfervations made in the Presence of M.ʳ Phillips a Mafter in his Majefty's Navy, & feveral able Pilots in December *1736*. The variation at that Time being nearly 15 Deg.ˢ Wefterly.

16.4 Title with acknowledgement. Charles Labelye, n.p., 1737.

and original documents' published before 1806 and, more importantly, 'manuscript maps & surveys made between the years 1771 and 1806'.[36] In other cases attack might be the most effective means of defence. The following is not part of a pamphlet or newspaper advertisement but an extract from a published map by Herman Moll:

> Among all the cheats that the world are daily abused with, none have lately been more scandalous than that of maps, sometimes new ones are put out by ignorant pretenders, sometimes mean and imperfect foreign maps are copied and published by them as their own, and having no judgement or knowledge of what is good or bad, correct or incorrect, they basely impose on the public with pompous titles, and pretend they are countenanced and assisted by those who either never saw, or despise their wretched performances.[37]

The same author complained – also on one of his own maps – that his *Pocket companion* to the roads of England and Wales (1717) had been 'copied four times very confused and scandalously' by rival cartographers.[38]

Chronological contexts

By the sixteenth century there were many kinds of document that could be expected to carry a date. For the consumer this practice could do nothing but good; for the producer it was a mixed blessing, helpful at first as a guarantee of topicality, inconvenient thereafter as an admission of staleness. A definite time-reference was most useful when the mutability of man's or God's creations formed an essential part of a map-maker's story. This makes it strange that the first European maps to be dated were portolan charts, which are notorious for having been copied and recopied over long periods with comparatively little alteration.[39] One likely vehicle for chronological precision was an overtly historical map: where temporal statements could be ambiguous as between past and present an author had good reason to be as definite as possible, so it was not altogether paradoxical that modern dates should be more common in Ortelius's antiquarian *Parergon* than among his regular maps. Time frames were also involved in the correct representation of physically unstable coastlines or floodplains, well illustrated by the Roman and medieval insets to the same publisher's maps of Flanders and Friesland (Fig. 16.5).[40]

Another incentive for dating was when maps had been added to freshen up an aging atlas in its later editions: it would be a pity in this situation for the new material to pass unnoticed among the old. Altogether Ortelius provided dates for about thirty per cent of his maps, excluding the *Parergon* for which the corresponding figure is nearer one half. Compared with Mercator and Blaeu this was a good score. Perhaps the strongest motive for historical precision was to warn readers that the publisher's copyright had not yet run out: this explains why a date and the words 'cum privilegio' often appear close together on early printed maps. In English cartography the issue was brought to a head when copyright became legally restricted to maps that bore a date of publication.[41] In all these manoeuvrings there was one clear danger for future historians. Exactly what event was being dated? An editor should at least know when he himself acquired a map; but the time at which it was surveyed might be as inaccessible to him as the name of the original author. Otherwise Speed could have told readers that much of his information for Ireland in '1610' had actually been collected thirty or forty years earlier.[42]

Apart from carrying numerical dates, early maps often congratulated themselves on being 'new', 'latest' or 'modern', a salutary reminder that belief in progress did not originate with nineteenth-century whig historians. Such was the value of topicality that a degree of subterfuge might sometimes seem necessary to assert it. The temptation was for an engraver to doctor the original footnote by changing an earlier to a later year. The new date might be defended, with a touch of sophistry, as relating to publication rather than composition, but this excuse was not available when the words 'Survey'd in the Winter, 1766' became 'Survey'd in the Winter, 1775' in John Montresor's otherwise unaltered plan of New York city.[43] Editors interested

16.5 Abraham Ortelius, East Friesland in Roman times, by Joachim Hopper, inset to *Frisia occidentalis* by Sibrandus Leonis Leouardiensis ([Antwerp], 1579).

16.6 Armorial bearings of English and Irish nobility. John Speed, The British Isles (London, *c.*1603). See also Fig. 1.15.

in topicality would have to watch out for loose ends, if necessary redirecting a dedication from a dead to a living recipient. By the mid-eighteenth century, when readers were expected to be capable of appreciating good cartography, an over-confident publisher might tolerate such anachronisms as no longer worth correcting.[44] On the whole, however, publishers were becoming more careful about chronology. It was certainly wise to drop the expression 'géographe du roi' when the French monarchy became a republic. In the next generation an Arrowsmith map would sometimes carry several successive dates recording 'additions', though often in very small script as if warning readers not to expect too much change.

Titles and authors were not the only marginalia that could make a good impression. In a manuscript map drawn for private use salesmanship might be promoted by exhibiting the customer's coat of arms or, alongside an estate plan, a picture of his principal residence. A publisher could pursue the same end with a fulsome dedication, or by printing the heraldic insignia of potential purchasers (Fig. 16.6) who might then be expected to pay an additional fee.[45] This kind of personalisation may seem rather ineffective as a marketing device until one counts the 724 coats of arms that John Warburton managed to squeeze into a single map of Middlesex, Essex and Hertfordshire.[46] Less well-connected or less foresighted cartographers could suffer the embarrassment of decorating a map with shields that they were unable to fill.[47]

There was one device for claiming credit which differed in character from all those considered above, and which affected the interior of a map rather than the margins. With the growth of scientific self-consciousness, some authors were emboldened to compare their own new positions for coasts, rivers, towns and other features with those they wished to controvert, an effect achievable by superimposing the later outline upon the earlier. The best-known example is a much-reproduced map of 1693 by the Paris Académie de Sciences in which the Academy's recent astronomically-based version of France was contrasted with that of Nicolas Sanson (1679).[48] The new coast was accentuated with shading, the old one drawn as a single fine line; the additional town names were in roman lettering, the original names in italics. In Scotland John Cowley tried to improve on this idea by comparing no less than six outlines on a single sheet.[49] Meanwhile the same technique had been used more extensively in John Senex's *A new map of the world* (1721) to show the effect of modern longitude reckonings in eastern Asia and elsewhere.[50] Another way of distinguishing different representations on a common base was double printing from two plates: in 1736 this made it possible for the opinions of Philippe Buache (black) and Henry Popple (red) to appear without confusion on the same map of eastern Canada.[51] Here too the author was careful to show both latitude and longitude; otherwise the process of superimposition would have been too arbitrary to serve much purpose.

Guides to understanding

It is now time to review some of the genuinely technical information that post-fifteenth-century editors might be expected to provide around the edges of a map. The first distinction to be drawn is between an acquaintance with the region depicted and a knowledge of geography and cartography in general. The former could reasonably be considered an authorial preserve, the latter might sometimes inspire the participation of a well-educated outsider. Suppose for instance that the original cartographer had neglected to provide a linear scale. A devoted publisher could, it is true, attempt to make good the omission by comparing a sample of his author's distances with those on a scaled map of the same area, but that was probably asking too much. He might however supplement an existing scale with further information about units of measurement, for instance by subdividing miles into furlongs, by specifying the number of paces in each mile, or by translating miles into hours of travel time. He might also notice longer and shorter units with the same name, and units belonging to particular regions, though these ran the risk of being overdone when twenty scale-bars appeared on the same sheet.[52] Perhaps the most logical policy was to quote standards from two sources, the country being mapped and the country publishing the map. Besides specifying linear quantities a diligent editor might sometimes 'refer the geographic reason unto the celestial' (as Mercator expressed it) by saying how many terrestrial distance-units made up a degree of latitude.[53] One enthusiast even threw in a numerical value (9000 leagues) for the earth's circumference.[54]

Another fact that might be less accessible to publishers than to authors was a map's approximate north-south alignment, though a non-expert should have been able to supply this information without too much trouble by studying earlier sources. In a compass-indicator the counterpart of alternative distance-units was the angle between true and magnetic north. To put this difference on a map would certainly be commendable, but even if the local magnetic variation was on record it could not be added to an author's north point without a knowledge of which north he had chosen in the first place. As we have seen when discussing protraction, this was a subject that most early cartographers preferred to avoid, and editors could not generally be expected to make amends. In Ortelius's atlas of 1606, for example, there were only four maps that showed a pair of north arrows diverging from a single point, and in no case are we told which line was magnetic and which geographical.

A comparable problem often arose in relation to numerical longitudes. The reader might expect degrees to be counted eastwards or westwards from a familiar datum-point, but unless the original map had specified this there was no way of replacing an unsuitable prime meridian by anything better. Similarly, even if the margins of a map were scaled for latitude and longitude it would often be impossible to add a graticule without being told the projection. Like magnetic north, this was a neglected subject. Mercator's version of Ptolemy gave numerical definitions for the central meridians and standard parallels on every map – all in a thoroughly Alexandrian spirit.[55] Seventeen years later his own atlas did the same, but in its later editions these useful notes were generally omitted: one is tempted to characterise Mercator's posthumous publishers (Jodocus Hondius, Henry Hondius and Jan Jansson) as being more interested in decoration than mathematics.

Thenceforth, for many decades, mathematical geography received little overt attention from surveyors, draughtsmen, engravers or map-publishers. In a collection of more than four hundred English advertisements for maps and atlases published between 1660 and 1720 only one item was credited with any particular projection (Mercator's, needless to say).[56] Such reticence may be partly excused by the lack of accepted names for several important projections until the eighteenth century or later: 'sinusoidal', for example, was apparently not recorded until 1863.[57] On the other hand if earlier cartographers had really wanted to identify a projection there was nothing to stop them inventing a name for it. Explicable or not, this state of self-censorship persisted well beyond the end of our period, so completely that any reference to a non-Mercatorian projection on a pre-Victorian map ran the risk of branding its author as an amateur. It was not until 1895, apparently, that any publisher took the trouble to identify the graticules he had put in his atlas, let alone give reasons for his choice.[58] Those withholding guidance on this subject included nearly all the most scientifically-minded publishers of the eighteenth and nineteenth centuries, among them the directors of the British Ordnance Survey – who at one stage came near to forgetting which projection had been used on some of their own maps.[59] In these circumstances readers could hardly expect to be instructed on such refinements as the earth's ellipticity or the exact length of a foot.

The one service that an editor could rely on performing was to diffuse an atmosphere of professionalism over both the map and its margins. For stating a scale or a north point, in particular, there have always been orthodox modes of presentation, though like other aspects of map style these have naturally been subject to fashion. Thus wind-heads (cherub-like faces expelling visible breath across sea and land) were commonly used for direction until the mid-sixteenth century, afterwards giving way to compass roses and later to simple meridian lines.[60] At any period styles could also vary according to the cartographer's exact purpose: on a sixteenth-century land map the scale divisions were probably short vertical lines in a horizontal block; on sea charts they were often simply dots – a smaller target for criticism where precision was regarded as important.[61] Some of these customs could be explained functionally. On a plan meant for determining exact distances the first division of the scale-line would often be subdivided and numbered 'backwards' from right to left, a practice that became common on English military maps from the 1660s onwards.[62] The reason for this odd-looking layout became clear as soon as one started actually measuring from such a scale (Fig. 4.19) – a practice that many cartographers encouraged by drawing a pair of open dividers above the bar.

However, not every element of editorial policy could be explained by reference to either artistic fashion or practical map-use. For any specialist in-group there are certain rules of procedure, usually unwritten and unspoken, that bring no particular benefit to anyone except as a means of exposing the amateur or imposter. Most map conventions served a 'territorial' purpose in this pseudo-ecological sense as well as making a statement about the world. An unusually clear example of a useless convention was the marginal referencing of distant towns where a road ran off the map. In the late eighteenth and early nineteenth centuries it was correct to label roads 'From X' in the left-hand margin and 'To Y' in the right-hand margin.[63] In one sense this was a trap for the half-knowledgeable, embodying a kind of serialism foreign to the essential character of cartography.

Arts of presentation

Deferring styles of cartographic ornament for later consideration, our focus may now be shifted from margins to interiors. Some maps called for quasi-editorial

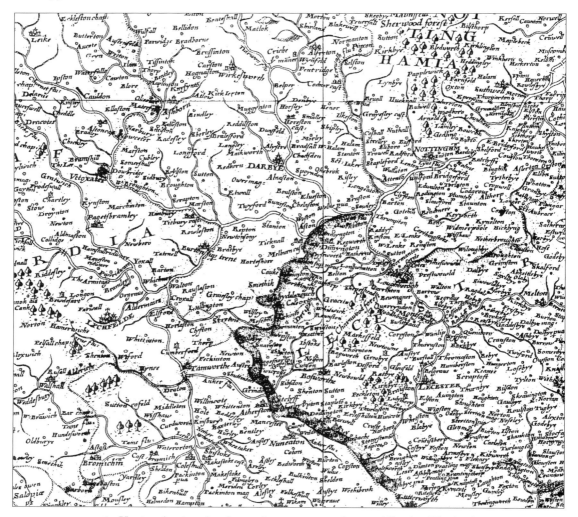

16.7 Varying density of settlement between east and west midlands of England,
original scale c.1:440,000. Christopher Saxton, *Britannia insularum* … (London, 1583).

attention because they were incomplete, others because they were overcrowded and
needed thinning out. This was especially likely where the original cartographer had
applied an inflexible 'thematic' rule of selection regardless of its local effects.
A principle of wayfinding at stake here is that however widely a traveller has
strayed, he should find something on the map to help him back on course: hence
the need for a more even spread of cartographic detail than might seem to exist on
the ground (Fig. 16.7). Here was sufficient justification for reinforcing a map with
ostensibly unimportant facts. The arguments for the opposite process of editing-out
were not so strong, and in this connection it would be worth knowing when the
seventeenth-century word 'clutter' first penetrated the cartographer's vocabulary.
The idea it expresses was certainly clear to Joel Gascoyne when he complained of

16.8 England: arms of King Philip and Queen Mary, the former scratched out. Diogo Homem, MS map of the British Isles, 1558. See note 66.

too much map-detail being 'burthensome to the eye'.[64] On the whole, clutter is a notion that comes more naturally to producers than consumers: in seeking value for money the average reader has probably always been willing to risk a certain amount of eye-strain. A cynic might suggest that map-makers filled up some blank spaces to rebut accusations of laziness, and made a virtue of avoiding clutter elsewhere because they really were lazy. But of course there was more to it than that – as witness the cartographer who created more work for himself by removing dubiously useful hachures that he had already allowed to reach proof stage.[65]

Overcrowding aside, there would probably be little in the interior content of a non-thematic map that a pre-modern editor would feel willing or able to reject on grounds of undesirability. Maps were not often controversial in the same sense as books. An author might of course make a political point by expressing territorial claims as if they were accomplished facts. This was common enough in the post-medieval cartography of European colonialism, but there it was probably more usual for a publisher to start afresh than to change an existing map that gave offence. The most spectacular case of political editing in British cartographic history comes as a surprise: it is the partial erasure of a Spanish heraldic device from one of Diogo Homem's charts, an act of vandalism that some scholars have attributed to an anonymous interventionist better known as Queen Elizabeth I (Fig. 16.8).[66]

More recently the most common excuse for removing or withholding information has been military, rather than political in any broader sense. In his map of Canada the French hydrographer Jacques Nicolas Bellin omitted soundings from harbours owned by France, 'that foreigners might not know how to attack these ports', while at the same time telling all he knew about coasts belonging to other countries.[67] In the 1870s the British Ordnance Survey was instructed to censor all its maps by erasing every fort, together with nearby contour lines and spot heights.[68] The results would obviously arouse suspicion as well as being open to ridicule, and some twentieth-century revisers in the same predicament chose to hide the blank spaces behind a screen of partly fictitious non-military detail – the latter an even more reprehensible practice though admittedly harder to detect.[69] An earlier parallel for this kind of obscurantism was not very close: it came in Braun and Hogenberg's sixteenth-century urban map-views, where *in-situ* pictures of local residents were allegedly meant to discourage Turkish readers who might have found out too much about infidel fortifications if their religion had not forbidden the portrayal of the human figure.[70]

An editor's most important reason for altering the draft supplied to him was simple enough: he would want to make the final map more likeable. Lacking this particular gift, Sir John Narborough 'scandalised' some eminent readers by the 'rudeness' with which he had charted the Strait of Magellan: they could hardly be expected to know that as a geographical statement the chart in question would not be outdone for another ninety years.[71] What Narborough needed, in the words of

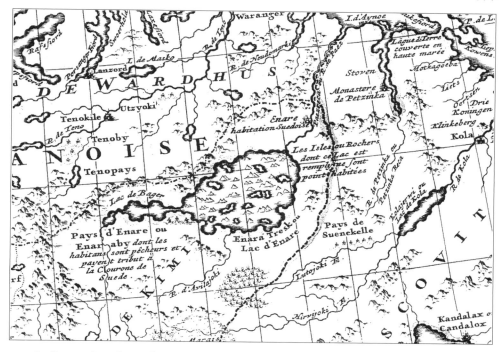

16.9 Restoration of worn line-work by re-engraving. Johannes Covens and Cornelis Mortier, *Carte des courones du nord* (Amsterdam, *c.*1721). University Library, Amsterdam. Marco van Egmond, 'The secrets of a long life: the Dutch firm of Covens and Mortier (1685–1866) and their copper plates', *Imago Mundi*, liv (2002), p. 73.

Willem Blaeu, was to have his map put 'into more finished form'.[72] Some aspects of this process have already been mentioned as appertaining to the copyist. Others were matters of general policy: avoiding superfluous lines and visible corrections, for example, and keeping both words and point symbols as small and compact as legibility allowed. 'Retouching' a worn plate could doubtless be placed in the same category (Fig. 16.9).

Further analysis is not so easy. In the first place, individual signs and marks should be good to look at, and also legible without effort. They should avoid monotony and blandness by showing bold differences in visual character, with linear, punctiform and areal elements each making their own impact. The resulting blend must be balanced and harmonious, no one kind of symbol too forcefully distracting attention from the others. The proportions that compose the mixture could – and, aesthetically speaking, should – vary quite markedly from one region to another, but taken together a map's contents look better if more or less evenly distributed across its surface. The qualities to aim at were completeness, plausibility and a general air of self-confidence: assume a virtue if you have it not, would be the critic's advice to the editor. In matters of presentation, a map might deviate from established usage, but only if it succeeded in indoctrinating posterity with its own style: an author immediately identifiable from an unsigned map might earn the gratitude of future cartobibliographers on that account but deserves no special credit for editorial flair, whatever his individual talents as a designer. It hardly needs pointing out that none of these virtues had anything to do with geographical merit: there are numerous medieval *mappaemundi* that must surely be accepted as master-works of editorial skill.

Before trying to apply the above-mentioned standards we must acknowledge the influence of scale. In small atlas maps an editor's opportunities for decision-making were inevitably reduced. Custom demanded the inclusion of coasts, rivers, mountains, countries and cities, and once these were in place there was probably little room for anything else. By contrast, at the upper end of the topographical scale-range two key editorial processes, selection and conventionalisation, would both be severely restricted: nothing important need be crowded off a cadastral-style map and almost every feature could be represented in plan rather than symbolically. In well-settled countries it was somewhere within these extremes, say between 1:1,000,000 and 1:10,000, that the editor had most to offer. His highest achievements must probably be sought in mid sixteenth-century Italy and then soon afterwards in the Low Countries, reaching a peak with Ortelius's *Theatrum orbis terrarum*. The best English expression of the Ortelian ideal was in the county maps of Saxton (Fig. 16.10) and Speed.

From about 1660 onwards, surprisingly, the effect of improved surveying technology was to depress the presentational standard of European maps. Two possible reasons suggest themselves. Scales were growing larger as surveys became

16.10 Successful cartographic design, original scale 1:313,783. From Christopher Saxton, *Lancastriae comitatus* (London, 1577).

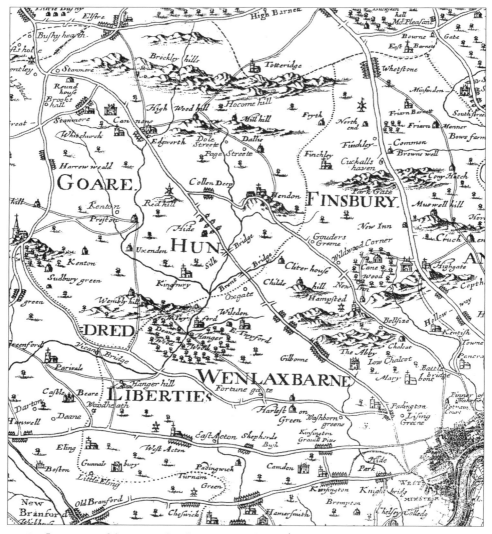

16.11 Less successful cartographic design, original scale 1:94,567. From John Ogilby, *An actuall survey of Midlesex* (London, 1673).

more accurate, but at the same time poor-quality field data were increasingly rejected on scientific grounds, giving many maps a scrappy, unfinished appearance and anticipating later suggestions of an 'inverse relationship between the best scientific surveys and those with the greatest topographical detail' (Fig. 16.11).[73] Then there was the change in symbol-design from profile to plan, to be further considered in the next chapter. This was a slow and sometimes painful process. It might almost be said, for instance, that no map featuring both profile hills and a fairly close road network could be counted an editorial success. Even when the new conventions were accepted it took time to find the most effective ways of drawing and combining them. In this case it was the French who took the lead, notably with

16.12 18th-century map design. Routes from London to Luton Park, Bedfordshire, 1767. British Library, Add. MS 74215.

Cassini's map of France in 1756. By about 1770 a new ideal was emerging in many draughtsmen's studios and engravers' workshops (Fig. 16.12). England for instance had begun to nurture a generation of engravers, among them Garnet Terry, William Palmer, Samuel Neele and Benjamin Baker, whose county maps were finally getting the balance right: it continued to be right throughout the period of the Old Series British one-inch Ordnance Survey maps from 1801 to 1872.

The perils of scale-reduction

One aspect of the editorial process absorbed enough time and energy to justify separate discussion, and that was the enhancement of a map's potential popularity by changing its scale. Among atlas-makers Nicolas Sanson gets credit for standardising a whole sequence of maps in this respect, no doubt in the interests of inter-regional comparability.[74] Another reason for interfering with scales was to maintain a fair balance between the more and less important items in a collection. Robert Hooke took up a characteristically doctrinaire position by advising that the 202 maps in Moses Pitt's *English atlas* should all be based on one or other of three globes with diameters measuring fourteen inches, eight feet and ninety feet.[75]

Consistency in this situation might sometimes require a map to be made larger than before. Enlargement, as a purely graphic operation unaccompanied by re-surveying, re-plotting or re-compiling, is not generally acceptable to modern authorities, partly perhaps because the original surveyor is assumed to have first chosen the smallest viable scale and then sought to economise by not measuring more precisely than this required – in which case enlarging his map would reveal errors that were previously too small to be noticeable.[76] This argument has not always been thought conclusive. Wall maps, temporary or permanent, are obviously a special case, their magnification justifiable on aesthetic and heuristic grounds.[77] Enlargements in more conventional format included those of Giovanni Magini's Italian maps by Joan Blaeu,[78] much of Sanson's output by Alexis-Hubert Jaillot,[79] and possibly the maps of Nicolas de Fer by de Fer himself.[80] Matteo Ricci enlarged his own maps for distribution in China on the reasonable ground that Chinese writing took up more space than European.[81] Other cases are harder to verify, but sometimes the look of a map may itself be sufficient proof of enlargement. In the north-west corner of Oronce Finé's *Nova totius Galliae descriptio* (1538) southern England displays a number of geometrical embayments that look more like gross exaggerations of a typical portolan coastline than any kind of original drawing (Fig. 16.13).[82] Other maps arouse suspicion by having fewer placenames than one would normally expect.

It was much more common for scales to be changed in a contrary direction – in the words of Laurent Fries, avoiding 'superfluous length and width'.[83] In a territory divided by internal administrative or political boundaries, this advice could be taken by individually redrawing each subdivision, otherwise unaltered, on a smaller sheet than before, which is how Blaeu treated Pont's Scottish maps. Alternatively a large map could be cut into arbitrary rectangular blocks, again without further modification. Usually, however, reduction involved diminishing scale as well as size. Indeed from the sixteenth century onwards making small maps out of large ones was probably the most important cartographic task that could be described as editorial. Economy was a powerful motive for engraving manuscript maps at less

16.13 Evidence of over-enlargement. From Oronce Finé, *Nova totius Galliae descriptio* (Paris, 1553).

than their original scale, often by a factor of half.[84] Also with a view to profitability, a map might seek two markets by appearing in two formats, full size and 'pocket' size.[85] The process of reduction might be thought to benefit from being carried out in stages: taken too far and too fast it was more in danger of offending knowledgeable readers by an injudicious selection and rejection of detail. With a complex hierarchy of larger and smaller scales covering the same area, as in the publications of the British Ordnance Survey, it was therefore a useful practice for each map to be derived from the scale immediately above it.

Down-scaling had two components, generalisation and omission. For intelligibility to survive reduction, line-work might have to be made simpler by straightening minor curves and omitting short branches. A small tract of land or water would look unconvincingly simplistic if reduced too far and might even be mistaken for an accidental blemish. In that case it would have to be treated like a superfluous point-symbol and omitted altogether. Some kind of value judgement ought to have been involved in choosing from a thematic category that had been made too numerous by a change of scale. In the case of settlements, for instance, the likeliest criteria of eligibility were size, population and economic or political importance; another ground for selection was a terminal position on an important routeway, though this could be misleading if the routes were subsequently edited out.[86] Such information was not always available to an editor. Suppose he had room for some but not all of the villages in his chosen territory, and suppose every village, large or small, had been represented on the original map by the same kind of symbol. Without a demographic census or some other extraneous statistical source, how could winners be distinguished from losers? Here is a cartographic paradox not often stated in plain terms: a small map may require more knowledge from its author than a large one.

An editor who lacked such knowledge (and most of them did lack it) might conceivably be influenced not so much by the places themselves as by their names. Toponymic characteristics with most appeal seem easy enough to guess at: intelligibility, congeniality – in either an aesthetic or linguistic sense – and of course brevity. Intelligible names were those that carried a connotative meaning, or that were known to stand for people as well as places. Congenial names were easily pronounceable or pleasantly evocative: Worthy in Hampshire looks better from this standpoint than Wormwood in Middlesex. Some cartographers may have been half-consciously resistant to names derived from a language that was not their own. If an original map featured both Celtic and Anglo-Saxon toponymy, would a smaller version of it perhaps betray the nationality of the editor? Along the Welsh border a good test of un-Englishness is the prefix 'Llan'. When John Speed, himself a Londoner, reduced his *Countye of Monmouth* for inclusion in a general map of Wales, he chose only 12 per cent of the 'Llan' names compared with 21 per cent of the other names. Unintelligible in a different way were 'floating' regional names with no attachment to either a point symbol or a boundary. Compare for example

16.14 Schematic river meanders. Nicolas Sanson, *Partie meridio^le du royaume d'Irlande* (Paris, 1665).

Speed's small map of Ireland with the larger maps by Robert Lythe from which it was partly derived. In County Clare, Speed used 33 per cent of Lythe's non-regional names and only 8 per cent of his floaters. Another editorial policy, perhaps more easily authenticated, was to favour settlement names that could serve as locational markers for nearby rivers or territorial boundaries in preference to names that stood aloof from other detail.

Although posterity must allow the early editor a good deal of freedom in such matters, a thoroughly eccentric choice may nonetheless be stigmatised as unwise or even positively erroneous: how else can one describe the inclusion of Egremont, Cumberland, in a non-thematic representation of northern England that finds no place for Carlisle, Durham, Hull, Lancaster or Newcastle?[87] In general, name-selection is a subject that has attracted little inquiry, and some of the foregoing theories may prove too facile to stand the test of serious research. What does seem highly probable is that the main influence on editorial behaviour in this respect was the need for an even spread of names across each map.

Descriptive matter was usually even more vulnerable than nomenclature to editorial blue-pencilling. When Speed copied Saxton's county maps, he could hardly be blamed for disregarding anomalies peculiar to a single site like the Roman road between Warwickshire and Leicestershire, and the hounds and huntsmen near Abergavenny in Monmouthshire. These deletions reflected a widespread editorial preference for the uniform and the anonymous that had nothing to do with scale. But besides suppressing an author's individuality, some editors felt obliged to impose their own – not just on borders, decoration, scripts and symbols but also, through 'personal curvature', in the treatment of linear geographical features. In 1587, for instance, Joan Martines added imaginary portolan-style capes and bays to an outline of Ireland that was otherwise based on Mercator's simplistic image of 1564.[88] Later Nicolas Sanson found it hard to copy other map-makers' river systems without introducing numerous schematic meander loops (Fig. 16.14). As with so many distinctions in cartography, the line between editing and authorship is not always easy to draw.

CHAPTER 17

Forbear much writing

ACONVENTION IS ANYTHING on a map, other than words and realistic pictures, that forms part of a factual statement about the world. There is a wide range of features that satisfy this definition. Point conventions are exemplified by circles for towns, line conventions by arrows for tidal currents. Area conventions include most kinds of shading or colouring for land and water surfaces, while script conventions govern the appearance of names and descriptive phrases as opposed to their linguistic content. Nor should we forget those properties of a map – scale, north-alignment and projection, for example – which can hardly count as realistic and which are too abstract to be visualised as separate marks or signs.[1] Another way of defining conventions might be as symbolic devices whose significance needs to be learned, though this raises the awkward problem of whether there are any signs that everyone understands by instinct. Let us simply add that map conventions, once acquired, eventually lose their capacity to arouse our interest, like 'dead' metaphors in verbal language, though in this case sleep or unconsciousness may be a more suitable analogy than death.

A symbol is 'unconscious' in the foregoing sense when it has become too well known for its conventionality to be recognised. Not all the viewers of a map achieve this degree of familiarity at the same time: a few initiates may understand without noticeable effort a symbol that would baffle everyone else. This is why the makers of portolan charts never bothered to explain themselves: their clientele was already 'in the know' and among the medieval map-using community there was still no such person as a general reader whose ignorance might need correcting. As the printing press made maps more widely available, this comfortable state of togetherness could no longer be taken for granted and cartographic symbolism now had to be formally elucidated, a process for which word-of-mouth communication was clearly insufficient. It therefore seems fitting that the advent of map-keys or 'characteristic sheets' should coincide with the early age of print, their first known appearance being in 1533 as an adjunct to Peter Apian's published map of Franconia.[2] After that the very presence of such explanatory matter probably inspired some cartographers to enlarge their repertoire of symbols (Fig. 17.1): a tabulated list looks feeble unless its length exceeds a certain minimum.

Meanwhile there was growing complexity in the world to be mapped and among the peoples who had created that world. Conventions were needed for new (or newly discovered) categories of settlement, newly exploited minerals, new

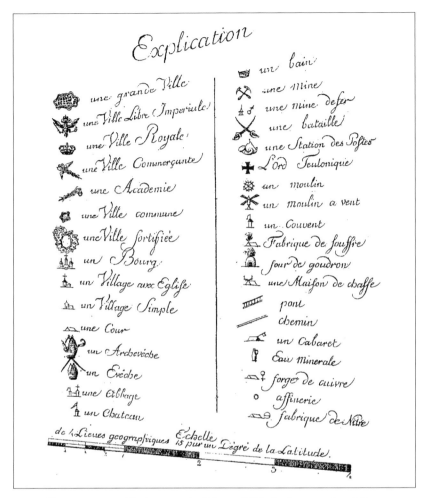

17.1 Key to topographical map, 1769. Johann Wilhelm Jaeger, atlas of Germany (Frankfurt am Main, 1789).

modes of travel and transport and new kinds of industrial enterprise. All these additional data might theoretically have been explained through the medium of words, supplemented where necessary by contractions and initials.[3] Aaron Rathborne advised cartographers to 'forbear much writing'[4] but the practical case against verbalisation has never been wholly conclusive and even in the sophisticated twentieth century non-realistic map symbols could be seen as an evil.[5] At all events it can sometimes be only too evident, 'intuitively', if a map has fewer words than it ought to have (Fig. 17.2). Prejudice apart, two arguments suggest themselves. In the first place conventions inevitably added to somebody's work-load. The author had to invent them; the reader had to learn them, or at least to track them down in a marginal explanation. Symbols also conveyed less meaning than could be packed

17.2 Low-density placenames, original scale *c.*1:130,000. G.L. le Rouge, Barbados (Paris, [1748]).

into a verbal statement. Their effect was to replace the individual by the general, and in doing so to give cartography an unwarranted degree of precedence over geography. Taken together, these two weaknesses had the further disadvantage of being complementary. Thus the idea of urbanism could be made less abstract by devising individual signs for towns of different types, but only at the cost of increasing the strain on the reader's memory. Generalisation could also be an effective way to mask an author's lack of knowledge, but this was hardly something to be pleased about: it is better to banish ignorance than to hide it.

On the other hand non-verbal signs could succeed in a strictly cartographic sense by occupying a more definite location than a line of script, which on small-scale maps might consume the equivalent of several hundred miles; and conventions, by saving space, left room for further information that would otherwise be squeezed off the page. Unlike words and phrases with their beginnings, middles and ends, a symbol could be apprehended almost instantaneously. It was also accessible to members of every language group, and being created for its own medium rather than imported from elsewhere it could avoid the kind of awkwardness sometimes found in verbal communication, where 'hill' and 'fort' might look too much like 'mill' and 'port'. Finally, conventional signs, unlike realistic images, were easy to draw. In that respect their generalising quality was put to good use, and on a printer's copper plate this effect could be enhanced by reproducing identical symbols with a stamp or punch instead of cutting each one individually by guiding a blade across the metal.

As a designer of conventional signs an author's task was to minimise the foregoing disadvantages and maximise the advantages. Certain requirements of cartographic symbolism were common to any semiotic system, one of the most elementary being that different signs should have different meanings. But some map-makers' applications of this principle, though undeniably effective, might still be hard to justify in strictly theoretical terms. Take the distinction between roman and italic script, increasingly common on maps after italic had passed the peak of its popularity in about 1570: when a century later John Adams chose roman for 'Rydall' as a village name and italic for the same word as a house name, it would have been difficult to say why those significations should not replace each other.[6] Such arbitrariness among map symbols was probably unavoidable to some degree, but there were three general objectives that merited a symbolist's attention: recognisability, proportionality and 'locatability'.

The role of cartographic mimesis

If symbols are to do their work with the minimum of help from a key, as seems desirable, there must be some pre-existing association between sign and subject-matter, independent of any particular map. Over the centuries cartographers have

17.3 From John Harding, Scotland, mid 15th century. See note 9.

17.4 Pictorial symbolism. From Olaus Magnus, Scandinavia (Venice, 1539).

generally grown more adept at forging this kind of link. Its most common basis, sometimes known as 'mimetic depiction', is a visual resemblance between icon and object.[7] It must be admitted that Paul Harvey, reviewing the course of both history and prehistory around the world, found a widespread preference among the earliest map-makers for abstract compositions totally unlike their referents, almost to the extent of supporting a sequential law that 'symbols' developed earlier than 'pictures'. Perhaps Harvey's primitive symbol-maps should be seen as bridges between oral and visual communication in which an otherwise incomprehensible relationship was explained by word of mouth – itself a reason why signs should be quickly and easily drawn so as not to keep the listener-viewer waiting. In some places, at any rate, Harvey's pictures started so early that most of the previous phase is irrecoverable: his only European example of a pre-pictorial tradition was in the

17.5 Indo-China, with ibis, Indian goats, giraffe and lions. From Euphrosynus Ulpius, globe, 1542. See note 12 and Fig. 10.11.

Valcamonica of northern Italy, and even there visual realism had already begun to gain ground in the middle of the first millennium BC.[8] Given the Eurocentric predilections of the present study, an overall dominance of mimetic symbolism can therefore be accepted without too much misgiving.

In medieval cartography the kinship between map and picture was particularly close. On the Hereford and Ebstorf world maps, for instance, decoration pure and simple was almost totally absent. Signs were drawn to simulate their referents in as much detail as the parchment could accommodate: sometimes, it might almost be said, in more detail, as when a quarter of John Harding's Scotland was buried under enormous castles and churches (Fig. 17.3).[9] If necessary a resemblance could be created by taking a verbal analogy at face value, as in the representation of evocatively named features like the Pillars of Hercules or the Caspian Gates. The objections to such literal-minded imagery would soon become apparent, but residual *in-situ* pictures were still plentiful in the sixteenth century. Olaus Magnus's *Carta marina* (1539) combines them with cartographic art-work to striking effect and on almost equal terms (Fig. 17.4).[10] Thirty years later, and less obtrusively, Philipp Apian's map of Bavaria included soldiers, cannon, a charcoal burner, miners, furnaces or kilns, horses, deer and goats.[11] Elsewhere, however, pictorialism by this time was losing much of its appeal. As more maps originated with an *ad hoc* survey, matter that might be thought superfluous was driven from their interiors by the accumulation of strictly planiform detail. Pictures survived to better advantage on the margin of settlement, particularly where they had reference to areas rather than points. Animals especially (Fig. 17.5) – elephants for want of towns, in Jonathan Swift's overworked phrase – were still to be found occupying empty spaces well after 1800.[12] Another refuge of visual 'realism' was the sea, for monsters and mermaids until the seventeenth century, for ships until considerably later. This kind of artist's licence could sometimes be geographically informative: ships of different sizes were useful for distinguishing degrees of navigability while exotic wildlife in a narrow sea at the edge of a map might helpfully suggest the presence of an unseen ocean beyond the neat line. Human figures cavorting on the water rather than in it would immediately indicate the presence of ice.[13]

As long as signs remained more or less self-explanatory there could be difficulties of separating portraiture from classification. A cartographer's conventions were influenced by his own cultural inheritance as much as by what he saw around him: few medieval buildings featured domes in the shape of pepper-pots, despite the testimony of the Hereford and Ebstorf maps; nor were tall church spires as common in English villages as some town-based Tudor cartographers might have thought.[14] It was also understandable for eighteenth-century British surveyors to map the coniferous forests on the shores of Hudson Bay with deciduous-looking tree symbols.[15] On the other hand convention could sometimes be enlivened with a certain admixture of realism. Thus Ortelius's anomalous tower at Cremona, measuring a full inch from top to bottom at a map-scale of about 1:200,000, was convincingly justified in his accompanying letterpress commentary as 'a turret of a wonderful height, far exceeding all the rest of this country'. In other maps, like those of Timothy Pont (according to some authorities), conventional and realistic symbols were applied indiscriminately to a single category of subject-matter.[16]

But the general trend in post-medieval European point symbols was for individuality to be disregarded. The nearest we can get to a statement of principle on this subject is a passage in Mercator's atlas:

> The little rounds do note the true site of every place, and from thence their distance is to be taken. Then there are marks whereby places may be known, bare villages by bare rounds, when there is nothing worth the note. The castles and forts of note are marked by a little hook upon the round; monasteries by a cross. The towns have commonly two towers, the hamlets of nobles one.[17]

This passage makes a number of valid points, some of which we shall reconsider in later paragraphs. The most relevant at the moment is that similarity could be extended by being made more abstract: thus 'bare' was matched with 'bare' in the foregoing reference to circle-symbols, which incidentally could also be crossed out with a diagonal stroke as an appropriate marker for 'places ruinated'.[18] At the same time mimesis could be brought under control, visually, by applying the rhetorical

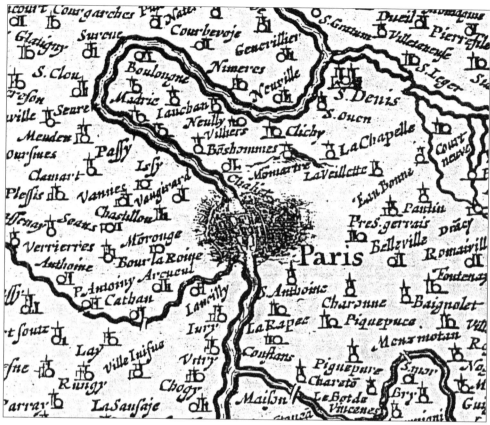

17.6 Planiform town representation (Paris), original scale *c.*1:265,000. Gerard Mercator, Henricus Hondius and Joannes Janssonius, *Atlas or a geographicke description of the world*, ii (London, 1636), p. 299.

device of synecdoche where the whole is represented by the part, as with the cross for a monastery in Mercator's example. Other such substitutions in contemporary atlas maps are a church for a village, a flag for a castle, a crozier for a bishop's seat, a horn for a hunting lodge, a hammer for an iron forge, a wheel for a water mill, what looks like a bath-tub for a bathing place[19] and, perhaps most famous of all, crossed swords for the site of a battle.[20] Some conventional signs came ready-made from non-cartographic codes of symbolism, for instance an eagle for a German imperial city or various kinds of metallurgist's abbreviation for minerals. Occasionally inspiration failed, as when a university had to be shown non-mimetically by a star and a market by what looked like an ordinary town symbol.

More lightly hinted at ('a little hook upon the round') was the possibility of applying two symbols to the same site. Thus on Godfried Mascop's map of the bishopric of Münster a village (*pagus*) appears as a profiled hemispherical cap, a notable house (*pretorium, sive villa*) as an inverted triangle, and a combination of these two objects as something like an icecream cornet.[21] Another example, from early British Ordnance Survey maps, was the alternation of dots and dashes in a single boundary that separated both counties (dashes) and parishes (dots), an idea that had already occurred to the cartographic amateur Daniel Beaufort.[22] However, to interpret each little tower or 'annexe' in a normal composite urban symbol as a single real-life building within the town – church, mansion or municipal hall – seems to be stretching probability too far.[23] On the whole, the process of schematisation appears to have been especially fast during the seventeenth century. As part of the second transition – from 'pictures' to 'surveys' – in Harvey's three-term developmental theory, it is much better documented than his previous shift from symbols to pictures.[24]

In general, renaissance pictorial conventions acknowledged that most items of landscape furniture were more familiar to earthbound observers in elevation than in plan. Indeed some early attempts at planiformity can only be described as gallant failures, especially the representation of large cities and (worse) their suburbs on small-scale maps (Fig. 17.6).[25] Only in the eighteenth century, when map-scales in general were becoming larger, did simulated plans begin to make rapid progress as a substitute for profiles. This tendency may have originated in military cartography, but by the 1720s it was already affecting civil maps. Samuel Wyld recommended to estate surveyors that 'the ground plot of buildings ought in all cases to be expressed by the same scale that the rest of the plot was laid down by ... [N]ever go about to draw the representation of a house or barn in the midst of the plot, so big as will cover an acre or two of land'.[26] This criticism was surely directed at elevations rather than exaggerated plan-forms. In fact there was a certain range of scales and subjects, including those of the typical European estate map, where it was a pity to sacrifice the opportunity that profile symbolism gave for the functional classification of buildings. To that extent planiformity could sometimes be seen a triumph of dogma over common sense.

By mid-century the drift towards planiform symbolism on regional maps was affecting the portrayal of hills, trees, fences, bridges and cliffs. Another example was the growing use of concentric rather than horizontal waterlines along sea and lake coasts.[27] The benefits of a planiform image were inversely proportional to the height of the object symbolised. Thus while low-profile earthen fortifications were already appearing two-dimensionally in Tudor times, lighthouses and windmills have remained in side-view even to the present day. On some nineteenth-century maps the third dimension survived mainly in the principle of shadowing by reference to an imaginary light.[28] This effect was achieved by a slight thickening of the appropriate lines, so that a rectangular building-block would be shadowed on its right-hand and lower edges, a rectangular body of water on its left-hand and upper edges, both rectangles being left empty unless there was some other reason for filling them.[29] In choosing which edges of a building to shadow some designers insisted on following the points of the compass even on an obliquely orientated

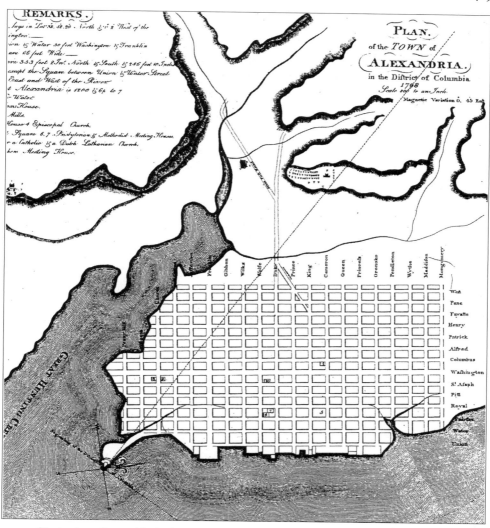

17.7 Shadowing of street blocks (more commonly done when north is at the top), original scale *c.*1:8800. George Gilpin, Alexandria, District of Columbia, USA (Alexandria, 1798).

map: but such maps were now unusual enough to make the results look unacceptably strange (Fig. 17.7).[30]

The representation of plurality by point-symbolism posed an awkward dilemma. A given number of buildings on a map would seem at first sight to denote the same number on the ground, but sometimes this interpretation was manifestly impossible, as when nine dwelling houses on Thomas Nairne's map of South Carolina were said in an accompanying note to be occupied by 800 people.[31] At small scales mimesis might consist rather of plurality signifying plurality and nothing further, the exact number of referents being deliberately mis-stated – in most cases understated by a wide if indefinite margin. The truth is that buildings, tightly or loosely clustered, were often functioning *en masse* as a kind of area-characteristic, like the tree symbols in a patch of forest, as we have already discovered by comparing map copies with originals. It might be thought better in such circumstances for every settlement, agglomerated or otherwise, to be given just one symbol and no more, a form of synecdoche rather obscurely characterised in a modern textbook as 'classification'.[32] This kind of total or ultimate generality was anticipated in the fourteenth-century Gough map of Britain, where the Forest of Dean appears as a single tree.

Area symbols

The most attractive kind of area symbol was colour.[33] Techniques of map-colouring have played little part in the literature on cartographic history, though lip service is sometimes paid to their supposed importance by paraphrasing or quoting early instructions for colour-manufacture, a practice that fails to establish whether maps differed significantly from other graphic media in the demands their colours made on industrial chemistry.[34] Defying the trend of modern map-philosophical opinion, we can usefully distinguish between functional and decorative colour. In the conveying of purely geographical information, shortage of colours has seldom been a matter for complaint. Mathematically speaking, four categories are known to suffice, but this rule applied only to the differentiation of adjacent territories.[35] When non-spatial characteristics were in question there could be no such constraint, though in practice half a dozen colours would often have been enough. The interior of a stately home might be painted in twelve shades of white, but few cartographers of any period would wish to take this as a precedent.[36]

Perhaps the first clear textual reference to a cartographer's choice of colour is an inscription in a papyrus map of *c.*1300 BC, explaining that certain hills had been coloured red.[37] Ranging more widely, the tenth-century Palestinian geographer al-Muqaddasi used different colours for routes, sands, seas, rivers and mountains. Clearly his inspiration was to some extent mimetic.[38] In later European maps of land cover there would likewise be a rough approximation to the prevailing tone of actual soil or vegetation: dark green for woods, light green for pasture or meadow and brown or yellow for arable, the last perhaps accompanied by a 'similitude of ridges and furrows'.[39] With the passage of time some conventions became detached from their mimetic roots. Two especially persistent traditions are thought to date from remote antiquity in Mediterranean latitudes. One was the use of blue for water, an exception usually being made for the Red Sea.[40] The other was the representation of buildings in red, which to an English surveyor might suggest brick[41] but which probably originated as an allusion to Roman roof-tiles.[42]

For planiform buildings to be distinguished with either carmine or grey had become a widespread convention by the eighteenth century, though the exact

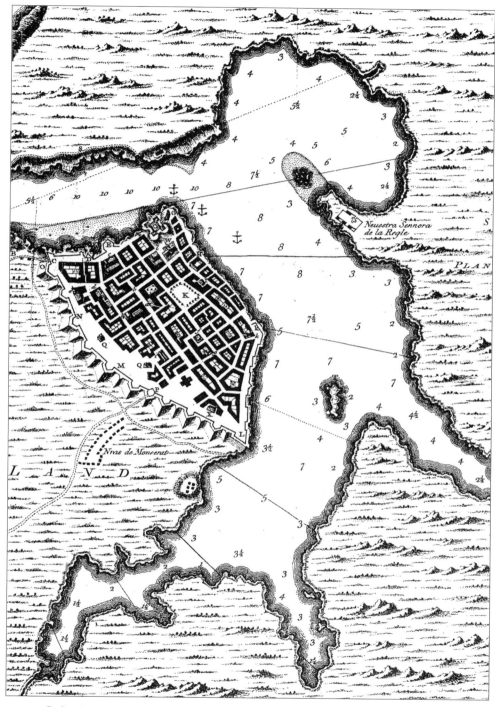

17.8 Indiscriminate use of grass and bush symbols, mid 18th century, original scale *c.*1:16,000.
J. Hinton, *An exact plan of the city, fortifications & harbour of Havana in the island of Cuba*, in
Universal Magazine, 1762.

significance of these colours seems to have varied: one possibility was carmine for stone or slate, grey for wood, mud or thatch; another was carmine for dwellings and grey for non-residential buildings. In general colour-coding became more rigorously imitative as more cartographers picked up the message. On Saxton's county maps as coloured by contemporary artists, for example, some hills were brown and others, for no particular reason, green. Later, relief features were more consistently shown in brown or grey to symbolise the absence of grass or trees at higher altitudes. With the rise of thematic cartography colours necessarily became more versatile: in twentieth-century Ireland a map showing land use in green and orange might have to be accompanied by a note disclaiming political intentions.[43]

Even before the sixteenth century different colours were occasionally printed on the same sheet from separate copper plates, but until the advent of lithography the easiest way to colour a printed surface was by painting individual black and white impressions with a brush.[44] In Elizabethan maps, especially, much fine engraving was almost obliterated by over-enthusiastic brushwork, but thenceforth colours generally grew lighter on both print and manuscript until in the course of the eighteenth century they often became restricted to a gentle tint along territorial boundaries. Already in 1720 Thomas Hearne could report having 'heard some knowing men say that the copies of Saxton's maps that are not coloured are preferable to such as are coloured'.[45] Then, with the revival of colour in the age of thematic cartography and the consequent need for clarification, it became common to anticipate future manuscript additions by engraving an explanatory key as a series of appropriately labelled empty boxes which a colourist would later fill in by hand.[46] An error in these messages could play havoc with a map's colour scheme.[47]

A monochrome area symbol could take several forms. The simplest was for closely-spaced small dots or lines to simulate an unbroken tint. Where the dots were just distinguishable, such a stipple could be used for a sandy desert, as in Matteo Ricci's representation of north-west China (apparently a rare case of western symbolism being influenced by Chinese maps)[48] and later by John Ogilby in the 'pine barrens' of Carolina.[49] Furrows, grass, trees, bushes and waves or ripples could also be drawn to resemble a continuous surface.[50] For marshland or rough pasture the continuity was broken to reveal a scatter of localised water-surfaces or tussocks of coarse grass (Fig. 17.8). Where every area on the map was characterised in some way (roads being excluded as essentially linear rather than two-dimensional features) the resultant impression of comprehensive portraiture could be both plausible and aesthetically satisfying – most of all, perhaps, in town plans (Fig. 17.9) and in what is regrettably an unusual genre, the large-scale engraved estate map.[51] On the other hand, when not carried through consistently the same effect could look woefully unkempt.[52] And even where partial 'carpeting' was done neatly it might be difficult to keep the map design in balance: see for example Ortelius's over-emphatic treatment of the Crau in southern France.[53] The one characteristic

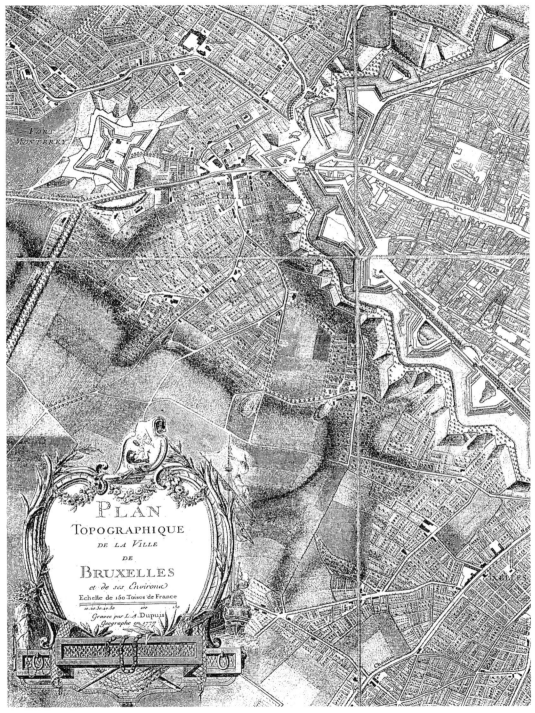

17.9 'Carpeting' of land cover, urban and rural. Louis-André Dupuis, *Plan topographique de la ville de Bruxelles et de ses environs* (Brussels, 1777).

for which a heavier hand might seem excusable was woodland, but this could easily be disturbed by the intrusion of placenames, as with the 'labels' hollowed out of woods in Jan Jansson's map of Luneburg.[54]

On the face of it, 'carpeting' might be taken to imply that something was known about the whole of the earth's surface. Many cartographers were prepared to act on this assumption, covering virtually every acre above sea level with a pattern of meaningless grass-tufts or bushes growing out of horizontal or very slightly tilted profile-lines. Ortelius's map of the country round Seville in 1579 is a good illustration of this rather fussy approach.[55] Another example, one is tempted to suggest, was the map featuring shadowy forests, 'champains' and 'wide-skirted meads' imagined by Shakespeare for King Lear's threefold division of Britain. Such indiscriminate landscaping marked a debasement of cartographic symbolism, for if water was coloured or shaded, white space by a process of elimination became sufficiently representative of land. Even the British Ordnance Survey broke this austere but logical rule by stippling the entire land surface of its beautiful one-inch 'Fifth Relief' map, though only with the palest and finest of light-brown dots, it is true.

More difficult to interpret is the kind of land cover that purports to distinguish single parcels, whether furlongs of so-called open field tillage or modern-style hedged enclosures. Under the heading of 'field patterns' these features have attracted a degree of attention in the historical study of the rural landscape that most early cartographers would have found alarming. We are not here concerned with estate and cadastral cartography, where (with some exceptions) individual field boundaries will be found as correct as anything else.[56] The problem arises with other kinds of map – with almost every other kind of map, it sometimes seems. Military surveys are perhaps the most familiar case (Fig. 17.10), already discussed in chapter 10. From them the 'fieldscape' spread to British and Irish county maps, apparently under the influence of John Rocque, and also to road maps at similar scales.[57] In addition, and at first sight unexpectedly, fields were sometimes shown by maritime surveyors, though against the background of a chart, it must be admitted, few of them can ever be said to carry much conviction. Implausibility in this situation is manifested in two ways. Firstly, the drawing itself may lack assurance and bravura – the lines thin and wavy, the enclosures unrealistically incomplete, a condition well exemplified by William Barents's 'hydrographic' map of Majorca in 1595.[58] Secondly, the shapes and sizes of the fields themselves may simply look improbable. For instance, there has surely never been a real landscape comprising row upon row of parallelograms with successive rows leaning in different directions and every third row sloping the same way as the first.[59] Conventionalisation is also indicated when the same field lay-out appears to embody a random mixture of vertical and oblique views.[60]

In most fieldscapes, however, the mapped outlines might just conceivably approximate to the truth. Then one can only report that the enclosures shown on eighteenth-century European topographical maps are generally considered too large

17.10 Conventional field boundaries in military cartography, original scale *c*.1:28,000. From *Boston and its environs and harbour, with the rebels works raised against that town in 1775* (London, 1778).

to satisfy historical common sense, and that this opinion has several times been verified by reference to contemporary estate surveys.[61] But perhaps the most effective argument is based on simple economics. In practice, field boundaries are seldom very useful in finding one's way, as any English pedestrian can discover with the help of the 1:25,000 Ordnance Survey map. Why should a topographical surveyor undertake the heavy additional labour of measuring these features when so few of his customers were likely to benefit from them? In summary, pre-nineteenth-century fields were usually a way of defining a whole landscape as 'improved' rather than a record of particular fences.

The foregoing example draws attention to a serious weakness in the cartographic language – its inability to express the psychological phenomenon of doubt. The problem is well illustrated by Vincenzo Coronelli's representation of the south polar regions.[62] Like a fictitious but better-known example in the works of Lewis Carroll, Coronelli's map is totally empty. What was he trying to tell us? Perhaps white paper meant a confession of ignorance, like the blank pages in next year's diary. Or perhaps it was making a positive geographical statement to the effect that Antarctica consists entirely of sea, or entirely of land. Ortelius had tried to solve this problem a century earlier in a map of the world as recorded by the ancient Greeks and Romans (Fig. 17.11). Known sea was stippled, known land was left white except

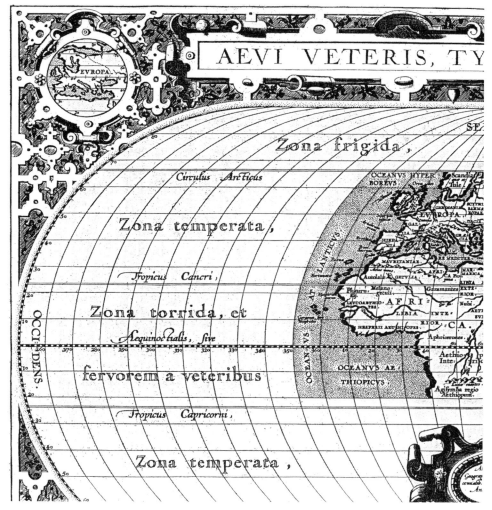

17.11 The ancient world, known and unknown. From Abraham Ortelius, *The theatre of the whole world, Parergon* (London, 1606), p. vi.

for its mountains, rivers and names. Unknown areas were also white apart from a graticule that covered the rest of the world. In some cases, Ortelius suggested without actually saying so, the frontier of knowledge coincided with a line of latitude or longitude.[63] He knew that there was something unusual about this map as acknowledged by its curious title, in English 'a draught and shadow of the ancient geography'. An alternative approach, recorded many years later, was for an unaccompanied graticule to assert a known lack of mappable geographical features, with entirely undiscovered lands or waters beginning where the graticule gave place to blank paper. It is an idea said to have been developed by J.B. Bourguignon d'Anville in a map of North America.[64] Like some other potentially useful cartographic conventions, this one suffered from not being known to enough cartographers.

17.12 Main roads with 'shadowing', original scale *c.*1:180,000. From J.Chilcott, *Guide to Bristol, Clifton, Hotwells and its environs* (Bristol, 1826). See also Fig. 9.14.

Conventionalising lines and strokes

Lines were the most difficult of conventions from a mimetic standpoint, being by nature too narrow for their appearance to vary without some sacrifice of locatability. The medieval solution was to show roads as single threads and to exaggerate the width of rivers, sometimes grossly.[65] When, reversing this distinction, a double line was adopted for roads, it became possible to make highways quasi-realistically wider than lanes, but there was still only limited scope for differentiation on a normal-sized topographical map. (In eighteenth-century England, turnpike gates, signposts and milestones might be sufficient indicators of importance.) The only other possibility on an uncoloured map was to treat major roads in the same way as buildings and 'shadow' their right-hand and lower edges on the mimetic principle that they were more likely than minor roads to be raised by artificial surfacing above the level of the adjacent land (Fig. 17.12). This was a common practice from the 1770s onwards, but started going out of fashion when lithography came to take the place of engraving.[66] Meanwhile coloured fillings had often been used to emphasise roads as such, especially in manuscript cartography, but not much for separating better roads from worse: that function of colour became popular only in the late nineteenth and the twentieth centuries, mainly as a service to cyclists and motorists. For other kinds of line perhaps the most effective method of characterisation was by tinting one side with a band of transparent colour through which adjoining non-linear features could remain legible. This process was easily overdone, however, especially on complicated small-scale maps; in print the same charge could have been brought against hatched or stippled boundary 'zones' if more maps had made use of these.

One particularly successful linear application of mimesis is credited to the fifteenth-century cartographer Nicolaus Germanus.[67] This was the pecked or interrupted line for a real-world boundary that could be crossed without meeting any physical obstruction, just as a pointer could be carried through a broken line on paper without touching any of the dots and dashes. Such lines were self-evidently appropriate for political and other legal boundaries, for the edges of unfenced roads as in John Ogilby's *Britannia*, for intermittent or underground streams, for contour lines, for routes of journeys unmarked by a normal road symbol, for proposed walls, roads or canals that had not yet been built, and for linear features of any kind whose existence was regarded as doubtful.[68] Experience shows, however, that different kinds of broken line become irritating when there are more than three or four of them on the same map: cartographic historians have discovered this for themselves by trying to plot the routes of different explorers in one colour on a common base.[69] For territorial boundaries, dot size was a useful measure of rank in a hierarchical structure: as Herman Moll expressed it, 'divisions of countries' could appear as 'prick-lines', with 'larger pricks for provinces, and smaller for subdivisions', while 'divisions of [i.e. between] nations, were often shown by chain-lines'.[70]

In conventionalising script the main categories were majuscules or capitals and miniscules or 'lower case', a difference related to variations in letter-size (discussed below) though obviously not the same. The use of capitals for area-names and miniscules for points and lines is at least as old as the Catalan Atlas of 1375. In the

17.13 Part of Roman Gaul with 'stone letters'. Abraham Ortelius, *The theatre of the whole world*, *Parergon* (London, 1606), p. xiii.

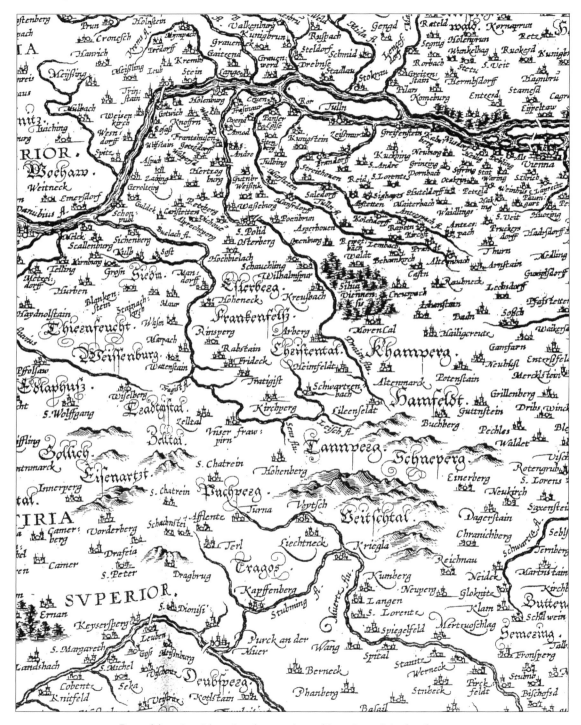

17.14 Part of Austria with regional names in gothic script, original scale *c.*1:1,150,000.
Abraham Ortelius, *The theatre of the whole world* (London, 1606), p. 63.

earliest printed lettering some artists chose capitals throughout because they were easier to engrave, but once this difficulty had been overcome at the end of the fifteenth century the Catalan system quickly became almost universal.[71] Otherwise the only common early use of script that might be called mimetic was in distinguishing ancient from modern. In Ortelius's maps of the classical world there are many open capitals of a kind that he hardly ever applied to his own time. Perhaps his intention was to simulate carving on masonry in the manner of a Roman inscription, an hypothesis supported by the later draughtsmen's term 'stone letters' for this kind of script (Fig. 17.13).[72] Gothic writing seems first to have been used for any class of feature that was comparatively rare, for instance territorial names as opposed to settlement names in sixteenth-century central Europe (Fig. 17.14). In map titles of the same period gothic could be used to identify a name as German rather than Latin. Its semi-mimetic use for antiquities, familiar today, has first been found on British Ordnance Survey maps of 1812.[73] Perhaps the most useful function of individualised script, however, would be to show whether the relation of word to object was connoting or denoting. We need to know that Cook's 'Doubtful Island' was a name and not a confession of ignorance and luckily his choice of lettering solves what might otherwise be an awkward semantic problem.[74]

A sense of proportion

Apart from being easily recognised, a good map-symbol must be in some sense 'proportionate', its visual impact in harmony with the attention conferred on it by the reading public. Importance has naturally depended on historical context. When ships were small and fragile, it was appropriate for sea shading to be dominated by tumultuous waves.[75] As naval architecture improved, a calmer-looking stipple came into vogue and eventually the sea was left blank.[76] Proportionality posed an increasing problem as societies, economies and artifacts became more complicated. The rule adopted by the Ordnance Survey in 1824 – 'names are to be so written as to be legible at distances proportional to the importance of the places'[77] – was applicable not just to script but to every other feature of a map. The significance of a point-symbol could be expressed by size, which would be best conceived as the diameter of an imaginary enclosing circle rather than either the height or breadth of what was actually drawn. (Of course 'point' in the phrase 'point-symbol' refers not to the symbol itself but to the amount of space that would be occupied by a planimetric representation of the same feature on a small-scale map.) For lines the most easily apprehended variable was thickness, and for area-fillings it was ratio of inked to ink-free paper. The letters of the alphabet, not often seen as conventional signs in the ordinary cartographic sense, turned out to be the most iconically adaptable symbols in the map-maker's repertoire, because style, size, lateral extension, thickness, colour and underlining could all be adjusted to define the significance of a name or other inscription.

The most difficult of these aids to symbolism in proportionate terms was colour, which unlike length or breadth did not change continuously along a single dimension and which on that account was often employed without regard to relative magnitude, notably in differentiating the various countries shown on a political map. It was most vigorously exploited for this purpose by Johann Baptist Homann and certain other German and Dutch cartographers of the early eighteenth century.[78] Where colours need to be made proportional in modern maps this has generally been achieved by differences in the density of a single hue, increasing its apparent 'value' or 'saturation' with a series of abstract line and dot patterns.[79] This method is most useful for thematic cartography, where there are few other symbols or inscriptions to disturb its visual effect, but it also became familiar in the layer-colouring of relief features on topographical maps.

In traditional map-making the colour most often used for emphasis was red. Some artists, including the author of the Gough map, would have allowed an even higher place to gold as an appropriate medium for particularly important names.[80] Gold had mimetic associations with rarity and material value, it is true, but suffered on paper from a certain lack of visual impact, especially when occurring in small quantities: a gold placename can never shine as brightly as a gold crown. Red on the other hand could sometimes be too powerful, especially in a non-quantitative 'chorochromatic' context. According to George Orwell 'all' cartographers indulge this propensity in political maps by reserving red for their own homeland and its dependencies,[81] though some members of former subject nations have been known to characterise this colour as a symbol for blood. In reality it may have been only the British who adopted Orwell's formula, but their early twentieth-century empire was large enough to make his opinion look like a universal truth.

The significance of location

The one attribute of a cartographic sign that seems in no way conventional (ignoring the effects of scale and projection) is the exact position it occupies on the map. With a point-symbol the area covered is almost certainly too large in planimetric terms. So if a town appeared as a cluster of multi-storey buildings in profile, exactly where within the limits of that image was the town itself? At the centre of the symbol, or somewhere on its lower edge where the houses were shown resting on the ground? Seen in this light the best kind of point-symbol would be disc-shaped, preferably with a dot placed in the centre to resolve any lingering locational doubts, and small enough to fit the space left by other detail in the immediate neighbourhood. In the thirteenth century Roger Bacon represented cities by 'little red circles'.[82] As we have seen, many subsequent cartographers achieved a double semiotic purpose by combining Bacon's symbols with a stylised architectural profile.

On the same principle, the precise location represented by a strip-symbol was a line mid-way between its edges – except in the case so often quoted by modern teachers of cartography where a river, canal, road and railway ran side by side along a narrow valley on a small-scale map. Locatability was not an issue with area symbols because their extent depended on the planimetric boundaries enclosing them. The only topic that calls for comment here is a kind of point-symbolism (in a purely graphic sense of the term) that actually refers to a whole region, as where for instance the national output of a commodity is represented by a single proportional circle somewhere within the borders of the nation concerned. In modern statistical maps this kind of symbol may pose a genuine puzzle for the uninstructed novice. Examples from traditional cartography were flags, coats of arms or enthroned kings proclaiming the political allegiance of a territory, ships indicating that a body of water was navigable, and individual plants or animals epitomising the flora and fauna of the surrounding area.[83] In theory, these imperfectly located icons should doubtless have occupied a central position within their territory. In practice a map-maker might try to avoid confusion by placing them as far as possible from features whose location was meant to be taken literally. In general, however, we should hesitate to interpret icons as randomly scattered over a map; even with a medieval *mappamundi* analogies like 'confetti' or 'buckshot' are probably best avoided.[84]

One solution to the problem of locatability deserves notice for its sheer flamboyance, if not for its long-term success. This was to fill an entire land-mass with a single enormous picture, or montage of pictures, representing phenomena to be found in at least a part of the area covered. Such displays occurred especially in sixteenth-century charts of Africa and the Americas where there was as yet little true geographical detail to show behind the portolan-style coastline but where plants, animals and people could be portrayed as instructively exotic. Originally a Portuguese speciality, this technique later became a favourite with chart-makers based in Dieppe.[85]

A rule with an opposite tendency that seems too simple to need learning was that script should be placed as near as possible to what it denoted. This justified the spacing-out of national and continental names as illustrated by the treatment of 'Africa', 'Asia' and 'Europa' on the Hereford world map, and the bending or fragmentation of inscriptions to fit them inside odd-shaped territories. Names of mountain ranges, rivers and streets could deviate from straight alignments for the same reason, and those of individual hills were sometimes arranged in a tight circle around the summit. The demands of spatial flexibility in such cases no doubt helped to popularise scripts whose individual characters were separate rather than con-joined. A more general influence in the same direction was the letterpress printing of books from moveable type, a practice which as we have seen was soon extended to the inscriptions on many woodcut maps. Here we may notice in passing that among west European cartographers the new trend towards discrete letter-forms was especially well exemplified by successive hand-written words and phrases of the 1560s and 1570s, a transition period when many influential printed maps and atlases were becoming available as models of design. Thereafter cursive and 'secretary' styles were rather quickly abandoned by serious cartographic draughtsmen except on rare occasions as a special 'display' feature.[86] In England the change to separate characters was sufficiently pronounced to give later scholars a method of very roughly dating Elizabethan manuscript maps – a point well illustrated, though never explicitly spelt out, in R.A. Skelton's catalogue of the undated Cecil maps at Hatfield House.[87]

For a map-maker to keep all his writing horizontal (or vertical, for that matter) was unusual enough to provoke comment from a modern historian.[88] Where congestion made the rule of proximity unworkable it might be necessary to invent some other principle. One suggestion was that other things being equal a name should stand on the right-hand side of its referent, though we may wonder how many professional cartographers, past or present, have actually heard of this advice.[89] On pre-renaissance maps the framing of an inscription might be used to link it with its subject, as in John Harding's map of Scotland, where each ribbon enclosing a town name was neatly brought into contact with the appropriate buildings.[90] Less elegantly, name and symbol could be connected by an arrow or tie-line. A cartouche, somewhat paradoxically, was a device for blocking off a title from every individual region on a map and associating it by default with the entire display. This forms one of many areas of overlap between convention and our next subject, which is decoration.

CHAPTER 18

To deck and beautify your plot

To A MODERN CONNOISSEUR and collector, what most distinguishes early maps is an abundance of decoration that present-day cartographers would find both intellectually unjustified and artistically embarrassing. On this subject, historically-minded map philosophers have taken a more lenient view, opposing the distinction between necessary and superfluous detail with the argument that every dot and line on every map fulfils a definite non-aesthetic purpose, if not scientific and topographic then in some way social or political. Thus one unchallenged or almost unchallenged authority has insisted that emblems and ornament, far from being 'inconsequential marginalia', should count as basic to the way a map conveys its cultural meaning.[1]

In fact it was only at certain periods and in certain cultures, most notably in Europe between the sixteenth and nineteenth centuries, that artistic adornment could be easily disentangled from other aspects of cartography. Even then, the more utilitarian kinds of map were often left with little or no decorative garnish, especially those meant for navigational, engineering, legal or military use or as contributions to specialised scientific research. Ornamental display worked better in maps catering for the general reader, a class that might on this account become more numerous if the decorator's embellishments were not too expensive. Also susceptible to an artist's ministrations were maps designed for wealthy individuals, who with any luck would make them known to other possible patrons of cartography in the same social set. This high-profile category included estate maps, not so much for corporate bodies (whose members might disapprove of unnecessary expenditure) as for private owners happy to be recognised as persons of wealth and refinement.[2] Estate surveyors were encouraged by their textbooks 'to deck, and beautify your plot',[3] though advice on this subject was often almost disdainfully vague, as in 'according as you shall see convenient'[4] or 'with such other ornaments as are commonly introduced in works of this kind'.[5] It would certainly be odd to find a writer from this period regarding such beautification as an 'integral' part of any map. Which is not to deny that, on a wider canvas, lavish ostentation could proclaim the power and glory of principalities, kingdoms and empires. It could likewise celebrate the merits of the aristocracy, the 'plantocracy', the bourgoisie, the white race, the cartographer's fellow-countrymen or the educated classes in general.

18.1 Cartouche with hammer and quadrant. Gerard Mercator, Henricus Hondius and Joannes
Janssonius, *Atlas or a geographicke description of the world*, i (London, 1636), p. 121.

Decoration as gesture

To a clever modern or post-modern investigator, the messages conveyed on the
outskirts of an early map can be endlessly subtle. Take for instance the rather
inconspicuous sea creature in Oronce Finé's *La terre sainte* (1517), who on a second
glance can be seen to carry the initial letters of the cartographer's name. This, we
are told, is a dolphin, which, wearing as it does a crown of fleurs-de-lys, is an
allusion to Francis I, dauphin of France. It is also an allusion to Finé himself, whose
birthplace was in the Dauphiné. The reading 'd'eau fine' makes a play on 'dauphin',

which can also be pronounced punningly as 'd'O.Fine'.[6] (Not everyone spells Finé's name with an accent.) A less transparent example is the dog in the lower margin of Laurence Nowell's 'General description of England and Ireland' in c.1564.[7] This ferocious creature has usually been assigned a purely canine role, snapping at the empty purse of an impecunious cartographer seated nearby as if in allusion to the lines 'Hark, hark, the dogs do bark, the beggars are coming to town'. But he has also been held to represent those sixteenth-century Irishmen, 'perceived as savage and barbaric', who were so often engaged in defending their country against the English.[8] After which it is no surprise to read that in a Nottinghamshire estate survey of 1635 pictures of ladybirds and butterflies might stand for courtiers of King Charles I, while snails on the same map depict the landowner's reluctant withdrawal from the metropolis and its problems into a haven of rural peace.[9] Sometimes such pleasantly fantastical chains of elucidation are hard to get started. A burly figure alongside a seventeenth-century map of Germany looks for all the world as if poised to smash a surveyor's quadrant with a large hammer (Fig. 18.1). Was this a premonitory salute to those later cartographic historians who dislike the idea of 'accuracy'?

One class of people has often been omitted by cartographic scholars from their ideological analyses and that is the map-making community itself. Elaborate embellishments, whatever their impersonal significance, could be justified as proving an author's mastery, technical or financial, of any problem the graphic arts could put to him. On this self-promotional level aesthetics might indeed be almost as important as geography. Joan Blaeu, for instance, made no distinction between 'supply' and 'decore', requesting 'ornament, titles, mile scales, and some other details' from his Scottish correspondents with equal importunity.[10] There is certainly one sense in which decoration and information can make a genuine if not particularly agreeable synthesis, and that is where whole tracts of terrestrial space turn out to look like something that a non-cartographic artist might take pleasure in drawing for its own sake. Among astronomers this had been a familiar theme since ancient times.[11] Popular identifications for the geographical synthesist were with animate objects, as when Belgium appeared as a lion, Europe as a woman (Fig. 18.2), and the whole world as an eagle, lily or clover leaf,[12] illustrating thought-processes that would all too easily degenerate into those of the modern 'cartoon map'.[13]

More often artistic satisfaction lay in bringing a touch of style to details inseparable from the map-maker's geographical message. Mercator's italic lettering was an obvious case in point.[14] So were some of the more smoothly curvilinear map projections of the same period. (The Latin language could almost be interpreted as a literary parallel for this kind of 'built-in' decoration on post-renaissance maps.) A less obtrusive example was the shading of field boundaries in certain late eighteenth-century estate surveys – by which means, according to one textbook, 'the colour from the outside of the field will seem by degrees to vanish, and to look the more beautiful to the eye'.[15] 'Vanishing' colour washes differed from the political and

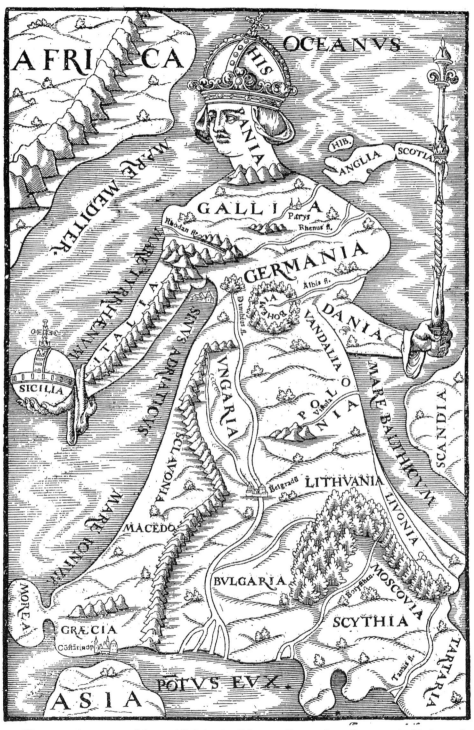

18.2 'Europa regina, queen of the world'. Sebastian Münster, *Europa*, from *Cosmography* (Basle, 1580).

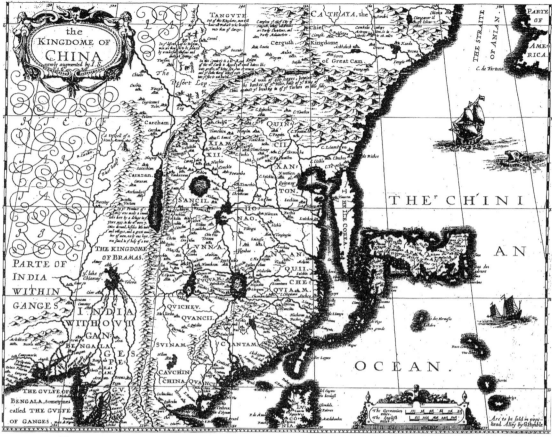

18.3 Territorial name ('Tartaria') with flourishes. John Speed, China, London, 1626, from *Prospect of the most famous parts of the world* (London, 1627).

legal boundary-tints of more complicated maps in having no obvious demonstrative function, but on the best estate plans the shading was so discreet and so well blended with other detail that the term 'decoration' seems too crude.[16] It was also possible for clarity and beauty to fight against each other: on some maps – John Speed's China, for instance – the script is so overburdened with loops and flourishes that it takes a perceptible effort to decipher its meaning (Fig. 18.3).[17]

In this chapter, however, our concern is mainly with extraneous features more amenable to separate discussion. We may first delimit antique from modern with the comment that opportunities for 'decking' diminished in the age of government surveys as taxpayers saw how much of their money could be spent on even the plainest of maps. A distinctive trend of the early nineteenth century was foreshadowed in Lord William Bentick's advice that for maps of British India 'the plainest method and style will appear to be the best'.[18] The new spirit was also expressed in the visual sobriety of the first British Ordnance Survey publications and those of the admiralty at the same period. 'I should like to see your time better occupied than in beautifying', the hydrographer to the Royal Navy told a junior survey officer in 1831.[19] Not long afterwards, authors of official British tithe maps were urged to show lines, names and numbers 'with no other ornament or colour whatever'.[20] Among private cartographic publishers of this period, it is true, there was a certain reaction against visual puritanism, analogous to the contemporary or slightly later aversion from Georgian restraint in the world of architecture, but in the end the drift to simplicity reasserted itself, and by the nineteen-hundreds any display of art for art's sake would infallibly tag its author as a cartographic outsider. In an age of utilitarianism the only remaining scope for purely artistic motivation had to lie somewhere outside the map-maker's neat line. Early six-inch Ordnance Survey sheets for instance were totally plain except that customers who took them as a county atlas received a full-size title-page with the county name set out in lettering that can only be described as extravagant. (Engraving a one-word title in this style against a plain background is said to have taken seven days.) A later outlet for the aesthetic impulse was to fold the map, protect it with a cover, and decorate the cover.[21]

Having set this historical limit we may return to an age when draughtsmen seemed to enjoy supplying a quota of non-planiform graphics somewhere on the face of every map. The problem was that improved geographical accuracy and comprehensiveness tended to push such features into a marginal position and so to make them structurally more independent and in consequence artistically more aggressive. Outside the territorial boundaries and inside the neat line there were titles, scales, north-points, explanatory tables and various kinds of picture. Then there were the spaces between circular or oval world maps and their rectangular outer frames, an early example being the four corners guarded by angels in the twelfth-century Sawley map.[22] Thus externalised, decoration could now stand for anything or nothing instead of contributing to a strictly geographical statement.

Literal and other meanings

At this point some thought must be given to the place of visual realism in cartographic decoration. What features of a map are manifestly *not* decorative? The most likely answers are (a) planiform line-work and (b) conventional symbolism, including script, that includes no more pictorial elaboration than is needed to describe the two-dimensional lay-out of the earth's surface. Suppose we define decoration as anything on a map that lies outside these two categories. Clearly there need be no hesitation about admitting semi-abstract motifs like strapwork and acanthus foliage. But how do we deal with features inside the map-frame which were clearly informative rather than artistic but which for all their complexity of spatial structure could never be described as cartographic? (For present purposes an

18.4 English costume designs. From John Speed, *The kingdome of England* (London, 1610).

image is non-cartographic when, if it occurred in isolation, no normal person would call it a map or consider it to be part of a map.) Even after this effort at analysis, an individual item may be difficult to classify. Take the large tents and their important-looking occupants that straddle the northern Sahara in the Maggiolo chart of 1563.[23] Political geography was evidently a major issue here, but the most startling features of these vignettes, and indeed of the whole map, are the complex foliated designs that cover every tent from top to bottom and from side to side. Probably the patterns in question carried some heraldic significance of practical import to north African men of affairs (though they can hardly have meant much to the average European reader) but until some historian comes forward to pursue the matter we can only assume that form was taking priority over substance in the Maggiolos' colour schemes.

Marginal pictures deserve a few more words of their own as a matter of courtesy, however tempting it may be to dismiss them as intruders from some other medium of communication. Joan Blaeu described their subject matter as 'the graces of each region, the fruits which it bears, the metals it produces, the animals it begets or nourishes ... [and] ... the arms or insignia, with the names, whether of dukes or counts or barons or other notable men'.[24] Counting these as two categories, naturalistic and heraldic, we may add: (3) representations of planets, satellites, comets and other heavenly bodies; (4) diagrams elucidating problems of astronomy or mathematical geography; (5) images of temporal sequence: day and night, the months of the year, the four seasons, the signs of the zodiac, or perhaps just 'time' as a philosophical conception; (6) symbols for the study of cosmography or geography, including portraits of explorers, surveyors or draughtsmen equipped with the instruments of their craft; (7) personifications or icons of particular regions, for example the four continents; also of geographical phenomena such as winds from various quarters, the elements of fire, earth, water and air, and the seven wonders of the world; (8) pictures illustrating classical mythology; (9) vignettes with a religious theme, such as scenes from the Bible or of missionaries at work; (10) illustrations of local history; (11) views of towns, industrial sites, rural landscapes or stretches of water, though these perhaps would qualify as Blaeu's 'graces'; (12) 'fashion plates' of typical native inhabitants in appropriate costume (Fig. 18.4);[25] (13) portraits or genealogies of kings, princes or noblemen with a territorial interest in the map, and perhaps of Roman emperors or other imposing but dubiously relevant historical personages (Fig. 18.5); (14) figures representing the benefits of good government, such as 'justice', 'concord', 'peace' and 'counsel', and sometimes even more basic human experiences like desire and fortune and pleasure.

Cutting across this kind of thematic classification was the difference between general and particular, and between different degrees of particularity. A castle or mill might be either an imaginary vision or an authentic likeness of one individual site. Here too distinctions could sometimes be difficult to enforce: for all practical

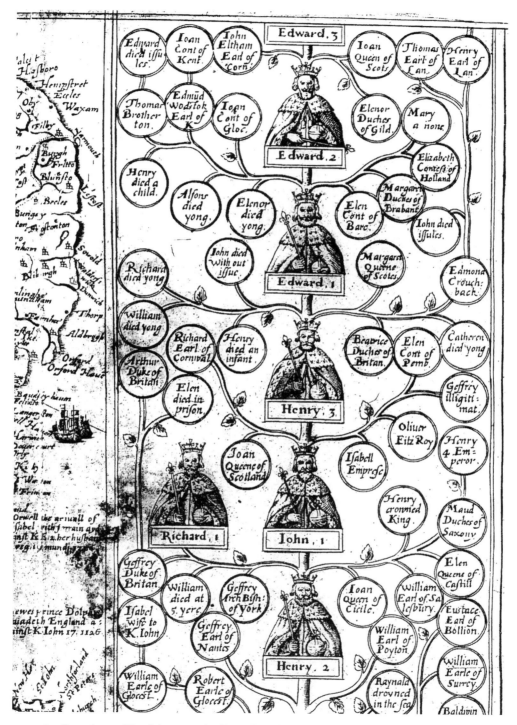

18.5 Genealogy of English monarchs. From John Speed, The British Isles (London, c.1603).

18.6 Surveyors at work. From John Rocque, Middlesex (London, 1757).

purposes a portrait of Ptolemy counted for no more than a portrait of Europa when nobody knew what Ptolemy looked like.[26] Nor were the surveying instruments depicted by a cartographer invariably the same as the ones he had used to make his map (Fig. 18.6): in this connection we may remember the unknown annotator who tried to tell us something by substituting a circumferentor for a theodolite in the margin of an estate map by John Rocque.[27]

To some tastes the foregoing categorisation may appear unduly simple. Today it would be widely agreed that 'cartography was always intended to mean more than it appears to show', even though without the gift of telepathy there may be no way of putting this proposition to the test.[28] What then were the meanings behind the meanings? One's first reaction to a large sample of decorative maps is that a normal reader would usually respond by feeling impressed, inspired, uplifted, respectful, appreciative (of what, may not be immediately obvious) and in general happy rather than sad. However, on this subject it may be better to follow a well-known map historian's analysis of Tudor English cartography, which is no doubt also applicable to many later maps. In Table 1, headings under 'map type' are simplified and a column for 'practical uses' is omitted.[29]

TABLE I

Map type	Symbolic meanings	Social function of image
Estate plans	Seignurial authority; proprietorship; class; pride, attitudes towards landscape and discovery of nature	Maintenance of social structure based on land; rise of absolutism; development of aesthetic consciousness
City and town plans	The ideal city; antiquity, fame and celebration of cities; mercantile wealth; the power of cities	Utopian city planning; differentiation of town and country in social and political terms
County maps	The county community	The identity of the county as a social unit with development of regional society and culture; the intellectual discovery of England
National maps	The nation and the crown; patriotism; the political state	Rise of the national consciousness; ethnocentrism; exercise of secular power
World maps	Empire; the New World science and the liberal arts	Attitudes to conquest and exploration; the creation of world views; the Renaissance 'discovery of space'
Celestial maps	Emblems for cosmographies and religious systems; the royal image	Reinforcement or refutation of religious, magical, and scientific beliefs; astrology; divine right of monarchy

The uses of ornament

To continue with more elementary kinds of map-reading, the best excuse for artistic licence was instructional rather than aesthetic: it drew the wayward reader's eye to some utilitarian feature of special importance, and especially to any message that involved the whole of the mapped area rather than one particular location. Among the most likely subjects for such treatment were titles, north points and scale bars; also the whole composition as incorporated in a single view. Seen at a glance, an ordinary undecorated map is complex, meaningless and not necessarily good to look at – in words attributed to Rigobert Bonne, 'so dry a subject that opportunities for treating it as a picture should not be lost'.[30] To satisfy Bonne, the cartographer would have to set a trap that needed priming with some more immediately attractive bait. Only when the bait was taken could the act of vision repeat itself,

18.7 Cartouche with scale bars, Hainault, from Gerard Mercator, Henricus Hondius and Joannes Janssonius, *Atlas or a geographicke description of the world*, ii (London, 1636), p. 239. See also Fig. 11.9.

advancing from the impressionistic to the analytical and then discriminating one map-component from another.

Scale bars like titles could be decorated by placing them in ornamental boxes (Figs 11.9, 18.7). Direction-indicators had a more complicated history. Functionally, all they needed was a single line with an arrowhead at its northern end. On navigational charts it would be helpful to draw another thirty-one lines radiating at equal intervals, and then legibility might benefit further from the addition of a contrasting symbol at a ninety-degree angle, typically a cross to mark the special Christian significance of the east (Fig. 18.8). To some cartographers, Robert Lythe in Elizabethan Ireland for instance, a pattern of simple wheel spokes seemed effective enough,[31] but by widening all the lines into daggers or arrowheads it was easier for contrasting colours to distinguish one compass point from another – and

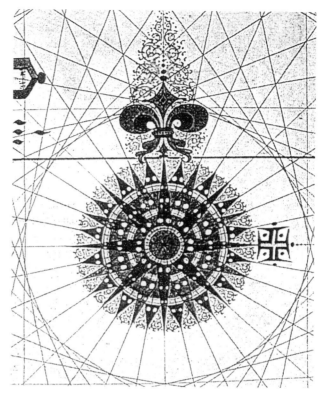

18.8 Portuguese compass rose with fleur de lys (north) and cross (east). From Sebastien Lopez, Atlantic Ocean, 1558. British Library, Add. MS 27303. See also Figs 11.8, 11.9.

perhaps for the whole design to evoke the idea of the pole star, an allusion unaffected by the discovery that magnetic north was different from true north.[32] Other features of the compass rose were more clearly decorative, sometimes including a small inner circle enclosing a pictorial vignette. The more ornate the composition as a whole, the greater the need for special emphasis on the actual north point. Such was the function of the schematised lily or fleur-de-lys. This motif was already serving various non-navigational purposes, especially in the heraldic art of the French monarchy, before its adoption on maps and compasses towards the end of the fifteenth century, and it long remained current in France as a cartographic point-symbol expressive of government authority.[33] One rather dubious reason for its adoption was that the initial 't' of the word 'tramontana', meaning 'north wind' on Italian and Spanish compass indicators, reminded map-makers of a three-pointed flower.[34] Later the flower migrated from the inside to the outside of the compass circle.[35] This made the middle point of the triad visually dominant as well as easily distinguishable from the symbols for other compass directions, a fact that did not prevent some artists from considerably over-enlarging and over-decorating the whole image.[36] On maps with only a single meridian arrow the fleur-de-lys was less useful (though kept alive from force of habit), but as part of a compass rose it remained one of Europe's most common cartographic symbols through both space and time – the only important exception being a short period of political incorrectness when French Revolutionary cartographers tipped their north lines with a so-called 'cap of liberty' instead.[37]

Scales and north points were seldom difficult to recognise, but in a crowded composition other features that ought to be peripheral might tend to merge into the rest of the map, with a risk that proximity on the page might become hard to tell from proximity in the landscape. A case in point, popular from the sixteenth century onwards, was the inset map, perhaps a large-scale town or harbour plan, perhaps a peripheral area of the main subject detached from its correct position as a way of saving space. Normally a single ruled line would be sufficient to separate the major from the minor map, though cartouches, discussed below, were occasionally made to serve the same purpose.[38] It was only if inset boundaries were left out that their value at once became evident, as when part of the west African coast was deliberately moved about 400 miles out of position by Jorge de Aguiar in 1492[39] or, worse still, when the whole world in miniature appeared without preamble occupying part of the Sahara desert on a map of 1561.[40] If a 'line-of-convenience' was in danger of being misinterpreted as some kind of geographical reality there was a practical case for adding a few unmistakably decorative touches to it, like the discreetly ornamental loops at the corners of labels enclosing regional and county names on the fourteenth-century Gough map of Britain.

Of equal interest were boundaries delimiting cartographic and non-cartographic information or, within the map, what philosophers might distinguish as language and meta-language.[41] Such was the principal justification of the cartouche. This word reportedly derives from the Italian 'cartoccio', an oval enclosure surrounding the heraldic insignia of a pope or prince, though in cartographic usage there is no need to insist on any particular shape or any particular content.[42] This innovation has sometimes been credited to engravers who, seeing their delicate line-work get lost beneath insensitive hand colour, chose to pre-empt the business of map-decoration by incising it on their own work surface[43] – though in fact the first occurrence of a cartouche, we are told, was in a fifteenth-century manuscript.[44] In the early sixteenth century, title-cartouches were found particularly in regional compilations of recent origin, as if to distinguish these from the more familiar untitled Ptolemaic atlas-pages; they also appeared in a number of rectangular world maps. (With conical and circular projections, written matter could conveniently occupy the corner spaces unaccompanied by any special frame.)

Besides delimiting titles, inset maps and scale bars, an internal decorative border might enclose a guide to conventional signs, an explanatory account of a map projection, or matter from any of the pictorial categories enumerated earlier in this chapter. With so many subjects to choose from it is understandable that cartouches were not infrequently left blank, doubtless on the assumption that somebody would think of something to put in them.[45] A contrary reason for their emptiness was that

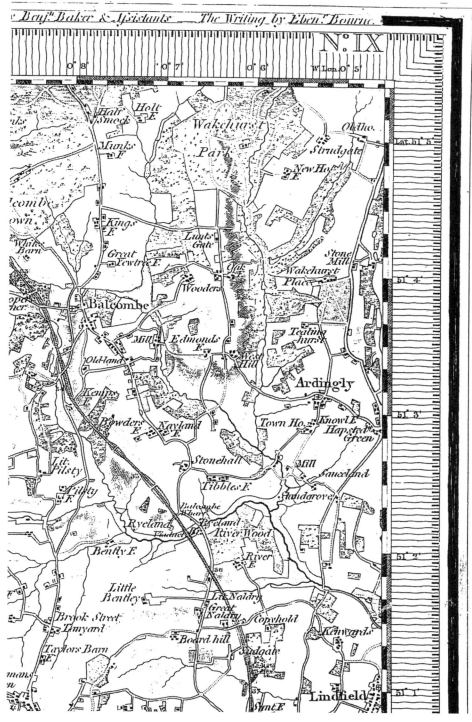

18.9 'Piano-key' border. From Ordnance Survey, one inch to one mile, sheet IX, central and west Sussex, published 1813 with later revisions.

an artist had plagiarised a different author's multi-compartmental cartouche without having enough inscriptions of his own to go round: this may explain the gaps in Francis Jobson's scale-cartouche for Ulster (1591), mentioned again below.[46] The same kind of hiatus may appear in print when an earlier inscription has been erased as obsolete and not replaced.[47] It is also conceivable that some draughtsmen and engravers, forgetting why the cartouche had been invented in the first place, saw it as an object worthy of admiration in its own right: there were certainly some blank enclosed panels in seventeenth-century wall decoration, by Grinling Gibbons for instance, that would have encouraged a suggestive map-maker to take this view.

On the argument put forward here, a border surrounding the whole map would help to mark off language from reality, but when these two realms of being were in any case physically separate and composed of different materials there was little point in labouring the distinction between them. It was mainly in an aesthetic sense that a map's external and internal boundaries needed different degrees of emphasis.[48] More particularly, by simulating a carved or moulded surround, a border might implicitly acknowledge the object inside it as valuable enough to deserve the protection of a real frame (Figs 1.6, 3.1, 3.5). It might seem to follow that the width of a border should be keeping with the size of the enclosed map: gross infringements of this rule are certainly difficult to regard with satisfaction.[49] In any case, as maps became more plentiful such presentational pomp and circumstance would begin to seem too heavy-handed, which perhaps explains why Blaeu's and Jansson's borders are generally narrower and less elaborate than those of say Ortelius. Another factor working in the same direction may have been the increasing use of marginal latitude and longitude divisions as encouraged by Mercator's *Atlas*: though not incompatible with an ornamental border they would at least slightly weaken its effect. One practical counter-argument remained operative, however. In a multi-sheet map the constituent parts could be more easily combined into a mosaic if the outermost borders of the whole assemblage were distinguished from the single-line internal borders dividing one sheet from another. Not that this excused the British Ordnance Survey's unpleasant 'piano key' motif for its early one-inch margins, because here the outer edge of the composite map was already differentiated by its scales of latitude and longitude (Fig. 18.9). Some customers must have liked the piano keys: to 'improve' early sheets where they had been absent the department's printers chose at one stage to produce separate strips of paper in this style which they then laboriously cut out and pasted into position.[50]

For all its variety of subject, representational art in a map margin had one property so obvious that it could easily be forgotten. Some early maps had no easily identifiable alignment *vis-à-vis* the reader, frustrating the modern expectation that a catalogue should say which cardinal compass point appears uppermost. Where names and other script were written at many different angles this lack of directional guidance must often have been deliberate: 'multiple, simultaneous orientations'

18.10 Up-dating of ship designs. Christopher Saxton, *Britannia insularum* ... , editions of 1583 (above) and *c.*1688 (below).

seem especially characteristic of seventeenth-century Russian maps, for example.[51] But as verbal texts became more familiar, so map readers increasingly expected the 'right way up' to be at once apparent.[52] Prominent titles would help to achieve this end. So would any pictorial addenda naturalistic enough to have a top and bottom, such as views of scenery or human figures.

In the sixteenth century and later, cartographic embellishment took its place among many minor art forms whose characteristic shapes acknowledged various kinds of functional imperative: tombstones, church memorials, door and window surrounds, architectural friezes and entablatures, fireplaces, picture frames and book titles. Many map designs were taken directly from these sources, though one would welcome more examples of this association than have yet been given by historians.[53] Other compositions were copied from earlier cartographers: in maps of Irish provinces drawn by or copied from Francis Jobson, for instance, the cartouches are strongly reminiscent of two chosen by Christopher Saxton's engravers for counties in the English midlands.[54] A copyist's models should not be too deeply rooted in tradition, however, because in cartography, as in so many other spheres, ornament had an important non-aesthetic role as a proof of up-to-dateness. This is why such fashion-sensitive appendages as ships and costumed figures were sometimes modernised by wholesale erasure and re-engraving (Fig. 18.10).[55] Not that naturalism was indispensable to the process of dating by style: a wholly abstract system of ornament can sometimes locate a map to within a few decades, as it is hoped to show in the course of the next few paragraphs.

The sequence of styles

Most of the developments now to be discussed will be familiar from a brief summary written many years ago by Edward Lynam with a lightness of touch that more recent semioticists have failed to emulate.[56] The subject is complicated and from a chronological standpoint sometimes frustrating. Fashion could vary from one country to another and in the same country from one cartographer to another; and sometimes it is a relatively inconspicuous or visually unexciting feature that gives the best clue to a map's date. Periodicity is not always well developed: some devices were common to several historical periods, for instance the blazoning of a title on a simulated ribbon or banner; and apart from issues of trendiness, a major brake on stylistic change in published maps was the longevity of copper plates. In manuscript cartography some artists – especially local estate surveyors – appear to have held aloof from stylistic influences of any era. Some were presciently forward-looking, others old-fashioned to the point of moribundity. Ireland, long a reputed haven of conservatism, may fittingly illustrate the last of these tendencies. At least two of its mid-seventeenth-century maps, and another from as late as 1694, were decorated with cartouches straight from the age of John Speed (Fig. 18.11).[57]

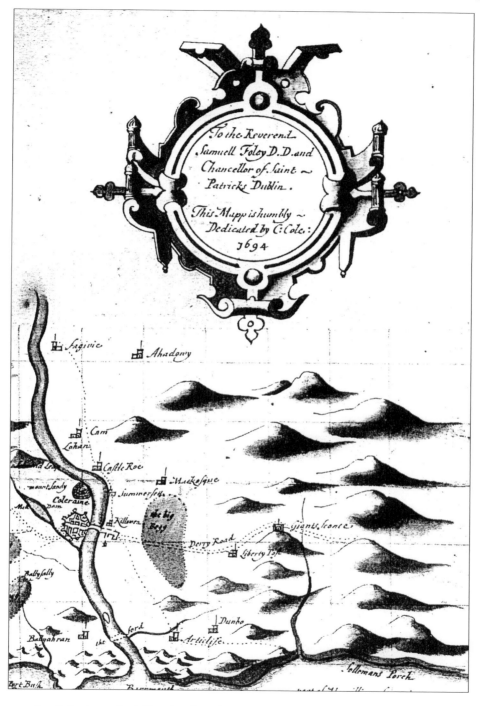

18.11 Coleraine and neighbourhood, northern Ireland, 1694. Christopher Cole's re-use of a cartouche-design from John Speed's map of Leinster (1610). Holkham Hall, John Innys Collection, vol. 98, fol. 121a.

J. da Poluara

Delineacio Orarum maritimarum, Terræ vulgo indigetatæ Terra do Natal, jtem Sofalæ, Mozambitæ, & Melindæ, Jnsulæq Sancti Laurentij, Jnsularum Maldiuicarum, Seylon insulæ, & Promon:torij Comorini, ad Jndiam siti unà cum Jnsulis, Scopulis, Puluinis, Vadis, veris Ventorum tractibus, & genuino sin:gulorum locorum situ, ad exac:tis:simas schnographicas Jndica:rum tabulas recognita atq emensata.

Affbeeldinghe der custen des landts genaempt Terra do Natal, jtem van alle de custen van Çoffala, Mozambique, Melinde, ende t'eylandt van S. Loreazo: met alle haere ey:landen, clippen, droochten, ende ondiepten. Jtem d'eylan:den van Maldiua tot het eylandt Çeylon, ende den hoeck van Comori toe, aende custen van Jndien ligghende, met de waerachtighe streckinghe ende gheleghentheyt der zeluer, alles seer correctelijck naer d'allerbeste Jndiaensche Pas ende Lees-caerten, ouersien ende verbeetert.

Arnoldus F. à Lan:gren, delintauit & sculpsit.

18.12 Dutch strapwork cartouche. Arnold van Langeren, East Africa and Indian Ocean, from Jan Huyghen van Linschoten, *Itinerario* (Amsterdam, 1595–6). See also Fig. 11.9.

In the earliest phase of cartouche-design the prime consideration was to look as different from geographical reality as possible, an effect best achieved by incorporating recognisable objects at a scale obviously far larger than that of any map – fruit, vegetables, leaves, animals, human beings. These would form part or all of an enclosure, usually elongated from left to right, that might also be diversified with scrolls or folds. Such features seem to have been characteristically German. Italian styles were generally simpler, sometimes no more than a plain rectangle. Contrary to national stereotypes, it was the supposedly undemonstrative Flemings who took most satisfaction in display for its own sake and who from the 1560s began to elaborate the basic rectangular outline. To this end its frame was made wider and cut or moulded to simulate wood, metal, stone or leather, with tabs, flaps and serrations projecting in complex patterns of which the most distinctive was a perforated trefoil or loop, often with another 'strap' threaded through it. From the frame various subordinate components would protrude upwards, downwards or sideways to give an irregular, spiky outline in which the most popular motifs were hanging cords, tassels, ribbons, fronds, flowers (often in vases or urns), fruit, birds, butterflies, fish, snakes, turtles and snails, though seldom any animal larger than a monkey. Human figures were usually of classical inspiration. As in many later maps, a more direct message might be conveyed by a prominent coat of arms, but the general effect was bizarre and even slightly sinister, recalling the mannerist style in vogue among contemporary painters. These displays were large, colourful and domineering, detached from both the border and the body of the map: in Mercator's map of Sardinia and Sicily the largest land mass was smaller than the cartouche.[58] It was a style that reached its zenith in the 1590s, with circles or ovals then becoming a more common alternative to the rectangle (Fig. 18.12).[59] Towards the 1620s the 'straps' seemed to get longer, narrower and more clearly separate one from another.[60]

An aesthetic watershed might seem to have been crossed when engravers took the trouble to rework a plate by deleting and replacing the whole of a large decorative compartment for reasons that had nothing to do with geography. This is what happened to many of Mercator's cartouches after his atlas had been taken over by Jan Jansson and Henry Hondius in 1630 (Fig. 18.13).[61] Yet the idea of the scrolled or moulded frame continued to provide a basis for variation, with fruit and foliage remaining much in evidence (now fitted more snugly into the design) and with acanthus leaves challenging the leather or wooden strips of the previous generation. The border of the cartouche might incorporate a single grotesquely distorted face as a centrepiece, but other human figures would often stand alongside or above the main framework; likewise animals that were larger but generally more companionable than in the weird-looking menageries of the Elizabethans. Meanwhile the frames themselves were becoming smaller and more compact, their outlines undisturbed by obtrusive projections, their inner and outer edges more often oval or bean-shaped than rectangular or circular.

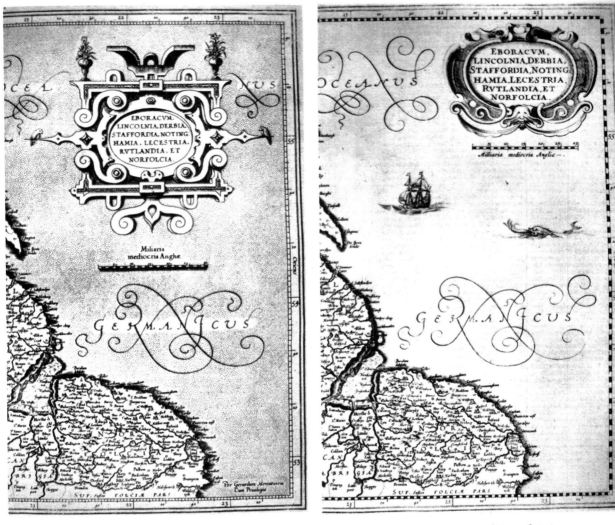

18.13 From Gerard Mercator's map of eastern England. *Left*: edition of 1595. *Right*: edition of 1636, with marginal features re-engraved.

From about 1630 first Willem and later Joan Blaeu helped to popularise a new kind of cartouche by arranging their human figures in pairs, of contrasting appearance though equal in size, one on each side of the main frame as if appointed protectors of its interior (Fig. 18.14). Sometimes these 'guardians' had a mythological or Biblical look, but often their costumes and hair-styles were refreshingly modern and homely. It soon became customary for one or both sidepieces to expand into a scene of diversity and animation invading the irregular space between the map and its border. In the 1640s Blaeu's rival Jan Jansson began to follow his example.[62] On some maps the cartouche proper became nearly lost among adventitious pictorial

18.14 Cartouche with 'guardians'. Isle of Man, from Johannes Janssonius, *Theatrum orbis terrarum sive atlas novus* (Amsterdam, 1646). See also Fig. 2.8.

matter which in extreme cases could push the title itself into a semi-detached position strung out along the upper border.

By 1680–1730 the more extended late seventeenth-century 'compartments' were coming to resemble contemporary European baroque architecture. Many of their ingredients remained familiar – scrolled brackets, acanthus leaves, swags of fruit, paired 'guardian' figures and grotesque face-masks. Newer fashions were a curtain or scalloped shell as background to the words of the title, and a display of any objects that could be made to frame those words within a 'trophy' of radiating lines: swords, spears, muskets, flags, tools of agriculture, stalks or sheaves of corn, musical instruments or rays of the sun (Fig. 18.15). However realistic the individual elements in a design, their juxtaposition was often improbable or impossible: sometimes they just hovered in the air, connected by a buoyant mass of cloud. The effect was striking but hardly comfortable: timid viewers may have felt restless and uneasy as banners were waved, weapons brandished, wings flapped, horses whipped, trumpets sounded and drums beaten.

All this extravagance naturally had its price. The more complex the designs, the harder they were to draw or engrave. It was common in the eighteenth century for a cartographic title-piece to be signed by a different artist from its accompanying geographical outline, and sometimes a map was printed as a proof or first edition

18.15 Baroque cartouche. Richard Budgen, map of Sussex (London, 1724).

before this specialist got to work on it.[63] A converse expedient in manuscript maps was for the graphic portion of the cartouche to be printed in the corner of a blank sheet, leaving the main body of the map and the words of the title to be filled in by hand. North points were sometimes treated in the same way. Mass-produced cartouches sometimes broke the unwritten rule that each page in an atlas should be individually decorated; and any kind of 'sub-contractual' marginalia went against the modern dogma that all peripheral ornament must be accepted as an essential and irremovable part of its host map. It is fair to add that separately printed cartouches were especially associated with the baroque and cognate eighteenth-century styles; and that the engraved compartment for manuscript estate maps was always something of a rarity – not known to the present writer except in Ireland.[64]

Baroque designs in the fullest sense were most enthusiastically embraced in the Netherlands and Germany. French title-panels, though featuring similar motifs, were for the most part more compact, more delicately engraved, and less grandiose in appearance. It was the same national sensibility that originated the rococo style of the period 1740 to 1770. Rococo cartouches had no straight lines, either internally

18.16 Rococo cartouche. Estate map, MS, Dodderhill, Worcestershire, 1770. Edward Lynam, *British maps and map-makers* (London, 1944), opposite p. 32. See also Fig. 18.6.

or externally, but rather a series of narrow curved segments in which the characteristic forms of shell, foliage, wood and possibly china were teasingly combined into a single imaginary substance. The curves enclosed an upright shape, often broader at the top than the bottom in the manner of a vase (Fig. 18.16). Colours were seldom obtrusive, and the overall feeling was one of elegance and refinement. Although pure rococo had a relatively short life, its birth helps to divide Europe's cartographic history into two periods, separated by what contemporaries apparently saw as a rather abrupt increase of sophistication among map users. Henceforth, instead of having his gaze initially directed towards over-emphatic marginal explanations, the reader was encouraged to look first at the interior of a map – in the belief that he would need no preparation to understand what he found there.

Artists who found the new style insipid reacted by twisting the cartouche out of its symmetrical shape[65] and by merging some of the carved and moulded side-pieces with tree branches or tendrils in a rococo interpretation of the seventeenth-century scenic view.[66] In fact after about 1770 many map-makers abandoned the formal cartouche in favour of an almost wholly naturalistic title-frame. This might be built up as a series of loosely connected segments – perhaps part of the cartographer's neat line, an underlying strip of ground and associated landscape furniture, and

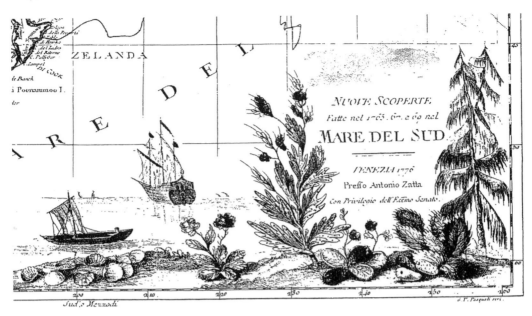

18.17 Naturalistic cartouche. Antonio Zatta, map of south-west Pacific (Venice, 1776).

finally a tree with near-vertical trunk and spreading crown whose branches could be continued if necessary by wisps of cloud en route for a point higher on the neat line, thus fortuitously demarcating a convenient block of marginal space (Fig. 18.17). Within this apparently accidental boundary the title itself might appear carved on a natural rock-face or the side of an ancient-looking stone monument. As in earlier periods, there would often be room for depictions of human activity. Local scenes were especially characteristic of independently published English county and town maps, whose authors could draw on their own experience for appropriate topographical allusions.[67] A different kind of 'realistic' gesture was to write titles and other inscriptions, *trompe l'oeil* style, on simulated cards or scraps of paper that lay on the map surface as if dropped by chance.[68]

Whatever the vagaries of fashion, map-titles in book illustrations and pocket atlases from every period were necessarily less elaborate than those to be found on larger sheets. But in the last decades of the eighteenth century maps of every size were increasingly subject to a process of decorative simplification. The least obtrusive enclosure for a legend was now a plain rectangle, or sometimes a single or double line describing a circle or oval that was completely without accompaniment except perhaps for being slightly 'shadowed' (Fig. 18.18).[69] Even more minimalistic was a reversion to the early days of the Ptolemaic printed atlas, the words of the title now being written by themselves in whatever part of the margin left room for them. The main change since the sixteenth century was that modern titles were a good deal longer and could be made more emphatic by using a different script for each line.[70]

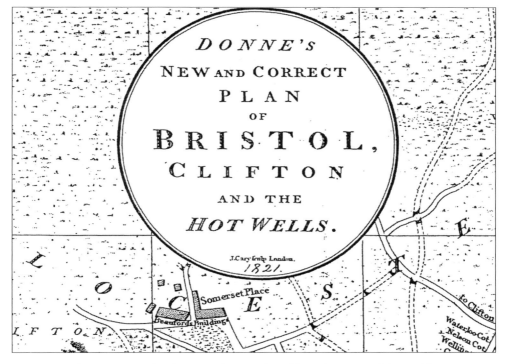

18.18 Circular cartouche. Benjamin Donne, plan of Bristol *c*.1800, with later additions.

The first moves towards a nineteenth-century resurgence of marginal decoration were tentative. Sometimes they took the form of a single free-standing architectural vignette, which in an English county map would probably feature a cathedral or large church.[71] The county name itself was still left unenclosed, or perhaps put in a centrally positioned label in or above the upper border of the map. In the 1830s this kind of title, sometimes now in a scroll or banner, became part of a minor aesthetic explosion involving almost all the popular styles of the previous three hundred years – strapwork, coats of arms, fruit and foliage, views of buildings and landscapes, and symbolic human figures, sometimes with a top dressing of medieval revivalism unknown in the age of Saxton.[72] Compared with authentic sixteenth-century ornament the individual elements were now often small, spaced-out, attenuated and rather niggling (Fig. 18.19). Another not very tasteful mid nineteenth-century trend was to continue the diversification of letter forms, distorting large capitals to simulate twigs or logs, or sometimes human figures frozen in a variety of uncomfortable poses.

Art in cartography: the historian's balance-sheet

Today, given the wholesale change in attitudes to cartographic ornament since *c*.1850, how far can we blame earlier map-makers for distracting us with what

18.19 Mock-Elizabethan cartouche. Thomas Moule, City of Bath, from *Moule's English counties* (London, 1830–7).

Samuel Wyld called 'pretty diversions'?[73] A simple argument immediately offers itself. If people want a map, they will draw a map. If they want a work of art, they have other ways of satisfying that desire – by painting a wholly abstract picture, for example. From which we can surely infer that where cartographic and aesthetic values were in conflict, it was undesirable (perhaps we should grit our teeth and say 'wrong') for a map-maker to let the aesthetic values prevail. But what is meant by 'conflict' here? In the first place, decorative themes should not be mistakable for geographical facts. Perhaps the clearest example, already encountered more than once in the foregoing pages, was the label enclosing a maritime placename misinterpreted by later copyists as the coast of a rectangular island. On a larger scale, purely decorative horticultural features such as herbaceous borders, parterres, lawns and paths in the body of a map should not be read as characterising any particular garden. Marginal illustrations were sometimes more or less misleading in the same way. It was ill-advised for Justus Danckerts to embellish his mid-seventeenth-century chart of western Europe with blatantly oriental figures copied from Willem Janszoon Blaeu's map of China.[74] Nor was an elephant the ideal vehicle for the title of an eighteenth-century English estate map.[75] Besides countries and peoples, the cartographic medium itself could be caught up in this kind of confusion: no useful purpose was served by checkered borders that looked like scales of distance without actually meaning anything.[76]

   We can also regret the misuse of decoration and other extraneous material as a shield for ignorance, a practice illustrated in Johannes Ruysch's world map of 1507 by scrolls covering the west coasts of South America and Cuba.[77] As it happened Ruysch's scrolls carried an apology and explanation, and it soon became widely agreed that wherever else an adventitious feature might appear it should not be allowed to intersect a coastline: such at least is the message of Pierre Desceliers's world chart, where this kind of collision is avoided by twenty-six text panels, eighteen compass roses, five scale bars, two coats of arms and one blank cartouche.[78]

All the same, in the later seventeenth century a less scrupulous map-maker could still place a cartouche over central Africa to avoid committing himself about the sources of the White Nile.[79]

It was not usual, one must admit, for a map's geographical meaning to be negated by its decorative framework. More often the conflict was between disparate claims on the reader's attention. In most cases the decision whether or not to add embellishments was simply a matter of taste. So where now are the modern scholar's grounds for approval or disapproval? One final reflection draws us back from history to historiography. Map-makers have seldom had the power to decide whether their creations would be preserved for any length of time beyond the onset of factual obsolescence. When a map-sheet got through what R.A. Skelton called the 'interval of vulnerability' its future might depend on the purely aesthetic preferences of dealers, collectors and interior decorators.[80] Without the frivolity admired by generations of non-geographers, a hard-headed map historian in a modern seat of learning could easily find himself with only a very short past to write about.[81]

CHAPTER 19

Maps and society

THE FINAL PAGES OF THIS book are devoted to rearranging and reconsidering a few ideas that have already been introduced. As in previous chapters, we begin with a quotation. 'Maps and society' is the title of a marathon series of highly successful lectures organised at the Warburg Institute in London by Catherine Delano-Smith and Tony Campbell which at the time of writing has occupied a total of more than hundred hours. Afloat on this tidal wave of fact and opinion, few members of the Warburg audience rehearsing the same theme could admit to any shortage of material. Nevertheless, when cartographic technique is claiming attention a social perspective may be difficult to sustain.

As a matter of logic, mapping does not have to be a social activity. A map could have been made by a single-handed Robinson Crusoe for no one else's benefit but his own. The Crusoe phenomenon is admittedly rare, best kept as a debating weapon against socially hyper-conscious cartographic philosophers. One reason for its rarity (apart from the low incidence of shipwrecks with only one survivor) is that we can get better value from our maps by paying specialists to make them than by trying to do the job ourselves. The specialist thus engaged may be no more skilful than his customers. They may indeed excel him – though by a narrower margin than in whatever employment actually brings them a living. Given a rationale for commercial map-making, the distribution of roles among surveyors, draughtsmen, engravers, publishers and others is dependent on technical factors considered earlier. The overriding general principles, however, are those of elementary economics – the division of labour and the law of comparative costs. This is all obvious enough. The fact remains that no admirer of Adam Smith's famous introductory chapter in *The wealth of nations* would respond to it by giving a lecture on 'pins and society'.

Why should maps be different from pins? One reason, surely, is that they purport to be better at telling us important facts about the world. In modern map-historical literature, and especially in the writings of J.B. Harley, social analysis of this communicative process has paid special attention to the phenomenon of 'interest'.[1] The map is a tool, a weapon even, in the relations between potentially antagonistic social groups – conquerors and conquered, governors and governed, soldiers and civilians, owners and slaves, employers and employed, landlords and tenants, rich and poor, them and us. In these dichotomies, the advantage of one group is served by channelling the flow of information to the other group. The most blatantly 'interest-dominated' way to do this is by telling lies, which on a map means showing geographical features in places where they do not exist – as for example when the Moluccas were allegedly moved by early Portuguese cartographers from outside to inside the Portuguese sphere of influence defined by their country's agreement with Spain.[2] Maps meant to deceive can indeed claim to be the most social of all maps, having no conceivable value for the castaway on his desert island; but their importance in the present discussion is limited by being necessarily parasitic on maps that try to tell the truth.

The other choice for the cartographic manipulator is between speech and silence. Unfortunately, the withholding of facts, like the falsification of facts, has little significance for our present purpose. How did they make maps in those days? To reply 'By deciding what to leave out' does not take us very far. Here again an appeal to elementary economics seems best to meet the case, with the cartographer's selection of material reflecting the presumptive wishes of those prepared to buy his map. A less drastic means of enforcing authorial preferences has been the allotment of emphasis among topics already chosen for inclusion, a subject broached in an earlier chapter under the heading of 'proportionality'. We can agree that this idea has various social ramifications: it does not need much in the way of commentary here, though, because map-makers' techniques for giving emphasis have often been described.

How then can the business of map production be brought within the range of socially-focused cartographic thought? Perhaps we should consider the producers as an interest-group distinct from those who employ them and from those others who are either benefited or disadvantaged by their work. For most of our chosen period, map-making was not recognised as a single occupation: this much is suggested by the late arrival of the term 'cartography' in the world's dictionaries. In the absence of any strong unifying force, different parts of the cartographic process have been structured to varying degrees. In some of them, group standards and customs have been maintained by methods familiar in other occupations. Many map-making firms have been family businesses, their skills transmitted from father to son. Elsewhere pupils have studied the foundations of cartography in mathematical schools. Masters have trained apprentices in surveying, engraving or draughtsmanship. In some countries civilian surveyors have been formally organised in societies and institutes; in a different sense, the same has been true of their military and naval colleagues. Sometimes the map-maker has been a collective rather than an individual performer, notably in national cartographic agencies embodying an internal division of labour, such as the British Ordnance Survey. In many of these spheres the economic dominance of the post-medieval publisher must have exercised a standardising influence on various aspects of map production. But the status of cartographers in society has never been sufficiently awe-inspiring to exclude the enthusiastic amateur. There have always been self-taught and part-time map-makers, some of them earning high repute and a certain amount of money.

Note however that within the ambit of European culture such individualists, even the most original and creative, are likely to have drawn part of their inspiration from a broader and more conformist cartographic mainstream.

These complexities make it understandable that no single socio-economic specification can be found to embrace the whole of our subject. The most we can say is that map-making has sometimes resembled a business, sometimes a profession. On the one hand, returning to Adam Smith, the map historian may be inclined to agree that 'people of the same trade seldom meet together, even for merriment and diversion, but the conversation ends in a conspiracy against the public, or in some contrivance to raise prices'.[3] Smith's combative term 'against' has an anticipatory flavour of J.B. Harley. Here, then, interest consists of transferring as much wealth as possible from buyer to seller. But at this point we may remember the words of another eminent map historian whose generation overlapped with Harley's. Complaining that cartographers are less readily believed than prose writers, Ir.C. Koeman attributed this inequitable attitude to historians' 'lack of understanding of the professional moral code of map authors'.[4] In other words mapping, for all its commercial dynamics, may also have some kinship with a more socially elevated range of occupations – not with the law, medicine or the church, perhaps, but at any rate with engineering, accountancy and teaching. Here the producer is paid partly in money and partly in respect, and different interest groups compete to climb a ladder of prestige.

We may next briefly recall the effect of these social influences on the face of the map itself. One pervasive impression is that truthfulness and accuracy have never been the only reasons for cartographic success. Testing a surveyor's work against experience on the ground is normally a slow and intermittent operation occupying years or decades. (Perhaps this helps to explain why the publication of critical reviews has never played much part in cartographic history.) Yet many maps have won high praise before this process has been finished or even begun. 'It is a passing fair piece of work' was how one of Queen Elizabeth I's lord justices described a brand-new map of an Irish province.[5] How could he have known this? He surely did no measuring or protracting himself. Perhaps, like so many grateful recipients, he was impressed by a neat and artistic appearance, a sufficiency of familiar names, and an absence of large empty spaces and obvious loose ends.

Apart from sheer amiability, an important source of reassurance in the initial impact of a map is its decorative content. Visual extravagance plays the same part in cartography as in the architecture of banking or insurance. The bank says: Look, I have too much wealth of my own to feel any necessity for stealing yours. The map says: Look, if I can afford to waste money on useless ornament why should I need to economise by skimping on accuracy? Not perhaps the most convincing of arguments; and in periods when ostentation might seem counter-productive the cartographer would have to proceed with more finesse. As we have found, a map can hold many clues that help to classify its author as insider or outsider. In themselves such signals may have no significance or value, but suspicions will be aroused if they are left out. For instance, there is no good reason why a map of England and Wales should have north at the top: it is just how for several centuries the cartographic establishment has chosen to behave. This aspect of vocational image-building is familiar in many spheres of gainful employment. 'I have always thought the manner would be the easiest part of a profession,' says a long-forgotten fictional character. 'I could belong to any of them, if nothing else was needed. I expect that is how the manners became established. There had to be something that was within people's power.'

Not all such ritual behaviour could be deemed innocuous. Some of it has positively impeded the expression of geographic truth, just as doctors' illegible handwriting has sometimes proved an obstacle to medical efficiency. In cartography the worst examples have been the omission of essential details from a map – the unit of distance in a scale-bar, the location of a prime meridian for reckoning longitude, the difference between true and magnetic north and, most incorrigible of deficiencies, the identity of an author's chosen projection. Unexplained conventional signs are another common failing: it may be true that Jean Rotz used a dotted line to delimit the area off Labrador 'where men goeth a-fishing', but in that case why did he not say so?[6] Sometimes it would have been easier to state such facts than to withhold them: a modern example is the absence from so many isopleth maps of the 'stations' between which the isopleths have been interpolated.

Certain bad habits can be explained by sheer unwillingness to break ranks, a characteristic of professional 'manners' that probably arose from fear of being mistaken for an amateur. But inertia was not just a matter of style: substance could suffer just as badly. In the opinion of Didier Robert de Vaugondy, it was 'better to follow a received sentiment than to forge a chimera in order to bring something new to geography'.[7] In fact many erroneous outlines were disseminated in print long after their falsity had been made plain. This phenomenon is probably better known than anything else in cartographic history, but it can easily be given too much emphasis. Error often stemmed from failures of communication that were nobody's fault. And of course most of the errors were eventually rectified, thanks to the ultimate triumph of professional virtue. But the process has nearly always been impeded and delayed by the operation of professional vice in its familiar guises of conservatism, pride and complacency.

Reluctance to change was only one reason for suppressing some or all of the information needed to understand a map. A clue here is the word 'mystery' as a synonym for 'craft' or 'business', indicating the practitioner's desire to maintain a certain distance from his clientele, lest they should try to usurp his function. In fact, as map historians know only too well, secrecy pervades almost every department of their chosen subject matter. Survey records were seldom made available to accompany the plans that derived from them, and the lines representing measured angles and distances were nearly always erased from a map before it left its editor's custody. However verbose the text of an atlas, it hardly ever explained how the accompanying maps had actually come into existence. A pointer in the same direction, at any rate before about 1930, is the lack of teaching manuals on any aspect of cartography apart from surveying, this exception being mainly attributable to the links that connected land admeasurement with more obviously useful occupations like engineering and property valuation.

So much for Adam Smith's conspiracy theory. We must now switch categories from trade to profession. It should already be clear that these two concepts have certain points in common – the 'manner', for example, and the impulse to mystification. The difference lies in the professional's reasons for seeking a high standard of performance – not just a desire for profit but also a genuine love of truth. Although in practice the two motives can sometimes be hard to separate, there is one case where the distinction seems unmistakable. A cartographer may have some affinity with other recognised professions – a lawyer in his treatment of boundaries, a soldier in his assessment of heights and slopes, a linguist in his anxieties about the spelling of names, a teacher in his concern for clarity of exposition. But his closest links have been with the least clandestine of all professions, namely natural and experimental science as practised by university dons and before that by leisured churchmen or by gentlemen amateurs. Among the scientists who have both taught and learned from the map-making fraternity are geologists, hydrologists, geomagnetists and archaeologists. Even more directly connected with cartography are geodesy (with which in this context we may bracket metrology) and geophysics. Within these fields one problem assumed importance in the seventeenth century and remained unsolved until the twentieth: it was to calculate the length of a degree of latitude in terrestrial units, and in doing so to determine the exact size and shape of the geoid, a task that benefited cartography by allowing astronomers and land surveyors to provide data for the same map. Prerequisites for a solution were base measurement and triangulation at the highest level of accuracy.

What distinguished science from other methods of data collection was that all its procedures were public enough to be replicated by any independent observer. The talent and integrity needed to achieve this end were sufficiently rare to command a gratifying measure of deference and thus to define a superior caste. Map historians have been especially conscious of such elitism where their favourite themes of class (in the Marxian sense) and ethnicity can be built into the record. A well-documented example is pre-Victorian India, whose British occupiers – rational, knowledgeable and gentlemanlike – could assert an all-round superiority over the native population by making cartography as difficult as possible.[8] In many cases this quest for status carried scientifically-minded geometers well beyond the requirements of simple land measurement. The accuracy of a graphic representation, unlike that of a numerical statement, is limited by the physical instability of the materials composing it. The control surveys of the eighteenth and nineteenth centuries, dominated as they so often were by standards appropriate to geodesy, would often transcend the needs of any cartographic enterprise that was likely to be based on them. A map could even suffer positive harm if too much energy was diverted from detail to control.

This last danger was well illustrated by the experience of Colonel Thomas Colby. In 1827–8 a base for the first Ordnance Survey triangulation of Ireland was measured by Colby's subordinates with all possible care through the parish of Magilligan and adjoining parishes in County Londonderry. To judge from a later investigation, its length of nearly eight miles was correct to about the nearest inch, an accuracy more than a hundred times greater than was required by the maps under contemplation in the 1820s.[9] Not long afterwards the landscape features of the Survey's new Magilligan sheet were pronounced too inaccurate to show their face outside the departmental office.[10] Thus widely did the gap between control and detail lie open within the limits of a single parish.

On a more general view of 'western' map history, the cartographic businessman's attitude to science was respectful if not always well-informed, altogether a more productive relationship than the unconformity separating different intellectual strata in the worlds of modern art, music and poetry. But for a final summary we must briefly return to the historical literature of 'traditional' societies outside Europe. Here, in thousands of recently-published pages, there is an almost total silence about how early non-European map-makers were recruited, trained, supervised and remunerated.[11] In the absence of more direct enlightenment, one's strongest impression is of a certain parallelism with European antiquity and with medieval Christendom. In these mainly Asian and African communities, maps were made by people of wide-ranging talents and achievements – civil servants, teachers, priests, soldiers. Their livelihood came from salaries, rewards or endowments, and was not directly related through the medium of market forces to the execution of individual cartographic tasks within any particular sphere of employment. In these societies no one person, and no department, is described as a full-time professional map-maker; and whether in print or manuscript no one is found selling a specified map for a specified sum of money.

Setting these extra-European environments against the 'western' socio-economic framework taken for granted in the middle chapters of the present book, suppose we were to gather up all the maps produced under western and non-western conditions throughout the history of the world, and that we were then to compare the merits of the resultant collections. Early maps can certainly be studied without attempting a definition of cartographic merit, but given our terms of reference it would be cowardly to finish without confronting this somewhat loaded question.

What is it a map does that nothing else can do? Merit is the capacity to give satisfaction by doing just that. On this criterion, when map-making is treated as a trade or profession the results are better than when it is not. This is one reason why our curiosity about early cartographic methods may be worth trying to satisfy.

Notes

Chapter 1: Map history miniaturised

1 J.H. Andrews, 'What was a map? The lexicographers reply', *Cartographica*, xxxiii, 4 (1996), pp. 1–11.
2 Peter Whitfield, *The mapping of the heavens* (London, 1995).
3 J.B. Harley (ed. Paul Laxton), *The new nature of maps* (Baltimore, 2001); Denis Wood with John Fels, *The power of maps* (London, 1993).
4 W.S.W. Vaux (ed.), *The world encompassed by Sir Francis Drake, being his next voyage to that of Nombre de Dios* (London, 1854), p. 79.
5 Catherine Delano Smith, 'Cartography in the prehistoric period in the old world: Europe, the Middle East, and North Africa' in J.B. Harley and David Woodward (eds), *The history of cartography, volume one: cartography in prehistoric, ancient, and medieval Europe and the Mediterranean* (Chicago, 1987), pp. 54–101.
6 O.A.W. Dilke, *Greek and Roman maps* (London, 1985), pp. 146–7, reviewed by Richard J.A. Talbert, *Journal of Roman Studies*, lxxvii (1987), p. 210 and by Christian Jacob, *Imago Mundi*, xxxviii (1986), pp. 106–7.
7 Robert Baldwin, *Globes* (Greenwich, 1992).
8 J.B. Harley and David Woodward (eds), *The history of cartography, volume two, book one: cartography in the traditional Islamic and south Asian societies* (Chicago, 1992); J.B. Harley and David Woodward (eds), *The history of cartography, volume two, book two: cartography in the traditional east and southeast Asian societies* (Chicago, 1994); David Woodward and G. Malcolm Lewis (eds), *The history of cartography, volume two, book three: cartography in the traditional African, American, Arctic, Australian, and Pacific societies* (Chicago, 1998).
9 Ken Garland, *Mr Beck's underground map* (Harrow Weald, Middlesex, 1994): 1931 et seq.
10 John P. Snyder, *Flattening the earth: two thousand years of map projections* (Chicago, 1993).
11 Matthew Edney, 'Mathematical cosmography and the social ideology of British cartography, 1780–1820', *Imago Mundi*, xlvi (1994), pp. 101–4.
12 Norman J.W. Thrower, 'Doctors and maps', *The Map Collector*, lxxi (1995), pp. 10–14; Gimpel and Eva Wajntraub, 'Physician, map thyself: men of medicine and cartography', *Mercator's World*, v, 2 (2000), pp. 48–53.
13 J. Lennart Berggren and Alexander Jones (eds), *Ptolemy's* Geography: *an annotated translation of the theoretical chapters* (Princeton, NJ, 2000), p. 59.
14 Christian Jacob, 'Mapping in the mind: the earth from ancient Alexandria' in Denis Cosgrove (ed.), *Mappings* (London, 1999), pp. 29–30.
15 O.A.W. Dilke, *The Roman land surveyors: an introduction to the agrimensores* (Newton Abbot, 1971); Helen M. Wallis and Arthur H. Robinson (eds), *Cartographical innovations: an international handbook of mapping terms to 1900* ([Tring], 1987), 'plan', pp. 45–51.
16 O.A.W. Dilke, 'Itineraries and geographical maps in the early and late Roman empires' in Harley and Woodward, *History of cartography, volume one*, pp. 238–42: the Peutinger Table; Benet Salway, 'The nature and genesis of the Peutinger map', *Imago Mundi*, lvii, 2 (2005), pp. 119–35; Emily Albu, 'Imperial geography and the medieval Peutinger map', ibid., pp. 136–48.
17 H. Rackham (ed.), *Pliny: natural history*, i (London, 1944), passim; J.J. Tierney, 'The map of Agrippa', *Proceedings of the Royal Irish Academy*, lxiii C, 4 (1963), pp. 151–66.

18 Cordell D.K. Yee, 'Cartography in China' in Harley and Woodward, *History of cartography, volume two, book two*, pp. 35–202, 228–31.

19 Gari Ledyard, 'Cartography in Korea', ibid., pp. 235–345; Kazutaka Unno, 'Cartography in Japan', ibid., pp. 346–477.

20 Ahmet T. Karamustafa et al., 'Islamic cartography' in Harley and Woodward, *History of cartography, volume two, book one*, pp. 3–292.

21 P.D.A. Harvey, *Medieval maps* (London, 1991).

22 P.D.A. Harvey, *The history of topographical maps: symbols, pictures and surveys* (London, 1980), p. 10; C. Koeman, 'Hoe oud is het woord kartografie?', *Geografisch Tijdschrift*, viii, 3 (1974), pp. 230–1.

23 J.B. Mitchell, 'The Matthew Paris maps', *Geographical Journal*, lxxxi (1933), pp. 27–34; P.D.A. Harvey, 'Matthew Paris's maps of Britain' in P.R. Coss and S.D. Lloyd (eds), *Thirteenth century England IV: proceedings of the Newcastle upon Tyne conference* (Woodbridge, Suffolk, 1992), pp. 109–21; Daniel Birkholz, *The king's two maps: cartography and culture in thirteenth century England* (New York, 2004).

24 R.A. Skelton and P.D.A. Harvey (eds), *Local maps and plans from medieval England* (Oxford, 1986).

25 E.J.S. Parsons, *The map of Great Britain* circa *AD 1360 known as the Gough map* (Oxford, 1958); David Buisseret, *The mapmakers' quest: depicting new worlds in renaissance Europe* (Oxford, 2003), pp. 4–5; Nick Millea, *The Gough map: the earliest road map of Great Britain* (Oxford, 2007).

26 Tony Campbell, 'Portolan charts from the late thirteenth century to 1500' in Harley and Woodward, *History of cartography, volume one*, pp. 371–463.

27 David Woodward, 'Medieval *mappaemundi*' in Harley and Woodward, *History of cartography, volume one*, pp. 314–18.

28 John Larner, *Marco Polo and the discovery of the world* (New Haven, CT, 1999).

29 Georges Grosjean (ed.), *Mapamundi: the Catalan atlas of the year 1375* (Dietikon–Zürich, 1978), reviewed by Tony Campbell, *Imago Mundi*, xxxiii (1981), pp. 115–16; Piero Falchetta, *Fra Mauro's world map with a commentary and translation of the inscriptions* (Turnhout, 2006).

30 Helen Wallis, 'Cartographic knowledge of the world in 1492', *Mariner's Mirror*, lxxviii (1992), pp. 407–18.

31 Henry N. Stevens, *Ptolemy's Geography: a brief account of all the printed editions down to 1730* (London, 1908).

32 David Buisseret (ed.), *Monarchs, ministers and maps: the emergence of cartography as a tool of government in early modern Europe* (Chicago, 1992); Peter Barber, 'Maps and monarchs in Europe 1550–1800' in Robert Oresko, G.C. Gibbs and H.M.Scott (eds), *Royal and republican sovereignty in early modern Europe* (Cambridge, 1997), pp. 75–124.

33 George Best in Richard Collinson (ed.), *The three voyages of Martin Frobisher* (London, 1867), p. 30.

34 R.A. Skelton, *Looking at an early map* (Lawrence, KS, 1965), pp. 14–27.

35 Andrew Sharp, *The voyages of Abel Janszoon Tasman* (Oxford, 1968); Michael Ross, 'The mysterious Eastland revealed', *The Globe*, liii (2002), pp. 1–22.

36 R.A. Skelton, *Captain James Cook after two hundred years* (London, 1969).

37 Barber, 'Maps and monarchs in Europe', p. 93.

38 J.N.L. Baker, *A history of geographical discovery and exploration* (new ed., London, 1937); Daniel B. Baker (ed.), *Explorers and discoverers of the world* (Detroit, 1993). Ayer's Rock is now called Uluru.

39 Robert W. Karrow Jr, *Mapmakers of the sixteenth century and their maps: bio-bibliographies of the cartographers of Abraham Ortelius, 1570* (Winnetka, IL, 1993), pp. 568–83: 1507–16; John W. Hessler, *The naming of America: Martin Waldseemüller's 1507 world map and the* Cosmographiae introductio (Washington, 2008).

40 Bibliothèque Royale Albert 1ᵉʳ, *Le cartographe Gerard Mercator, 1512–1594* (Brussels, 1994).

41 Karrow, *Mapmakers of the sixteenth century*, pp. 120–2: c.1520.

42 R.V. Tooley, 'Maps in Italian atlases of the sixteenth century', *Imago Mundi*, iii (1939), pp. 12–47; Edward Lynam, *The Carta marina of Olaus Magnus, Venice 1539 and Rome 1572* (Jenkintown, PA, 1949), p. 29.

43 Eila M.J. Campbell, 'The early development of the atlas', *Geography*, xxxiv (1949), pp. 187–95; R.A. Skelton, 'Early atlases', *Geographical Magazine*, xxxii, 11 (1960), pp. 529–43.

44 B.van't Hoff, *Gerard Mercator's map of the world (1569)* (Rotterdam, 1961).

45 Ir. C. Koeman, *Joan Blaeu and his grand atlas: introduction to the facsimile edition of* Le grand atlas, *1663* (Amsterdam, 1970).

46 Mario M. Witt, 'Vincenzo Coronelli as cartographer', *Mariner's Mirror*, lx (1974), pp. 143–52; Helen Wallis, 'Bibliographical note', *Vincenzo Coronelli, Libro dei globi (Venice, 1693, 1701)* (Amsterdam, 1969), pp. v–xxii.

47 Lucas Jansz. Waghenaer, *Spieghel der zeevaerdt* (Leyden, 1584–5; Amsterdam, 1964); Günter Schilder, 'Lucas Janszoon Waghenaer's nautical atlases and pilot books' in John A.Wolter and Ronald E. Grim (eds), *Images of the world: the atlas through history* (Washington, 1997), pp. 135–59; Wallis and Robinson, *Cartographical innovations*, 'island atlas', pp. 320–3.

48 Georg Braun and Frans Hogenberg, *Civitates orbis terrarum* (Cologne and Antwerp, 1572–1618; Amsterdam, 1965).

49 R.A. Skelton, *County atlases of the British Isles, 1579–1703* (London, 1970), pp. 30–44: 1612.

50 John Ogilby (ed. J.B. Harley), *Britannia, London 1675* (Amsterdam, 1970).

51 [Nicolas] Tassin, *Les plans et profils de toutes les principales villes et lieux considerables de France* (Paris, 1638).

52 Augustin Lubin, *Orbis Augustinianus sive conventuum ordnis eremitarum Sancti Augustini chorographica et topographica descriptio* (Paris, 1659).

53 Josef W. Konvitz, *Cartography in France, 1660–1848: science, engineering, and statecraft* (Chicago, 1987).

54 Christian Sandler, *Die Reformation der Kartographie um 1700* (Munich, 1905), pp. 6–8.

55 Mary Sponberg Pedley, *Bel et utile: the work of the Robert de Vaugondy family of mapmakers* (Tring, 1992); J.B. Harley, 'The bankruptcy of Thomas Jefferys: an episode in the economic history of eighteenth century map-making', *Imago Mundi*, xx (1966), pp. 27–48.

56 Geoffrey King, *Miniature antique maps: an illustrated guide for the collector* (Tring, 1996).

57 Helen Wallis, 'The map collections of the British Museum Library' in Helen Wallis and Sarah Tyacke (eds), *My head is a map: essays and memoirs in honour of R.V. Tooley* (London, 1973), p. 11; Peter Barber, *The map book* (London, 2005), pp. 164–5.

58 J.B. Harley, 'The re-mapping of England, 1750–1800', *Imago Mundi*, xix (1965), p. 56–67.

59 David Smith, 'The enduring image of early British townscapes', *Cartographic Journal*, xxviii, 2 (1991), pp. 163–75; David Smith, 'Inset town plans on large-scale maps of Great Britain', *Cartographic Journal*, xxix, 2 (1992), pp. 118–36.

60 J.B. Harley, Barbara Bartz Petchenik and Lawrence W. Towner, *Mapping the American revolutionary war* (Chicago, 1978).

61 Yolande O'Donoghue, *William Roy, 1726–1790: pioneer of the Ordnance Survey* (London, 1977).

62 Marc Duranthon, *La carte de France: son histoire 1678–1978* (Paris, 1978), pp. 31, 34.

63 William Ravenhill (ed.), *Christopher Saxton's 16th century maps: the counties of England and Wales* (Shrewsbury, 1992), pp. 16–17: 1579.

64 R.A. Skelton, 'Cartography' in Charles Singer (ed.), *A history of technology*, iv (Oxford, 1958), pp. 605–11.

65 Konvitz, *Cartography in France*, p. 60; Lloyd A. Brown, *The story of maps* (New York, 1949, 1979), pp. 241–79.

66 W.A. Seymour (ed.), *A history of the Ordnance Survey* (Folkestone, 1980); Tim Owen and Elaine Pilbeam, *Ordnance Survey: map makers to Britain since 1791* (Southampton, 1992).

67 Peter Whitfield, *The charting of the oceans: ten centuries of maritime maps* (London, 1996).

68 Roger J.P. Kain and Elizabeth Baigent, *The cadastral map in the service of the state: a history of property mapping* (Chicago, 1992).

69 Arthur H. Robinson, *Early thematic mapping in the history of cartography* (Chicago, 1982).

Chapter 2: A science of facts?

1   Yolande Jones, 'Aspects of relief portrayal on 19th-century British military maps', *Cartographic Journal*, xi, 1 (1974), p. 3.

2   R.J. Coggins and M.A. Knibb, *The first and second books of Esdras* (Cambridge, 1979), pp. 154, 157; Cecil Jane (ed.), *Select documents illustrating the four voyages of Columbus*, ii (London, 1933), p. 42.

3   C.E. Heidenreich, 'Explorations and mapping of Samuel de Champlain, 1603–1632', *Cartographica*, Monograph 17 (1976), plate 1.

4   Arthur H. Robinson, *Early thematic mapping in the history of cartography* (Chicago, 1982).

5   Barry Cunliffe, *The extraordinary voyage of Pytheas the Greek* (London, 2001), p. 128; H.F. Tozer, *A history of ancient geography* (Cambridge, 1897), p. 162.

6   Gerhard Friedrich Müller, quoted in J.C. Beaglehole, *The life of Captain James Cook* (London, 1974), p. 487.

7   John Stuart Mill, *A system of logic, ratiocinative and inductive* (London, 1895), p. 201.

8   Paul Carter, *The road to Botany Bay: an essay in spatial history* (London, 1987), ch. 2, pp. 34–68.

9   John Leighly, *California as an island: an illustrated essay* (San Francisco, 1972); Dora Beale Polk, *The island of California: a history of the myth* (Spokane, 1991); below, Fig. 2.12.

10   Numa Broc, *La géographie des philosophes, géographes et voyageurs français au xviiie siècle* (Paris, 1974), p. 48.

11   Richard Hakluyt, world map, reproduced from *Principal navigations* (1598) in W.P. Cumming, R.A. Skelton and D.B. Quinn, *The discovery of North America* (London, 1971), p. 224.

12   Günter Schilder, *Australia unveiled: the share of the Dutch navigators in the discovery of Australia* (Amsterdam, 1976), p. 81: John Brooke, 1622.

13   Dee Longenbaugh, 'Dating query on Russian map', *The Map Collector*, xxxix (1987), p. 55.

14   J.H. Andrews, *Shapes of Ireland: maps and their makers, 1564–1839* (Dublin, 1997), p. 31.

15   Reproduced in Max Leinekugel le Cocq, *Premieres images de la terre* (Paris, 1977), pp. 62–3.

16   Bertrand Russell, *Human knowledge, its scope and limits* (London, 1948), pp. 475–6, 506–7.

17   A.H.W. Robinson, *Marine cartography in Britain: a history of the sea chart to 1855* (Leicester, 1962), pp. 34–46.

18   Tozer, *A history of ancient geography*, p. 167.

19   Horace Leonard Jones (ed.), *The geography of Strabo*, i (Cambridge, MA, 1969), pp. 41–3.

20   Herbert George Fordham, *John Cary: engraver, map, chart and print-seller and globe-maker, 1754 to 1835* (1925; Folkestone, 1976), p. 115.

21   V. Coronelli, *Corso geografico* (Venice, 1693); Herman Moll, *The compleat geographer: or, The chorography and topography of all the known partes of the earth* (London, 1723), p. xii.

22   David Woodward, 'Medieval *mappaemundi*' in J.B. Harley and David Woodward (eds), *History of cartography: volume one, cartography in prehistoric, ancient, and medieval Europe and the Mediterranean* (Chicago, 1987), pp. 294–9; Helen M. Wallis and Arthur H. Robinson (eds), *Cartographical innovations: an international handbook of mapping terms to 1900* ([Tring], 1987), 'zone map', pp. 159–60.

23   Gerald R. Tibbetts, 'The Balkhi school of geographers', p. 128 and 'Later cartographic developments', p. 140 in J.B. Harley and David Woodward (eds), *The history of cartography, volume two, book one: cartography in the traditional Islamic and south Asian societies* (Chicago, 1992).

24   E.G.R. Taylor, 'A letter dated 1577 from Mercator to John Dee', *Imago Mundi*, xiii (1956), p. 59.

25   Gerard Mercator, Henry Hondius and Joannes Janssonius (ed. R.A. Skelton), *Atlas or a geographicke description of the world*, i (Amsterdam, 1636, 1968), p. 19.

26   Evelyn Edson, *Mapping time and space: how medieval mapmakers viewed their world* (London, 1997), p. 5; Michel Mollat du Jourdin and Monique de la Roncière, with Marie-Madeleine Azard, Isabelle Raynaud-Nguyen and Marie-Antoinette Vamerau, *Sea charts of the early explorers, 13th to 17th century* (New York, 1984), p. 8.

27   Aubrey de Sélincourt (ed.), *Herodotus: the histories* (London, 1959), pp. 253–4; Christian Jacob, 'Lectures antiques de la carte', *Études Françaises*, xxi, 2 (1985), pp. 26–30.

28 Edson, *Mapping time and space*, p. 125.

29 Gordon L. Davies, *The earth in decay: a history of British geomorphology, 1578–1878* (London, 1968), passim.

30 K.J. Tinkler, 'Worlds apart: eighteenth-century writing on rivers, lakes and the terraqueous globe' in K.J. Tinkler (ed.), *History of geomorphology from Hutton to Hack* (Boston, 1989), p. 65.

31 H. Rackham (ed.), *Pliny: natural history*, ii (London, 1942), p. 6: 'anfractu'; Miller Christy (ed.), *The voyages of Captain Luke Fox of Hull, and Captain Thomas James of Bristol in search of a north-west passage, in 1631–32, with narratives of ... earlier north-west voyages* (London, 1894), p. 199: Thomas Button.

32 E.G.R. Taylor, *Tudor geography, 1485–1583* (London, 1930), p. 158.

33 Mollat du Jourdain et al., *Sea charts of the early explorers*, pp. 48–51: Guillaume Le Testu, 1556; above, Fig. 1.15.

34 Frank Debenham (ed.), *The voyage of Captain Bellingshausen to the Antarctic seas, 1819–1821* (London, 1945), p. 410.

35 Alexander Dalrymple (ed. Andrew Cook), *An account of the discoveries made in the South Pacifick Ocean* (Sydney, 1996), p. 78.

36 John Leighley, 'Error in geography' in Joseph Jastrow (ed.), *The story of human error* (New York, 1936), pp. 105–14.

37 O.A.W. Dilke, *Greek and Roman maps* (London, 1985), p. 36.

38 Samuel Purchas, *Hakluytus Posthumus or Purchas his pilgrimes, contayning a history of the world in sea voyages and lande travells by Englishmen and others*, i (Glasgow, 1905), pp. 346–7.

39 Mercator-Hondius-Janssonius, *Atlas*, i, p. 20; Joseph de Acosta, *History of the Indies, 1608* (London, 1880), pp. 19–28; Alfred Hiatt, *Terra incognita: mapping the antipodes before 1600* (London, 2008).

40 Tibbetts, 'The Balkhi school of geographers', pp. 121–2.

41 G.E. Manwaring (ed.), *A cruising voyage round the world, by Woodes Rogers, 1712* (London, 1928), p. 237; Charles de Brosses, *Histoire des navigations aux terres australes*, 1756, cited in Beaglehole, *Life of Cook*, pp. 119–20; Dalrymple, *Discoveries in the South Pacifick Ocean*.

42 Edson, *Mapping time and space*, p. 161.

43 Günter Schilder, 'Willem Jansz. Blaeu's wall map of the world, on Mercator's projection, 1606–07 and its influence', *Imago Mundi*, xxxi (1979), p. 38.

44 William P. Cumming, 'Early maps of the Chesapeake Bay area: their relation to settlement and society' in David B. Quinn (ed.), *Early Maryland in a wider world* (Detroit, 1982), p. 270: 1670.

45 Glyndwr Williams, *The British search for the northwest passage in the eighteenth century* (London, 1962), p. 93.

46 Oskar Peschel, quoted by R.A. Skelton in R.A. Skelton, Thomas E. Marston and George D. Painter, *The Vinland map and the Tartar relation* (New Haven, CT, 1965), p. 118.

47 Derek Howse and Michael Sanderson, *The sea chart: an historical survey based on the collections in the National Maritime Museum* (Newton Abbot, 1973), p. 51: Antonio Sanches, 1633.

48 Carter, *The road to Botany Bay*, pp. 103–13, pl. 5: map of Australia, 1827, from T.J. Maslen, *Friend of Australia*; J. Wreford Watson, *Mental images and geographical reality in the settlement of North America* (Nottingham, 1967), pp. 8–9.

49 Edward Grey (ed.), *The travels of Pietro della Valle in India* (London, 1892), p. 15: 1623.

50 Tony Campbell, *Early maps* (New York, 1981), p. 60, pl. 26: Pierre Mortier, map of Central America and West Indies (Amsterdam, *c*.1705).

51 J.C. Beaglehole (ed.), *The journals of Captain James Cook on his voyages of discovery: i, The voyage of the* Endeavour, *1768–1771* (Cambridge, 1955), app. 2, p. 516.

52 P.D.A. Harvey, *The history of topographical maps: symbols, pictures and surveys* (London, 1980), p. 49; A.R. Millard, 'Cartography in the ancient Near East' in Harley and Woodward, *History of cartography: volume one*, pp. 113–14.

53 Vicomte de Santarem, *Atlas composé de mappemondes, de portulans et de cartes hydrographiques et historiques depuis le xie jusqu'ai xviie siècle* (Paris, 1849), plates 8, 12, 22; below, Fig. 18.11; Bernhard Varenius, *Cosmography and geography in two parts* (London, 1683), pp. 46–8.

54 Reproduced with commentaries in Walter W. Ristow (compiler), *A la carte: selected papers on maps and atlases* (Washington, 1972), pp. 40–41, 43 and Rodney W. Shirley, *The mapping of the world: early printed world maps, 1472–1700* (London, 1983), plate 99, p. 134.

55 Beaglehole, *Voyage of the* Endeavour, p. 270: 1770; below, Fig. 11.16; Mary Blewitt, *Surveys of the seas: a brief history of British hydrography* (London, 1957), p. 87: James Cook, 1770.

56 John L. Allen, 'Patterns of promise: mapping the plains and prairies, 1800–1860', *Great Plains Quarterly*, iv, 1 (1984), p. 18.

57 Quoted in Richard J. Chorley, Robert P. Beckinsale and Antony J. Dunn, *The history of the study of landforms or the development of geomorphology*, i (London, 1964), p. 6.

58 C.G.C. Martin, *Maps and surveys of Malawi* (Rotterdam, 1980), p. 20: Loopata Mountains.

59 Kazutaka Unno, 'The geographical thought of the Chinese people: with special reference to ideas of terrestrial features', *Memoirs of the Research Department of the Toyo Bunko*, xli (Tokyo, 1983), p. 88.

60 C.T. Smith, 'The drainage basin as an historical basis for human activity' in Richard J.Chorley (ed.), *Water, earth and man* (London, 1969), p. 102.

61 C.R.D. Bethune (ed.), *The discoveries of the world from their first original unto the year of our lord 1555 by Antonio Galvano* (London, 1862), p. 215; I.A. Wright (ed.), *Documents concerning English voyages to the Spanish Main, 1569–1580* (London, 1932), p. 110.

62 D. Graham Burnett, *Masters of all they surveyed: exploration, geography and a British El Dorado* (Chicago, 2000), p. 228.

63 Armando Cortesao (ed.), *The suma oriental of Tomé Pires* (London, 1944), p. 200: 1512–15.

64 Christopher Lloyd (ed.), *The voyages of Captain James Cook round the world* (London, 1949), p. 218, n. 7.

65 J.W. Frezier, 1717, quoted in Robert E. Gallagher (ed.), *Byron's journal of his circumnavigation, 1764–1766* (Cambridge, 1964), p. 171.

66 Thomas Suarez, 'A 2,000-year premonition: maps of the great southern continent – before its discovery', *Mercator's World*, i, 2 (1996), p. 16.

67 Mansel Longworth Dames (ed.), *The book of Duarte Barbosa: an account of the countries bordering on the Indian Ocean and their inhabitants*, ii (London, 1921), map opposite p. 134; Valerie A. Kivelson, 'Cartography, autocracy and state powerlessness: the uses of maps in early modern Russia', *Imago Mundi*, li (1999), p. 86: 1701.

68 A.E. Nordenskiöld, *Facsimile atlas to the early history of cartography* (1889; New York, 1973), plate XXII: 1490.

69 Edward Lynam, *The Carta marina of Olaus Magnus, Venice 1539 and Rome 1572* (Jenkintown, PA, 1949), p. 18.

70 D.B. Quinn (ed.), *The voyages and colonising enterprises of Sir Humphrey Gilbert*, i (London, 1940), p. 164.

71 Williams, *The British search for the northwest passage*, pp. 12, 35, 64; William Barr and Glyndwr Williams (eds), *Voyages to Hudson Bay in search of a northwest passage, 1741–1747*, i (London, 1994), p. 16 and passim.

72 Beaglehole, *Voyage of the* Endeavour, p. 62: 1769.

73 Beaglehole, *Life of Cook*, p. 601; G.S. Ritchie, *The Admiralty chart: British naval hydrography in the nineteenth century* (London, 1967), p. 77.

74 J.C. Beaglehole (ed.), *The journals of Captain James Cook on his voyages of discovery: ii, The voyage of the* Resolution *and* Adventure, *1772–1775* (Cambridge, 1961), p. 190.

75 Brian Hooker, 'James Cook's secret search in 1769', *Mariner's Mirror*, lxxxvii (2001), p. 301; Blewitt, *Surveys of the seas*, p. 116.

76 Albert Hastings Markham (ed.), *The works and voyages of John Davis, the navigator* (London, 1880), pp. 215–18.

77 Charles de Brosses, 1756, and John Callender, quoted by G.R. Crone and R.A. Skelton, 'English collections of voyages and travels, 1625–1846' in Edward Lynam (ed.), *Richard Hakluyt and his successors* (London, 1946), p. 120.

78 Jonathan Williams, 'Memoir on the use of the thermometer in discovering banks, soundings etc.', *Transactions of the American Philosophical Society*, iii (1793), p. 83.

79 Hildegard Binder Johnson, 'Science and historical truth on French maps of the 17th and 18th centuries', *Proceedings of the Minnesota Academy of Science*, xxv–xxvi (1957–8), p. 329.

80 J.M. Cohen (ed.), *The four voyages of Christopher Columbus* (London, 1969), pp. 222–4.

81 Matthew Flinders, *A voyage to Terra Australis*, i (London, 1814), p. lxxiii.

82 George Edmundson (ed.), *Journal of the travels and labours of Father Samuel Fritz in the River of the Amazons between 1686 and 1723* (London, 1922), p. 149.

83 Saul Jarcho, 'Christopher Packe (1686–1749), physician-cartographer of Kent', History of Cartography Conference, Washington, 1977, typescript, pp. 4–5; William Ravenhill, 'The Honourable Robert Edward Clifford, 1767–1817: a cartographer's response to Napoleon', *Geographical Journal*, clx (1994), p. 164.

84 Abraham Ortelius (ed. R.A. Skelton), *The theatre of the whole world, London, 1606* (Amsterdam, 1968), p. 4: Africa; p. 9: Peru; p. 105: Tartary.

85 Guy Meriwether Benson with William R. Irwin and Heather Moore Riser, *Lewis and Clark: the maps of exploration, 1507–1814* (Charlottesville, VA, 2002), pp. 27, 30, 40, 66.

86 Varenius, *Cosmography and geography*, pp. 48–9; G. Malcolm Lewis, 'Misinterpretation of Amerindian information as a source of error on Euro-American maps', *Annals of the Association of American Geographers*, lxxvii, 4 (1987), p. 543.

87 John Mitchell, *A map of the British colonies in North America, with the roads, distances, limits, and extent of the settlements* (London, 1755).

88 Gordon Manley, 'Saxton's survey of northern England', *Geographical Journal*, lxxxiii (1934), p. 310.

89 Edson, *Mapping time and space*, p. 28: twelfth-century copy of fourth- or fifth-century map of Asia attributed to St Jerome; p. 124: Matthew Paris, world map, *c.*1250; below, Fig. 12.1.

90 Clements R. Markham (ed.), *The natural and moral history of the Indies, by Father Joseph de Acosta*, i (London, 1880), pp. 152–3.

91 E.G.R. Taylor (ed.), *The original writings and correspondence of the two Richard Hakluyts*, ii (London, 1935), p. 494.

92 Samuel Dunn, *A map of the British empire in North America* in Thomas Jefferys, *The American atlas* (London, 1776; Amsterdam, 1974).

93 Quinn, *Gilbert*, i, pp. 141–2.

94 J. Lennart Berggren and Alexander Jones (eds), *Ptolemy's* Geography: *an annotated translation of the theoretical chapters* (Princeton, NJ, 2000), pp. 69, 70.

95 Map of North and South America, 1775, Jeffreys, *American atlas*, nos 1, 2, 3; Samuel Hearne, *A journey from Prince of Wales's Fort in Hudson's Bay, to the Northern Ocean* (London, 1795), p. vii.

96 Cohen, *Four voyages of Columbus*, pp. 43–52; C.V. Sölver and G.J. Marcus, 'Dead reckoning and the ocean voyages of the past', *Mariner's Mirror*, xliv (1958), pp. 31–2; P.C. Fenton, 'The navigator as natural historian', *Mariner's Mirror*, lxxix (1993), pp. 51–3.

97 Williams, *The British search for the northwest passage*, p. 33.

98 Beaglehole, *Voyage of the* Endeavour, p. 290.

99 Burnett, *Masters of all they surveyed*, p. 81.

100 Berggren and Jones, *Ptolemy's* Geography, pp. 67, 72.

101 P.D.A. Harvey and Harry Thorpe, *The printed maps of Warwickshire, 1576–1900* (Warwick, 1959), plates 2 and 4.

102 J.B. Harley and P. Laxton (eds), *A survey of the county palatine of Chester: P.P. Burdett: 1777* (Liverpool, 1974), p. 21.

103 P.D.A. Harvey, *Medieval maps* (London, 1991), pp. 24, 26, 29, 34, 36; below, Figs 8.20, 16.1.

104 Andreas Stylianou and Judith A.Stylianou, *The history of the cartography of Cyprus* (Nicosia, 1980), pp. 204, 218, 219, 220, 223.

105 Cordell D.K. Yee, 'Chinese maps in political culture' in J.B. Harley and David Woodward (eds), *The history of cartography: volume two, book two: cartography in the traditional east and southeast Asian societies* (Chicago, 1994), p. 86.

106 R.A. Skelton, *Explorers' maps* (London, 1958), p. 89: Theodore de Bry, map of Guiana, 1599.

107 Linen Hall Library, *The Hyberniae Novissima Descriptio by Jodocus Hondius*, introductory note by J.H. Andrews (Belfast, 1983); R.W. Shirley, *Early printed maps of the British Isles, 1477–1650, a bibliography* (London, 1973), no. 175, p. 69; no. 177, pp. 69–70.

108 Jean Pierre Sanchez, 'Myths and legend in the old world and European expansionism on the American continent' in Wolfgang Haase and Meyer Reinhold (eds), *The classical tradition and the Americas*, i (Berlin and New York, 1994), pp. 189–240; F.J. Manasek, 'Frislant; phantom island of the North Atlantic', *Mercator's World*, ii, 1 (1997), pp. 14–18.

109 Reproduced in Cumming, Skelton and Quinn, *The discovery of North America*, p. 206.

110 Catherine Delano Smith, 'Maps as art *and* science: maps in sixteenth century Bibles', *Imago Mundi*, xlii (1990), p. 69.

111 Hartmann Schedel, *Liber cronicarum* (Nuremberg, 1493), ff. xlviii, lxviii, lxxx, lxxxiiii; Richard L.Kagan, '*Urbs* and *civitas* in sixteenth- and seventeenth-century Spain' in David Buisseret (ed.), *Envisioning the city: six studies in urban cartography* (Chicago, 1998), pp. 79–80.

112 Ibid., pp. 94–6; G. and E. Wajntraub, 'An illustrated history of the Holy City', *Mercator's World*, i, 4 (1996), pp. 26–8; John A.Pinto, 'Origins and development of the ichnographic city plan', *Journal of the Society of Architectural Historians*, xxxv, 1 (1976), p. 49.

113 Elizabeth Rodger, 'An eighteenth-century collection of maps connected with Philippe Buache', *Bodleian Library Record*, vii, 2 (1963), pp. 96–106.

114 R.A. Skelton, *Charts and views drawn by Cook and his officers and reproduced from the original manuscripts* (Cambridge, 1969), LVIII: Henry Roberts, *A general chart exhibiting the discoveries made by Captn James Cook in this and his two preceeding voyages; with the tracks of the ships under his command*, 1784.

115 Quoted by L.P. Kirwan, *A history of polar exploration* (London, 1962), p. 74.

116 M.M. Sweeting, 'The enclosed depression of Carran, County Clare', *Irish Geography*, ii, 5 (1953), pp. 218–24.

Chapter 3: *So many guess plots*

1 Robert E. Gallagher (ed.), *Byron's journal of his circumnavigation, 1764–1766* (Cambridge, 1964), pp. 29–30.

2 Ibid., pp. 40–2, 164–76.

3 Herman Moll, *The compleat geographer: or, The chorography and topography of all the known partes of the earth* (London, 1723), p. 272; B.M. Chambers, 'Where was Pepys Island? A problem in historical geography', *Mariner's Mirror*, xix (1933), pp. 446–54.

4 Fergus Fleming, *Barrow's boys* (London, 1998), pp. 48, 67, 349.

5 Derek Howse and Michael Sanderson, *The sea chart: an historical survey based on the collections in the National Maritime Museum* (Newton Abbot, 1973), p. 98.

6 William Ravenhill, 'Bird's-eye view and bird's-flight view', *The Map Collector*, xxxv (1986), pp. 36–8.

7 View of the earth from 100 miles above Berlin, J.E. Bode, *Anleitung zur allgemeinen Kentniss der Erdkugel* (Berlin, 1786).

8 Helen M. Wallis and Arthur H. Robinson (eds), *Cartographical innovations: an international handbook of mapping terms to 1900* ([Tring], 1987), p. 137.

9 Hugh Carrington (ed.), *The discovery of Tahiti: a journal of the second voyage of H.M.S. Dolphin round the world, under the command of Captain Wallis, R.N., in the years 1766, 1767 and 1768* (London, 1948), pp. 126, 242.

10 Gerald Strauss, *Sixteenth-century Germany, its topography and topographers* (Madison, WI, 1959); Karen Severud Pearson, 'Germania illustrata: the discovery of the German landscape by 16th century geographers, cartographers and artists', History of Cartography Conference, Washington, 1977, typescript; David Buisseret, 'The estate map in the old world' in David Buisseret (ed.), *Rural images: estate maps in the old and new worlds* (Chicago, 1996), pp. 8–22; Roger J.P. Kain and Elizabeth Baigent, *The cadastral map in the service of the state: a history of property mapping*

(Chicago, 1992), pp. 123–4; Lucia Nutti, 'Mapping places: chorography and vision in the renaissance' in Denis Cosgrove (ed.), *Mappings* (London, 1999), pp. 90–108.

11 Dorothy Sylvester, *Map and landscape* (London, 1952), part iii, especially ch. 18.

12 Juergen Schulz, 'Jacopo de' Barbari's view of Venice: map making, city views, and moralized geography before the year 1500', *The Art Bulletin*, lx (1978), p. 438.

13 Martin Clayton, *Leonardo da Vinci: a curious vision* (London, 1996), p. 97.

14 Dan Pedoe, *Geometry and the liberal arts* (London, 1976), p. 52; plate 13, following p. 148.

15 Maya Hambly, *Drawing instruments: their history, purpose and use for architectural drawings* (London, 1982), pp. 38–9; Philip Steadman, *Vermeer's camera* (Oxford, 2001), ch. 1.

16 Fred Dubery and John Willats, *Drawing systems* (London, 1972), pp. 60–1, 83–5.

17 Arthur Lovat Higgins, *Phototopography: a practical manual of photographic surveying methods* (Cambridge, 1926), pp. ix–xi.

18 J. Lennart Berggren and Alexander Jones (eds), *Ptolemy's* Geography: *an annotated translation of the theoretical chapters* (Princeton, NJ, 2000), p. 58.

19 William Cuningham, *The cosmographical glasse* (London, 1559), f. 8 and view of Norwich.

20 R[obert] N[orton], *A mathematicall apendix ... for mariners at sea, and for cherographers and surveyors of land* (London, 1604), p. 22.

21 William Barr and Glyndwr Williams (eds), *Voyages to Hudson Bay in search of a northwest passage, 1741–1747*, i (London, 1994), pp. 283–4.

22 Quoted in W.A. Seymour (ed.), *A history of the Ordnance Survey* (Folkestone, 1980), p. 364: 1785.

23 Diana C.F. Smith, 'The progress of the *Orcades* survey, with biographical notes on Murdoch Mackenzie Senior (1712–1797)', *Annals of Science*, xliv (1987), p. 285.

24 Gordon Manley, 'Saxton's survey of northern England', *Geographical Journal*, lxxxiii (1934), pp. 308–16; below, Fig. 12.5.

25 R.W. Bremner, 'Mount Etna and the distorted shape of Sicily on early maps', *The Map Collector*, xxxii (1985), p. 28.

26 J.C. Beaglehole (ed.), *The journals of Captain James Cook on his voyages of discovery: The voyage of the* Resolution *and* Discovery, *1776–1780* (Cambridge, 1967), pp. 338–9.

27 F.J. North, *Humphrey Lhuyd's maps of England and Wales* (Cardiff, 1937), p. 37.

28 R.A. Skelton, 'Tudor town plans in John Speed's *Theatre*', *Archaeological Journal*, cviii (1951), pp. 116–17.

29 J.H. Andrews, 'The maps of the escheated counties of Ulster, 1609–10', *Proceedings of the Royal Irish Academy*, lxxiv C (1974), pp. 149–50.

30 J.C. Stone, 'Timothy Pont and the mapping of sixteenth-century Scotland: survey or chorography?', *Survey Review*, xxxv, 276 (2000), pp. 425–6.

31 John Dunmore (ed.), *The journal of Jean-François de Galaup de La Pérouse, 1785–1788* (London, 1995), ii, p. 405.

32 Ian R. Stone, 'W.R. Broughton and the insularity of Sakhalin', *Mariner's Mirror*, lxxxii (1996), p. 77.

33 Peter Barber, 'The Evesham world map: a late medieval English view of God and the world', *Imago Mundi*, xlvii (1995), pp. 25–7.

34 *Journal of a 2nd voyage for the discovery of a north west passage from the Atlantic to the Pacific; performed in the years 1821–22–23 in His Majesty's ships* Fury *and* Hecla, *under the orders of Captain William Edward Parry, R.N., F.R.S., and commander of the expedition* (New York, 1969), p. 251.

35 Berggren and Jones, *Ptolemy's* Geography, p. 70.

36 John Ogilby (ed. J.B. Harley), *Britannia, London 1675* (Amsterdam, 1970), preface.

37 David Beers Quinn, *The Roanoke voyages, 1584–1590*, ii (London, 1955), p. 828.

38 British Library, Index to original drawings of first survey of England and Wales, 1784–1841, no. 299B.

39 William P. Cumming, *British maps of colonial America* (Chicago, 1974), p. 86, n. 16: John Mitchell on Nathaniel Blackmore's chart of Nova Scotia, 1714–15.

40 John Smith, *A description of New England* (London, 1616), pp. 4–5, quoted in Norman J.W. Thrower (ed.), *The compleat plattmaker: essays on chart, map, and globe making in England in the seventeenth and eighteenth centuries* (Berkeley, CA, 1978), p. 106.

41 Gary T. Moore and Reginald G. Colledge (eds), *Environmental knowing: theories, research, and methods* (Stroudsburg, PA, 1976), p. 113.

42 A.W. Siegal and S. White, 'The development of spatial representations of large-scale environments' in H.W. Reese (ed.), *Advances in child development and behavior*, x (New York, 1975), pp. 10–55; Erika L. Ferguson and Mary Hegarty, 'Properties of cognitive maps constructed from texts', *Memory and Cognition*, xxii (1994), pp. 455–73.

43 Robert Lloyd, 'The estimation of distance and direction from cognitive maps', *The American Cartographer*, xvi (1989), p. 119; R.W. Byrne, 'Memory for urban geography', *Quarterly Journal of Experimental Psychology*, xxxi (1979), pp. 147–54.

44 Edward K. Sadalla and Lorin J. Staplin, 'The perception of traversed distance', *Environment and Behaviour*, xii (1980), pp. 65–80; Edward K. Sadalla and Stephen J. Magel, 'The perception of traversed distance: intersections', *Environment and Behaviour*, xii (1980), pp. 167–82; Edward K. Sadalla and Lorin J. Staplin, 'An information storage model for distance cognition', *Environment and Behaviour*, xii (1980), pp. 183–93; Christopher Spencer, Mark Blades and Kim Morsley, *The child in the physical environment: the development of spatial knowledge* (Chichester, 1989), p. 126.

45 'Remarks by M. Bellin, in relation to his maps drawn for P. Charlevoix's history of New France', *Gentleman's Magazine*, xvi (1746), p. 73.

46 Berggren and Jones, *Ptolemy's Geography*, p. 118.

47 Albert Stevens and Patty Coupe, 'Distortions in judged spatial relations', *Cognitive Psychology*, x (1978), pp. 422–37.

48 Michael Blakemore, 'From way-finding to map-making: the spatial information fields of aboriginal peoples', *Progress in Human Geography*, v (1981), p. 13; Byrne, 'Memory for urban geography', pp. 147–54; Spencer et al., *The child in the physical environment*, p. 30.

49 Jack L. Nasar, Hugo Valencia, Zainal Abidin Omar, Shan-Chy Chueh and Ji-Hyuan Hwang, 'Out of sight, further from mind', *Environment and Behaviour*, xvii, 5 (1985), pp. 627–39.

50 J.W. Watson, 'Mental distance in geography: its identification and representation' in J. Keith Fraser (ed.), *22nd IGU Congress proceedings* (Ottawa, 1979), pp. 38–50; Spencer et al., *The child in the physical environment*, p. 126; Christopher Board, 'Maps as models' in R.J. Chorley and P. Haggett (eds), *Models in geography* (London, 1967), p. 675.

51 Barbara Tversky, 'Distortions in memory for maps', *Cognitive Psychology*, xiii (1981), pp. 407–33; Rudyard Kipling, *A book of words* (London, 1928), pp. 107–8.

52 Stevens and Coupe, 'Distortions in judged spatial relations', pp. 422–37.

53 Byrne, 'Memory for urban geography', pp. 147–54; Spencer et al., *The child in the physical environment*, p. 29.

54 Lynn S. Liben and Roger M. Downs, 'Understanding person-space-map relations: cartographic and developmental perspectives', *Developmental Psychology*, xxix, 4 (1993), p. 746.

55 Waldo R. Tobler, 'Computation of the correspondence of geographical patterns', *Papers of the Regional Science Association*, xv (1965), pp. 131–9; W.R. Tobler, 'Medieval distortions: the projections of ancient maps', *Annals of the Association of American Geographers*, lvi, 2 (1966), pp. 351–60; Andrews, 'The maps of the escheated counties of Ulster', pp. 149–50.

56 View from Erwin Raisz, *Mapping the world* (New York, 1956), p. 91.

57 David G. Kendall, 'Construction of maps from "odd bits of information"', *Nature*, ccxxxi (1971), p. 158.

58 J.H. Andrews, *Shapes of Ireland: maps and their makers, 1564–1839* (Dublin, 1997), p. 12; W.R. Tobler, 'Bidimensional regression', typescript, Santa Barbara, CA, 1977, p. 27; Tobler, 'Medieval distortions: the projections of ancient maps', pp. 355, 360.

59 T. Hennigan, *Mastering statistics* (Basingstoke, 1988), p. 212; J. Crawshaw and J. Chambers, *A concise course in A-level statistics* (Cheltenham, 1989), p. 493; Alan Graham, *Statistics* (London, 1994), p. 195.

Chapter 4: Strict dimensuration

1   John Ogilby (ed. J.B. Harley), *Britannia, London 1675* (Amsterdam, 1970), preface, p. [i].

2   N[icholas] Person, *Novae archiepiscopatus Moguntini tabulae* (Mainz, n.d.): *Bischoffsheim*.

3   Albert Hastings Markham (ed.), *The voyages and works of John Davis the navigator* (London, 1880), p. 275.

4   Samuel Eliot Morison, *Admiral of the ocean sea, a life of Christopher Columbus* (1942; Boston, 1970), p. 190.

5   Anita McConnell, 'La Condamine's scientific journey down the River Amazon, 1743–1744', *Annals of Science*, xlviii (1991), p. 7.

6   W.E. May, *A history of marine navigation* (Henley-on-Thames, 1973), pp. 108–18; Günter Schilder, *Australia unveiled: the share of the Dutch navigators in the discovery of Australia* (Amsterdam, 1976), p. 59.

7   Richard Copeland, *An introduction to the practice of nautical surveying and the construction of sea-charts, … translated from the French* [1808] *of C.F. Beautemps-Beaupre* (London, 1823), p. 8.

8   N.J.W. Thrower (ed.), *The three voyages of Edmond Halley in the* Paramore, *1698–1701* (London, 1981), pp. 338–40; [Alexander] Dalrymple, *Essay on nautical surveying* (2nd ed., London, 1786), pp. 13–14.

9   D. Graham Burnett, *Masters of all they surveyed: exploration, geography, and a British El Dorado* (Chicago, 2000), p. 221.

10 Florian Cajori, 'History of determinations of the heights of mountains', *Isis*, xii (1929), pp. 498–514.

11 E.G.R. Taylor, 'Robert Hooke and the cartographical projects of the late seventeenth century (1666–1696)', *Geographical Journal*, lxxxii (1937), p. 538; Anne E. Baily-Kahn, 'The little story of the contour line', *UCLA Map Library: Newsletter and Selected Acquisitions*, v, 2 (1985), pp. 22–3.

12 Arnold Horner, *Wicklow and Dublin Mountains in 1812: Richard Griffith's map for the bogs commissioners of 1809–1814* (Dublin, 2004), p. 9.

13 Mary Pedley, '"I Due Valentuomini Indefessi": Christopher Maire and Roger Boscovich and the mapping of the papal states (1750–1755)', *Imago Mundi*, xlv (1993), pp. 64–5; Christopher Maire, *Nuova carta geografica dello stato ecclesiastico* (Rome, [1755]).

14 'Utriusque Frisiorum regionis noviss: descriptio, 1568' in Abraham Ortelius (ed. R.A. Skelton), *The theatre of the whole world London 1606* (Amsterdam, 1968), p. 48.

15 Philip Grierson, *English linear measures, an essay in origins* (Reading, 1972), pp. 20–4.

16 Ogilby, *Britannia*, preface, p. [iii].

17 'A breviate of Monsieur Picart's account of the measure of the earth', *Philosophical Transactions of the Royal Society of London*, x (1675), p. 268.

18 Horace Leonard Jones (ed.), *The geography of Strabo* (Cambridge, MA), v (1969), p. 287.

19 D.R. Dicks, *The geographical fragments of Hipparchus* (London, 1960), pp. 42–6.

20 Thomas McGreevy, *The basis of measurement, volume 1: historical aspects* (Chippenham, Wilts, 1995), pp. 140–58; Josef W. Konvitz, *Cartography in France, 1660–1848: science, engineering, and statecraft* (Chicago, 1987), pp. 46–7.

21 Edward Belcher, *A treatise on nautical surveying* (London, 1835), p. 265.

22 J. Lee, *Maps and plans of Manchester and Salford, 1650 to 1843* (Altringham, Cheshire, 1957), p. 13; Manchester Public Libraries, *Maps of Manchester, 1650–1848* (Manchester, 1969), no. 3: Charles Laurent, engineer and geographer, 1793.

23 J. Wartnaby, *Surveying: instruments and methods* (London, 1968), no. 6.

24 Tony Campbell, 'Portolan charts from the late thirteenth century to 1500' in J.B. Harley and David Woodward (eds), *The history of cartography, volume one: cartography in prehistoric, ancient, and medieval Europe and the Mediterranean* (Chicago, 1987), p. 387, n. 152.

25 Wartnaby, *Surveying: instruments and methods*, no. 5.

26 Charles Close (ed. J.B. Harley), *The early years of the Ordnance Survey* (1926; Newton Abbot, 1969), p. 38.

27 J.B. Harley, 'The Society of Arts and the surveys of English counties, 1759–1809', *Journal of the Royal Society of Arts*, cxii (1963), pp. 44–5.

28 Joseph Lindley, *Memoir of a map of the county of Surrey, from a survey made in the years 1789 and 1790* (London [1793]), p. 66.

29 May, *History of marine navigation*, p. 115.

30 B. Talbot, *The new art of land surveying; or, a turnpike road to practical surveying* (Wolverhampton, 1779), p. 404.

31 Uta Lindgren, 'Astronomische und geodätische Instrumente zur Zeit Peter und Philipp Apians' in Hans Wolff et al., *Philipp Apian und die Kartographie der Renaissance* (Munich, 1989), p. 53: 1550; Samuel Sturmy, *The mariners magazine* (London, 1669), v, pp. 3–4; A. Sarah Bendall, *Maps, land and society: a history, with a carto-bibliography of Cambridgeshire estate maps, c.1600–1836* (Cambridge, 1992), p. 130, quoting John Reid, *The Scots gard'ner*, 1683.

32 Richard Benese, *This boke sheweth the maner of measurynge of all maner of lande* (Southwark, 1537); Valentine Leigh, *The most profitable and commendable science of surveying* (London, 1577), unpaginated.

33 G.L'E. Turner, 'Introduction: some notes on the development of surveying and the instruments used', *Annals of Science*, xlviii, 4 (1991), p. 316.

34 Uuno Varjo, 'Observations on the mapping of the village of Nummi in the parish of Nousiainen in 1786', *Fennia*, clxiv, 1 (1986), p. 80.

35 A.W. Richeson, *English land measuring to 1800: instruments and practices* (Cambridge, MA, 1966), pp. 108–9.

36 J.B. Harley, 'The contemporary mapping of the American revolutionary war' in J.B. Harley, Barbara Bartz Petchenik and Lawrence W. Towner, *Mapping the American revolutionary war* (Chicago, 1978), p. 9.

37 John Holwell, *A sure guide to the practical surveyor, in two parts* (London, 1678), p. 175; John Hammond, *The practical surveyor* (3rd ed., enlarged by Samuel Warner, London, 1750), p. 11; W. Emerson, *The art of surveying, or, measuring land* (London, 1770), p. 5.

38 William Yolland, *An account of the measurement of the Lough Foyle base in Ireland* (London, 1847).

39 Ronald C. Cox, 'Engineering surveying instrumentation: 250 years of development', *Transactions of the Institution of Engineers of Ireland*, ciii (1979), p. 35.

40 William Ford Stanley, *Surveying and levelling instruments* (London, 1895), pp. 394–400; E.G.R. Taylor, *The mathematical practitioners of Hanoverian England, 1714–1840* (Cambridge, 1966), p. 37; J.A. Bennett, *The divided circle: a history of instruments for astronomy, navigation and surveying* (Oxford, 1987), p. 153; Charles H. Cotter, *A history of nautical astronomy* (London, 1968), pp. 91–6; Richard I. Ruggles, *A country so interesting: the Hudson's Bay Company and two centuries of mapping, 1670–1870* (Montreal, 1991), p. 15.

41 Bennett, *Divided circle*, pp. 151–2.

42 O.A.W. Dilke, *Greek and Roman maps* (London, 1985), p. 32; Klaus Geus, 'Measuring the earth and *oikoumene*: zones, meridians, *sphragides* and some other geographical terms used by Eratosthenes of Cyrene' in Richard Talbert and Kai Brodersen (eds), *Space in the roman world: its perception and presentation* (New Brunswick, NJ, 2004), pp. 11–26.

43 May, *History of marine navigation*, pp. 46–9; below, Figs 4.22, 18.10.

44 T.M. Perry, *The discovery of Australia: the charts and maps of the navigators and explorers* (London, 1982), plate 50: Louis de Freycinet, *Carte générale de la Terre Napoléon (à la Nouvelle Hollande)*, 1808.

45 O.A.W. Dilke, *The Roman land surveyors: an introduction to the agrimensores* (Newton Abbot, 1971), pp. 66–7.

46 O.A.W. Dilke and Margaret S. Dilke, 'Perception of the Roman world', *Progress in Geography*, ix (1976), pp. 46–51.

47 Bennett, *Divided circle*, p. 150; George Adams, *Geometrical and graphical essays* (London, 1791), pp. 209–13; Stanley, *Surveying and levelling instruments*, pp. 416–19.

48 R.T. Gunther, *Early science at Oxford* (Oxford, 1922), pp. 331–2.

49 Germaine Aujac, 'The foundations of theoretical cartography in archaic and classical Greece' (prepared by the editors) in Harley and Woodward (eds), *The history of cartography, volume one*, p. 134.

50 E.G.R. Taylor, *The haven-finding art: a history of navigation from Odysseus to Captain Cook* (London, 1956); H.O. Hill and E.W. Paget-Tomlinson, *Instruments of navigation: a catalogue of instruments at the National Maritime Museum with notes upon their use* (London, 1958), pp. 8–11; Richeson, *English land measuring to 1800*, p. 58; May, *History of marine navigation*, pp. 123–7; Bennett, *Divided circle*, pp. 32–6; John Roche, 'The cross-staff as a surveying instrument in England 1500–1640' in Sarah Tyacke (ed.), *English map-making, 1500–1650* (London, 1983), pp. 107–11.

51 Alan Neale Stimson and Christopher St John Hume Daniel, *The cross-staff: historical development and modern use* (London, 1977).

52 Ernest R. Cooper, 'The Davis back-staff or English quadrant', *Mariner's Mirror*, xxx (1944), pp. 59–64.

53 D.W. Waters (ed.), *The planispheric astrolabe* (Greenwich, 1979).

54 R.T. Gunther, 'The astrolabe: its uses and derivatives', *Scottish Geographical Magazine*, xliii (1927), pp. 141–3; Alan Stimson, *The mariner's astrolabe: a survey of known, surviving sea astrolabes* (Utrecht, 1988).

55 J.A. Bennett and Olivia Brown, *The compleat surveyor* ([Cambridge], 1982), pp. 4–7; J.A. Bennett, 'Geometry and surveying in early-seventeenth-century England', *Annals of Science*, xlviii (1991), p. 347.

56 Turner, 'Introduction', p. 316; Bennett, *Divided circle*, pp. 49–50.

57 Markham, *Works and voyages of John Davis*, p. 333.

58 Mary Terrall, 'Representing the earth's shape: the polemics surrounding Maupertius's expedition to Lapland', *Isis*, lxxxiii (1992), p. 225; Rob Iliffe, '"Aplatisseur du monde et de Cassini": Maupertius, precision measurement, and the shape of the earth in the 1730s', *History of Science*, xxxi (1993), pp. 335–75.

59 Bennett, *Divided circle*, p. 132.

60 Benjamin Martin, *The new art of surveying by the goniometer* (London, 1766), p. 2.

61 Bennett, *Divided circle*, pp. 136–8.

62 E.A. Reeves, *Maps and map-making* (London, 1910), pp. 20–2; Taylor, *Mathematical practitioners of Hanoverian England*, p. 32.

63 Richeson, *English land measuring*, pp. 9–10.

64 Bennett, *Divided circle*, pp. 145–9.

65 Martin, *New art of surveying*, p. 15.

66 Richeson, *English land measuring*, pp. 181–3.

67 James E. Kelley, Jnr, 'Perspectives on the origin and uses of the portolan charts', *Cartographica*, xxxii, 3 (1995), p. 2; May, *History of marine navigation*, ch. 2; Amir D. Aczel, *The riddle of the compass: the invention that changed the world* (New York, 2001), pp. 29–33.

68 Sydney Chapman and Julius Bartels, *Geomagnetism*, ii (Oxford, 1940), pp. 898–937.

69 Norman J.W. Thrower, 'Edmond Halley as a thematic geo-cartographer', *Annals of the Association of American Geographers*, lix (1969), pp. 661–9.

70 Thrower, *Three voyages of Edmond Halley*, pp. 366–7.

71 Bennett, *Divided circle*, pp. 40, 149; Adams, *Geometrical and graphical essays*, pp. 213–18.

72 Taylor, *Mathematical practitioners of Hanoverian England*, p. 374.

73 Eduard Imhof, 'Beiträge zur Geschichte der topographischen Kartographie', *International Yearbook of Cartography*, iv (1964), pp. 133–40; Jim Bennett and Stephen Johnston, *The geometry of war, 1500–1750* (Oxford, 1996), pp. 58–63.

74 E.G.R. Taylor, 'The plane table in the sixteenth century', *Scottish Geographical Magazine*, xlv (1929), pp. 205–11.

75 Stanley, *Surveying and levelling instruments*, pp. 188–93; Eric G. Forbes, 'Tobias Mayer's new astrolabe (1759): its principles and construction', *Annals of Science*, xxvii (1971), pp. 110, 116; James R. Smith, *From plane to spheroid: determining the figure of the earth from 3000 B.C. to the eighteenth century Lapland and Peruvian survey expeditions* (Rancho Cordova, CA, 1986), pp. 34–5.

76 Reeves, *Maps and map-making*, pp. 25–6; Smith, *From plane to spheroid*, pp. 29–33.

77 Maurice Daumas, *Les instruments scientifiques aux xviie et xviiie siècles* (Paris, 1953), pp. 249–55.

78 Albert van Helden, 'The telescope in the seventeenth century', *Isis*, lxv (1974), pp. 38–58.

79 Stanley, *Surveying and levelling instruments*, pp. 36–40; Richeson, *English land measuring*, pp. 149, 164–5; J.C. Beaglehole (ed.), *The journals of Captain James Cook on his voyages of discovery: the voyage of the* Resolution *and* Discovery, *1776–1780* (Cambridge, 1967), p. 144.

Chapter 5: Remarkable objects

1 Frank Debenham, *Exercises in cartography* (London, 1937), p. 3.

2 Helen M. Wallis and Arthur H. Robinson (eds), *Cartographical innovations: an international handbook of mapping terms to 1900* (Tring, 1987), nusquam.

3 W.A. Seymour (ed.), *A history of the Ordnance Survey* (Folkestone, 1980), p. 37.

4 Thomas Breaks, *A complete system of land-surveying, both in theory and practice* (Newcastle and London, 1771), p. 376.

5 Frank Debenham (ed.), *The voyage of Captain Bellingshausen to the Antarctic seas, 1819–1821* (London, 1945), p. 197.

6 National Library of Ireland, MS 3242: William Edgeworth's survey of Co. Roscommon, 1813.

7 National Archives of Ireland, Dublin: Ordnance Survey, general correspondence, file 774 (1851).

8 Charles Close and H.StJ.L. Winterbotham, *Text book of topographical and geographical surveying* (3rd ed., London, 1925), plates II–IV, pp. 16–18.

9 Seymour, *History of the Ordnance Survey*, pp. 16–17, 37.

10 Matthew H. Edney, *Mapping an empire: the geographical construction of British India, 1765–1843* (Chicago, 1997), pp. 237–8.

11 H.R. Calvert, *Astronomy: globes, orreries and other models* (London, 1967), p. 19.

12 W.E. May, *A history of marine navigation* (Henley-on-Thames, 1973), pp. 7–8.

13 Charles H. Cotter, *The astronomical and mathematical foundations of geography* (London, 1966), pp. 108–9.

14 J. Lennart Berggren and Alexander Jones (eds), *Ptolemy's* Geography: *an annotated translation of the theoretical chapters* (Princeton, NJ, 2000), p. 10; O.A.W. Dilke, *Greek and Roman maps* (London, 1985), p. 76; O.A.W. Dilke, 'The culmination of Greek cartography in Ptolemy' in J.B. Harley and David Woodward (eds), *The history of cartography, volume one: cartography in prehistoric, ancient, and medieval Europe and the Mediterranean* (Chicago, 1987), pp. 182–3; above, Figs 2.10, 2.11; below, Fig. 8.14.

15 Berggren and Jones (eds), *Ptolemy's* Geography, pp. 84–5.

16 Gerard Mercator, Henry Hondius and Joannes Janssonius (ed. R.A. Skelton), *Atlas or a geographicke description of the world*, i (Amsterdam, 1636, 1968), p. 34.

17 Abraham Ortelius (ed. R.A. Skelton), *The theatre of the whole world, London, 1606* (Amsterdam, 1968), pp. 3–4: *Asiae nova descriptio*.

18 Horace Edward Jones (trans.), *The geography of Strabo*, viii (London, 1967), p. 129.

19 Egnatio Danti, *Le scienze matematiche ridotte in tavole* (Bologna, 1577), quoted in Thomas Frangenberg, 'Chorographies of Florence: the use of city views and city plans in the sixteenth century', *Imago Mundi*, xlvi (1994), p. 58.

20 Richard Carnac Temple (ed.), *The travels of Peter Mundy, in Europe and Asia, 1608–1667*, iii (London, 1919), p. 347.

21 Edward Grey (ed.), *The travels of Pietro della Valle in India* (London, 1892), pp. 11–12: 1623; Bertha S. Phillpotts (ed.), *The life of the Icelander Jón Ólafsson, traveller to India*, ii (London, 1932), pp. 58–9: 1623.

22 W.S.W. Vaux (ed.), *The world encompassed by Sir Francis Drake, being his next voyage to that to Nombre de Dios* (London, 1854), pp. 78–9.

23 Pasfield Oliver (ed.), *The voyage of François Leguat of Bresse to Rodriguez, Mauritius, Java, and the Cape of Good Hope*, i (London, 1891), p. 21: 1690; J.C. Beaglehole (ed.), *The journals of Captain James*

Cook on his voyages of discovery, i: the voyage of the Endeavour, *1768–1771* (London, 1955), p. 12; G.S. Ritchie, *The Admiralty chart: British naval hydrography in the nineteenth century* (London, 1967), p. 73.

24 Ernst C. Abbe and Frank J. Gillis, 'Henry Hudson and the early exploration and mapping of Hudson Bay, 1610–1631' in John Parker (ed.), *Merchants and scholars: essays in the history of exploration and trade collected in memory of James Ford Bell* (Minneapolis, 1965), p. 96.

25 E.G.R. Taylor, *The mathematical practitioners of Hanoverian England, 1714–1840* (Cambridge, 1966), p. 86: *c.*1818; A. Chapman, 'The accuracy of angular measuring instruments used in astronomy between 1500 and 1850', *Journal for the History of Astronomy*, xiv (1983), pp. 133–7.

26 Derek Howse, *Nevil Maskelyne: the seaman's astronomer* (Cambridge, 1989), pp. 129–41.

27 Seymour, *History of the Ordnance Survey*, pp. 40, 143–5; Clements R. Markham, *A memoir on the Indian surveys* (2nd ed., London, 1878; rpr. Amsterdam, 1968), pp. 139–42; A.R. Hinks, *Maps and survey* (5th ed., Cambridge, 1944), pp. 216–30.

28 A.R. H[inks], 'Nautical time and civil date', *Geographical Journal*, lxxxvi (1935), pp. 153–7.

29 William Brooks Greenlee (ed.), *The voyage of Pedro Alvares Cabral to Brazil and India* (London, 1938), p. 155.

30 George T. Staunton (ed.), *The history of the great and mighty kingdom of China and the situation thereof compiled by the Padre Juan Gonzales de Mendoza*, ii (London, 1854), pp. 253–4.

31 William Foster (ed.), *The voyage of Thomas Best to the East Indies, 1612–14* (London, 1934), pp. 86–7; William Foster (ed.), *The voyage of Nicholas Downton to the East Indies, 1614–15* (London, 1939), pp. 46–7; Samuel Purchas, *Hakluytus Posthumus; or, Purchas his pilgrimes, contayning a history of the world in sea voyages and lande travells by Englishmen and others*, iv (Glasgow, 1905), p. 567: Martin Pring, 1615.

32 Ursula S. Lamb, 'The Spanish cosmographic juntas of the sixteenth century', *Terrae Incognitae*, vi (1974), pp. 51–64.

33 Lloyd A. Brown, *The story of maps* (New York, 1949, 1979), p. 209; Derek Howse, *Greenwich time and the discovery of the longitude* (Oxford, 1980), p.12; W.F.J. Mörzer Bruyns, 'Longitude in the context of navigation', p. 44 and A.J. Turner, 'In the wake of the act, but mainly before', p. 122, both in William J.H. Andrewes (ed.), *The quest for longitude* (Cambridge, MA, 1996).

34 Howse, *Greenwich time*, p. 79.

35 Richard Hakluyt (ed. D.B. Quinn and R.A. Skelton), *The principall navigations voiages and discoveries of the English nation*, ii (Cambridge, 1965), p. 610.

36 C.E. Heidenreich, 'Explorations and mapping of Samuel de Champlain, 1603–1632', *Cartographica*, Monograph 17 (1976), pp. 55–67.

37 James Boswell, *The life of Samuel Johnson, LL.D.* (Everyman's Library, London, 1949), i, pp. 183–4; Albert J. Kuhn, 'Dr Johnson, Zachariah Williams, and the eighteenth-century search for the longitude', *Modern Philology*, lxxxii (1984), pp. 40–52; E.G.R. Taylor, 'A reward for the longitude', *Mariner's Mirror*, xlv (1959), pp. 59–66.

38 Norman J.W. Thrower, 'Edmond Halley and thematic geo-cartography' in Norman J.W. Thrower (ed.), *The compleat plattmaker: essays on chart, map, and globemaking in England in the seventeenth and eighteenth centuries* (Los Angeles, 1978), pp. 209–19; E.G.R. Taylor, 'The early navigator', *Geographical Journal*, cxiii (1949), p. 61.

39 Howse, *Greenwich time*, pp. 55, 56; Owen Gingerich, 'Cranks and opportunists: "nutty" solutions to the longitude problem' in Andrewes, *Quest for longitude*, pp. 134–48.

40 T.R. Robinson, 'Notice of determination of the arc of longitude between the observatories of Armagh and Dublin', British Association report, ninth meeting (1840): *Reports on the state of science*, pp. 19–22; V.L. Bosazza and C.G.C. Martin, 'Geographical methods of exploration surveys in the nineteenth century' in C.G.C. Martin (ed.), *Maps and surveys of Malawi* (Rotterdam, 1980), pp. 28–9.

41 Leo Bagrow (ed. Henry W. Castner), *A history of Russian cartography up to 1800* (Wolfe Island, Ontario, 1975), app. 3, p. 263.

42 Robert W. Karrow Jr, *Mapmakers of the sixteenth century and their maps: bio-bibliographies of the cartographers of Abraham Ortelius, 1570* (Winnetka, IL, 1993), p. 207.

43 Clements R. Markham (ed.), *Tractatus de globis et eorum usu; a treatise descriptive of the globes constructed by Emery Molyneux, and published in 1592, by Robert Hues* (London, 1889), p. 97.

44 Dava Sobel, *Longitude: the true story of a lone genius who solved the greatest scientific problem of his time* (London, 1996); Rupert T. Gould, 'John Harrison and his timekeepers', *Mariner's Mirror*, xxi, 2 (1935), pp. 1–24.

45 Berggren and Jones, *Ptolemy's* Geography, pp. 29–30, 63.

46 Observatoire de Paris, *Jean Picard et la mesure sur la terre* (Paris, 1982).

47 Andrew A. Lipscomb (ed.), *The writings of Thomas Jefferson*, xiv (Washington, 1905), p. 477; Edney, *Mapping an empire*, pp. 87–91.

48 Derek Howse, 'The lunar distance method of measuring longitude' in Andrewes, *Quest for longitude*, pp. 149–62.

49 Miller Christy (ed.), *The voyages of Captain Luke Foxe of Hull, and Captain Thomas James of Bristol … in 1631–32* (London, 1894), pp. 211–12: voyage of Robert Bilot, 1615.

50 Charles Coulston Gillispie (ed.), *Dictionary of scientific biography*, xiv (New York, 1976), p. 275; Karrow, *Mapmakers of the sixteenth century*, p. 186: Oronce Finé, 1544.

51 Clements R. Markham (ed.), *The voyages of William Baffin, 1612–1622* (London, 1881), pp. 21–2.

52 D.H. Sadler (ed.), *Man is not lost: a record of two hundred years of astronomical navigation with the Nautical Almanac, 1767–1967* (London, 1968); Eric G. Forbes, *The birth of scientific navigation: the solving in the 18th century of the problem of finding longitude at sea* (Greenwich, 1974).

53 Randolph Cock, 'Precursors of Cook: the voyages of the *Dolphin*, 1764–8', *Mariner's Mirror*, lxxxv (1999), pp. 45–6.

54 Sadler, *Man is not lost.*

55 Howse, *Nevil Maskelyne*, pp. 93–4.

56 E.G.R. Taylor, *Navigation in the days of Captain Cook* (Greenwich, 1975), p. 7.

57 D.W. Waters, 'Seamen, scientists, historians, and strategy', *British Journal for the History of Science*, xiii (1980), p. 191.

58 Howse, 'The lunar distance method', p. 156.

59 William A. Spray, 'British surveys in the Chagos archipelago and attempts to form a settlement at Diego Garcia in the late eighteenth century', *Mariner's Mirror*, lvi (1970), pp. 69, 73; John Dunmore (ed.), *The journal of Jean-François de Galaup de la Pérouse, 1785–1788* (London, 1995), passim.

60 Andrew David, 'Vancouver's survey methods and surveys' in Robin Fisher and Hugh Johnston (eds), *From maps to metaphors: the Pacific world of George Vancouver* (Vancouver, 1993), p. 66.

61 Ritchie, *Admiralty chart*, p. 218.

62 Marcel Watelet (ed.), *Gérard Mercator cosmographe: le temps et l'espace* (Antwerp, 1994), pp. 251–5: 1541; Edward Luther Stevenson, *Terrestrial and celestial globes, their history and construction*, i (New Haven, CT, 1921), pp. 128–9.

63 Geoff Armitage, *The shadow of the moon: British solar eclipse mapping in the eighteenth century* (Tring, 1997).

64 Rodney Shirley, 'A neglected map of the world', *The Map Collector*, xlvi (1989), pp. 34–8.

65 Rodney W. Shirley, *The mapping of the world: early printed world maps, 1472–1700* (London, 1983), plate 15, p. 4; plate 63, pp. 79, 80–1.

66 Ibid., plate 247, p. 348: Isaac Habrecht, Strasbourg, 1628; Max Leinekugel le Cocq, *Premieres images de la terre* (Paris, 1977), pp. 70–1: Jehan Cossin, 1570.

67 W.P. Cumming, R.A. Skelton and D.B. Quinn, *The discovery of North America* (London, 1971), pp. 18–19: Guillaume le Testu, world map, 1566.

Chapter 6: A survey in its literal sense

1 J. Lennart Berggren and Alexander Jones (eds), *Ptolemy's* Geography*: an annotated translation of the theoretical chapters* (Princeton, NJ, 2000), p. 63.

2 E.H. Bunbury, *A history of ancient geography*, ii (1883; New York, 1959), p. 555.

3   Royal Institution of Chartered Surveyors, *Five centuries of maps and map-making: an exhibition* (Westminster, 1953).

4   J.C. Beaglehole, *The life of Captain James Cook* (London, 1974), p. 634.

5   Richard Benese, *This boke sheweth the maner of measurynge of all maner of lande* ... (Southwark, 1537); E.G.R. Taylor, 'The surveyor', *Economic History Review*, xvii (1947), pp. 129–30.

6   O.A.W. Dilke, *Mathematics and measurement* (London, 1987), pp. 8, 18.

7   O.A.W. Dilke, *The Roman land surveyors: an introduction to the agrimensores* (Newton Abbot, 1971), pp. 201–11; Ir. C. Koeman, *Geschiedenis van de kartografie van Nederland* (Canaletto-Alphen aan den Rijn, 1983), pp. 135–48; Walter W. Ristow, 'Dutch polder maps', *Quarterly Journal of the Library of Congress*, xxxi, 3 (1974), pp. 136–49; Amelia Clewly Ford, *Colonial precedents of our national land system as it existed in 1800* (1910; Philadelphia, 1976); William Pattison, *Beginnings of the American rectangular land survey system, 1784–1800* (Chicago, 1957); Hildegard Binder Johnson, *Order upon the land: the U.S. rectangular land survey and the upper Mississippi country* (New York, 1976); Don W. Thomson, 'The history of surveying and mapping in Canada', *Surveying and Mapping*, xxvii, 1 (1967), pp. 69, 70–1; Daniel Hopkins, 'An extraordinary eighteenth-century map of the Danish sugar plantation island St Croix', *Imago Mundi*, xli (1989), pp. 44–58.

8   Johnson, *Order upon the land*, p. 58; James Corner and Alex S. MacLean, *Taking measures: across the American landscape* (New Haven, CT, 1996), p. 56.

9   Eduard Imhof, 'Beiträge zur Geschichte der topographischen Kartographie', *International Yearbook of Cartography*, iv (1964), p. 132; Robert W. Karrow Jr, *Mapmakers of the sixteenth century and their maps; bio-bibliographies of the cartographers of Abraham Ortelius* (Winnetka, IL, 1993), p. 322.

10 Aaron Rathborne, *The surveyor in foure bookes* (London, 1616), p. 152.

11 Samuel Wyld, *The practical surveyor, or the art of land-measuring, made easy* (London, 1725), p. 84.

12 Imhof, 'Beiträge zur Geschichte der topographischen Kartographie', p. 133; William Leybourn, *The compleat surveyor* (London, 1679), p. 257.

13 J.H. Andrews, *A paper landscape: the Ordnance Survey in nineteenth-century Ireland* (Oxford, 1975), pp. 60–1, 82.

14 L. Gallois, 'Les origines de la carte de France: la carte d'Oronce Finé', *Bulletin de Géographie Historique*, 1891, pp. 18–34; F. de Dainville, 'How did Oronce Fine draw his large map of France', *Imago Mundi*, xxiv (1970), p. 53; Numa Broc, 'Quelle est la plus ancienne carte "moderne" de la France', *Annales de Géographie*, dxiii (1983), pp. 513–30.

15 Berggren and Jones, *Ptolemy's Geography*, pp. 96–107.

16 William Cuningham, *The cosmographicall glasse* (London, 1559).

17 Charles Coulston Gillispie (ed.), *Dictionary of scientific biography*, viii (New York, 1973), pp. 135–6; J.Robson, *A treatise on geodesic operations, or county surveying, land surveying and levelling* (Durham, 1821), pp. 135–6.

18 William Bowie, 'Long lines of triangulation', *Geographical Review*, xvi (1926), pp. 638–9.

19 J.B. Harley and P. Laxton, *A survey of the county palatine of Chester P.P. Burdett 1777* (Liverpool, 1974), introduction, pp. 15–16; T. Pilkington White, *The Ordnance Survey of the United Kingdom* (1886; Amsterdam, 1975), p. 34.

20 N.D. Haasbroek, *Gemma Frisius, Tycho Brahe and Snellius and their triangulations* (Delft, 1968), p. 14.

21 E.G.R. Taylor, 'The earliest account of triangulation', *Scottish Geographical Magazine*, xliii (1927), pp. 341–5; A. Pogo, 'Gemma Frisius, his method of determining differences of longitude by transporting timepieces (1530), and his treatise on triangulation (1533)', *Isis*, xxii (1934–5), pp. 469–85; Haasbroek, *Gemma Frisius, Tycho Brahe and Snellius*, p. 11.

22 Imhof, 'Beiträge zur Geschichte der topographischen Kartographie', pp. 133–6.

23 A.W. Richeson, *English land measuring to 1800: instruments and practices* (Cambridge, MA, 1966), pp. 47–8.

24 British Library, Sloane MS 3651; E.G.R. Taylor (ed.), *A regiment for the sea and other writings on navigation by William Bourne* (Cambridge, 1963), pp. xxv–xxvi, Fig. 5.

25   R.T. Gunther (ed.), *The theodelitus and topographical instrument of Leonard Digges ... described by his son Thomas Digges in 1571, reprinted from* Longimetra, *the fyrst booke of* Pantometria (Oxford, 1927), pp. 27–34.

26   Haasbroek, *Gemma Frisius, Tycho Brahe and Snellius*, p. 12.

27   William Ravenhill, *Christopher Saxton's 16th century maps: the counties of England and Wales* (Shrewsbury, 1992), pp. 22–3.

28   J.H. Andrews, 'The Irish surveys of Robert Lythe', *Imago Mundi*, xix (1965), pp. 22–31.

29   J.B. Harley, 'The Society of Arts and the survey of English counties, 1759–1809', *Journal of the Royal Society of Arts*, cxii, 5089 (1963), p. 45; cxii, 5090 (1964), p. 120.

30   J.H. Andrews, 'An Ordnance Survey playlet of 1842', *Sheetlines*, xxxiii (1992), p. 10.

31   Frank Debenham, *Map making* (2nd ed., London, 1940), pp. 127–86.

32   A. Sarah Bendall, *Maps, land and society: a history, with a carto-bibliography of Cambridgeshire estate maps, c.1600–1836* (Cambridge, 1992), pp. 131–2.

33   J. Eyre, *The exact surveyor: or, The whole art of surveying of land* (London, 1654), p. 5; John Wing, *Geodaetes practicus redivivus. The art of surveying: formerly publish'd by Vincent Wing, math. now much augmented and improved* (London, 1700), pp. 128–9; Thomas Breaks, *A complete system of land surveying both in theory and practice* (Newcastle and London, 1771), p. 292.

34   Sarah Bendall, 'Draft town maps for John Speed's *Theatre of the empire of Great Britaine*', *Imago Mundi*, liv (2002), p. 37.

35   Charles Close (ed. J.B. Harley), *The early years of the Ordnance Survey* (1926; Newton Abbot, 1969), p. 38.

36   Imhof, 'Beiträge zur Geschichte der topographischen Kartographie', p. 140.

37   Helmet Häuser, 'Zum kartographischen Werk des Mainzer Kupferstechers und Ingenieurs Nikolaus Person' in Elisabeth Geck and Guido Pressler (eds), *Festschrift für Josef Benzing zum sechzigsten Geburtstag* (Wiesbaden, 1964), pp. 170–86.

38   J.H. Andrews, 'Sir William Petty: a tercentenary reassessment', *The Map Collector*, xli (1987), p. 36.

39   Imhof, 'Beiträge zur Geschichte der topographischen Kartographie', p. 135.

40   John A. Pinto, 'Origins and development of the ichnographic city plan', *Journal of the Society of Architectural Historians*, xxxv, 1 (1976), p. 41; Hilary Ballon and David Friedman, 'Portraying the city in early modern Europe: measurement, representation, and planning' in David Woodward (ed.), *The history of cartography, volume three, part one: cartography in the European renaissance* (Chicago, 2007), pp. 682–3.

41   J.H. Andrews, *Plantation acres: an historical study of the Irish land surveyor and his maps* (Belfast, 1985), pp. 36, 60–1, 65–6; B.W. Higman, *Jamaica surveyed* (Kingston, 2001), pp. 51–3.

42   Richard Norwood, *The sea-man's practice, contayning a fundamentall problem in navigation, experimentally verified* (London, 1637).

43   Uuno Varjo, 'Observations on the mapping of the village of Nummi in the parish of Nousiainen in 1786', *Fennia*, clxiv, 1 (1986), p. 82.

44   Close, *Early years of the Ordnance Survey*, p. 120.

45   Debenham, *Map making*, pp. 90–1.

46   Joan Gadol, *Leon Battista Alberti: universal man of the early renaissance* (Chicago, 1969), pp. 167–92; Pinto, 'Origins and development of the ichnographic city plan', pp. 36–8.

47   Martin Kemp et al., *Leonardo da Vinci* (London, 1989), p. 176; Martin Clayton, *Leonardo da Vinci: a curious vision* (London, 1996), pp. 90–1; Samuel Y. Edgerton Jr, 'From mental matrix to *mappamundi* to Christian empire: the heritage of Ptolemaic cartography in the renaissance' in David Woodward (ed.), *Art and cartography: six historical essays* (Chicago, 1987), pp. 39–40.

48   Herbert George Fordham, *Some notable surveyors & map-makers of the sixteenth, seventeenth, and eighteenth centuries and their work* (Cambridge, 1929), p. 47.

49   B. Szczesniak, 'A note on the studies of longitudes made by M. Martini, A. Kircher, and J.N. Delisle from the observations of travellers to the Far East', *Imago Mundi*, xv (1961), pp. 89–93.

50   Andrew A. Lipscomb (ed.), *The writings of Thomas Jefferson*, xiv (Washington, 1905), pp. 471–85.

51 Nevil Maskelyne, 'Concerning the latitude and longitude of the Royal Observatory at Greenwich; with remarks on a memorial of the late M. Cassini de Thury', *Philosophical Transactions of the Royal Society of London*, lxxvii, pt 1 (1787), pp. 151–87; William Roy, 'An account of the mode proposed to be followed in determining the relative situation of the Royal Observatories of Greenwich and Paris', *Philosophical Transactions of the Royal Society of London*, lxxvii, pt 1 (1787), pp. 188–226; William Roy, 'An account of the trigonometrical operation, whereby the distance between the meridians of the Royal Observatories of Greenwich and Paris has been determined', *Philosophical Transactions of the Royal Society of London*, lxxx, pt 1 (1790), pp. 111–270.

52 François de Dainville, 'La levée d'une carte en Languedoc à l'entour de 1730' in *La cartographie reflet de l'histoire* (Paris, 1986), p. 369; Walter W. Ristow, 'The French-Smith map and gazetteer of New York state', *Quarterly Journal of the Library of Congress*, xxxv, 1 (1979), p. 69, quoting E.B. Hunt, 1851.

53 Breaks, *Complete system of land-surveying*, p. 461.

54 Stephen M. Stigler, *The history of statistics: the measurement of uncertainty before 1900* (Cambridge, MA, 1986), pp. 12–15, 55–61, 145–6.

55 Gillispie, *Dictionary of scientific biography*, ii (1970), p. 102; v (1972), p. 304; viii (1973), p. 137.

56 Matthew H. Edney, *Mapping an empire: the geographical construction of British India, 1765–1843* (Chicago, 1997), p. 29; Ferdinand Hassler, *Survey of the coast of the United States: further rectification of facts alleged in the discussion of Congress, in December, 1842* (Washington), p. 5.

57 Leo Bagrow, 'Ivan Kirilov, compiler of the first Russian atlas, 1689–1737', *Imago Mundi*, ii (1937), pp. 78–82.

58 J.B. Harley and Yolande O'Donoghue, *The old series Ordnance Survey maps of England and Wales, scale 1 inch to 1 mile*, i (Lympne, 1975), p. xi.

59 A. Arrowsmith, *Outlines of the physical and political divisions of South America* (London, 1811, additions to 1814).

60 W.A. Seymour (ed.), *A history of the Ordnance Survey* (Folkestone, 1980), p. 143.

Chapter 7: Laid down in our drafts

1 R.H. Major (ed.), *Early voyages to Terra Australis, now called Australia* (London, 1859), p. 97.

2 Ivor Waters, *The town of Chepstow, i, Riverside* (Chepstow, Mon., 1972), p. 8: plan of Chepstow, 1835.

3 Samuel Wyld, *The practical surveyor, or, The art of land-surveying made easy* (London, 1725), p. 109.

4 R. Herbin and A. Pebereau, *Le cadastre français* (Paris, 1953), p. 19: Bréveaux, 1803.

5 H.W. Dickinson, 'A brief history of draughtsmen's instruments', *Transactions of the Newcomen Society*, xxvii (1949–50, 1950–1), pp. 73–84.

6 D.W. Waters, *The art of navigation in Elizabethan and early Stuart times* (London, 1958), p. 212.

7 William Emerson, *The art of surveying or measuring land* (London, 1770), pp. 74–5.

8 Waters, *Art of navigation*, pp. 63–4.

9 Wyld, *Practical surveyor*, p. 48.

10 A.H.W. Robinson, *Marine cartography in Britain: a history of the sea chart to 1855* (Leicester, 1962), pp. 64–70; David Baxandall, 'The inventor of the station pointer', *Empire Survey Review*, iii (1933), pp. 19–21.

11 George Adams, *Astronomical and geographical essays* (London, 1789), containing *Catalogue of mathematical and philosophical instruments*.

12 George Adams the younger, *Geometrical and graphical essays* (London, 1797), p. 130.

13 Paul Ballard, 'Cartographic drawing instruments – the eighteenth and nineteenth centuries', *Cartography* (Australian Institute of Cartographers), viii, 3 (1974), pp. 140–6.

14 T. Baker, *Elementary treatise on land and engineering surveying* (London, 1857), pp. 158–9.

15 Maya Hambly, *Drawing instruments: their history, purpose and use for architectural drawings* (London, 1982), p. 32; T. Baker (ed. F.E. Dixon), *A rudimentary treatise on land and engineering surveying* (London, 1909), p. 176.

16 Hambly, *Drawing instruments*, p. 32; Arthur H. Robinson, *Elements of cartography* (New York, 1953), p. 82.

17 M.L. Ryder, 'Parchment', *History Today*, xxiv, 10 (1974), pp. 716–20; Public Record Office, *An introduction to parchment* (London, 1996).

18 Lloyd A. Brown, *The story of maps* (1949; New York, 1979), p. 161; Marc Duranthon, *La carte de France: son histoire, 1678–1978* (Paris, 1978), p. 34.

19 Wyld, *Practical surveyor*, p. 154; Joseph Lindley, *Memoir of a map of the county of Surrey, from a survey made in the years 1789 and 1790* (London, [1793]), p. 71.

20 William Hume, *Geodaesia accurata: or, Surveying made easy, by the chain only* (London, 1754), p. 441.

21 T.A. Larcom (ed.), *The history of the survey of Ireland commonly called the Down Survey* (Dublin, 1851), p. 324.

22 Emerson, *Art of surveying,* pp. 74–5.

23 J.E. Portlock, *Memoir of the life of Major General Colby* (London, 1869), p. 232; William Edgeworth's diary, 19 January 1822 (Naples), National Library of Ireland, MS 14124, p. 67.

24 Frank Debenham, *Exercises in cartography* (London, 1937), p. 12.

25 Arthur Hopton, *Speculum topographicum; or, The topographicall glasses* (London, 1611), p. 44.

26 John Holwell, *A sure guide to the practical surveyor* (London, 1678), pp. 54–5.

27 Wyld, *Practical surveyor*, p. 69.

28 Richard Norwood, *The seaman's practice* (London, 1637).

29 Adams, *Geometrical and graphical essays*, p. 333.

30 J.B. Mitchell, 'The Matthew Paris maps', *Geographical Journal*, lxxxi (1933), p. 28.

31 Juergen Schulz, 'Jacopo de' Barbari's view of Venice: map making, city views, and moralized geography before the year 1500', *The Art Bulletin*, lx (1978), p. 425; Rodney W. Shirley, *The mapping of the world: early printed world maps, 1472–1700* (London, 1983), p. 28; Siegmund Günther, *Peter und Philipp Apian, zwei deutsche Mathematiker und Kartographen* (Prague, 1882; Amsterdam, 1967), p. 120.

32 Josef W. Konvitz, *Cartography in France, 1660–1848: science, engineering, and statecraft* (Chicago, 1987), p. 25.

33 Ralph Agas, *A preparative to plotting of landes and tenements for surveigh* (London, 1596), p. 5.

34 Emerson, *Art of surveying*, pp. 74–5.

35 Monique Pelletier, *La carte de Cassini* (Paris, 1990), p. 113.

36 Evelyn Edson, *Mapping time and space: how medieval mapmakers viewed their world* (London, 1997), p. 15; http://www.maphist@geo.uu.nl, 19–21 April 2002, 'When was map orientation changed to north at top'; 20–3 April 2002, 'Map orientation'; 22 April 2002, 'North at top'; 5–7 August 2003, 'Why is north up?'; 24, 27 December 2003, 'North point convention'.

37 J. Lennart Berggren and Alexander Jones (eds), *Ptolemy's* Geography: *an annotated translation of the theoretical chapters* (Princeton, NJ, 2000), p. 94.

38 Robert W. Karrow Jr, *Mapmakers of the sixteenth century and their maps: bio-bibliographies of the cartographers of Abraham Ortelius, 1570* (Winnetka, IL, 1993), p. 321.

39 George Atwell, *The faithful surveyour* (Cambridge, 1662), p. 10.

40 R.A. Skelton, with A.D. Baxter and S.T.M. Newman (ed. J.B. Harley), *Saxton's survey of England and Wales, with a facsimile of Saxton's wall map of 1583* (Amsterdam, 1974), p. 10.

41 Gerard Mercator, Henry Hondius and Joannes Janssonius (ed. R.A. Skelton), *Atlas or a geographicke description of the world*, ii (Amsterdam, 1636, 1968), p. 301.

42 John Norden, *Preparative to his Speculum Britanniae* (1596), pp. 8–9.

43 P.D.A. Harvey and Harry Thorpe, *The printed maps of Warwickshire, 1576–1900* (Warwick, 1959), p. 22; Oxford Historical Society (ed.), *Remarks and collections of Thomas Hearne*, vii (Oxford, 1906), p. 284.

44 'A survey of certain estates belonging to Benjamin Hyett Esq.', by John Merrett, 1780, Gloucestershire Record Office, D6 E4.

45 H.G. Fordham, *Maps, their history, characteristics and uses* (Cambridge, 1921), p. 41.

46 G.A. Hayes-McCoy, *Ulster and other Irish maps, c.1600* (Dublin, 1964), pp. 18–19.

47 H. Threlfall, *A textbook on surveying and levelling* (2nd ed., London, 1929), pp. 194–5; John R. Hébert, 'Vicente Sebastián Pintado, surveyor general of Spanish West Florida, 1805–17: the man and his maps', *Imago Mundi*, xxxix (1987), pp. 56–69.

48 A.H.W. Robinson, *Marine cartography in Britain: a history of the sea chart to 1855* (Leicester, 1962), plate 32: 1748.

49 Angela Fordham, *Maps of the Falkland Islands*, Map Collectors' series, xi (London, 1964), plate IXb.

50 Philippa Glanville, *London in maps* (London, 1972), plates 26, 27, pp. 122–5.

51 R.A. Skelton, 'Cartography' in Charles Singer (ed.), *A history of technology*, iv (Oxford, 1958), p. 600.

52 Duranthon, *Carte de France*, p. 31.

53 Edward Lynam, 'Woutneel's map of the British Isles, 1603', *Geographical Journal*, lxxxii (1933), p. 536; Berggren and Jones, *Ptolemy's Geography*, p.14; Conrad E. Heidenreich, 'Measures of distance employed on 17th and early 18th century maps of Canada', *Canadian Cartographer*, xii (1975), p. 123.

54 Waters, *Art of navigation*, p. 65.

55 William Bedwell, *A briefe description of the town of Tottenham High Crosse in Middlesex* (London, 1631), ch. 3.

56 Philip Grierson, *English linear measures, an essay in origins* (Reading, 1972), p. 25: 1593.

57 Norwood, *Seaman's practice*; James R. Smith, *From plane to spheroid: determining the figure of the earth from 3000 BC to the 18th century Lapland and Peruvian survey expeditions* (Rancho Cordova, CA, 1986), p. 68.

58 W. Ravenhill, *Two hundred and fifty years of map-making in the county of Surrey … 1597–1823* (Lympne, 1974); W. Ravenhill, 'As to its position in respect to the heavens', *Imago Mundi*, xxviii (1976), p. 91; J.H. Andrews, Introduction to William Petty, *Hiberniae delineatio* (Shannon, 1969), p. 18.

59 E.G.R. Taylor, 'A regional map of the early 16th century', *Geographical Journal*, lxxi (1928), p. 476: 1513; F.J. North, *Humphrey Lhuyd's maps of England and of Wales* (Cardiff, 1937), p. 32: 1573.

60 J.R. Smith, 'The pear-shaped earth', *Geographical Magazine*, lviii, 11 (1986), pp. 572–7; James R. Smith, *Introduction to geodesy: the history and concepts of modern geodesy* (New York, 1997); Ken Alder, *The measure of things: the seven-year odyssey that transformed the world* (London, 2002).

61 D.G. Moir in Royal Scottish Geographical Society, *The early maps of Scotland to 1850* (Edinburgh, 1973), p. 87.

62 Mary Sponberg Pedley, *Bel et utile: the work of the Robert de Vaugondy family of mapmakers* (Tring, 1992), pp. 109–12: Didier Robert de Vaugondy, 1775.

63 'An advise for the use of maps', Mercator-Hondius-Janssonius, *Atlas*, ii, p. 275.

64 Alfred E. Lemmon, John T. Magill and Jason R. Wiese (eds), *Charting Louisiana: five hundred years of maps* (New Orleans, 2003), p. 73: Jacques Nicolas Bellin, 'Cours du fleuve Saint Louis' [1763].

65 George H.T. Kimble (ed.), *Esmeraldo de situ orbis, by Duarte Pacheo Pereira* (London, 1937), p. 26: 1505–8; Pierre du Val, *Carte des Indes Orientales*, 1677, reproduced in Robert Clancy and Alan Richardson, *So they came south* (Silverwater, New South Wales, 1988), pp. 106–7.

66 W.G. Perrin, 'The prime meridian', *Mariner's Mirror*, xiii (1927), pp. 109–24.

67 F.J. North, *Maps: their history and uses, with special reference to Wales* (Cardiff, 1933), p. 8: Martin Cortes, 1556.

68 Berggren and Jones, *Ptolemy's Geography*, p. 71.

69 Lucie Lagarde, 'Le passage du nord-ouest et la Mer de l'Ouest dans la cartographie française du 18ᵉ siècle, contribution à l'étude de l'oeuvre des Delisle et Buache', *Imago Mundi*, xli (1989), Fig. 5, p. 32: Guillaume Delisle, map of America, 1722.

70 Christian Sandler, *Die Reformation der Kartographie um 1700* (Munich, 1905), p. 19; *A new introduction to the knowledge and use of maps; rendered easy and familiar to any capacity* (3rd ed., London, 1774), p. 27; Perrin, 'The prime meridian', p. 120.

71 Mireille Pastoureau, *Les atlas français xviᵉ–xviiᵉ siècles: répertoire bibliographique et étude* (Paris, 1984), Fig. 123, *Côtes de Bretagne*, from *Neptune françois*, 1773.

72 Robert Sayer, *West Indian atlas* (London, 1775), p. ii.

73 [Herman Moll], *The compleat geographer: or, The chorography and topography of all the known parts of the earth* (London, 1723), p. x.

74 Reproduced in Thomas R. Adams, 'The map treasures of the John Carter Brown Library', *The Map Collector*, xvi (1981), p. 5.

75 Shirley, *The mapping of the world*, plates 59, 61, 62, pp. 70–8.

76 Edward Harrison, *Idea longitudinis* (London, 1696), p. 4.

77 Derek Howse, *Greenwich time and the discovery of the longitude* (Oxford, 1980), pp. 138–51.

Chapter 8: So strangely distorted

1 D.H. Maling, *Coordinate systems and map projections* (London, 1973), p. 83.

2 Hildegard Binder Johnson, *Carta marina: world geography in Strassburg, 1525* (Westport, CT, 1963), p. 59.

3 David Woodward, 'Roger Bacon's terrestrial coordinate system', *Annals of the Association of American Geographers*, lxxx (1990), p. 116.

4 Maling, *Coordinate systems and map projections*, p. 203.

5 Information kindly supplied by Iain Donovan.

6 Brian Adams, 'Projections of the Ordnance Survey ten-mile maps' in Roger Hellyer, *The 'ten-mile' maps of the Ordnance Surveys* (London, 1992), p. 178.

7 R.W. Shirley, *Early printed maps of the British Isles, 1477–1650: a bibliography* (London, 1973), plate 13, pp. 36–7; plates 15, 16, 17, 19, pp. 66–9.

8 Andrew Sharp, *The voyages of Abel Janszoon Tasman* (Oxford, 1968), pp. 318–19; Günter Schilder, *Australia unveiled: the share of the Dutch navigators in the discovery of Australia* (Amsterdam, 1976), pp. 148, 342.

9 Rodney W. Shirley, *The mapping of the world: early printed world maps, 1472–1700* (London, 1983), plate 162, p. 220: Gennaro Picicaro, 1597.

10 Johannes Keuning, 'The history of geographical map projections until 1600', *Imago Mundi*, xii (1955), pp. 5–9.

11 Richard Uhden, 'An equidistant and a trapezoidal projection of the early fifteenth century', *Imago Mundi*, ii (1937), pp. 8–9; François de Dainville, *Le langage des géographes* (Paris, 1964), pp. 34–6; Rodney W. Shirley, 'All the world within a circle: some unusual world maps on a single polar projection', *The Map Collector*, x (1980), pp. 2–11.

12 Bernhard Varenius, *Cosmography and geography in two parts* (London, 1683), pp. 330–1; A.E. Nordenskiöld, *Facsimile atlas to the early history of cartography* (1889; New York, 1973), p. 92.

13 John P. Snyder, *Flattening the earth: two thousand years of map projections* (Chicago, 1993), pp. 17–18.

14 Robert W. Karrow Jr, 'The cartographic collections of the Newberry Library', *The Map Collector*, xxxii (1985), p. 15: Franz Ritter, 1610.

15 Keuning, 'The history of geographical map projections', pp. 7–8.

16 C. Broekema, 'Notes on the history of map projections' in Ton Croiset van Uchelen, Koert van der Horst and Günter Schilder (eds), *Theatrum orbis librorum: liber amicorum presented to Nico Israel on the occasion of his seventieth birthday* (Utrecht, 1989), p. 7.

17 Snyder, *Flattening the earth*, p. 20.

18 Geoffrey King, *Miniature antique maps* (Tring, 1996), p. 135: James Moxon, 1691.

19 Arthur R. Hinks, *Map projections* (Cambridge, 1912), p. 70.

20 Woodward, 'Roger Bacon's terrestrial coordinate system', pp. 112–15.

21 Keuning, 'History of geographical map projections', p. 20; Walters Art Gallery, *The world encompassed* (Baltimore, 1952), plate XXXV: Michael Tramezinus, western hemisphere, Rome, 1554.

22 John A. Wolter and Ronald E. Grim (eds), *Images of the world: the atlas through history* (Washington, 1997), p. 15: Jean Rotz, *Boke of idrography*, 1542, world map.

23 Keuning, 'History of geographical map projections', p. 20; Dainville, *Langage des géographes*, p. 36.

24 A. Arrowsmith [junior], *Outlines of the world* (London, 1825), p. 1.

25 J. Lennart Berggren and Alexander Jones (eds), *Ptolemy's* Geography: *an annotated translation of the theoretical chapters* (Princeton, NJ, 2000), p. 82.

26 Ibid., pp. 88–93.

27 E.L. Stevenson's translation quoted by Snyder, *Flattening the earth*, p. 13; differently rendered by O.A.W. Dilke, 'The culmination of Greek cartography in Ptolemy' in J.B. Harley and David Woodward (eds), *The history of cartography, volume one: cartography in prehistoric, ancient, and medieval Europe and the Mediterranean* (Chicago, 1987), p. 188, and by Berggren and Jones, *Ptolemy's Geography*, p. 93.

28 Keuning, 'History of geographical map projections', p. 15: 'Kuntsmann iv'.

29 Reproductions in Shirley, *The mapping of the world*, plate 33, p. 33; plate 43, pp. 48–9; plate 47, p. 54; plate 159, p. 216.

30 Oswald A.W. Dilke and Margaret S. Dilke, 'The adjustment of Ptolemaic atlases to feature the new world' in Wolfgang Haase and Meyer Reinhold (eds), *The classical tradition and the Americas*, i (Berlin and New York, 1994), pp. 117–34.

31 George Kish, 'The cosmographic heart: cordiform maps of the 16th century', *Imago Mundi*, xix (1965), pp. 13–21; Frank Canters and Hugo Decleir, *The world in perspective: a directory of world map projections* (Chichester, 1989), p. 151; Ruth Watson, 'Cordiform maps since the sixteenth century: the legacy of nineteenth-century classificatory systems', *Imago Mundi*, lx, 2 (2008), pp. 182–94.

32 Snyder, *Flattening the earth*, p. 34; Shirley, *The mapping of the world*, plate 35, p. 36: 1511.

33 Hinks, *Map projections*, p. 53; Snyder, *Flattening the earth*, pp. 34–7.

34 Max Leinekugel le Cocq, *Premieres images de la terre* (Paris, 1977), pp. 70–1: 1570.

35 Snyder, *Flattening the earth*, p. 50: 1606, 1609.

36 Ibid., pp. 34, 52.

37 Norman J.W. Thrower, 'New geographical horizons: maps' in Fredi Chiappelli (ed.), *First images of America: the impact of the new world on the old*, ii (Berkeley, CA, 1976), p. 662.

38 Snyder, *Flattening the earth*, p. 38.

39 Keuning, 'History of geographical map projections', pp. 21–2.

40 Shirley, *The mapping of the world*.

41 Giorgio Mangani, 'Abraham Ortelius and the hermetic meaning of the cordiform projection', *Imago Mundi*, l (1998), p. 63.

42 'The case against rectangular world maps', *Cartographic Journal*, xxvi, 2 (1989), pp. 156–7.

43 Quoted by Snyder, *Flattening the earth*, p. 46.

44 Shirley, *The mapping of the world*, plate 43, pp. 48–9; plate 53, pp. 62–3; plate 75, p. 100; plate 95, p. 127.

45 Ibid., plates 47, 48, p. 54; plate 159, p. 216.

46 Quoted by W.W. Jervis, *The world in maps: a study in map evolution* (London, 1938), p. 27.

47 Snyder, *Flattening the earth*, p. 43.

48 Ibid., pp. 5, 16, 18, 49.

49 Ingrid Kretschmer, 'Kartenprojektionen in Gerhard Mercators Atlas' in Hans H. Blotevogel and Rienk Vermij (eds), *Gerhard Mercator und die geistigen Strömungen des 16. und 17. Jahrhunderts* (Duisburger Mercator-Studien, Band 3, Bochum, 1995), pp. 68–9.

50 Snyder, *Flattening the earth*, p. 33.

51 William Ravenhill, 'John Adams, his map of England, its projection, and his *Index villaris* of 1680', *Geographical Journal*, cxliv (1978), pp. 424–37.

52 Mireille Pastoureau, 'The 1569 world map' in Marcel Watelet (ed.), *The Mercator atlas of Europe* (Antwerp, 1998), pp. 79–87; Andrew Taylor, *The world of Gerard Mercator, the mapmaker who revolutionised geography* (London, 2004), pp. 223–8; Mark Monmonier, *Rhumb lines and map wars: a social history of the Mercator projection* (Chicago, 2004).

53 Richard I. Ruggles, *A country so interesting: the Hudson's Bay Company and two centuries of mapping, 1670–1870* (Montreal, 1991), plates 9 (p. 133), 12 (p. 136), 14 (p. 138): 1772–91; Samuel Hearne, *A journey from Prince of Wales's Fort in Hudson's Bay, to the Northern Ocean* (London, 1795), route map.

54 Snyder, *Flattening the earth*, pp. 156–7.

55 J.B. Harley (ed. Paul Laxton), *The new nature of maps* (Baltimore, 2001), pp. 66–7.

56 Brigitte Englisch, 'Erhard Etzlaub's projection and methods of mapping', *Imago Mundi*, xlviii (1996), p. 103.

57 B. Van 'T Hoff, *Gerard Mercator's map of the world* (Rotterdam, 1961), p. 46.

58 Edward Wright, *On certaine errors in navigation* (London, 1599); Leslie B. Cormack, '"Good fences make good neighbours": geography as self-definition in early modern England', *Isis*, lxxxii (1991), p. 649.

59 Shirley, *The mapping of the world*, plate 214, p. 292: from the 1610 edition of Wright's *Certain errors*.

60 Snyder, *Flattening the earth*, pp. 65–73: Philippe de la Hire, Antoine Parent, John Green, Patrick Murdoch.

61 *Notes and comments on the composition of terrestrial and celestial maps (1772) by Johann Heinrich Lambert*, translated and introduced by Waldo R. Tobler (Ann Arbor, MI, 1972); Snyder, *Flattening the earth*, p. 76.

62 Varenius, *Cosmography and geography*; [John Green], *The construction of maps and globes* (London, 1717).

63 Tobler in Lambert, *Notes and comments*, p. x.

64 Norman J.W. Thrower, *Maps and civilization: cartography in culture and society* (Chicago, 1996), p. 122.

65 William Alingham, *A short account of the nature and use of maps* (London, 1703), p. 22.

66 Berggren and Jones, *Ptolemy's Geography*, pp. 95, 121.

67 Green, *Construction of maps and globes*, p. 108.

68 R.W. Shirley, *Early printed maps of the British Isles 1477–1650: a bibliography* (London, 1973), plate 85: Nicolas Sanson, *Isles Britanniques*, 1648; Rodney W. Shirley, *Printed maps of the British Isles, 1650–1750* (Tring, 1988), p. 58: Nicolas de Fer, *Isles Britaniques*, c.1705; Edward Lynam, *The Carta marina of Olaus Magnus, Venice 1539 and Rome 1572* (Jenkintown, PA, 1949), p. 5; Agustin Hernando, *El mapa de Espana, siglos xv–xviii* (Madrid, 1995), p. 201: Nicolas Sanson, Spain, 1658; R.V. Tooley and Charles Bricker, *A history of cartography: 2500 years of maps and mapmakers* (London, 1969), p. 129: Jean Baptiste d'Anville, *Royaume de Corée*, 1737.

69 Gerard Mercator, Henry Hondius and Joannes Janssonius (ed. R.A. Skelton), *Atlas or a geographicke description of the world*, i (Amsterdam, 1636, 1968), pp. 163–4.

70 E.Hatton, *A mathematical manual: or delightful associate* (London, 1728), p. 103.

71 Alexander Jamieson, *A treatise on the construction of maps* (London, 1814), p. 121.

72 Snyder, *Flattening the earth*, p. 105.

73 Dainville, *Langage des géographes*, p. 31.

74 J.A. Steers, *An introduction to the study of map projections* (7th ed., London, 1949), p. 248.

75 Brian Adams, '"Parallel to the meridian of Butterton Hill" – do I laugh or cry?', *Sheetlines*, xxxviii (1994), pp. 15–19; Yolande Hodson, *Popular maps: the Ordnance Survey Popular Edition one-inch map of England and Wales 1919–1926* (London, 1999), pp. 65–8; Charles Close (ed. J.B. Harley), *The early years of the Ordnance Survey* (1926; Newton Abbot, 1969), p. 149.

76 Maling, *Coordinate systems and map projections*, p. 216.

77 M.J. Collet, *La carte de France dite de l'état-major* (Paris, 1887), pp. 19–27; *Dictionnaire de biographie française*, vi (Paris, 1954), p. 990: Bonne, Charles-Rigobert-Marie.

78 Florian Cajori, *The chequered career of Ferdinand Rudolph Hassler, first superintendent of the United States Coast Survey* (Boston, 1929), pp. 114–17; Mark Monmonier, 'Practical and emblematic roles of the American polyconic projection', History of Cartography Conference, Portland, Maine, 2003, typescript; Ferdinand Hassler, 'On the mechanical organisation of a large survey, and the particular application to the Survey of the Coast', *American Philosophical Society Transactions*, new series, ii (1825), pp. 385–408; Walter Satzinger, 'The Great Atlas of Germany edited by Johann Wilhelm Jaeger, Frankfurt am Main, 1789', History of Cartography Conference, Greenwich, 1975, typescript, p. 10.

79 Maling, *Coordinate systems and map projections*, p. 200; Mark Monmonier, *Rhumb lines and map wars: a social history of the Mercator projection* (Chicago, 2004), ch. 7.

Chapter 9: One fair card or map

1 Valerie A. Kivelson, 'Cartography, accuracy and state powerlessness: the uses of maps in early modern Russia', *Imago Mundi*, li (1999), pp. 92–4; Valerie Kivelson, *Cartographies of Tsardom: the land and its meanings in seventeenth-century Russia* (Ithaca, 2006), pp. 60–6.

2 B. Talbot, *The new art of land measuring; or, A turnpike road to practical surveying* (Wolverhampton, 1779), p. 243.

3 Aaron Rathborne, *The surveyor in foure bookes* (London, 1616), p. 172; J. Eyre, *The exact surveyor: or, The whole art of surveying of land* (London, 1654), p. 190; Adam Martindale, *The country-survey-book: or Land-meter's vademecum* (8th ed., London, 1711), pp. 96–7; Samuel Wyld, *The practical surveyor, or The art of land-measuring, made easy* (London, 1725), p. 81; Talbot, *New art of land measuring*, pp. 236–7.

4 J.H. Andrews, *Plantation acres: an historical study of the Irish land surveyor and his maps* (Belfast, 1985); A. Sarah Bendall, *Maps, land and society: a history, with a carto-bibliography of Cambridgeshire estate maps, c.1600–1836* (Cambridge, 1992); David H. Fletcher, *The emergence of estate maps: Christ Church, Oxford, 1600–1840* (Oxford, 1995); David Buisseret (ed.), *Rural images: estate maps in the old and new worlds* (Chicago, 1996); Sarah Bendall, *Dictionary of land surveyors and local map-makers of Great Britain and Ireland*, 2 vols (London, 1997).

5 D. Hodson, *Maps of Portsmouth before 1801* (Portsmouth, 1978); Elisabeth Stuart, *Lost landscapes of Plymouth: maps, charts and plans to 1800* (Stroud, 1991); Paul Kerrigan, *Castles and fortifications in Ireland, 1485–1945* (Cork, 1995).

6 Marcus Merriman, 'Italian military engineers in Britain in the 1540s' in Sarah Tyacke (ed.), *English map-making, 1500–1650: historical essays* (London, 1983), pp. 57–67; Bendall, *Dictionary of land surveyors*, i, pp. 13, 15.

7 Emile van der Vekene, *Les plans de la forteresse de Luxembourg* (Luxembourg, [1987]), p. [11]; Elizabeth [sic] A. Stuart, 'Armada maps of Plymouth', *The Map Collector*, xlii (1988), p. 8; Martha Pollak, 'Military architecture and cartography in the design of the early modern city' in David Buisseret (ed.), *Envisioning the city: six studies in urban cartography* (Chicago, 1998), pp. 120–1.

8 Robert W. Karrow Jr, *Mapmakers of the sixteenth century and their maps: bio-bibliographies of the cartographers of Abraham Ortelius, 1570* (Winnetka, IL, 1993), pp. 297–8; Catherine Delano-Smith and Roger J.P. Kain, *English maps: a history* (London, 1999), p. 210: Bath, 1735.

9 James Elliot, *The city in maps: urban mapping to 1900* (London, 1987); David Smith, 'The earliest printed maps of British towns', *Bulletin of the Society of Cartographers*, xxvii, 2 (1993), pp. 25–45. For earlier town plans see chapter 2, pp. 57–9.

10 Frank Kitchen, 'Cosmo-choro-poly-grapher: an analytical account of the life and works of John Norden, 1547?–1625' (University of Sussex, D.Phil. thesis, 1993), p. 34, quoting Norden's description of Northamptonshire; Delano-Smith and Kain, *English maps*, plate 15, pp. 178–9; Sarah Bendall, 'Draft town maps for John Speed's *Theatre of the empire of Great Britaine*', *Imago Mundi*, liv (2002), p. 40.

11 Thomas Frangenberg, 'Chorographies of Florence: the use of city views and city plans in the sixteenth century', *Imago Mundi*, xlvi (1994), p. 56.

12 David Buisseret, *The mapmakers' quest: depicting new worlds in renaissance Europe* (Oxford, 2003), p. 23; Martha D. Pollak, *Military architecture, cartography and the representation of the early modern European city* (Chicago, 1991), pp. 31, 42, 49, 83.

13 P.D.A. Harvey, 'The Portsmouth map of 1545 and the introduction of scale maps into England' in John Webb, Nigel Yates and Sarah Peacock (eds), *Hampshire studies* (Portsmouth, 1981), pp. 33–49.

14 Ralph Agas, *A preparative to plotting of landes and tenements for surveigh* (London, 1596), p. 18.

15 Philippa Glanville, *London in maps* (London, 1972), plates 14, 15, pp. 98–101.

16 A.A. Horner, 'Two eighteenth-century maps of Carlow town', *Proceedings of the Royal Irish Academy*, lxxviii C (1978), pp. 119–21.

17 Johannes Mejer, *Newe Landesbeschreibung der zweii Hertzogthümer Schleswich und Holstein* (1652; Hamburg-Bergedorf, 1963).

18 Cyprian Lucar, *A treatise named Lucar solace* (London, 1590), pp. 9–10, 44, 48; Talbot, *New art of land surveying*, pp. ix, 403.

19 John Varley, 'John Rocque: engraver, surveyor, cartographer and map-seller', *Imago Mundi*, v (1948), p. 85.

20 Herbert George Fordham, *John Cary: engraver, map, chart and print-seller and globe-maker, 1754 to 1835* (1925; Folkestone, 1976), p. xxi.

21 John Norden, *Preparative to his Speculum Britanniae* (London, 1596), p. 14.

22 David Buisseret, 'Monarchs, ministers and maps in France before the accession of Louis XIV' in David Buisseret (ed.), *Monarchs, ministers and maps: the emergence of cartography as a tool of government in early modern Europe* (Chicago, 1992), p. 106; Karrow, *Mapmakers of the sixteenth century*, p. 439; Monique Pelletier, *La carte de Cassini* (Paris, 1990), pp. 21–4.

23 Donald Hodson, *County atlases of the British Isles published after 1703, volume 1: atlases published 1704 to 1742* (Welwyn, Herts, 1984), pp. 169–79; Varley, 'John Rocque', pp. 90–1; J.B. Harley, 'The bankruptcy of Thomas Jefferys: an episode in the economic history of eighteenth century map-making', *Imago Mundi*, xx (1966), pp. 42–4.

24 Andrews, *Plantation acres*, pp. 356–7; Arnold Horner, *Mapping Offaly in the early nineteenth century, with an atlas of William Larkin's map of King's County, 1809* (Bray, Co. Wicklow, 2006), pp. 15–19.

25 Donald Hodson, 'Dating county maps through mapsellers' advertisements', *The Map Collector*, xxvi (1984), pp. 16–18.

26 G.R. Crone, *Maps and their makers* (3rd ed., London, 1966), pp. 106–7.

27 P.D.A. Harvey and Harry Thorpe, *The printed maps of Warwickshire, 1576–1900* (Warwick, 1959), p. ix; R.A. Skelton, 'The mapping of Sussex, 1575–1825' in Harry Margary (ed.), *Two hundred and fifty years of map-making in the county of Sussex* (Lympne, 1970), unpaginated.

28 Ifor M. Evans and Heather Lawrence, *Christopher Saxton, Elizabethan map-maker* (Wakefield, 1979), pp. 38–9.

29 J.B. Harley, 'Christopher Saxton and the first atlas of England and Wales, 1579–1979', *The Map Collector*, viii (1979), p. 7; William Ravenhill, *Christopher Saxton's 16th century maps: the counties of England and Wales* (Shrewsbury, 1992), p. 18.

30 Kitchen, 'Cosmo-choro-poly-grapher', Figs. 6, 13, 14, 15.

31 Joan Blaeu, *Le grand atlas*, iii (Amsterdam, 1663, 1967), p. 69: Bremae & Furdae.

32 J.H. Andrews and Rolf Loeber, 'An Elizabethan map of Leix and Offaly: cartography, topography and architecture' in William Nolan and Timothy P. O'Neill (eds), *Offaly, history and society: interdisciplinary essays on the history of an Irish county* (Dublin, 1998), pp. 262–3; Blaeu, *Grand atlas*, iii, pp. 50–1: *Marchionatus Brandenburgi*; Catherine Delano-Smith, 'Milieus of mobility: itineraries, route maps, and road maps' in James R. Akerman (ed.), *Cartographies of travel and navigation* (Chicago, 2006), pp. 61–4.

33 Gerard Mercator, Henry Hondius and Joannes Janssonius (ed. R.A. Skelton), *Atlas or a geographicke description of the world*, ii (Amsterdam, 1636, 1968), pp. 237–65; J. Keuning, 'The history of an atlas: Mercator-Hondius', *Imago Mundi*, iv (1947), pp. 52, 56.

34 William Ravenhill, 'Christopher Saxton's surveying: an enigma' in Tyacke, *English map-making, 1500–1650*, pp. 112–19, quotation on p. 115; Karrow, *Mapmakers of the sixteenth century*, p. 441.

35 Paul Pfinzing, *c.*1590, quoted in Fritz Bönisch, 'The geometrical accuracy of 16th and 17th century topographical surveys', *Imago Mundi*, xxi (1967), p. 64.

36 See above, chapter 4, pp. 100–4.

37 Felix E. Schelling (ed.), *The complete plays of Ben Jonson*, i (London, 1910), p. 545: *Epicoene*, 1609.

38 Norden, *Preparative to his Speculum Britanniae*, pp. 12–13.

39 R.A. Skelton, *County atlases of the British Isles, 1579–1850* (London, 1970), p. 35.

40 R[obert] N[orton], *A mathematicall apendix for mariners at sea, and for cherographers and surveyors of land* (London, 1604), p. 22.

41 John Love, *Geodaesia: or, The art of surveying and measuring of land made easie* (London, 1688), p. 145; Wyld, *Practical surveyor*, p. 148; William Gardiner, *Practical surveying improved: or, Land measuring according to the most correct methods* (London, 1737), p. 98.

42 John Wing, *Geodaetes practicus redivivus. The art of surveying: formerly publish'd by Vincent Wing, math. now much augmented and improved* (London, 1700), p. 242.

43 E.G.R. Taylor (ed.), *A regiment for the sea and other writings on navigation by William Bourne* (Cambridge, 1963), introduction, pp. xxix–xxxii.

44 Nathanael Carpenter, *Geography delineated forth in two bookes* (Oxford, 1635), p. 193.

45 John Gregorie, *The description and use of the terrestrial globe* in *Gregorii posthuma; or, Certain learned tracts written by John Gregorie* (London, 1649), p. 316.

46 Venterus Mandey, *Synopsis mathematica universalis* (London, 1729), p. 628.

47 [John Green], *The construction of maps and globes* (London, 1717), p. 87.

48 'An account of the astronomical, geographical, and physical observations made at the Cape of Good Hope, in 1751, 1752 and 1753, by order of the French king, by the Abbe de la Caille', *Gentleman's Magazine*, xxv (1755), pp. 511–12.

49 J. Lennart Berggren and Alexander Jones (eds), *Ptolemy's Geography: an annotated translation of the theoretical chapters* (Princeton, NJ, 2000), pp. 96–107; Arthur Hopton, *Speculum topographicum; or, The topographicall glasses* (London, 1611), pp. 85–6.

50 Abraham Ortelius (ed. R.A. Skelton), *The theatre of the whole world, London, 1606* (Amsterdam, 1968), p. 68: Lake Constance; Mercator-Hondius-Janssonius, *Atlas*, ii, p. 309: Lake Geneva.

51 Karrow, *Mapmakers of the sixteenth century*, pp. 469–70: Upper Lusatia, by Bartholomaeus Scultetus, 1593, copied in Blaeu, *Grand atlas*, iii, p. 46.

52 Love, *Geodaesia*, p. 144.

53 Bönisch, 'The geometrical accuracy of 16th and 17th century topographical surveys', p. 63.

54 'A pamphelett conteiginge the description of Mylford havon …' in George Owen (ed. H. Owen), *Description of Pembrokeshire*, ii (London, 1892), pp. 545, 547, 569.

55 Quoted by R.A. Gardiner, 'Philip Symonson's "New description of Kent", 1596', *Geographical Journal*, cxxxv (1969), p. 136.

56 John Ogilby, *Britannia, London 1675* (Amsterdam, 1970).

57 Sarah Tyacke, *London map-sellers, 1660–1720* (Tring, 1978), p. 65: 1694.

58 J.B. Harley and William Ravenhill, 'Proposals for county maps of Cornwall (1699) and Devon (1700)', *Devon and Cornwall Notes and Queries*, xxxii (1971), pp. 34–5.

59 *Proposals by way of subscription for making a new survey, and publishing a most correct map, of the county of Sussex*, reproduced in Tony Campbell, 'Laying bare the secrets of the British Library's map collections', *The Map Collector*, lxii (1993), p. 38.

60 J.B. Harley, 'The re-mapping of England, 1750–1800', *Imago Mundi*, xix (1965), pp. 56–67.

61 J.B. Harley, 'The Society of Arts and the survey of English counties, 1759–1809', *Journal of the Royal Society of Arts*, cxii, 5089 (1963), pp. 43–6; cxii, 5090 (1964), pp. 119–24; cxii, 5092 (1964), pp. 269–75; cxii, 5095 (1964), pp. 538–43.

62 Paul Laxton, 'The geodetic and topographical evaluation of English county maps 1740–1840', *Cartographic Journal*, xiii, 1 (1976), pp. 52–3, n. 18.

63 Moses Pitt, *The English atlas*, ii (London, 1681): 'Vera totius Marchionatus Badensis et Hochbergensis …'.

64 Laxton, 'Geodetic and topographical evaluation of English county maps', p. 53, n. 28; above, Fig. 6.9.

65 Quoted by Harley, 'The bankruptcy of Thomas Jefferys', p. 44.

66 J.B. Harley (ed.), *A map of the county of Lancashire, 1786, by William Yates* (Liverpool, 1968), p. 12.

67 National Library of Ireland, MSS 2016, 3242.

68 E.J.S. Parsons, *The map of Great Britain circa AD 1360 known as the Gough map* (Oxford, 1958), pp. 7–10.

69 Sarah Tyacke and John Huddy, *Christopher Saxton and Tudor map-making* (London, 1980), p. 7.

70 E.G.R. Taylor, *The mathematical practitioners of Tudor and Stuart England* (Cambridge, 1968), p. 165.

71 Peter M. Barber, 'The British Isles' in Marcel Watelet (ed.), *The Mercator atlas of Europe* (Antwerp, 1998), p. 68.

72 David Marcombe, 'Saxton's apprenticeship: John Rudd, a Yorkshire cartographer', *Yorkshire Archaeological Journal*, l (1978), p. 173.

73 William Cuningham, *The cosmographical glasse* (London, 1559), f. 135.

74 Lloyd A. Brown, *The story of maps* (New York, 1949, 1979), p. 245; L. Gallois, 'Sur les progrès de la cartographie de la region parisienne jusqu'à la carte de Cassini' in *Regions naturelles et noms de pays* (Paris, 1908), p. 311.

75 P.D.A. Harvey, *The history of topographical maps: symbols, pictures and surveys* (London, 1980), pp. 58–62.

76   Gonzalo de Reparaz Ruiz, 'The topographical maps of Portugal and Spain in the 16th century', *Imago Mundi*, vii (1950), pp. 75–82; Geoffrey Parker, 'Maps and ministers: the Spanish Habsburgs' in Buisseret, *Monarchs, ministers and maps*, pp. 130–4.

77   J.H. Andrews, 'The Irish surveys of Robert Lythe', *Imago Mundi*, xix (1965), pp. 22–31.

78   Jeffrey C. Stone, *The Pont manuscript maps of Scotland: sixteenth century origins of a Blaeu atlas* (Tring, 1989).

79   Herman Richter, 'Willem Jansz. Blaeu with Tycho Brahe on Hven, and his map of the island: some new facts', *Imago Mundi*, iii (1939), pp. 53–60.

80   R.T. Gunther (ed.), *The theodelitus and topographical instrument of Leonard Digges of University College, Oxford, described by his son Thomas Digges in 1571* (Oxford, 1927), p. 34.

81   Parker, 'Maps and ministers', pp. 130–1; Ricardo Padrón, *The spacious world: cartography, literature, and empire in early modern Spain* (Chicago, 2004), pp. 51, 63.

82   Ravenhill, 'Saxton's surveying: an enigma', p. 118.

83   William Petty, *Hiberniae delineatio* (London, 1685; Shannon, 1969).

84   Parker, 'Maps and ministers', Figs. 5.2, 5.3, pp. 131–2; David Buisseret, 'Spanish peninsular cartography, 1500–1700' in David Woodward (ed.), *The history of cartography, volume three, part one: cartography in the European renaissance* (Chicago, 2007), pp. 1083–5.

85   J.H. Andrews, 'Science and cartography in the Ireland of William and Samuel Molyneux', *Proceedings of the Royal Irish Academy*, lxxx C (1980), p. 250.

86   J.B. Harley, 'William Yates and Peter Burdett: their role in the mapping of Lancashire and Cheshire during the eighteenth century', *Transactions of the Historic Society of Lancashire and Cheshire*, cxv (1963), p. 116.

87   William Borlase, quoted by William Ravenhill, 'The Lizard as a landfall', *Journal of Navigation*, xxxv, 1 (1982), p. 84.

88   Edward Williams, William Mudge and Isaac Dalby, 'An account of the trigonometrical survey carried on in the years 1791, 1792, 1793, and 1794, by order of his grace the duke of Richmond, late master of the ordnance', *Philosophical Transactions of the Royal Society of London*, lxxxv, pt 2 (1795), p. 415; W.A. Seymour (ed.), *A history of the Ordnance Survey* (Folkestone, 1980), pp. 21–2; John Robert Seeley, *The expansion of England* (London, 1883), p. 8. See above, chapter 6, p. 146.

89   J.B. Harley, 'English county map-making in the early years of the Ordnance Survey: the map of Surrey by Joseph Lindley and William Crosley', *Geographical Journal*, cxxxii (1966), p. 373.

90   Quoted in J.B. Harley, 'Cheshire maps, 1787–1831', *Cheshire Round*, i, 9 (1968), p. 297.

91   J.B. Harley, *Christopher Greenwood county map-maker, and his Worcestershire map of 1822* (Worcester, 1962), p. viii.

Chapter 10: Going in the dark

1   R.A. Skelton, *The military survey of Scotland, 1747–1755* (Edinburgh, 1967), p. 7.

2   R.A. Skelton, 'The military surveyor's contribution to British cartography in the 16th century', *Imago Mundi*, xxiv (1970), pp. 77–8.

3   M.H. Edney, 'British military education, mapmaking and "map-mindedness" in the later Enlightenment', *Cartographic Journal*, xxxi (1994), p. 17.

4   Peter J. Guthorn, *British maps of the American revolution* (Monmouth Beach, NJ, 1972), p. 30; J.B. Harley, 'The contemporary mapping of the American revolutionary war' in J.B. Harley, Barbara Bartz Petchenik and Lawrence W. Towner, *Mapping the American revolutionary war* (Chicago, 1978), pp. 33, 39, 49; Douglas W. Marshall, 'The British engineers in America: 1755–1783', *Journal of the Society for Army Historical Research*, li (1973), pp. 155–63.

5   Gerard Hayes-McCoy (ed.), *Ulster and other Irish maps, c.1600* (Dublin, 1964); Joan Murphy, 'Measures of map accuracy assessment and some early Ulster maps', *Irish Geography*, xi (1978), p. 97.

6   D.G. Moir in Royal Scottish Geographical Society, *The early maps of Scotland to 1850* (Edinburgh, 1973), p. 104.

7 Harley, 'Contemporary mapping of the American revolutionary war', p. 4.

8 Ibid., p. 20: 1775; p. 33: *c.1777*.

9 J.H. Andrews, 'Charles Vallancey and the map of Ireland', *Geographical Journal*, cxxxii (1966), pp. 48–61.

10 Yolande Hodson and Alan Gordon, *An illustrated history of 250 years of military survey* ([London], 1997), pp. 4–5.

11 Yolande Hodson, 'Maps in the field' in Peter Barber and Christopher Board (eds), *Tales from the Map Room: fact and fiction about maps and their makers* (London, 1993), p. 115.

12 Charles Close (ed. J.B. Harley), *The early years of the Ordnance Survey* (1926; Newton Abbot, 1969), p. 3.

13 Skelton, *Military survey of Scotland*, plate 2; Jessica Christian, 'Paul Sandby and the military survey of Scotland' in Nicholas Alfrey and Stephen Daniels (eds), *Mapping the landscape: essays on art and cartography* (Nottingham, 1990), pp. 18–22.

14 Skelton, *Military survey of Scotland*, pp. 3–5.

15 Don W. Thomson, *Men and meridians: the history of surveying and mapping in Canada*, i (Ottawa, 1966), pp. 94, 102; W.P. Cumming, 'The Montresor-Ratzer-Sauthier sequence of maps of New York City, 1766–76', *Imago Mundi*, xxxi (1979), p. 55.

16 Quoted in Moir, *Early maps of Scotland*, p. 104.

17 British Library, *The American war of independence, 1775–83* (London, 1975), p. 77: Stillwater, 1777; p. 93: Newport, 1775; p. 114: Philadelphia, 1779; Michael Swift, *Historical maps of Europe* (London, 2000), p. 41: Fort Louis, Alsace, 1711; p. 46: La Rochelle, 1718; p. 55: Prague, 1742; p. 89: Luxembourg, 1801; Pro Civitate, *Carte de cabinet des Pay-Bas Autrichiens: Waterloo* (Brussels, 1965): 1777–8.

18 A.C. O'Dell, 'A view of Scotland in the middle of the eighteenth century', *Scottish Geographical Magazine*, lxix (1953), p. 59.

19 J.B. Harley, 'The map user in the revolution' in Harley, Petchenik and Towner, *Mapping the American revolutionary war*, pp. 98–9, quoting Douglas W. Marshall and Howard Henry Peckham, *Campaigns of the American revolution: an atlas of manuscript maps* (Ann Arbor, MI, 1976), p. 130.

20 Graeme Whittington, 'The Roy map: the protracted and fair copies', *Scottish Geographical Magazine*, cii (1986), part 1, pp. 18–28, part 2, pp. 66–73; G. Whittington and A.J.S. Gibson, *The military survey of Scotland, 1747–1755: a critique*, Historical Geography Research Group, no. 18 (1986).

21 James R. Coull, 'The district of Buchan as shown on the Roy map', *Scottish Geographical Magazine*, xcvi (1980), pp. 70–3.

22 Quoted by Whittington and Gibson, *Military survey of Scotland*, p. 10.

23 Peter Barber, 'The eyes of the general abroad', in Barber and Board, *Tales from the Map Room*, pp. 102–3: manuscript map of Dauphiné, Provence and Savoy, *c.1708*.

24 Roger J.P. Kain and Elizabeth Baigent, *The cadastral map in the service of the state: a history of property mapping* (Chicago, 1992), p. 195.

25 Yolande Hodson, 'Wartime difficulties – peacetime mapping' in Barber and Board, *Tales from the Map Room*, p. 126.

26 Andrews, 'Charles Vallancey and the map of Ireland', p. 52.

27 Harley, 'The map user in the revolution', pp. 84–5.

28 Gwyn Walters, 'Thomas Pennant's map of Scotland, 1777: a study in sources, and an introduction to George Paton's role in the history of Scottish cartography', *Imago Mundi*, xxviii (1976), p. 126.

29 Aaron Arrowsmith, *Map of Scotland constructed from original materials obtained under the authority of the parliamentary commissioners for making roads and building bridges in the Highlands of Scotland* (London, 1807).

30 J.H. Andrews, *Shapes of Ireland: maps and their makers, 1564–1839* (Dublin, 1997), ch. 9, pp. 248–76.

31 Thomson, *Men and meridians*, p. 99.

32 Ibid., p. 101; Brooke Hindle, *The pursuit of science in revolutionary America, 1735–1789* (Chapel Hill, NC, 1956), p. 175.

33 Julian P. Boyd (ed.), *The papers of Thomas Jefferson*, x (Princeton, NJ, 1954), p. 212.

34 Leo Bagrow (ed. Henry W. Castner), *A history of Russian cartography to 1800* (Wolfe Island, Ontario, 1975), app. 3, p. 263; E. Varep, *The prime meridian of Dagö and Osel* (Tartu, Estonia, 1975), p. 9; A.V. Postnikov, 'Materials, accompanying surveys and mapping of Russia in xviii–xix centuries, as memorials of history of cartography and as historic-geographic sources', History of Cartography Conference, Berlin, 1979, typescript, pp. 4–5.

35 Hodder and Stoughton, *Rudyard Kipling's verse: definitive edition* (London, 1940), p. 104.

36 C.E. Heidenreich, 'Explorations and mapping of Samuel de Champlain, 1603–1632', *Cartographica*, Monograph 17 (1976).

37 Frank Debenham, *Discovery and exploration: an atlas-history of man's journeys into the unknown* (London, 1960); Gail Roberts, *Atlas of discovery* (London, 1973); Felipe Fernandez-Armesto, *The Times atlas of world exploration* (London, 1991).

38 Ibid., p. 55; Patricia Galloway (ed.), *The Hernando de Soto expedition: history, historiography and 'discovery' in the southeast* (Lincoln, NE, 1997).

39 Samuel Hearne, *A journey from Prince of Wales's Fort in Hudson's Bay, to the Northern Ocean* (London, 1795), p. xliii.

40 David Beers Quinn and Raleigh Ashlin Skelton (eds), *The principall navigations voiages and discoveries of the English nation by Richard Hakluyt*, i (Cambridge, 1965), pp. 459, 461.

41 James P. Ronda, '"A chart in his way": Indian cartography and the Lewis and Clark expedition', *Great Plains Quarterly*, iv (1984), p. 50.

42 Fergus Fleming, *Barrow's boys* (London, 1998), pp. 196–7.

43 G. Malcolm Lewis, 'Misinterpretation of Amerindian information as a source of error on Euro-American maps', *Annals of the Association of American Geographers*, lxxvii, 4 (1987), p. 545: 1762; Barbara Belyea, 'Inland journeys, native maps' in G. Malcolm Lewis (ed.), *Cartographic encounters: Perspectives on native American mapmaking and map use* (Chicago, 1998), pp. 135–55.

44 *Journal of a 2nd voyage for the discovery of a north west passage from the Atlantic to the Pacific; performed in the years 1821–22–23 in His Majesty's ships* Fury *and* Hecla, *under the orders of Captain William Edward Parry, R.N., F.R.S., and commander of the expedition* (New York, 1969), pp. 196–7.

45 Barbara Belyea, 'Amerindian maps: the explorer as translator', *Journal of Historical Geography*, xviii, 3 (1992), p. 275.

46 Hearne, *Journey to the Northern Ocean*.

47 Hildegard Binder Johnson, 'Science and historical truth on French maps of the 17th and 18th centuries', *Proceedings of the Minnesota Academy of Science*, xxv (1957), xxvi (1958), pp. 332–3; Thomas Jefferys, *The American atlas* (London, 1776; Amsterdam, 1974), passim; Richard I. Ruggles, *A country so interesting: the Hudson's Bay Company and two centuries of mapping, 1670–1870* (Montreal, 1991), plates 27 (p. 151), 40 (p. 164), 49 (p. 173): 1819–45.

48 Rodney W. Shirley, *The mapping of the world: early printed world maps, 1472–1700* (London, 1983), plate 104, pp. 144–5; Günter Schilder, 'Willem Jansz. Blaeu's wall map of the world, on Mercator's projection, 1606–07 and its influence', *Imago Mundi*, xxxi (1979), pp. 38, 53.

49 A. Arrowsmith, *Outlines of the countries between Delhi and Constantinople* (London, 1814).

50 Robert Sayer, *A map of South America*, 1775, in Jefferys, *The American atlas*.

51 G. Malcolm Lewis, 'Travelling in uncharted territory', in Barber and Board, *Tales from the Map Room*, p. 41.

52 Alexander Keith Johnston, *The national atlas of historical, commercial and political geography constructed from the most recent and authentic sources* (Edinburgh, 1847).

53 Peter Whitfield, *New found lands: maps in the history of exploration* (London, 1998).

54 J.E. Delmar Morgan and C.H. Coote (eds), *Early voyages and travels to Russia and Persia by Anthony Jenkinson and other Englishmen* (London, 1886); J. Keuning, 'Jenkinson's map of Russia', *Imago Mundi*, xiii (1956), pp. 172–5; Samuel H. Baron, 'William Borough and the Jenkinson map of Russia', *Cartographica*, xxvi, 2 (1989), pp. 72–8; Krystyna Szykula, *The re-discovered Jenkinson map of Russia dated 1562: further investigation* (Wroclaw, [1989]). For other references see David Woodward (ed.), *The history of cartography, volume three: cartography in the European renaissance* (Chicago, 2007), pp. 1738, 1856–7.

55 John W. Webb, 'The Van Deutecum map of Russia and Tartary' in John Parker (ed.), *Merchants and scholars: essays in the history of exploration and trade collected in memory of James Ford Bell* (Minneapolis, 1965), pp. 78–9.

56 Morgan and Coote, *Early voyages and travels to Russia and Persia*, i, pp. 27–100.

57 William Camden, *Annales or, The history of the most renowned and victorious princesse Elizabeth, late queen of England* (3rd ed., London, 1635), p. 86.

58 Guy Meriwether Benson with William R. Irwin and Heather Moore Riser, *Lewis and Clark: the maps of exploration, 1507–1814* (Charlottesville, VA, 2002), p. 80.

59 Martin Plamondon II, *Lewis and Clark trail maps: a cartographic reconstruction*, i (Pullman, WA, 2000), p. 3.

60 Quoted by Carl I. Wheat, *Mapping the Transmississippi West, 1540–1861*, ii: *From Lewis and Clark to Fremont 1804–1845* (San Francisco, 1958), p. 48.

61 Bernard De Voto (ed.), *The journals of Lewis and Clark* (Boston, 1997), p. 482.

62 Silvio A. Bedini, 'The scientific instruments of the Lewis and Clark expedition', *Great Plains Quarterly*, iv (1984), p. 64; Wheat, *Mapping the Transmississippi West*, ii, pp. 45, 60.

63 W. Raymond Wood, 'Mapping the Missouri River: through the Great Plains, 1673–1895', *Great Plains Quarterly*, iv (1984), p. 37.

64 Dorothy Middleton, 'Francis Galton: travel and geography' in Milo Keynes (ed.), *Sir Francis Galton: the legacy of his ideas* (London, 1993), p. 49.

65 Francis Galton, *Tropical South Africa* (London, 1853).

Chapter 11: All marked with lines

1 Louis De Vorsey, Jr, 'The Gulf Stream on eighteenth century maps and charts', *The Map Collector*, xv (1981), pp. 2–10; xvi (1981), p. 46.

2 Tony Campbell, 'Portolan charts from the late thirteenth century to 1500' in J.B. Harley and David Woodward (eds), *The history of cartography, volume one: cartography in prehistoric, ancient, and medieval Europe and the Mediterranean* (Chicago, 1987), p. 373; Derek Howse and Michael Sanderson, *The sea chart: an historical survey based on the collections in the National Maritime Museum* (Newton Abbot, 1973), p. 53; Tony Campbell, *Early maps* (New York, 1981), p. 138.

3 Campbell, 'Portolan charts', pp. 383–4, 388.

4 E.G.R. Taylor, *The haven-finding art: a history of navigation from Odysseus to Captain Cook* (London, 1956), pp. 112–13.

5 Robert W. Karrow Jr, *Mapmakers of the sixteenth century and their maps: bio-bibliographies of the cartographers of Abraham Ortelius, 1570* (Winnetka, IL, 1993), p. 211.

6 George Percy Badger (ed.), *The travels of Ludovico di Varthema … AD 1503 to 1508* (London, 1863), p. 249.

7 Hermann Wagner, 'The origin of the mediaeval Italian nautical charts', *Report of the Sixth International Geographical Congress* (London, 1895), p. 700; J.E. Kelley Jr, 'Non-Mediterranean influences that shaped the Atlantic in the early portolan charts', History of Cartography Conference, Washington, 1977, typescript.

8 Heinrich Winter, 'A late portolan chart at Madrid and late portolan charts in general', *Imago Mundi*, vii (1950), p. 45.

9 J.E. Kelley Jr, 'The oldest portolan chart in the new world', *Terrae Incognitae*, ix (1977), p. 48.

10 Mary Blewitt, *Surveys of the seas: a brief history of British hydrography* (London, 1957), p. 18.

11 Kelley, 'Non-Mediterranean influences that shaped the Atlantic in the early portolan charts', p. 1.

12 David Woodward, 'Roger Bacon's terrestrial coordinate system', *Annals of the Association of American Geographers*, lxxx (1990), pp. 109–22.

13 Jonathan T. Lanman, 'On the origin of portolan charts', History of Cartography Conference, Ottawa, 1985, typescript, p. 12.

14 R.W. Bremner, 'The outline of the Mediterranean in "Lo conpasso de navegare" and the Pisane chart', *Proceedings of the VIII International Reunion for the History of Nautical Science and*

Hydrography, 1998, pp. 97–106; R.W. Bremner, 'Written portulans and charts from the 13th to the 16th century', *Proceedings of the IX International Reunion for the History of Nautical Science and Hydrography*, 2000, pp. 345–62.

15 Mateus Prunes, *Chart of the Mediterranean Sea and Western Europe*, introduction by John A. Wolter (Washington, 1981), p. [1].

16 Jonathan T. Lanman, *On the origin of portolan charts*, The Hermon Dunlap Smith Center for the History of Cartography, Occasional Publications, no. 2 (1987), p. 34.

17 A.E. Nordenskiöld, *Periplus: an essay on the early history of charts and sailing-directions* (Stockholm, 1897), pp. 16–17.

18 Scott A. Loomer, 'Mathematical analysis of medieval sea charts', American Congress of Surveying and Mapping, annual conference, technical papers, i (1986), p. 128.

19 Lanman, *Portolan charts*, pp. 4–18, 40.

20 Ernst Steger, *Untersuchungen über italienische Seekarten des Mittelalters auf Grund der kartometrischen Methode* (Göttingen, 1896), pp. 18–26; Wagner, 'Origin of the mediaeval Italian nautical charts', pp. 697–9.

21 Christopher Terrell, *The evolution of the sea chart* (Nicosia, 1999), p. 10.

22 James E. Kelley, Jr, 'Perspectives on the origin and uses of the portolan charts', *Cartographica*, xxxii, 3 (1995), p. 2.

23 Badger, *The travels of Ludovico di Varthema*, p. 249.

24 Taylor, *Haven-finding art*, pp. 92–6.

25 W.E. May, *A history of marine navigation* (Henley-on-Thames, 1973), pp. 45, 104–5.

26 Kelley, 'Origin and uses of the portolan charts', p. 4.

27 Campbell, 'Portolan charts', pp. 395–6; Heinrich Winter, 'The true position of Hermann Wagner in the controversy of the compass chart', *Imago Mundi*, v (1948), p. 24; Winter, 'A late portolan chart at Madrid', pp. 37–40.

28 Lanman, *Portolan charts*, pp. 27–8.

29 P.T. Pelham, 'The portolan charts: their construction and use in the light of contemporary techniques of marine survey and navigation' (MA thesis, Manchester, 1980), p. 84.

30 Campbell, 'Portolan charts', p. 388; Kelley, 'Origins and uses of the portolan charts', pp. 5–6.

31 John Dunmore (ed.), *The journal of Jean-François de Galaup de la Pérouse, 1785–1788*, ii (London, 1995), pp. 395, 430, 539–40.

32 Ramsay Cook (ed.), *The voyages of Jacques Cartier* (Toronto, 1993), p. 57: 1535.

33 E.G.R. Taylor (ed.), *The troublesome voyage of Captain Edward Fenton, 1582–1583* (Cambridge, 1959), p. 97.

34 C. Koeman, 'The chart trade in Europe from its origin to modern times', *Terrae Incognitae*, xii (1980), pp. 49–64.

35 E.G.R. Taylor, 'Hudson's strait and the oblique meridian', *Imago Mundi*, iii (1939), p. 49. But see Maria Fernanda et al., 'Portegese cartography in the renaissance' in David Woodward (ed.), *The history of cartography, volume three, part one: cartography in the European renaissance* (Chicago, 2007), pp. 1005–7 and Alison Sandman, 'Spanish nautical cartography in the renaissance', ibid., pp. 1137–8.

36 Richard Collinson (ed.), *The three voyages of Martin Frobisher in search of a passage to Cathaia and India by the north-west, AD 1576–8* (London, 1867), p. 319.

37 Helen Wallis (ed.), *Carteret's voyage round the world, 1766–1769*, ii (Cambridge, 1965), p. 238: 1772.

38 William Robert Broughton, *A voyage of discovery to the north Pacific Ocean in the years 1795, 1796, 1797, 1798* (London, 1804), p. 272.

39 William Herbert Hobbs, 'Verrazano's voyage along the North American coast in 1524', *Isis*, xli (1950), p. 272; Wallis, *Carteret's voyage*, pp. 168–9; Jan Wilson, *The Columbus myth: did men of Bristol reach America before Columbus?* (London, 1991), p. 149.

40 Michael Roe (ed.), *The journal and letters of Captain Charles Bishop on the north-west coast of America, in the Pacific and in New South Wales, 1794–1799* (Cambridge, 1967), p. 45: 1795; J.A. Schüller, 'A history of the Dutch nautical chart', *Journal of the Institute of Navigation*, vi, 2 (1953), pp. 195–6.

41 G.S. Ritchie, *The Admiralty chart: British naval hydrography in the nineteenth century* (London, 1967), pp. 163–4; Olivier Chapuis, 'Hydrographical departments' in John B. Hattendorf (ed.), *Oxford encyclopedia of maritime history*, ii (Oxford, 2007), pp. 162–7.

42 Howse and Sanderson, *The sea chart*, pp. 29, 37.

43 C. Koeman, *The sea on paper: the story of the Van Keulens and their 'Sea torch'* (Amsterdam, 1972).

44 E. Roukema, 'A discovery of Yucatan prior to 1503', *Imago Mundi*, xiii (1956), p. 34.

45 C.R.D. Bethune (ed.), *The discoveries of the world, from their first original unto the year of our lord 1555, by Antonio Galvano* (London, 1862), p. 82.

46 'Narrative of a journey from Aleppo to Basra in 1751 by John Carmichael' in Douglas Carruthers (ed.), *The desert route to India … 1745–1751* (London, 1928), pp. 135–79.

47 Michel Mollat du Jourdin and Monique de la Roncière with Marie-Madeleine Azard, Isabelle Raynaud-Nguyen and Marie-Antoinette Vamereau, *Sea charts of the early explorers, 13th to 17th century* (New York, 1984), p. 216: Nicolaus de Caverio, *c*.1505.

48 William Borough, chart of the north Atlantic, 1576, Hatfield House, CPM I.69; R.A. Skelton and John Summerson, *A description of maps and architectural drawings in the collection made by William Cecil, first Baron Burghley now at Hatfield House* (Oxford, 1971), p. 69 and plate 6; Howse and Sanderson, *The sea chart*, p. [38]: William Borough, chart of the North and Baltic Seas, *c*.1580; above, Fig. 8.11; below, Fig. 11.10.

49 Johannes and Gerard van Keulen, *De nieuwe groote ligtende Zee-fakkel* (Amsterdam, 1716–53; Amsterdam, 1970).

50 J.S. Cummins, *The travels and controversies of Friar Domingo Navarrete 1618–1686*, i (Cambridge, 1962), p. 46.

51 Commission to James Bassendine and others, 1588, in David Beers Quinn and Raleigh Ashlin Skelton (eds), *The principall navigations voiages and discoveries of the English nation by Richard Hakluyt*, i (Cambridge, 1965), p. 408.

52 Howse and Sanderson, *The sea chart*, pp. 46–7: 1596; Skelton and Summerson, *Maps and architectural drawings at Hatfield*, plate 15: 1604; Donald Wigal, *Historic maritime maps used for historic exploration, 1290–1699* (New York, 2000), map 78, p. 167: Jean Guérard, 1628.

53 Peter Whitfield, *The charting of the oceans: ten centuries of maritime maps* (London, 1996), pp. 99–101: Augustin de Ortiz, chart of the Caribbean, 1747.

54 Tony Campbell, 'The Drapers' Company and its school of seventeenth century chart-makers' in Helen Wallis and Sarah Tyacke (eds), *My head is a map: essays and memoirs in honour of R.V. Tooley* (London, 1973), p. 101.

55 Quoted by R.A. Skelton, *Explorers' maps* (London, 1958), p. 234: 1770.

56 Andrew David, 'Vancouver's survey methods and surveys' in Robin Fisher and Hugh Johnston (eds), *From maps to metaphors: the Pacific world of George Vancouver* (Vancouver, 1993), p. 53.

57 Ritchie, *Admiralty chart*, pp. 22–5.

58 Andrew S. Cook, 'Surveying the seas: establishing the sea routes to the East Indies' in James R. Akerman (ed.), *Cartographies of travel and navigation* (Chicago, 2006), pp. 88–9.

59 Blewitt, *Surveys of the seas*, p. 36: 1831 and 1837; Ritchie, *Admiralty chart*, p. 160: *c*.1827.

60 Terrell, *The evolution of the sea chart*, pp. 15–16.

61 D. Gernez, 'The works of Lucas Janszoon Wagenaer', *Mariner's Mirror*, xxiii (1937), p. 350.

62 Murdoch Mackenzie, *A maritim survey of Ireland and west coast of Great Britain* (London, 1776); Joseph Frederick Wallet Des Barres, *The Atlantic Neptune* (London, 1784).

63 Quoted by Olwen Caradoc Evans, *Marine plans and charts of Wales* (Map Collectors' Series, liv, London, 1969), pp. 15–16. See also Cook, 'Surveying the seas', pp. 70–1.

64 R.S.J. Clarke, 'The Irish charts in *Le Neptune François*: their sources and influence', History of Cartography Conference, Dublin, 1983, typescript, p. 1.

65 Blewitt, *Surveys of the seas*, p. 18; Alec Macdonald, 'Plans of Dover harbour in the sixteenth century', *Archaeologia Cantiana*, xlix (1937), pp. 111–12.

66 Gaspar Correa, *The three voyages of Vasco da Gama* (London, 1869), pp. 44–5.

67 Florence E. Dyer, 'The journal of Grenvill Collins', *Mariner's Mirror*, xiv (1928), pp. 197–219; William Bray (ed.), *The diary of John Evelyn*, ii (London, 1952), p. 175.

68   David, 'Vancouver's survey methods', p. 53; Matthew Flinders, *A voyage to Terra Australis*, i (London, 1814), p. 144; Ritchie, *Admiralty chart*, p. 78.

69   G.N.D. Evans, 'Hydrography: a note on eighteenth-century methods', *Mariner's Mirror*, lii (1966), p. 249.

70   Ritchie, *Admiralty chart*, p. 112.

71   Ibid., p. 24.

72   J.C. Beaglehole, *The life of Captain James Cook* (London, 1974), pp. 33–4.

73   Susanna Fisher, *Old sea charts*, list no. 27, n.d., item 56.

74   William Bell Clark (ed.), *Naval documents of the American revolution*, i (Washington, 1964), p. 1166; David, 'Vancouver's survey methods', p. 53.

75   Blewitt, *Surveys of the seas*, p. 30.

76   W.A. Seymour (ed.), *A history of the Ordnance Survey* (Folkestone, 1980), p. 102.

77   Andrew David, Felipe Fernandez-Armesto, Carlos Nori, Glyndwr Williams, *The Malaspina expedition, 1789–1794* (London and Madrid, 2001), p. 325.

78   William A. Spray, 'British surveys in the Chagos archipelago and attempts to form a settlement at Diego Garcia in the late eighteenth century', *Mariner's Mirror*, lvi (1970), p. 73.

79   David, 'Vancouver's survey methods', p. 63.

80   Blewitt, *Surveys of the seas*, pp. 86–7.

81   Dunmore, *Journal of La Pérouse*, p. 197.

82   William Bourne (ed. E.G.R. Taylor), *A regiment for the sea and other writings on navigation* (Cambridge, 1963), p. 57: 1571.

83   Luciana de Lima Martins, 'Mapping tropical waters: British views and visions of Rio de Janeiro' in Denis Cosgrove (ed.), *Mappings* (London, 1999), pp. 157–9.

84   Blewitt, *Surveys of the seas*, p. 35.

85   *Carte piana del Mare Mediterraneao*, reproduced in John A. Wolter and Ronald E. Grim (eds), *Images of the world: the atlas through history* (Washington, 1997), p. 282.

86   G.S. Ritchie, '500 years of graphical and symbolical representation on marine charts', History of Cartography Conference, Greenwich, 1975, typescript, p. 3.

87   *Abbreviations adopted in the [admiralty] charts* (London, 1835).

88   *Explanation of signs and abbreviations adopted in the charts issued by the Hydrographic Office, Admiralty* (London, 1866).

89   Norman J.W. Thrower, 'Edmond Halley as a thematic geo-cartographer', *Annals of the Association of American Geographers*, lix (1969), pp. 669–73; J. Proudman, 'Halley's tidal chart', *Geographical Journal*, c (1942), pp. 174–6; Howse and Sanderson, *The sea chart*, pp. 80–1; Helen M. Wallis and Arthur H. Robinson (eds), *Cartographical innovations: an international handbook of mapping terms to 1900* ([Tring], 1987), 'tidal map', pp. 154–6.

90   W.A. Spray, 'The surveys of John McCluer', *Mariner's Mirror*, lx (1974), p. 241.

91   Ritchie, '500 years of graphical and symbolical representation on marine charts', p. 8.

92   Flinders, *A voyage to Terra Australis*, i, p. v.

Chapter 12: The true features of the ground

1   Arthur H. Robinson and Barbara Bartz Petchenik, *The nature of maps: essays toward understanding maps and mapping* (Chicago, 1976), pp. 61–5.

2   Eduard Imhof (ed. H.J. Steward), *Cartographic relief presentation* (Berlin, 1982), pp. 1–3; Leo Bagrow (ed. R.A. Skelton), *History of cartography* (London, 1964), plate V: Mesopotamia, 4th millennium BC; J.B. Harley and David Woodward (eds), *The history of cartography, volume one: cartography in prehistoric, ancient, and medieval Europe and the Mediterranean* (Chicago, 1987), plate 5: Peutinger Table, 4th century; plate 7: Madaba mosaic map, 6th century; plate 22: Anglo-Saxon world map, 10th century; J.B. Harley and David Woodward (eds), *The history of cartography, volume two, book one: cartography in the traditional Islamic and south Asian societies* (Chicago, 1992), plate 11: Al Idrisi, world map, 15th century; Harley and Woodward, *History of cartography, volume two, book two:*

cartography in the traditional east and southeast Asian societies (Chicago, 1994), p. 60: China, 15th century; p. 362: Japan, 12th century; plate 17: Korea, 15th century; David Woodward and G. Malcolm Lewis (eds), *The history of cartography, volume two, book three: cartography in the traditional African, American, Arctic, Australian, and Pacific societies* (Chicago, 1998), p. 189: Mexico, 16th century; Denis Wood, 'Now and then: comparisons of ordinary Americans' symbol conventions with those of past cartographers', *Prologue: The Journal of the National Archives*, ix (1977), pp. 151–61; Denis Wood with John Fels, *The power of maps* (London, 1992), ch. 6. See above, Figs 1.4, 1.8, 2.7.

3 Alice Garnett, 'Insolation, topography and settlement in the Alps', *Geographical Review*, xxv (1935), pp. 601–17; Alice Garnett, 'Insolation and relief: their bearing on the human geography of Alpine regions', *Publications of the Institute of British Geographers* (1937), pp. 1–71; Dorothy Sylvester, *Map and landscape* (London, 1952), p. 207.

4 G.E. Morris, 'The profile of Ben Loyal from Pont's map entitled *Kyntail*', *Scottish Geographical Magazine*, cii (1986), pp. 74–9.

5 Günter Schilder, 'Organization and evolution of the Dutch East India Company's hydrographic office in the seventeenth century', *Imago Mundi*, xxviii (1976), p. 71.

6 Franz Grenacher, 'Die kartographische Erschliessung des Jura', *Regio Basiliensis*, xiii, 1 & 2 (1972), Abb. 10, p. 103: manuscript atlas, 1592.

7 William Alingham, *A short account of the nature and use of maps* (London, 1698), p. 24.

8 Barbara E. Mundy, *The mapping of New Spain: indigenous cartography and the maps of the Relaciones Geograficas* (Chicago, 1996), p. 101.

9 Coolie Verner, *Smith's Virginia and its derivatives* (Map Collectors' Series, xlv, London, 1968), plates I and III; Christopher W. Lane, 'Whose map is it anyway?', *The Map Collector*, xxxvi (1986), pp. 16–17.

10 Henry N. Stevens, *Lewis Evans: his map of the middle British colonies in America* (London, 1920), p. 37, quoting Thomas Pownall.

11 P.D.A. Harvey and Harry Thorpe, *The printed maps of Warwickshire, 1576–1900* (Warwick, 1959), plate 2: 1603; Gerard Mercator, Henry Hondius and Joannes Janssonius (ed. R.A. Skelton), *Atlas or a geographicke description of the world*, ii (Amsterdam, 1636, 1968), p. 297: Beauce; p. 313: Berry.

12 Mary Blewitt, *Surveys of the seas: a brief history of British hydrography* (London, 1957), p. [41]: John Narborough, Magellan Straits, 1670; Derek Howse and Michael Sanderson, *The sea chart: an historical survey based on the collections in the National Maritime Museum* (Newton Abbot, 1973), p. 104: Murdoch Mackenzie, 1776; above, Fig. 2.11.

13 William M. Ivins, Jr, *How prints look: photographs with a commentary* (Boston, 1958).

14 Brian M. Ambroziak and Jeffrey R. Ambroziak, *Infinite perspectives: two thousand years of three-dimensional mapmaking* (New York, 1999), pp. 42–3: Johann Heinrich Weiss and Joachim Eugen Müller, *Atlas de la Suisse* (1802).

15 R.A. Skelton and John Summerson, *A description of maps and architectural drawings in the collection made by William Cecil, first Baron Burghley now at Hatfield House* (Oxford, 1971), plate 1: anonymous plan of Dover Harbour, 1532.

16 Elizabeth Clutton, 'Some seventeenth century images of Crete: a comparative analysis of the manuscript maps by Francesco Basilicata and the printed maps by Marco Boschini', *Imago Mundi*, xxxiv (1982), p. 49.

17 William Jones, plan of Cork, 1602, Trinity College, Dublin, MS 1209/45.

18 John Dunstall, map of Carrickfergus, 1612, British Library, Cotton MS Aug. I, ii, 41.

19 Basil Jackson, *A course of military surveying* (London, 1838), pp. 70–2.

20 Ibid., p. 76.

21 J.H. Andrews, *A paper landscape: the Ordnance Survey in nineteenth-century Ireland* (Oxford, 1975), pp. 240–3.

22 E.A. Reeves, *Maps and map-making* (London, 1910), p. 129; H.S.L. Winterbotham, *A key to maps* (2nd ed., London, 1939), p. 97; Edward Lynam, 'Period ornament, writing, and symbols on maps, 1250–1800' in *The mapmaker's art: essays on the history of maps* (London, 1953), p. 41; above, Figs 10.10, 10.13.

23 Amato Pietro Frutaz, *Le piante di Roma*, ii (Rome, 1962), tav. 189–209 (Leonardo Bufalini), tav. 35 (Onofrio Panvinio, *Roma antica edita*, 1565); John A.Pinto, 'Origins and development of the ichnographic city plan', *Journal of the Society of Architectural Historians*, xxxv, 1 (1976), p. 44; Jessica Maier, 'Mapping past and present: Leonardo Bufalini's plan of Rome', *Imago Mundi*, lix, 1 (2007), p. 4, Fig. 5 et seq.

24 John Ogilby (ed. J.B. Harley), *Britannia, London 1675* (Amsterdam, 1970), no. 22, Daventry, Coleshill.

25 Abraham Ortelius (ed. R.A. Skelton), *The theatre of the whole world, London, 1606* (Amsterdam, 1968), p. 82.

26 *Civitatis imp. Lindaviensis territorum* (scale *c.*1:40,000) in Joan Blaeu, *Le grand atlas*, iii (Amsterdam, 1663, 1967), p. 137.

27 Ogilby, *Britannia*, no. 32, Bridgwater to Ascot; John Booth, *Antique maps of Wales* (Montacute, Somerset, 1977), pp. 82–3: G.Rollos, *An accurate map of Brecknockshire*, 1760.

28 Michael Swift, *Historical maps of Europe* (London, 2000), p. 33: battle of Fleurus, 1690; above, Fig. 10.1.

29 Merton College, Oxford, MS D.3.30; Sarah Bendall, 'Draft town maps for John Speed's *Theatre of the empire of Great Britaine*', *Imago Mundi*, liv (2002), pp. 30–45.

30 Edward Lynam, 'Maps of the Fenland' in *The Victoria history of the counties of England, Huntingdonshire*, iii (London, 1936), p. 294: Jonas Moore, *A true mapp of ye great Levell of the Fens*; Edward Lynam, *The Carta marina of Olaus Magnus, Venice 1539 and Rome 1572* (Jenkintown, PA, 1949), p. 33; Lynam, 'Period ornament', p. 41; Robert Macfarlane, *Mountains of the mind: a history of a fascination* (London, 2004), pp. 182–3.

31 David Buisseret, 'Modeling cities in early modern Europe' in David Buisseret (ed.), *Envisioning the city: six studies in urban cartography* (Chicago, 1998), pp. 125–43.

32 Yolande Jones, 'Aspects of relief portrayal on 19th century British military maps', *Cartographic Journal*, xi, 1 (1974), p. 7: quotation from 1832.

33 Mireille Pastoureau, *Les atlas français xvie–xviie siècles: répertoire bibliographique et étude* (Paris, 1984), Fig. 73: Nicolas de Fer, Girone, 1694.

34 British Library, King's Maps, 16.31.2: Medway; King's Maps, 16.46: Dover.

35 Sarah Bendall, *Dictionary of land surveyors and local map-makers of Great Britain and Ireland, 1530–1815*, i (London, 1997), plate 8.

36 F. Hull, *Catalogue of estate maps, 1590–1840, in the Kent county archives office* (Maidstone, Kent, 1973), plate 15.

37 John Varley, 'John Rocque: engraver, surveyor, cartographer and map-seller', *Imago Mundi*, v (1948), pp. 85, 89; Hugh Phillips, 'John Rocque's career', *London Topographical Record*, xx (1953), p. 10.

38 Alfred E. Lemmon, John T. Magill and Jason R. Wiese (eds), *Charting Louisiana: five hundred years of maps* (New Orleans, 2003), p. 58: Guillaume Delisle, 1718; Maud D. Cole, 'The cartographic treasures in the New York Public Library', *The Map Collector*, xliii (1988), p. 10: *A map of the countrey of the Five Nations*, 1724.

39 Walter Satzinger, 'The Great Atlas of Germany edited by Johann Wilhelm Jaeger, Frankfurt am Main, 1789', History of Cartography Conference, Greenwich, 1975, typescript, Beilage 5: *Carte topographique d'Allemagne*, Feuille XLIX, 1789.

40 A.M. Cubbon, *Early maps of the Isle of Man* (Douglas, 1967), p. 29; Roger Baynton-Williams, *Investing in maps* (London, 1969), p. 108; A. Arrowsmith, *Asia* (London, 1801).

41 Lemmon, Magill and Wiese, *Charting Louisiana*, p. 60: Bernard de la Harpe, 'Carte nouvelle de la partie de l'oüest de la Louisianne', 1723.

42 François de Dainville, 'From the depths to the heights', *Surveying and Mapping*, xxx, 3 (1970), p. 389; Charles Close (ed. J.B. Harley), *The early years of the Ordnance Survey* (1926; Newton Abbot, 1969), p. 45.

43 Josef W. Konvitz, *Cartography in France 1660–1848: science, engineering, and statecraft* (Chicago, 1987), p. 100.

44 Marc Duranthon, *La carte de France: son histoire 1678–1978* (Paris, 1978), p. 38.

45 John Holwell, *A sure guide to the practical surveyor, in two parts* (London, 1678), pp. 192–5.

46 Monique Pelletier, *La carte de Cassini: l'extraordinaire aventure de la carte de France* (Paris, 1990), pp. 98–9.

47 Jackson, *Military surveying*, p. 75.

48 J.B. Harley et al., *The old series Ordnance Survey maps of England and Wales, scale 1 inch to 1 mile*, i (Lympne, 1975), pp. xxx, xxxii; iv (Lympne, 1986), p. xviii; v (Lympne, 1987), pp. xiii, xiv.

49 Jackson, *Military surveying*, pp. 76–8.

50 W.E. May, *A history of marine navigation* (Henley-on-Thames, 1973), pp. 203–6; John Peter Oleson, 'Testing the waters: the role of sounding weights in ancient Mediterranean navigation' in Robert L. Hohlfelder (ed.), *The maritime world of ancient Rome* (Ann Arbor, MI, 2008), pp. 119–33.

51 G. de Boer and R.A. Skelton, 'The earliest English chart with soundings', *Imago Mundi*, xxiii (1969), pp. 9–16, citing M. Destombes, 'Les plus anciens sondages portés sur les cartes nautiques aux XVIᵉ et XVIIᵉ siècles', *Bull. de l'Inst. Océanographique*, special no. 2 (1968), pp. 199–222; above, Figs 2.3, 11.10–11.21, passim.

52 Jackson, *Military surveying*, p. 91; Jones, 'Relief portrayal on 19th century British military maps', pp. 4–6.

53 Skelton and Summerson, *Maps and architectural drawings in the collection made by William Cecil*, plate 8: 1581.

54 A.A. Horner, 'Some examples of the representation of height data on Irish maps before 1750, including an early use of the spot-height method', *Irish Geography*, vii (1974), pp. 68–80.

55 Dainville, 'From the depths to the heights', pp. 389–91.

56 Andrew A. Lipscomb (ed.), *The writings of Thomas Jefferson*, xiv (Washington, 1905), p. 480: 1816.

57 Derek J. Price, 'Medieval land surveying and topographical maps', *Geographical Journal*, cxxi, 1 (1955), p. 2.

58 Quoted in W.A. Seymour (ed.), *A history of the Ordnance Survey* (Folkestone, 1980), p. 363.

59 W. Whewell, 'Account of a level line measured from the Bristol channel to the English channel, during the year 1837–8 …', *British Association report, eighth meeting* (1839), vii, pp. 1–17.

60 Lipscomb, *Writings of Thomas Jefferson*, xiv, p. 480.

61 Seymour, *History of the Ordnance Survey*, p. 7.

62 Duranthon, *Carte de France*, p. 31.

63 See above, Fig. 12.6. Other examples reproduced in Bibliothèque Royale Albert Iᵉʳ, *Le cartographe Gérard Mercator 1512–1594* (Brussels, 1994), pp. 19–21; G.R. Crone, *Maps and their makers* (3rd ed., London, 1966), p. 70; Michel Mollat du Jourdin and Monique de la Roncière with Marie-Madeleine Azard, Isabelle Raynaud-Nguyen and Marie-Antoinette Vamereau, *Sea charts of the early explorers, 13th to 17th century* (New York, 1984), plate 12: Meccia de Viladestes, 1413; plates 13, 14: Cristoforo Buondelmonte, 1420.

64 David Buisseret (ed.), *Monarchs, ministers and maps: the emergence of cartography as a tool of government in early modern Europe* (Chicago, 1992), p. 38: John Rogers, map of Boulogne and vicinity, c.1546, British Library, Cotton MS Aug. I, ii, 77; Martin Clayton, *Leonardo da Vinci: a curious vision* (London, 1996), pp. 95, 102.

65 Ortelius, *Theatre*, p. 50: East Friesland; Edmund Dummer et al., 'A survey of the ports on the south west coast of England', 1698, British Library, Sloane MS 3233; Blewitt, *Surveys of the seas*, p. [39]: James Grant, plan of Boston harbour, 1775.

66 Robert Lythe, map of Carrickfergus, 1567, Trinity College, Dublin, MS 1209/26.

67 Konvitz, *Cartography in France*, pp. 67–71, 77–8, 95–102; Paul van den Brink, 'River landscapes: the origin and development of the printed river map in the Netherlands, 1725–1795', *Imago Mundi*, lii (2000), pp. 66–78, especially note 5.

68 John A. Wolter, 'The heights of mountains and the lengths of rivers', *Surveying and Mapping*, xxxii, (1972), pp. 317, 328 n. 13; Dainville, 'From the depths to the heights', p. 395.

69 Charles Hutton, 'An account of the calculations made from the survey and measures taken at Schiehallien, in order to ascertain the mean density of the earth', *Philosophical Transactions of the*

Royal Society of London, lxviii, 2 (1778), pp. 689–788; Derek Howse, *Nevil Maskelyne, the seaman's astronomer* (Cambridge, 1989), p. 136.

70   William Ravenhill, 'Churchman's contours?', *The Map Collector*, xxxiv (1986), pp. 22–5.

71   Richard Oliver, 'The Ordnance Survey in Great Britain, 1835–1870', PhD thesis, University of Sussex, 1985, p. 239.

72   Ordnance Survey, *Characteristic sheet for the six inch and one inch plans* (Southampton, 1847).

73   Bolton Glanvill Corney, *The quest and occupation of Tahiti by emissaries of Spain during the years 1772–1776*, i (London, 1913): plan of La Ysla de Amat, 1772, chart III; Aleksey K. Zaytsev, 'The three earliest charts of Akhtiar (Sevastopol) Harbour', *Imago Mundi*, lii (2000), p. 119: Ivan Bersenev, map of 1783; Richard W. Stephenson, *The cartography of northern Virginia: facsimile reproductions of maps dating from 1608 to 1915* (Fairfax County, 1981), plate 20, p. 34: Andrew Ellicott, map of the territory of Columbia, 1794; Michael Swift, *Historical maps of Europe* (London, 2000), p. 118: *Topografia dell isola di Zante*, 1821.

74   Yolande Hodson and Alan Gordon, *An illustrated history of 250 years of military survey* ([London], 1997), p. 8: H.B. Harris, map of battle of Corunna, 1809.

75   Ambroziak and Ambroziak, *Infinite perspectives*, p. 22; Erwin Raisz, *General cartography* (2nd ed., New York, 1948), p.103; Helen M. Wallis and Arthur H. Robinson, *Cartographical innovations: an international handbook of mapping terms to 1900* ([Tring], 1987), p. 219.

76   Oliver, 'The Ordnance Survey in Great Britain', pp. 128–9.

77   Arthur R. Hinks, *Maps and survey* (5th ed., Cambridge, 1944), p. 25.

78   Dainville, 'From the depths to the heights', p. 395; Wallis and Robinson, *Cartographical innovations*, 'hypsometric map', pp. 145–6, 'layer tints', pp. 229–30.

79   Harley, *Old series Ordnance Survey maps*, iv, p. xxviii.

80   [Yolande Hodson], 'Shaded contours', *Sheetlines*, iii (1982), pp. 10–11; Yolande Hodson, 'A little light relief: a personal view of collecting O.S. maps', *The Map Collector*, liv (1991), pp. 10–11; Peter Clark, '"A useful collection of maps and charts"', *The Map Collector*, xxxv (1986), p. 4; Raisz, *General cartography*, p. 116.

81   J.B. Harley and R.R. Oliver, *The old series Ordnance Survey maps of England and Wales*, vi (Lympne, 1992), passim.

Chapter 13: In no case arbitrary sounds

1   John Green, *Remarks, in support of the new chart of North and South America; in six sheets* (London, 1753), pp. 13, 15.

2   R.C.E. Quixley, *Antique maps of Cornwall and the Isles of Scilly* (Penzance, Cornwall, 1966), pp. 34–47: 1720–84.

3   C.C.A. Gosch (ed.), *Danish Arctic expeditions, 1605 to 1620*, ii (London, 1897), pp. 76–7; Kenneth Nebenzahl, *Atlas of Columbus and the great discoveries* (Chicago, 1990), pp. 158–9: Edward Wright, world map, 1599.

4   R.A. Skelton, *Explorers' maps: chapters in the cartographic record of geographical discovery* (London, 1958), p. 277; John Masefield (ed.), *Dampier's voyages*, i (London, 1906), p. 553.

5   Matthew H. Edney, *Mapping an empire: the geographical construction of British India, 1765–1843* (Chicago, 1997), p. 331.

6   Hans Wolff et. al., *Philipp Apian und die Kartographie der Renaissance* (Munich, 1989), p. 77: Philipp Apian, map of Bavaria, 1568; E. Heawood, 'Early map indexing', *Geographical Journal*, lxxx (1932), pp. 247–9.

7   'Stayshal', cartoon, *The Map Collector*, lxix (1994), p. 35.

8   Robin Flower, 'Laurence Nowell and the discovery of England in Tudor times', *Proceedings of the British Academy*, xxi (1935), pp. 3–29; F.J. North, *Humphrey Lhuyd's maps of England and of Wales* (National Museum of Wales, Cardiff, 1937).

9   George R. Stewart, *Names on the globe* (New York, 1975), pp. 15–17.

10 Edmund Gibson, *Britannia: or a chorographical description of Great Britain and Ireland ... by William Camden* (2nd ed., London, 1722), preface, p. [6].

11 Mary Blewitt, *Surveys of the seas: a brief history of British hydrography* (London, 1957), p. 35.

12 *Description géometrique de la France par M.Cassini de Thury* (Paris, 1783), p. 205.

13 J.C. Beaglehole (ed.), *The journals of Captain James Cook on his voyages of discovery, i: the voyage of the* Endeavour (London, 1955), pp. 291–4.

14 Quoted in Michael Charlesworth, 'Mapping, the body and desire: Christopher Packe's chorography of Kent' in Denis Cosgrove (ed.), *Mappings* (London, 1999), p. 116; Mary Sponberg Pedley, *The commerce of cartography: making and marketing maps in eighteenth-century France and England* (Chicago, 2005), pp. 170–1 (J.B. d'Anville), 182–3.

15 François de Dainville, 'La levée d'une carte en Languedoc à l'entour de 1730' in *La cartographie reflet de l'histoire* (Paris, 1986), p. 372.

16 Eilert Ekwall, *The concise dictionary of English place-names* (4th ed., Oxford, 1960), p. 331.

17 Bolton Glanvill Corney (ed.), *The quest and occupation of Tahiti by emissaries of Spain during the years 1772–1776*, i (London, 1913), p. 190.

18 R.A. Skelton, 'Catalogue of manuscript maps' in R.A. Skelton and John Summerson, *A description of maps and architectural drawings in the collection made by William Cecil, first Baron Burghley now at Hatfield House* (Oxford, 1971), no. 14, p. 41: map of English Channel coasts, late 16th century.

19 Francis W. Steer and Felix Hull, *Illustrated handbook to exhibition of Essex estate, county and official maps* (1947), plate vi: John Walker, 1615.

20 Thomas Porter, *The newest and exactest mapp of the most famous cities London and Westminster ...* (London, 1655).

21 J.B. Harley and Gwyn Walters, 'Welsh orthography and Ordnance Survey mapping, 1820–1905', *Archaeologia Cambrensis*, cxxxi (1982), p. 123: 1888.

22 Isaac Taylor, *Words and places, or etymological illustrations of history, ethnology and geography* (new edition, London, 1873), p. 311.

23 A.B. Taylor, 'The name "St Kilda"', *Scottish Studies*, xiii (1969), pp. 145–58.

24 Duncan A. Johnston, *Account of the methods and processes adopted for the production of the maps of the Ordnance Survey of the United Kingdom ... revised in 1901* (Southampton, 1902), p. 88; W.A.R. Richardson, 'The origin of place-names on maps', *The Map Collector*, lv (1991), pp. 18–23.

25 Hiob Ludolf, *A new history of Ethiopia* (London, 1682), p. 21.

26 R.A. Skelton, 'Ludolf's map of Abyssinia' in William Foster (ed.), *The Red Sea and adjacent countries at the close of the seventeenth century* (London, 1949), pp. 182–5.

27 G. Walters, 'The Morrises and the map of Anglesey', *Welsh History Review*, v, 2 (1970), p. 170.

28 A. Arrowsmith, *A map exhibiting the new discoveries in the interior parts of North America* (London, 1795); Carl I. Wheat, *Mapping the Transmississippi West 1540–1861, i: From Lewis and Clark to Fremont 1804–1845* (San Francisco, 1958), p. 57: *A map of Lewis and Clark's track, across the western portion of North America ... in 1804, 5 & 6* (1814).

29 M. Aurousseau, *The rendering of geographical names* (London, 1957), pp. 72–3; Derek Nelson, *Off the map: the curious histories of place-names* (New York, 1997), pp. 123–8.

30 Cécile Kruyfhooft, 'A recent discovery: Utopia by Abraham Ortelius', *The Map Collector*, xvi (1981), p. 12; Robert W. Karrow Jr, *Mapmakers of the sixteenth century and their maps: bio-bibliographies of the cartographers of Abraham Ortelius, 1570* (Winnetka, IL, 1993), p. 25; Peter Barber, *The map book* (London, 2005), pp. 132–3.

31 Kenneth Nebenzahl and Don Higginbotham, *Atlas of the American revolution* (Chicago, 1974), pp. 62–3.

32 Wolfgang Haase and Meyer Reinhold, *The classical tradition and the Americas, i: European images of the Americas and the classical tradition* (Berlin, 1994), p. 119.

33 Horace K. Mann, *The lives of the popes in the middle ages*, ix (London, 1925), p. 323.

34 Advice of Egnatio Danti in 1577, quoted by Thomas Frangenberg, 'Chorographies of Florence: the use of city views and city plans in the sixteenth century', *Imago Mundi*, xlvi (1994), pp. 50, 57.

35 Quoted by James Elliot, *The city in maps: urban mapping to 1900* (London, 1987), p. 26.

36   Masefield, *Dampier's voyages*, i, pp. 20–1.

37   R.V. Tooley, 'The early mapping of the Falkland Islands', *The Map Collector*, xx (1982), pp. 2–6; Angela Fordham, *Maps of the Falkland Islands*, Map Collectors' series, xi (London, 1964).

38   R.V. Tooley, Charles Bricker and G.R. Crone, *A history of cartography: 2500 years of maps and mapmakers* (London, 1969), p. 264: Daniel Djurberg, *Karta over Polynesien* (Stockholm, 1780).

39   Green, *Remarks, in support of the new chart*, p. 44.

40   Beaglehole, *Voyage of the* Endeavour, pp. 291–4; Michael E. Hoare (ed.), *The* Resolution *journal of Johann Reinhold Forster, 1772–1775*, iii (London, 1982), p. 480; Paul Carter, *The road to Botany Bay: an essay in spatial history* (London, 1987), p. 65: T.L. Mitchell, *c*.1838; Francis Beaufort, 'Karamania' (1817), quoted by Marion Hercock, 'Francis Beaufort, R.N., 1774–1857' in Patrick H. Armstrong and Geoffrey J. Martin (eds), *Geographers: biobibliographical studies*, xix (2000), p. 11.

41   John Dunmore (ed.), *The journal of Jean-François de Galaup de la Pérouse, 1785–1788*, ii (London, 1995), p. 434.

42   Don W. Thomson, *Men and meridians: the history of surveying and mapping in Canada, volume 1: prior to 1867* (Ottawa, 1966), p. 57.

43   Samuel Hearne, *A journey from Prince of Wales's Fort in Hudson's Bay, to the Northern Ocean* (London, 1795), p. 210.

44   D. Graham Burnett, *Masters of all they surveyed: exploration, geography and a British El Dorado* (Chicago, 2000), p. 114.

45   Fabian O'Dea, 'The wandering cape: the location of Pointe Riche, Newfoundland', *The Map Collector*, lxix (1994), pp. 3–8.

46   David Beers Quinn and Raleigh Ashlin Skelton (eds), *The principall navigations voiages and discoveries of the English nation by Richard Hakluyt*, i (Cambridge, 1965), pp. 408, 458.

47   R.H. Major (ed.), *Early voyages to Terra Australis, now called Australia* (London, 1859), p. 90.

48   'North-west Fox' in Miller Christy (ed.), *The voyages of Captain Luke Foxe of Hull, and Captain Thomas James of Bristol in search of a north-west passage, in 1631–32; with narratives of … earlier north-west voyages* (London, 1894), p. 86.

49   Charles Nicholl, *The creature in the map: a journey to El Dorado* (New York, 1995), p. 313.

50   'North-west Fox', p. 10.

51   Gosch, *Danish Arctic expeditions*, p. 75.

52   Frank Debenham (ed.), *The voyage of Captain Bellingshausen to the Antarctic seas, 1819–1821* (London, 1945), p. 107.

53   J.C. Beaglehole, *The life of Captain James Cook* (London, 1974), p. 590.

54   John Keay, *The great arc: the dramatic tale of how India was mapped and Everest was named* (London, 2000), pp. 166–71: 1856.

55   J.C. Beaglehole (ed.), *The journals of Captain James Cook on his voyages of discovery: ii: The voyage of the* Resolution *and* Adventure, *1772–1775* (Cambridge, 1961), p. 544.

56   H. Rackham (ed.), *Pliny: natural history* (London, 1942), p. 219.

57   J.C. Beaglehole (ed.), *The journals of Captain James Cook on his voyages of discovery: The voyage of the* Resolution *and* Discovery, *1776–1780* (Cambridge, 1967), p. 1103: David Samwell's journal.

58   E.E. Rich (ed.), *Copy book of letters outward, 1680–1687*, Champlain Society: Hudson's Bay Company series, xi (Toronto, 1948), pp. 281, 299, 322.

59   A. Teixeira da Mota, 'Some notes on the organization of hydrographical services in Portugal until the beginning of the 19th century', History of Cartography Conference, Greenwich, 1975, typescript, p. 4.

60   T.M. Perry, *The discovery of Australia: the charts and maps of the navigators and explorers* (London, 1982), p. 91; plate 50, pp. 94–5: *Carte générale de la Terre Napoléon*, 1808.

61   Perry, *Discovery of Australia*, plate 29, pp. 62–3: *A complete map of the southern continent*; Tooley, Bricker and Crone, *A history of cartography*, p. 264: Daniel Djurberg, *Karta over Polynesien* (Stockholm, 1780).

62   Alfred E. Lemmon, John T. Magill and Jason R. Wiese (eds), *Charting Louisiana: five hundred years of maps* (New Orleans, 2003), p. x: Emanuel Bowen, *A new and accurate map of Louisiana*, 1747.

63 Walter Raleigh, *The history of the world* (London, 1614), book 2, part 1, p. 574; Helen Wallis, 'The cartography of Drake's voyage' in Norman J.W. Thrower (ed.), *Sir Francis Drake and the famous voyage, 1577–1580: essays commemorating the quadricentennial of Drake's circumnavigation of the earth* (Berkeley, CA, 1984), pp. 152–3.

64 John Ogilby (ed. J.B. Harley), *Britannia, London 1675* (Amsterdam, 1970), no. 80.

65 Matthew Flinders, *A voyage to Terra Australis*, i (London, 1814), p. 142; Carter, *The road to Botany Bay*, pp. 184–5.

66 Reproduced in Iolo and Menai Roberts, 'Which John Evans are we talking about?', *The Map Collector*, xlvi (1989), p. 22.

67 J.M. Cohen, *The four voyages of Christopher Columbus* (London, 1969), p. 115.

68 Beaglehole, *Life of Cook*, p. 429.

69 Helen Wallis (ed.), *Carteret's voyage round the world, 1766–1769*, ii (Cambridge, 1965), p. 338.

70 A. Arrowsmith, *Asia* (London, 1801).

71 Fergus Fleming, *Barrow's boys* (London, 1998), pp. 312–13; Conrad E. Heidenreich, 'The fictitious islands of Lake Superior', *The Map Collector*, xxvii (1984), pp. 21–5; Seymour I. Schwartz, *The mismapping of America* (Rochester, NY, 2003), pp. 192–7.

72 Richard I. Ruggles, *A country so interesting: the Hudson's Bay Company and two centuries of mapping, 1670–1870* (Montreal, 1991), plate 35, p. 159.

73 G.M. Asher (ed.), *Henry Hudson the navigator* (London, 1860), p. 14.

74 Flinders, *A voyage to Terra Australis*, i, p. 124.

75 Major, *Early voyages to Terra Australis*, p. lxxv; Clements Markham (ed.), *The voyages of Pedro Fernandez de Quiros, 1595–1606* (London, 1904), p. xxviii: Luis Vaez de Torres, 1606.

76 Beaglehole, *Voyage of the* Resolution *and* Discovery, p. 363.

77 Jessie M. Sweet, 'Robert Jamieson and the explorers: the search for the north-west passage', *Annals of Science*, xxxi (1974), p. 24.

78 W. Kaye Lamb, *George Vancouver: a voyage of discovery to the north Pacific Ocean and round the world, 1791–1795*, i (London, 1984), p. 149.

79 Edward Tomkins, *Newfoundland's interior explored* (St John's, 1986), p. 5.

80 W.S.W. Vaux, *The world encompassed by Sir Francis Drake, being his next voyage to that to Nombre de Dios* (London, 1854), p. 132.

81 J. Wreford Watson, *Mental images and geographical reality in the settlement of North America*, (Nottingham, 1967), p. 6, quoting J.B. Brebner, *Explorers of North America, 1492–1806* (London, 1955), p. 203.

82 Francis Celoria, 'Delta as a geographical concept in Greek literature', *Isis*, lvii (1966), pp. 385–8.

83 Charles T. Beke (ed.), *True description of three voyages by the north-east towards Cathay and China, undertaken in the years 1594, 1595 and 1596* (London, 1853), p. 33.

84 Clements R. Markham (ed.), *The voyages of William Baffin, 1612–1622* (London, 1881), p. 99.

85 David Icenogle, 'The geographic and cartographic work of the American military mission to Egypt, 1870–1878', *The Map Collector*, xlvi (1989), p. 31.

Chapter 14: Shadowed and counterfeited

1 David Woodward, 'The woodcut technique' in David Woodward (ed.), *Five centuries of map printing* (Chicago, 1975), p. 43.

2 Quoted by Coolie Verner, 'Engraved title plates for the folio atlases of John Seller' in Helen Wallis and Sarah Tyacke (eds), *My head is a map: essays and memoirs in honour of R.V. Tooley* (London, 1973), p. 50.

3 Abraham Ortelius (ed. R.A. Skelton), *The theatre of the whole world, London, 1606* (Amsterdam, 1968), p. 55.

4 Mark Rose, *Authors and owners: the invention of copyright* (Cambridge, MA, 1993), p. 9.

5 Robert W. Karrow Jr, *Mapmakers of the sixteenth century and their maps: bio-bibliographies of the cartographers of Abraham Ortelius, 1570* (Winnetka, IL, 1993), pp. 67, 210, 350, 390; Nicholas Crane,

Mercator: the man who mapped the planet (London, 2002), pp. 235–7; Robert Haardt, 'The globe of Gemma Frisius', *Imago Mundi*, ix (1951), pp. 109–10; Mary Sponberg Pedley, *The commerce of cartography: making and marketing maps in eighteenth-century France and England* (Chicago, 2005), pp. 96–118.

6 8 Geo. II c. 13 ('Hogarth's act', 1734–5); 7 Geo. III c. 38 (1767); 17 Geo. III c. 57 (1777).

7 R.A. Skelton, 'Copyright and piracy in eighteenth-century chart publication', *Mariner's Mirror*, xlvi (1960), pp. 207–8; Mary Sponberg Pedley, *The commerce of cartography: making and marketing maps in eighteenth-century France and England* (Chicago, 2005), ch. 4.

8 David Hunter, 'Copyright protection for engravings and maps in eighteenth-century England', *The Library*, 6th ser., ix (1987), p. 147.

9 Gerald Strauss, *Sixteenth-century Germany: its topography and topographers* (Madison, WI, 1959), p. 82.

10 Quoted by D.G. Moir in Royal Scottish Geographical Society, *The early maps of Scotland to 1850* (Edinburgh, 1973), p. 47.

11 Don W. Thomson, *Men and meridians: the history of surveying and mapping in Canada, i, prior to 1867* (Ottawa, 1966), p. 65.

12 Richard Eden, *The arte of navigation* ([London], 1572).

13 J. Holwell, *A sure guide to the practical surveyor* (London, 1678), pp. 150–1.

14 H.W. Dickinson, 'A brief history of draughtsmen's instruments', *Transactions of the Newcomen Society*, xxvii (1949–50, 1950–1), p. 77.

15 Basil Jackson, *A course of military surveying* (London, 1838), p. 79.

16 Cennino d'Andrea Cennini (trans. Daniel V. Thompson Jnr), *Il libro dell'arte: the craftsman's handbook* (New Haven, CT, 1933), pp. 13–14.

17 J.C. Beaglehole, *The journals of Captain James Cook on his voyages of discovery, i: the voyage of the Endeavour, 1768–1771* (Cambridge, 1955), p. 617.

18 Aaron Rathborne, *The surveyor in foure bookes* (London, 1616), p. 175; John Love, *Geodaesia: or, The art of surveying and measuring of land made easie* (London, 1688), p. 139; John Wing, *Geodaetes practicus redivivus. The art of surveying: formerly publish'd by Vincent Wing, math. now much augmented and improved* (London, 1700), p. 183.

19 William Davis, *A complete treatise of land surveying by the chain, cross and offset staffs only* (London, 1798), pp. 204–5.

20 Robert Latham and William Matthews (eds), *The diary of Samuel Pepys*, ix (London, 1976), pp. 340, 437.

21 John Hammond (enlarged by Samuel Warner), *The practical surveyor: containing the most approved methods for surveying of lands and waters, by the several instruments now in use* (London, 1750), pp. 173–82; Maya Hambly, *Drawing instruments: their history, purpose and use for architectural drawings* (London, 1982), p. 29.

22 Ibid., p. 29; George G. André, *The draughtsman's handbook of plan and map drawing* (London, 1891), pp. 136–40 and plate 26.

23 *London encyclopedia* (London, 1829), sub. 'Camera lucida'; Hambly, *Drawing instruments*, p. 39.

24 Marcel Watelet (ed.), *The Mercator atlas of Europe* (Antwerp, 1998), p. 12; Crane, *Mercator*, pp. 171, 310.

25 W.S.W. Vaux (ed.), *The world encompassed by Sir Francis Drake, being his next voyage to that to Nombre de Dios* (London, 1854), p. xi.

26 J. Lennart Berggren and Alexander Jones (eds), *Ptolemy's* Geography: *an annotated translation of the theoretical chapters* (Princeton, NJ, 2000), p. 80.

27 Karrow, *Mapmakers of the sixteenth century*, p. 511.

28 Henry James, *Account of the process of engraving the Ordnance maps of the United Kingdom* ([Southampton], 1872), p. [2]; Hans W. Singer and William Strang, *Etching, engraving and other methods of printing pictures* (London, 1897), p. 27; Coolie Verner, 'Copperplate printing' in Woodward, *Five centuries of map printing*, pp. 52–3; http://www.maphis@geo.uu.nl, 24–6 September 2007, 'Image alignment & copper plate engravings'.

29 Robert Plot, *The natural history of Oxford-shire* (Oxford, 1705), p. (b).

30 François de Dainville, *Cartes anciennes de l'eglise de France* (Paris, 1956), p. 263: Jacques Nicolas Bellin.

31 Karrow, *Mapmakers of the sixteenth century*, pp. 132–4.

32 Andrew S. Cook, 'Alexander Dalrymple's *A collection of plans of ports in the East Indies* (1774–1775): a preliminary examination', *Imago Mundi*, xxxiii (1981), pp. 46–64.

33 Coolie Verner, 'Copperplate printing' in Woodward, *Five centuries of map printing*, ch. 3; Clifford H. Wood, 'Tonal reproduction processes in map printing from the 15th to the 19th centuries', *Cartographica*, xxii, 1 (1985), pp. 78–92.

34 R.A. Skelton, introduction, *Claudius Ptolemaeus Geographia Venice 1511* (Amsterdam, 1969), p. x; Mary Sponberg Pedley, *Bel et utile: the work of the Robert de Vaugondy family of mapmakers* (Tring, 1992), p. 109.

35 Roderick Barron (ed.), *The county maps of old England: Thomas Moule* (London, 1990), p. 11.

36 Elizabeth M. Harris, 'Miscellaneous map printing processes in the nineteenth century' in Woodward, *Five centuries of map printing*, p. 114, p. 157 n.7; Helen M. Wallis and Arthur H. Robinson (eds), *Cartographical innovations: an international handbook of mapping terms to 1900* ([Tring], 1987), 'steel engraving', pp. 298–9.

37 Woodward, 'The woodcut technique', pp. 46–7; David Woodward, 'Some evidence for the use of stereotyping on Peter Apian's world map of 1530', *Imago Mundi*, xxiv (1970), pp. 43–8; Skelton, *Claudius Ptolemaeus Geographia*, p. x.

38 Marcel P.R. van den Broecke, *Ortelius atlas maps* (Westrenen, 1996), pp. 28, 234–5.

39 H. Sankey, 'The maps of the Ordnance Survey', *Engineering*, xlv (1888), pp. 119–21.

40 Erwin Raisz, *General cartography* (2nd ed., New York, 1948), pp. 50, 161–2; David Woodward, *The All-American map: wax engraving and its influence on cartography* (Chicago, 1977).

41 William Edgeworth's diary, 3 June 1822, National Library of Ireland, MS 14124, p. 204.

42 National Archives of Ireland: Ordnance Survey, general correspondence, file 4668 (15 October 1885): C.W. Wilson.

43 P.D.A. Harvey and Harry Thorpe, *The printed maps of Warwickshire 1576–1900* (Warwick, 1959), p. 64 and passim.

44 Colonel T.F. Colby, 7 June 1843, National Archives (Public Record Office), London, WO 44/703.

45 Karrow, *Mapmakers of the sixteenth century*, p. 135.

46 F.R. Hassler, *Coast survey of the United States* (Philadelphia, 1842), pp. 1–2: Newark Bay, *c.*1816–18.

47 Coolie Verner, 'Mr Jefferson makes a map', *Imago Mundi*, xiv (1959), p. 102: 1787.

48 Dana Bennett Durand, *The Vienna-Klosterneuburg map corpus of the fifteenth century: a study in the transition from medieval to modern science* (Leiden, 1952), p. 197; Michael C. Andrews, 'Scotland in the portolan charts', *Scottish Geographical Magazine*, xlii (1926), pp. 141–2; Michael C. Andrews, 'The boundary between Scotland and England in the portolan charts', *Proceedings of the Society of Antiquaries of Scotland*, lx (1925–6), pp. 36–66; Bertha S. Phillpotts (ed.), *The life of the Icelander Jón Ólafsson, traveller to India*, i (London, 1923), p. 17: 1615.

49 Vicomte de Santarem, *Atlas composé de mappemondes, de portulans et de cartes hydrographiques et historiques depuis le xie jusqu'au xviie siècle* (Paris, 1849), plate 24: fifteenth-century world map, Florence, Biblioteca Medicea; above, p. 38.

50 Robert Lythe, map of central and southern Ireland, [1571], West Sussex Record Office, Chichester, PHA 9581.

51 James A. Welu, 'The sources and development of cartographic ornamentation in the Netherlands' in David Woodward (ed.), *Art and cartography: six historical essays* (Chicago, 1987), pp. 149, 153, 157, 163, 169.

52 R.A. Skelton, 'Tudor town plans in John Speed's *Theatre*', *Archaeological Journal*, cviii (1951), p. 111: Norwich, 1610.

53 P.E.H. Hair, 'A note on Thevet's unpublished maps of overseas islands', *Terrae Incognitae*, xiv (1982), p. 108.

54 D. Smith, 'The enduring image of early British townscapes', *Cartographic Journal*, xxviii (1991), p. 169.

55 Andrew Bonar Law, *John Speed: maps of Ireland* ([Dublin], 1979), pp. 8–9; http://www.maphist@geo.uu.nl, 4–20 Feburary 2007, 'Copper plate wear'.

56 Tony Campbell, 'Understanding engraved maps', *The Map Collector*, xlvi (1989), p. 9.

57 J.B. Harley, D.V. Fowkes and J.C. Harvey, *P.P. Burdett's map of Derbyshire, 1791 edition: an explanatory introduction* ([Derby], 1975), p. [3]: 5 June 1767.

58 A. Sarah Bendall, *Maps, land and society: a history, with a carto-bibliography of Cambridgeshire estate maps, c.1600–1836* (Cambridge, 1992), plates 6 and 7, facing p. 46; J.H. Andrews, 'Henry Pratt, surveyor of Kerry estates', *Journal of the Kerry Archaeological and Historical Society*, xiii (1980), p. 16.

59 David J. Butler, *The town plans of Chichester, 1595–1898* (Chichester, 1972), pp. 4–7.

60 Sarah Bendall, 'Draft town maps for John Speed's *Theatre of the empire of Great Britaine*', *Imago Mundi*, liv (2002), Figs 4, 5 (p. 34), 6, 7 (p. 36).

61 Wesley M. Stevens, 'Isidore's figure of the earth', *Isis*, lxxi (1980), p. 275; Scott D. Westrem, *The Hereford world map: a transcription and translation of the legends with commentary* (Turnhout, 2001), p. xviii.

62 Ortelius, *Theatre, Parergon*, p. xlii: Lhuyd's map of England and Wales, 1573.

63 Franck Cervoni, *Image de la Corse: 120 cartes de la Corse des origines à 1831* (Ajaccio, 1989).

64 G.R. Crone, 'The origin of the name Antillia', *Geographical Journal*, xci (1938), pp. 260–2; Evelyn Edson, *Mapping time and space: how medieval mapmakers viewed their world* (London, 1997), p. 62: 'Isidore' map.

65 A.C. Painter, 'Notes on some old Gloucestershire maps', *Transactions of the Bristol and Gloucester Archaeological Society*, li (1929), p. 86; Eugene Burden, 'Stedes – the Berkshire village that never was', *The Map Collector*, xl (1987), pp. 22–3.

66 Marcel Watelet (ed.), *Gérard Mercator, cosmographe: le temps et l'espace* (Antwerp, 1994), pp. 83–8.

67 Henry N. Stevens, *Lewis Evans: his map of the middle British colonies in America* (London, 1920), p. 37, quoting Thomas Pownall.

68 National Archives (Public Record Office), London, MPF 1/70.

69 J.H. Andrews, *Shapes of Ireland: maps and their makers, 1564–1839* (Dublin, 1997), ch. 3.

70 W.W. Greg, 'Bibliography – an apologia', *The Library*, 4th series, xiii, 2 (1932), p. 125.

71 David Woodward, 'Medieval *mappaemundi*' in J.B. Harley and David Woodward (eds), *The history of cartography, volume one: cartography in prehistoric, ancient, and medieval Europe and the Mediterranean* (Chicago, 1987), p. 325: Brunetto Latini, fourteenth-century world map.

72 W. Ravenhill, 'Joel Gascoyne, a pioneer of large-scale county mapping', *Imago Mundi*, xxvi (1972), p. 65.

73 Gerard Mercator, Henry Hondius and Joannes Janssonius (ed. R.A. Skelton), *Atlas, or a geographicke description of the world*, i (Amsterdam, 1636, 1968): 'The sixth map of England', following p. 65.

74 Lucas Jansz. Waghenaer, *De Spieghel der Zeevaerdt* (Leyden, 1584).

75 Peter M. Barber, 'Mapping Britain from afar', *Mercator's World*, iii, 4 (1998), pp. 22–3.

76 Vinton A. Dearing, *A manual of textual analysis* (Berkeley, CA, 1959); A.B. Taylor, 'Name studies in sixteenth century Scottish maps', *Imago Mundi*, xix (1965), pp. 81–99.

77 Sarah Tyacke and John Huddy, *Christopher Saxton and Tudor map-making* (London, 1980), p. 7; Daniel Birkholz, 'The Gough map revisited: Thomas Butler's *The Mape off Ynglonnd*, c.1547–1554', *Imago Mundi* lviii, 1 (2006), pp. 23–47.

78 Gerard Mercator, *Angliae Scotiae & Hibernie nova descriptio* (Duisburg, 1564).

79 David Buisseret, *The mapmakers' quest: depicting new worlds in renaissance Europe* (Oxford, 2003), p. 36: Leonardo da Vinci, Milan, c.1500.

80 J.H. Andrews, *Plantation acres: an historical study of the Irish land surveyor and his maps* (Belfast, 1985), p. 154.

Chapter 15: Cunningly compiled and made

1 Richard Gough, *British topography: or, an historical account of what has been done for illustrating the topographical antiquities of Great Britain and Ireland*, i (London, 1780), p. 109.

2 R.A. Skelton, *Looking at an early map* (Lawrence, KS, 1965), p. 1.

3 Abraham Ortelius (ed. R.A. Skelton), *The theatre of the whole world, London, 1606* (Amsterdam, 1968), p. 93: Illyria; p. 96: Hungary.

4 Numa Broc, *La géographie des philosophes, géographes et voyageurs français au xviiiᵉ siècle* (Paris, 1974), p. 168: Pieter van der Aa, *c.*1730.

5 Richard I. Ruggles, 'The cartographic lure of the northwest passage: its real and imaginary geography' in Thomas H.B. Symons (ed.), *Meta incognita: a discourse of discovery: Martin Frobisher's Arctic expeditions, 1576–1578* (Hull, Quebec, 1999), p. 221: John Dee, 1580.

6 E.G.R. Taylor (ed.), *The original writings and correspondence of the two Richard Hakluyts*, i (London, 1935), p. 81; Howard T. Fry, 'Alexander Dalrymple and Captain Cook: the creative interplay of two careers' in Robin Fisher and Hugh Johnston (eds), *Captain James Cook and his times* (Vancouver, 1979), p. 50.

7 Quoted in review of Matthew H. Edney and Irwin D. Novak (eds), *Reading the world: interdisciplinary perspectives on Pieter van den Keere's map Nova totius terrarum orbis geographica ac hydrographica tabula (Amsterdam, 1608/36)* (Portland, ME, 2001), *The Globe*, liii (2002), p. 69.

8 John Norden, 'A description of Ireland' [1608], National Archives (Public Record Office), London, MPF 1/117.

9 Matthew H. Edney, 'Reconsidering enlightenment geography and map making: reconnaissance, mapping, archive' in David N. Livingstone and Charles W.J. Withers (eds), *Geography and enlightenment* (Chicago, 1999), pp. 169–70.

10 Miles Harvey, *The island of lost maps: a true story of cartographic crime* (London, 2001), pp. 141–54.

11 Nicholas Crane, *Mercator: the man who mapped the planet* (London, 2002), pp. 184–5, 312.

12 John W. Blake, 'New light on Diogo Homem, Portuguese cartographer', *Mariner's Mirror*, xxviii (1942), p. 157.

13 John Green, *Remarks, in support of the new chart of North and South America; in six sheets* (London, 1753), p. 4.

14 Clements R. Markham, *Major James Rennell and the rise of modern English geography* (London, 1895), pp. 10–11.

15 Andrew A. Lipscomb (ed.), *The writings of Thomas Jefferson*, v (Washington, 1905), p. 286.

16 R.V. Tooley, 'Map making in France from the sixteenth to the eighteenth century', *Proceedings of the Huguenot Society of London*, xviii, 6 (1952), p. 475.

17 Broc, *La géographie des philosophes, géographes et voyageurs*, pp. 29–36; Anne Marie Claire Godlewska, *Geography unbound: French geographic science from Cassini to Humboldt* (Chicago, 1999), pp. 40–54; Mary Sponberg Pedley, *The commerce of cartography: making and marketing maps in eighteenth-century France and England* (Chicago, 2005), pp. 166–74.

18 [John Green], *The construction of maps and globes* (London, 1717), pp. 132, 136, 141.

19 Mireille Pastoureau, 'Maps at the Bibliothèque Nationale: a collection of collections', *The Map Collector*, xl (1987), pp. 11.

20 Bibliothèque Nationale, Paris, Ge. DD. 2987B; Andrew Bonar Law, *The printed maps of Ireland, 1612–1850* (Dublin, 1997), pp. 71–3: 1689.

21 P.D.A. Harvey, 'Cartography and its written sources' in F.A.C. Mantello and A.G. Rigg (eds), *Medieval Latin: an introduction and bibliographical guide* (Washington, 1996), p. 389.

22 J.B. Mitchell, 'The Matthew Paris maps', *Geographical Journal*, lxxxi (1933), p. 30; P.D.A. Harvey, 'Matthew Paris's maps of Britain' in P.R. Coss and S.D. Lloyd (eds), *Thirteenth century England IV: proceedings of the Newcastle upon Tyne conference* (Woodbridge, Suffolk, 1992), p. 117.

23 Horace Leonard Jones (ed.), *The geography of Strabo* (Cambridge, MA), i (1969), pp. 305, 435; ii (1969), pp. 5, 169, 213, 253; iii (1967), pp. 55, 333; iv (1968), pp. 13, 277, 345; v (1969), pp. 249, 291; vi

(1970), pp. 359, 377; vii (1966), pp. 15, 59, 129; viii (1967), pp. 3, 13, 143; H. Rackham (ed.), *Pliny: natural history*, ii (London, 1942), pp. 63, 89, 125, 175, 181, 255, 267, 365, 399, 421, 431.

24 Günter Schilder, *Australia unveiled: the share of the Dutch navigators in the discovery of Australia* (Amsterdam, 1976), map 56, p. 355: 'Bonaparte' map of Australia, 1644; James E. Kelley Jnr, 'Still lost in the Indies after all these years: interpreting early maps of America', Society for the History of Discoveries, Mackinic Island, Michigan, September 1994, p. 5.

25 Julian P. Boyd, *The papers of Thomas Jefferson*, x (Princeton, NJ, 1954), pp. 249–50.

26 Daniel Augustus Beaufort, MS map of Ireland, 1792, British Library, MS 53711A.

27 J.B. Harley (introduction), *The old series Ordnance Survey maps of England and Wales*, v (Lympne, 1987), p. xiii.

28 Green, *Remarks, in support of the new chart*, passim.

29 Matthew H. Edney, *Mapping an empire: the geographical construction of British India, 1765–1843* (Chicago, 1997), pp. 100–2.

30 David Bosse, '"To promote useful knowledge": *An accurate map of the four New England states* by John Norman and John Coles', *Imago Mundi*, lii (2000), pp. 147–8: 1785.

31 Donald Hodson, *County atlases of the British Isles published after 1703, volume one: atlases published 1704 to 1742 and their subsequent editions* (Tewin, Herts, 1984), p. 175: John Warburton, 1725.

32 Helen Wallis, 'The cartography of Drake's voyage' in Norman J.W. Thrower (ed.), *Sir Francis Drake and the famous voyage, 1577–1580: essays commemorating the quadricentennial of Drake's circumnavigation of the earth* (Berkeley, CA, 1984), p. 135: Richard Madox, 1582.

33 Green, *Remarks, in support of the new chart*, p. 4.

34 Rackham, *Pliny: natural history*, ii, pp. 3, 445.

35 Helen Wallis, 'Purchas's maps' in L.E. Pennington (ed.), *The Purchas handbook: studies of the life, times and writings of Samuel Purchas, 1577–1626*, i (London, 1997), p. 148.

36 Boleslaw Szczesniak, 'The seventeenth century maps of China: an inquiry into the compilations of European cartographers', *Imago Mundi*, xiii (1956), p. 126.

37 D.W. Waters, *The art of navigation in England in Elizabethan and early Stuart times* (London, 1958), p. 327.

38 *Calendar of state papers, colonial series, America and West Indies, 1669–1674*, p. 267: 1671; Jeanette D. Black, *The Blathwayt atlas, volume two, commentary* (Providence, 1975), pp. 192–5.

39 A.M. Cubbon, *Early maps of the Isle of Man* (Douglas, 1967), pp. 8–9.

40 Aaron Arrowsmith, *Memoir relative to the construction of a map of Scotland published by Aaron Arrowsmith in the year 1807* (London, [1809]), pp. 8–9.

41 John H. Andrews, 'The mapping of Ireland's cultural landscape, 1550–1630' in Patrick J. Duffy, David Edwards and Elizabeth FitzPatrick (eds), *Gaelic Ireland, c.1250–c.1650: land, lordship and settlement* (Dublin, 2001), p. 156.

42 Alexander Jamieson, *A treatise on the construction of maps* (London, 1814), p. 167.

43 Henry Bradley, 'Ptolemy's geography of the British Isles', *Archaeologia*, xlviii (1885), pp. 382–3.

44 Karel Kuchar, 'A map of Bohemia at the time of the thirty years war', *Imago Mundi*, ii (1937), p. 76.

45 Roland Chardon, 'A best-fit evaluation of De Brahm's 1770 chart of Northern Biscayne Bay, Florida', *The American Cartographer*, ix, 1 (1982), pp. 58–9.

46 Marcel Watelet (ed.), *The Mercator atlas of Europe* (Antwerp, 1998), pp. 76–7; Robert W. Karrow Jr, *Mapmakers of the sixteenth century and their maps: bio-bibliographies of the cartographers of Abraham Ortelius, 1570* (Winnetka, IL, 1993), p. 488.

47 James Rennell, *Memoir of a map of Hindoostan; or, The Mogul's empire* (London, 1788), p. 25.

48 Arthur Davies, 'The date of Juan de la Cosa's world map and its implications for American discovery', *Geographical Journal*, cxlii (1976), pp. 113–14.

49 William Hubbard, narrative of troubles with the Indians, quoted in John A. Wolter, 'Source materials for the history of American cartography', *American Studies*, xii, 3 (1974), p. 23, supplement to *American Quarterly*, May 1974.

50 James B. Caird, 'Early 19th century estate plans' in Finlay Macleod (ed.), *Togail tir, marking time: the map of the Western Isles* (Stornoway, 1989), pp. 49–61.

51 Ortelius, *Theatre, Parergon*, p. xix; Richard J.A. Talbert (ed.), *Barrington atlas of the Greek and Roman world* (Princeton, NJ, 2000), maps 39 and 40.

52 Gerard Mercator, Henry Hondius and Joannes Janssonius (ed. R.A. Skelton), *Atlas or a geographicke description of the world*, i (Amsterdam, 1636, 1968), p. 207.

53 A. Arrowsmith, *A map exhibiting the new discoveries in the interior parts of North America* (London, 1795, with additions to 1814).

54 Thomas O'Loughlin, 'An early thirteenth-century map in Dublin: a window into the world of Giraldus Cambrensis', *Imago Mundi*, li (1999), pp. 24–39.

55 A. Arrowsmith, *A map of the United States of North America* (London, 1796).

56 Mercator-Hondius-Janssonius, *Atlas*, ii, p. 441.

57 Harvey, 'Matthew Paris's maps of Britain', p. 120.

58 J.H. Andrews, *J. Lendrick map of Co. Antrim 1780 engraved by S. Pyle, London 1782* (Belfast, 1987).

59 Evelyn Edson, *Mapping time and space: how medieval mapmakers viewed their world* (London, 1997), pp. 17, 134.

60 Joan Blaeu, *Le grand atlas*, i (Amsterdam, 1663, 1967), 'Description de la terre', p. iii.

61 Gerald Strauss, *Sixteenth-century Germany, its topography and topographers* (Madison, WI, 1959), pp. 26–7; Crane, *Mercator*, pp. 160–1.

62 Broc, *La géographie des philosophes, géographes et voyageurs*, p. 34; Lucie Lagarde, 'Le passage du nord-ouest et la Mer de l'Ouest dans la cartographie français du 18e siècle, contribution à l'étude de l'oeuvre des Delisle et Buache', *Imago Mundi*, xli (1989), p. 24: map of North America, 1700; Mary Sponberg Pedley, *The commerce of cartography: making and marketing maps in eighteenth-century France and England* (Chicago, 2005), pp. 166–9.

63 Arrowsmith, *A map of the United States*; A. Arrowsmith, *Outlines of the world* (London, 1825), no. 37: William Burchell, Cape of Good Hope, 1811–15.

64 Ibid.

65 J.H. Andrews, 'Sir Richard Bingham and the mapping of western Ireland', *Proceedings of the Royal Irish Academy*, ciii C (2003), p. 92.

66 J.B. Harley, 'George Washington, map-maker', *Geographical Magazine*, xlviii, 10 (1976), pp. 592–3; Peter J. Guthorn, *American maps and map makers of the Revolution* (Monmouth Beach, NJ, 1966), pp. 8–9.

67 R.A. Skelton, 'Map compilation, production, and research in relation to geographical exploration' in Herman R. Friis (ed.), *The Pacific Basin: a history of its geographical exploration* (New York, 1967), p. 51.

68 Charles T. Beke (ed.), *True description of three voyages by the north east towards Cathay and China, undertaken in the years 1594, 1595 and 1596* (London, 1853), p. xxxii; Johannes Keuning, 'Nicolaas Witsen as a cartographer', *Imago Mundi*, xi (1954), p. 102: 1692.

69 *Journal of a 2nd voyage for the discovery of a north west passage from the Atlantic to the Pacific; performed in the years 1821–22–23 in His Majesty's ships* Fury *and* Hecla, *under the orders of Captain William Edward Parry, R.N., F.R.S., and commander of the expedition* (New York, 1969), pp. 252–3.

70 Green, *Remarks, in support of the new chart*, p. 5.

71 Ulla Ehrensvärd, 'Peter Gedda's maritime atlas of the Baltic, 1695', History of Cartography Conference, Greenwich, 1975, typescript, p. 6.

72 Mercator to Abraham Ortelius, 22 November 1570, quoted in *The Map Collector*, xxxvii (1986), p. 53.

73 Green, *Construction of maps and globes*, p. 136.

74 P.D.A. Harvey and Harry Thorpe, *The printed maps of Warwickshire, 1576–1900* (Warwick, 1959), p. 113 et seq.

75 J.H. Andrews, *Shapes of Ireland: maps and their makers, 1564–1839* (Dublin, 1997), pp. 135–6.

76 David Smith, 'The preparation of the county maps for Lysons' *Magna Britannia*', *Bulletin of the Society of University Cartographers*, xxv, 1 (1991), pp. 23–32; Pedley, *The commerce of cartography*, pp. 183–6.

77 William Camden (ed. Edmund Gibson), *Britannia: or, A chorographical description of Great Britain and Ireland*, i (London, 1722), preface.

78 J.H. Andrews, *History in the ordnance map: an introduction for Irish readers* (2nd ed., Newtown, Montgomeryshire, 1993), p. 38.

79 P.D.A. Harvey, *Mappa mundi: the Hereford world map* (London, 1996), p. 7.

80 Derek Howse and Michael Sanderson, *The sea chart: an historical survey based on the collections in the National Maritime Museum* (Newton Abbot, 1973), p. 103.

81 Blaeu, *Le grand atlas*, iii, p. 109: *Nassovia Comitatus* ('Hohe montes' and 'Feldberg').

82 Rodney W. Shirley, *The mapping of the world: early printed world maps, 1472–1700* (London, 1983), plate 179, pp. 240–1.

83 Ibid., plate 236., pp. 332–5.

84 C. Koeman, *The history of Abraham Ortelius and his* Theatrum orbis terrarum (Lausanne, 1964), pp. 36–7.

85 Gordon L. Herries Davies, *Sheets of many colours: the mapping of Ireland's rocks, 1750–1890* (Dublin, 1983), pp. 45, 60, 61.

86 J.H. Andrews, *A paper landscape: the Ordnance Survey in nineteenth-century Ireland* (Oxford, 1975), pp. 213–19, 268.

Chapter 16: Into more finished form

1 Reproduced in P.D.A. Harvey, *The history of topographical maps: symbols, pictures and surveys* (London, 1980), pp. 174–5: Charles Varle.

2 D.G. Moir and R.A. Skelton, 'New light on the first atlas of Scotland', *Scottish Geographical Magazine*, lxxxvi (1970), p. 152.

3 J.H. Andrews, *Shapes of Ireland: maps and their makers, 1564–1839* (Dublin, 1997), p. 128.

4 Thomas Jefferys, *The American atlas*, (London, 1776; Amsterdam, 1974), introduction by Walter W. Ristow, p. vii.

5 David Bosse, '"To promote useful knowledge": *An accurate map of the four New England states* by John Norman and John Coles', *Imago Mundi*, lii (2000), p. 144.

6 David Smith, 'The preparation of the county maps for Lysons' *Magna Britannia*', *Bulletin of the Society of University Cartographers*, xxv, 1 (1991), p. 30.

7 Andreas Stylianou and Judith A. Stylianou, *The history of the cartography of Cyprus* (Nicosia, 1980), pp. 60–1.

8 Edward Lynam, *The Carta marina of Olaus Magnus, Venice 1539 and Rome 1572* (Jenkintown, PA, 1949), p. 29; G.R. Crone and F. George, 'Olaus Magnus and his *Carta marina*: a problem in sixteenth-century cartography', *Geographical Journal*, cxiv (1949), p. 200; Carol Urness, 'Olaus Magnus: his map and his book', *Mercator's World*, vi, 1 (2001), pp. 26–33.

9 Coolie Verner, *Captain Collins'* Coasting pilot: *a carto-bibliographical analysis* (Map Collectors' Series, lviii, London, 1969), p. 24; Tony Campbell, 'For those in peril on the sea' in Peter Barber and Christopher Board (eds), *Tales from the Map Room: fact and fiction about maps and their makers* (London, 1993), p. 164.

10 Marco van Egmond, 'The secrets of a long life: the Dutch firm of Covens & Mortier (1685–1866) and their copper plates', *Imago Mundi*, liv (2002), p. 79.

11 R.V. Tooley, 'Maps in Italian atlases of the sixteenth century, being a comparative list of the Italian maps issued by Lafreri, Forlani, Duchetti, Bertelli and others, found in atlases', *Imago Mundi*, iii (1939), p. 13.

12 Tony Campbell, *Japan: European printed maps to 1800* (Map Collectors' Series, xxxvi, London, 1967), p. 3.

13 Edward Lynam, 'The early maps of Scandinavia', *Geographical Journal*, lxx (1927), p. 65.

14 Nicholas Crane, *Mercator: the man who mapped the planet* (London, 2002), p. 253.

15 Catherine Delano-Smith and Roger J.P. Kain, *English maps, a history* (London, 1999), p. 107.

16 C.G. Cash, 'Manuscript maps by Pont, the Gordons, and Adair, in the Advocates' Library, Edinburgh', *Scottish Geographical Magazine*, xxiii (1907), p. 584.

17 Nicolas Sanson, *Tables de la géographie ancienne et nouvelle* (Paris, 1667); Guilio Macchi (ed.), *Cartes et figures de la terre* (Paris, 1980), pp. 16, 447.

18 David Woodward, 'Paolo Forlani: compiler, engraver, printer, or publisher?', *Imago Mundi*, xliv (1992), p. 46.

19 Robert W. Karrow Jr, *Mapmakers of the sixteenth century and their maps: bio-bibliographies of the cartographers of Abraham Ortelius, 1570* (Winnetka, IL, 1993), p. 163.

20 Raymond Lister, *How to identify old maps and globes* (London, 1965), pp. 117–18; Paul Goldman, *Looking at prints; a guide to technical terms* (London, 1981), p. 6.

21 R.A. Skelton, *County atlases of the British Isles, 1579–1703* (London, 1970), p. 32.

22 Leo Bagrow, 'At the sources of the cartography of Russia', *Imago Mundi*, xvi (1962), p. 33.

23 Karrow, *Mapmakers of the sixteenth century*, pp. 54–5, 605.

24 Ibid., pp. 32, 283, 293, 327.

25 R.V. Tooley, *A sequence of maps of Africa* (Map Collectors' Series, lxxxii, London, 1972), p. 10.

26 Mary Sponberg Pedley, *Bel et utile: the work of the Robert de Vaugondy family of mapmakers* (Tring, 1992), p. 94: Robert de Vaugondy, map of the Americas, 1740.

27 A.H.W. Robinson, *Marine cartography in Britain: a history of the sea chart to 1855* (Leicester, 1962), p. 75.

28 W.P. Cumming, R.A. Skelton and D.B. Quinn, *The discovery of North America* (London, 1971), p. 279: 1625.

29 Alex Krieger and David Cobb with Amy Turner, *Mapping Boston* (Leventhal Family Foundation, 1999), p. 185: Henry Pelham, *A plan of Boston in New England and its environs* (London, 1777).

30 Jonathan D. Spence, *The memory palace of Matteo Ricci* (London, 1988), p. 96.

31 Richard W. Stephenson, 'Maps for the general public: commercial cartography of the American civil war', History of Cartography Conference, Ottawa, 1985, typescript, pp. 9–10: 1863.

32 G. Walters, 'Themes in the large scale mapping of Wales in the eighteenth century', *Cartographic Journal*, v, 2 (1968), p. 138.

33 Kenneth Nebenzahl and Don Higginbotham, *Atlas of the American revolution* (Chicago, 1974), pp. 14–15: *A general map of the middle British colonies* (London, 1776).

34 Ida Darlington and James Howgego, *Printed maps of London,* circa *1553–1850* (London, 1964), pp. 28–9; Matthew H. Edney, *Mapping an empire: the geographical construction of British India, 1765–1843* (Chicago, 1997), p. 320.

35 Rodney W. Shirley, *The mapping of the world: early printed world maps, 1472–1700* (London, 1983), p. 573: Jean-Dominique Cassini, *Planisphere terrestre ou sont marquées les longitudes de divers lieux de la terre, trouvées par les observations des eclipses des satellites de Jupiter dressé et presenté a sa majesté par Mr de Cassini le Fils de l'Académie Royale des Sciences*, 1696.

36 A. Arrowsmith, *Atlas to Thompson's Alcedo; or, Dictionary of America and West Indies* (London, 1816).

37 Herman Moll, *A new and exact map of Spain and Portugal* (London, 1711); Ashley Baynton-Williams, 'The world described: the life and times of Herman Moll, geographer (d. 1732)', *Map Forum*, i (2004), pp. 19–20.

38 Rodney W. Shirley, *Printed maps of the British Isles, 1650–1750* (Tring, 1988), p. 91.

39 Helen M. Wallis and Arthur H. Robinson, *Cartographical innovations: an international handbook of mapping terms to 1900* ([Tring], 1987), p. 12: Pietro Vesconti, 1311.

40 Abraham Ortelius (ed. R.A. Skelton), *The theatre of the whole world, London, 1606* (Amsterdam, 1968), pp. 42–5, 49, 50.

41 8 Geo. II c. 13 (1734–5); David Smith, *Antique maps of the British Isles* (London, 1982), p. 49.

42 Andrews, *Shapes of Ireland*, ch. 4.

43 W.P. Cumming, 'The Montresor-Ratzer-Sauthier sequence of maps of New York City, 1766–76', *Imago Mundi*, xxxi (1979), p. 57.

44 Peter Barber, 'Maps and monarchs in Europe 1550–1800' in Robert Oresko, G.C. Gibbs and H.M. Scott (eds), *Royal and republican sovereignty in early modern Europe: essays in memory of Ragnhild Hatton* (Cambridge, 1997), p. 120.

45 *A collection of several relations and treatises singular and curious, of John Baptista Tavernier* (London, 1680): advertisement for Moses Pitt's atlas.

46 Peter Barber, 'Finance and flattery' in Peter Barber and Christopher Board (eds), *Tales from the Map Room: fact and fiction about maps and their makers* (London, 1993), p. 140: 1725.

47 David Smith, 'Jansson versus Blaeu: a study in competitive response in the production of English county maps', *Cartographic Journal*, xxiii, 2 (1986), pp. 111–12; John H. Farrant, *Sussex depicted: views and descriptions, 1600–1800* (Lewes, Sussex, 2001), p. 78.

48 Josef W. Konvitz, *Cartography in France, 1660–1848: science, engineering and statecraft* (Chicago, 1987), pp. 7–8 (Fig. 1); R.A. Skelton, 'Cartography' in Charles Singer (ed.), *History of technology*, iv (Oxford, 1958), p. 606; above, Fig. 1.17.

49 Margaret Wilkes, *The Scot and his maps* (Motherwell, 1991), p. 8: 1734; Ulla Ehrensvärd, *The history of the Nordic map, from myths to reality* (Helsinki, 2006), p. 290: 1767.

50 Reproduced in Richard van de Gohm, *Antique maps for the collector* (Edinburgh, 1972), p. 88; Mead T. Cain, 'Unrecorded maps of the world and four continents – by Morden?', *The Map Collector*, lvii (1991), pp. 7–8; Bill Warren, 'A most curious map', *The Map Collector*, lxxiii (1995), p. 14: J.N. Buache, North Pacific, 1776.

51 Mark Babinski, *Henry Popple's 1733 map of the British Empire in America* (Garwood, NJ, 1998), p. 33.

52 Herbert George Fordham, *Maps: their history, characteristics and uses* (Cambridge, 1921), p. 38: Roch Joseph Julien, 1751.

53 Gerard Mercator, Henry Hondius and Joannes Janssonius (ed. R.A. Skelton), *Atlas or a geographicke description of the world*, ii (Amsterdam, 1636, 1968), p. 275.

54 Ortelius, *Theatre*, p. 25: unsigned map of Blois.

55 Józef Babicz, 'La résurgence de Ptolémée' in Marcel Watelet (ed.), *Gérard Mercator cosmographe: le temps et l'espace* (Antwerp, 1994), pp. 63, 65–8; Johannes Keuning, 'The history of an atlas: Mercator-Hondius', *Imago Mundi*, iv (1947), p. 39.

56 Sarah Tyacke, *London map-sellers, 1660–1720* (Tring, 1978).

57 John P. Snyder, *Flattening the earth: two thousand years of map projections* (Chicago, 1993), p. 49.

58 Ibid., p. 96.

59 J.H. Andrews, *A paper landscape: the Ordnance Survey in nineteenth-century Ireland* (Oxford, 1975), pp. 231, 233; Brian Adams, 'The projection of the original one-inch map of Ireland (and of Scotland)', *Sheetlines*, xxx (1991), pp. 12–15.

60 François de Dainville, *Le langage des géographes* (Paris, 1964), p. 92; Paul de F. Hicks Jr, 'Tracing the origins of wind heads to wind gods', *Mercator's World*, ii, 1 (1997), pp. 34–9.

61 *Vallard atlas*, Dieppe, 1547 (Petaluma, CA, 1991).

62 D. Hodson, *Maps of Portsmouth before 1801* (Portsmouth, 1978), pp. xxvii, 79 and passim.

63 Kenneth Nebenzahl and Don Higginbottom, *Atlas of the American revolution* (Chicago, 1974), pp. 116–17: *A plan of the city and environs of Philadelphia survey'd by N. Scull and G. Heap* (London, 1777).

64 J.B. Harley and William Ravenhill, 'Proposals for county maps of Cornwall (1699) and Devon (1700)', *Devon and Cornwall Notes and Queries*, xxxii (1971), p. 34; Mary Sponberg Pedley, *The commerce of cartography: making and marketing maps in eighteenth-century France and England* (Chicago, 2005), pp. 177–80.

65 Smith, 'The preparation of the county maps for Lysons' *Magna Britannia*', p. 26.

66 G.R. Crone, *Early maps of the British Isles, A.D.1000–A.D.1579* (London, 1961), pp. 25–6; Helen Wallis, 'Sixteenth-century maritime manuscript atlases for special presentation' in John A. Wolter and Ronald E. Grim (eds), *Images of the world: the atlas through history* (Washington, 1997), pp. 9, 25; Peter Barber (ed.), *The Queen Mary atlas*, 2 vols (London, 2005).

67 'Remarks by M. Bellin, in relation to his maps drawn for P. Charlevoix's history of New France', *Gentleman's Magazine*, xvi (1746), p. 73.

68 National Archives of Ireland: Ordnance Survey, general correspondence, file 3844 (1876–80).

69 Christopher Board, 'Falsification and security' in Barber and Board, *Tales from the Map Room*, pp. 106–7.

70 Simon Pointer, 'How art overcame adversity', *The Map Collector*, iv (1978), p. 11; Richard L. Kagan, '*Urbs* and *civitas* in sixteenth- and seventeenth-century Spain' in David Buisseret (ed.), *Envisioning the city: six studies in urban cartography* (Chicago, 1998), p. 105 n. 22.

71 Helen M. Wallis, 'Geographie is better than divinitie: maps, globes, and geography in the days of Samuel Pepys' in Norman J.W. Thrower (ed.), *The compleat plattmaker: essays on chart, map, and globe making in England in the seventeenth and eighteenth centuries* (Berkeley and Los Angeles, 1978), pp. 34–5: 1669–70.

72 D.G. Moir in Royal Scottish Geographical Society, *The early maps of Scotland to 1850* (Edinburgh, 1973), p. 43: 1631.

73 Paul Laxton, 'The geodetic and topographical evaluation of English county maps, 1740–1840', *Cartographic Journal*, xiii, 1 (1976), p. 44.

74 Mary Sponberg Pedley, 'Atlas editing in Enlightenment France', *Journal of Scholarly Publishing*, xxvii, 2 (1996), p. 112.

75 E.G.R. Taylor, '"The English atlas" of Moses Pitt, 1680–83', *Geographical Journal*, xcv (1940), p. 294.

76 A.H. Robinson and R.D. Sale, *Elements of cartography* (3rd ed., New York, 1969), p. 41; G.C. Dickinson, *Maps and air photographs* (London, 1969), pp. 147–8.

77 Lucio Gambi, *The gallery of maps in the Vatican* (New York, 1997); Francesco Fiorani, *The marvel of maps: art, cartography and politics in renaissance Italy* (New Haven, CT, 2005).

78 Giovanni Magini, *Italia* (Bologna, 1620); Joan Blaeu, *Atlas maior*, ix (Amsterdam, 1662).

79 F. Roland, 'Alexis-Hubert Jaillot, géographe du roi Louis XIV (1632–1712)', *Académie des Sciences, Belles Lettres et Arts de Besancon, Procès Verbaux et Mémoires*, 1919–20, p. 14.

80 Tony Campbell, *Weinreb + Douwma catalogue 7: maps* (London, 1971), p. 13.

81 Cordell D.K. Yee, 'Traditional Chinese cartography and the myth of westernisation' in J.B. Harley and David Woodward (eds), *History of cartography, volume two, book two: cartography in the traditional east and southeast Asian societies* (Chicago, 1994), p. 171.

82 Reproduced in Macchi, *Cartes et figures de la terre*, p. 254.

83 Hildegard Binder Johnson, *Carta marina: world geography in Strassburg, 1525* (Westport, CT, 1963), p. 51.

84 Yolande Hodson, Introductory notes, *Facsimile of the Ordnance Surveyors' drawings of the London area, 1799–1808* (London Topographical Society, Publication no. 144, 1991), p. [2]; W.A. Seymour (ed.), *A history of the Ordnance Survey* (Folkestone, 1980), pp. 47, 58; Mary Blewitt, *Surveys of the seas: a brief history of British hydrography* (London, 1957), p. 36.

85 R.A. Skelton, 'The first English world atlases' in K.-H. Meine, *Kartengeschichte und Kartenbearbeitung: Festschrift zum 80 Geburtstag von Wilhelm Bonacker* (Bad Godesberg, 1968), pp. 77–81; Geoffrey King, *Miniature antique maps* (Tring, 1996).

86 P.D.A. Harvey, 'Cartography and its written sources' in F.A.C. Mantello and A.G. Rigg (eds), *Medieval Latin: an introduction and bibliographical guide* (Washington, 1996), p. 389.

87 King, *Miniature antique maps*, pp. 141–4, especially p. 142: Arthur Hopton, 1611.

88 Andrews, *Shapes of Ireland*, p. 53, p. 56 n. 41.

Chapter 17: Forbear much writing

1 I. Robson, *A treatise on geodetic operations, or, County surveying, land surveying and levelling* (Durham, 1821), p. 249; Alan M. MacEachren, *How maps work: representation, visualization, and design* (New York, 1995).

2 Eila M.J. Campbell, 'The development of the characteristic sheet, 1533–1822', *Proceedings*, VIIIth General Assembly-XVIIth Congress, International Geographical Union (Washington, 1952), pp. 426–30; Catherine Delano Smith, 'Cartographic signs on European maps and their explanation before 1700', *Imago Mundi*, xxxvii (1985), pp. 9–29; François de Dainville, *Le langage des géographes* (Paris, 1964), plates XI-XXI.

3   Helen M. Wallis and Arthur H. Robinson (eds), *Cartographical innovations: an international handbook of mapping terms to 1900* ([Tring], 1987), 'letter symbol', pp. 231–2.

4   Aaron Rathborne, *The surveyor in foure bookes* (London, 1616), p. 175.

5   C.F. Close, 'The ideal topographical map', *Geographical Journal*, xxv (1905), p. 636.

6   Blake Tyson, 'John Adams's cartographic correspondence to Sir Daniel Fleming of Rydal Hall, Cumbria, 1676–1687', *Geographical Journal*, cli (1985), p. 28; [John Green], *The construction of maps and globes* (London, 1717), p. 10; Herman Moll, *The compleat geographer: or, The chorography and topography of all the known parts of the earth* (London, 1723), p. xiii.

7   Arthur H. Robinson and Barbara Bartz Petchenik, *The nature of maps: essays toward understanding maps and mapping* (Chicago, 1976), p. 61.

8   P.D.A. Harvey, *The history of topographical maps: symbols, pictures and surveys* (London, 1980), pp. 45–6.

9   Royal Scottish Geographical Society, *The early maps of Scotland to 1850* (Edinburgh, 1973), map 2, p. 5: 1457; C.E. Doble (ed.), *Remarks and collections of Thomas Hearne*, iii (Oxford, 1889), p. 16.

10 Edward Lynam, *The Carta marina of Olaus Magnus, Venice 1539 and Rome 1572* (Jenkintown, PA, 1949).

11 Hans Wolff, 'Die Bayerischen Landtafeln – das kartographische Meisterwerk Philipp Apians und ihr Nachwirken' in Hans Wolff et. al., *Philipp Apian und die Kartographie der Renaissance* (Munich, 1989), pp. 77–100: Augsburg, 1568; Robert W. Karrow Jr, *Mapmakers of the sixteenth century and their maps: bio-bibliographies of the cartographers of Abraham Ortelius, 1570* (Winnetka, IL, 1993), pp. 64–8.

12 Wilma George, *Animals and maps* (London, 1969), p. 25.

13 Lynam, *The Carta marina of Olaus Magnus*.

14 P.D.A. Harvey, 'A manuscript estate map by Christopher Saxton', *British Museum Quarterly*, xxiii, 3 (1961), p. 66.

15 Richard I. Ruggles, *A country so interesting: the Hudson's Bay Company and two centuries of mapping, 1670–1870* (Montreal, 1991), p. 35.

16 Jeffrey Stone, 'Writing and signs: an assessment of Pont's settlement signs' in Ian C. Cunningham (ed.), *The nation survey'd: essays on late sixteenth-century Scotland as depicted by Timothy Pont* (East Linton, East Lothian, 2001), pp. 50–3.

17 'In usum tabularum admonitio', Gerard Mercator, *Atlas* (Duisburg, 1595), English translation in Gerard Mercator, Henry Hondius and Joannes Janssonius (ed. R.A. Skelton), *Atlas or a geographicke description of the world*, ii (Amsterdam, 1636, 1968), p. 276.

18 Frank Kitchen, 'Cosmo-choro-poly-grapher: an analytical account of the life and work of John Norden, 1547?–1625' (University of Sussex, D.Phil. thesis, 1993), Fig. 15: Norden's map of Hampshire.

19 For example, Mercator-Hondius-Janssonius, *Atlas*, i, p. 183: Swabia; p. 191: Bohemia; p. 195: Moravia.

20 Edward Lynam, 'The character of England in maps' in *The mapmaker's art* (London, 1953), p. 10.

21 Abraham Ortelius (ed. R.A. Skelton), *The theatre of the whole world, London, 1606* (Amsterdam, 1968), p. 59.

22 Daniel Augustus Beaufort, *A new map of Ireland, civil and ecclesiastical* (London, 1792).

23 A.J. Bird, 'John Speed's view of the urban hierarchy in Wales in the early seventeenth century', *Studia Celtica*, x–xi (1975–6), p. 404.

24 Dainville, *Le langage des géographes*, p. 251.

25 Mercator-Hondius-Janssonius, *Atlas*, i, p.155: Cologne; p. 185: Vienna; ii, p. 299: Paris; p. 385: Genoa.

26 Samuel Wyld, *The practical surveyor, or, The art of land-measuring, made easy* (London, 1725), p. 112; William Gardiner, *Practical surveying improved: or, Land measuring according to the most correct methods* (London, 1737), p. 88; B.Talbot, *The new art of land measuring; or, A turnpike road to practical surveying* (Wolverhampton, 1779), p. 343.

27 Edward Lynam, 'Period ornament, writing, and symbols on maps, 1250–1800' in *The mapmaker's art* (London, 1953), pp. 45, 46.

28 National Archives of Ireland: Ordnance Survey, general correspondence, file 1416 (1854).

29 George G. André, *The draughtsman's handbook of plan and map drawing* (London, 1891), p. 48 and plate 13.

30 Richard W. Stephenson, *The cartography of northern Virginia: facsimile reproductions of maps dating from 1608 to 1915* (Fairfax County, 1981), plate 21, p. 35.

31 William P. Cumming, *British maps of colonial America* (Chicago, 1974), p. 8: [1711].

32 Arthur H. Robinson, Randall D. Sale, Joel L. Morrison and Phillip Muehrcke, *Elements of cartography* (5th ed., New York, 1984), pp. 130–1.

33 R.A. Skelton, 'Colour in mapmaking', *Geographical Magazine*, xxxii, 11 (1960), pp. 544–53; Ulla Ehrensvärd, 'Color in cartography: a historical survey' in David Woodward (ed.), *Art and cartography: six historical essays* (Chicago, 1987), pp. 123–46; Christopher Lane, 'The color of old maps', *Mercator's World*, i, 6 (1996), pp. 50–7; Wallis and Robinson, *Cartographical innovations*, 'colour', pp. 207–11.

34 Raymond Lister, *How to identify old maps and globes* (London, 1965), pp. 56–9.

35 Robin Wilson, *Four colours suffice: how the map problem was solved* (London, 2002).

36 James Lees-Milne, *Fourteen friends* (London, 1996), p. 161.

37 O.A.W. Dilke, *Greek and Roman maps* (London, 1985), p. 14.

38 Gerald R. Tibbetts, 'The Balkhi school of geographers' in J.B. Harley and David Woodward (eds), *The history of cartography, volume two, book one: cartography in the traditional Islamic and south Asian societies* (Chicago, 1992), p. 122.

39 William Davis, *A complete treatise of land surveying by the chain, cross and offset staffs only* (London, 1798), p. 197; J. Eyre, *The exact surveyor: or, The whole art of surveying of land* (London, 1654), p. 205.

40 David Woodward, 'Medieval *mappaemundi*' in J.B. Harley and David Woodward (eds), *History of cartography, volume one: cartography in prehistoric, ancient, and medieval Europe and the Mediterranean* (Chicago, 1987), p. 327.

41 Gardiner, *Practical surveying improved*, p. 88.

42 Skelton, 'Colour in mapmaking', p. 544.

43 *Report of departmental committee [on] housing conditions of the working classes in Dublin*, House of Commons sessional paper, 1914, xix, qq. 5937–42.

44 Ehrensvärd, 'Color in cartography', pp. 123–46.

45 H.E. Salter (ed.), *Remarks and collections of Thomas Hearne*, x (Oxford, 1915), p. 255.

46 J.H. Andrews, *Shapes of Ireland: maps and their makers, 1564–1839* (Dublin, 1997), p. 86: Baptista Boazio, *Irelande*, 1599; Vladimir E. Bulatov, Catherine Delano Smith and Francis Herbert, 'Andrew Dury's *Map of the present seat of war, between Russians, Poles, and Turks* (1769)', *Imago Mundi*, liii (2001), p. 75.

47 David Beaton, *Dorset maps* (Wimborne, 2001), p. 27.

48 Kazutaka Unno, 'The origin of the cartographical symbol representing desert areas', *Imago Mundi*, xxxiii (1981), pp. 82–7; Boleslaw Szczesniak, 'The seventeenth century maps of China: an inquiry into the compilations of European cartographers', *Imago Mundi*, xiii (1956), p. 122.

49 Cumming, *British maps of colonial America*, p. 4.

50 Johannes Keuning, 'The Van Langren family', *Imago Mundi*, xiii (1956), p. 108.

51 J.H. Andrews, *Plantation acres: an historical study of the Irish land surveyor and his maps* (Belfast, 1985), plate 24, p. 335.

52 Joan Blaeu, *Le grand atlas*, iii (Amsterdam, 1663, 1967), p. 97: *Descriptio agri civitatis Coloniensis*.

53 Ortelius, *Theatre*, p. 28: 1594; Mercator-Hondius-Janssonius, *Atlas*, ii, p. 231.

54 Ibid., i, p. 139.

55 Ortelius, *Theatre*, p. 18.

56 Andrews, *Plantation acres*, p. 172.

57 April Carlucci and Peter Barber, *Lie of the land: the secret life of maps* (London, 2001), pp. 32–3: 'Routes from London to Luton Hoo, Bedfordshire', 1767; G. Walters, 'Themes in the large-scale mapping of Wales in the eighteenth century', *Cartographic Journal*, v, 2 (1968), p. 142: C. Hassall and J. Williams, The road from the new port of Milford to the New Passage of the Severn and

Gloucester, 2" to 1 mile; Audrey M. Lambert, 'Early maps and local studies', *Geography*, xli (1956), pl. 1, p. 172: Richard Davis, map of Oxfordshire, 1797.

58 Peter Whitfield, *The charting of the oceans: ten centuries of maritime maps* (London, 1996), pp. 62–3: *Caertboeck van de Midlandtsche Zee*; above, Fig. 10.4.

59 Asher Rare Books, *Catalogue 30* (Haarlem, 1999), p. 51: J. Grodemetz, *Caerte van de baey en stadt van Gibraltar* (The Hague, c.1726–7).

60 Asher Rare Books, *Catalogue 30*, p. 46: Emanuel van Meteren, *Obsidio et expugnatio trajecti ad Mosam* (Amsterdam, 1632).

61 G.R. Crone et al., 'Landmarks in British cartography', *Geographical Journal*, cxxviii (1962), pp. 428–9; Paul Laxton, *John Rocque's map of Berkshire* (Lympne, 1971); William Ravenhill, *Two hundred and fifty years of map-making in the county of Surrey* (Lympne, 1974); Paul Laxton, 'The geodetic and topographical evaluation of English county maps, 1740–1840', *Cartographic Journal*, xiii (1976), pp. 44, 47; J.H. Andrews, *Two maps of eighteenth-century Dublin and its surroundings* (Lympne, 1977).

62 Vincenzo Coronelli, *Atlante Veneto* (Venice, 1691).

63 Ortelius, *Theatre, Parergon*, map vi: 1590.

64 Anne Marie Claire Godlewska, *Geography unbound: French geographic science from Cassini to Humboldt* (Chicago, 1999), p. 20.

65 E.J.S. Parsons, *The map of Great Britain circa AD 1360 known as the Gough map* (Oxford, 1958).

66 P.P. Burdett (ed. J.B. Harley and P. Laxton), *A survey of the county palatine of Chester 1777* (Chester, 1974), plate IX: 'Explanation'.

67 Karrow, *Mapmakers of the sixteenth century*, p. 257.

68 Mary Sponberg Pedley, *Bel et utile: the work of the Robert de Vaugondy family of mapmakers* (Tring, 1992), p. 90: Didier Robert De Vaugondy, 1774.

69 R.V. Tooley, Charles Bricker and G.R. Crone, *A history of cartography: 2500 years of maps and mapmakers* (London, 1969), p. 264: Daniel Djurberg, *Karta over Polynesien* (Stockholm, 1780); J.N.L. Baker, *A history of geographical discovery and exploration* (2nd ed., London, 1937), opposite p. 434; David Fletcher, 'The Ordnance Survey's nineteenth century boundary survey: context, characteristics and impact', *Imago Mundi*, li (1999), pp. 134, 140.

70 Moll, *The compleat geographer*, p. xiii.

71 Lynam, *The Carta marina of Olaus Magnus*, p. 32.

72 George G. André, *The draughtsman's handbook of plan and map drawing* (London, 1891), plate 5.

73 C.W. Phillips, *Archaeology in the Ordnance Survey, 1791–1965* (London, 1980), p. 6.

74 R.A. Skelton (ed.), *The journals of Captain James Cook on his voyages of discovery: Charts & views drawn by Cook and his officers and reproduced from the original manuscripts* (Cambridge, 1969), LI.

75 John Goss, *The mapmaker's art* ([Chicago], 1993), pp. 80–1: Hessel Gerritsz, *Mar del Sur, Mar Pacifico*, MS, 1622.

76 Green, *The construction of maps and globes*, p. 9; Lynam, 'Period ornament', p. 45; above, Figs 13.4, 14.4; below, Fig. 18.13.

77 J.H. Andrews, *A paper landscape: the Ordnance Survey in nineteenth-century Ireland* (Oxford, 1975), p. 319.

78 Goss, *The mapmaker's art*, pp. 144–5: Johann Baptist Homann, Livonia and Curland, 1714.

79 Robinson, Sale, Morrison and Muehrcke, *Elements of cartography*, pp. 165–7.

80 John Norden, map of Ireland, [1608], Trinity College, Dublin, MS 1209/1.

81 Sonia Orwell and Ian Angus (eds), *The collected essays, journalism and letters of George Orwell*, iii (London, 1970), p. 114.

82 David Woodward with Herbert M. Howe, 'Roger Bacon on geography and cartography' in Jeremiah Hackett (ed.), *Roger Bacon and the sciences: commemorative essays* (Leiden, 1997), p. 209.

83 Günter Schilder, 'Cornelis Claeszoon, founder and stimulator of Dutch maritime and colonial cartography', History of Cartography Conference, Dublin, 1983, typescript, p. 2: map of the Moluccas.

84 Naomi Reed Kline, *Maps of medieval thought: the Hereford paradigm* (Woodbridge, Suffolk, 2001), p. 98.

85 Amando Cortesao and Avelino Teixeira da Mota, *Portugaliae monumenta cartographica*, v (Lisbon, 1960), plates 568, 572, 574; Donald Wigal, *Historic maritime maps used for historic exploration 1290–1699* (New York, 2000), pp. 80, 83, 84, 86 (Miller atlas, 1519) and pp. 116, 120 (Guillaume Le Testu, 1556); Helen Wallis (ed.), *The maps and text of the Boke of Hydrography presented by Jean Rotz to Henry VIII now in the British Library* (Oxford, 1981).

86 Marcel Watelet (ed.), *Gérard Mercator, cosmographe: le temps et l'espace* (Antwerp, 1994), pp. 270–3: Mercator, map of Palestine, 1537; Ortelius, *Theatre*, p. 34: undated map of the Low Countries.

87 R.A. Skelton and John Summerson, *A description of maps and architectural drawings in the collection made by William Cecil, first Baron Burghley now at Hatfield House* (Oxford, 1971).

88 Karrow, *Mapmakers of the sixteenth century*, p. 110: Sebastian Cabot, world map, 1544; Mary Sponberg Pedley, *The commerce of cartography: making and marketing maps in eighteenth-century France and England* (Chicago, 2005), p. 171.

89 '[Thomas] Colby's instructions for the interior survey of Ireland', 1825, in Andrews, *Paper landscape*, p. 319.

90 Royal Scottish Geographical Society, *Early maps of Scotland*, map 2, p. 5; Alfred Hiatt, 'Beyond a border: the maps of Scotland in John Hardyng's *Chronicle*' in Jenny Stratford (ed.), *The Lancastrian court: proceedings of the 2001 Harlaxton symposium* (Donington, Lincs, 2001), pp. 78–94; above, Fig. 17.3.

Chapter 18: To deck and beautify your plot

1 J.B. Harley, 'Deconstructing the map' (1989) in J.B. Harley (ed. Paul Laxton), *The new nature of maps* (Baltimore, 2001), p. 160.

2 A. Sarah Bendall, *Maps, land and society: a history, with a carto-bibliography of Cambridgeshire estate maps, c.1600–1836* (Cambridge, 1992), pp. 46, 49.

3 J. Eyre, *The exact surveyor: or, The whole art of surveying of land* (London, 1654), p. 204.

4 John Love, *Geodaesia: or, the art of surveying and measuring of land made easie* (London, 1688), p. 144.

5 John Hammond (enlarged by Samuel Warner), *The practical surveyor: containing the most approved methods for surveying of lands and waters, by the several instruments now in use* (London, 1750), p. 94.

6 Tom Conley, *The self-made map: cartographic writing in early modern France* (Minneapolis, 1997), p. 94.

7 Peter Barber, 'A Tudor mystery: Laurence Nowell's map of England and Ireland', *The Map Collector*, xxii (1983), pp. 16–21.

8 Bernhard Klein, *Maps and the writing of space in early modern England and Ireland* (Basingstoke, Hants, 2001), pp. 115–16.

9 Sarah Bendall, *Dictionary of land surveyors and local map-makers of Great Britain and Ireland, 1530–1850*, i (London, 1997), p. 32.

10 D.G. Moir, 'A history of Scottish maps' in Royal Scottish Geographical Society, *The early maps of Scotland to 1850* (Edinburgh, 1973), pp. 46, 48; R.A. Skelton, *County atlases of the British Isles, 1579–1703* (London, 1970), p. 102.

11 Peter Whitfield, *The mapping of the heavens* (London, 1995), chapter 1.

12 Robert W. Karrow Jr, *Mapmakers of the sixteenth century and their maps: bio-bibliographies of the cartographers of Abraham Ortelius, 1570* (Winnetka, IL, 1993), pp. 32, 283, 447–8, 534; R.V. Tooley, *Leo Belgicus: an illustrated list of variants* (Map Collectors' Series, vii, London, 1963).

13 Gillian Hill, *Cartographical curiosities* (London, 1978), nos 52–8, pp. 45–9; John Goss, *The mapmaker's art: an illustrated history of cartography* ([Chicago], 1993), pp. 334–9; Helen M. Wallis and Arthur H. Robinson (eds), *Cartographical innovations: an international handbook of mapping terms to 1900* ([Tring], 1987), 'symbolic map', pp. 68–9.

14 A.S. Osley, *Mercator: a monograph on the lettering of maps, etc. in the 16th century Netherlands* (London, 1969).

15 William Davis, *A complete treatise of land surveying by the chain, cross and offset staffs only* (London, 1798), p. 229.

16 Edward Lynam, *British maps and map-makers* (London, 1944), opposite p. 32: lands belonging to Charles Henry Talbot, 1770.

17 Reproduced in Goss, *The mapmaker's art*, plate 5.51, p. 168: 'Pte of Tartaria'.

18 M.H. Edney, 'British military education, mapmaking, and military "map-mindedness" in the later Enlightenment', *Cartographic Journal*, xxxi, 1 (1994), p. 14.

19 Mary Blewitt, *Surveys of the seas: a brief history of British hydrography* (London, 1957), p. 36.

20 Quoted in Roger J.P. Kain and Hugh C. Prince, *The tithe surveys of England and Wales* (Cambridge, 1985), p. 81: 1837.

21 John Paddy Browne, *Map cover art: a pictorial history of Ordnance Survey cover illustrations* ([Southampton, 1990]).

22 Reproduced in Catherine Delano-Smith and Roger J.P. Kain, *English maps, a history* (London, 1999), p. 36: *c*.1200; above, Figs 8.5, 8.13.

23 Goss, *The mapmaker's art*, plate 2.17, pp. 48–9.

24 Quoted by D.G. Moir and R.A. Skelton, 'New light on the first atlas of Scotland', *Scottish Geographical Magazine*, lxxxvi (1970), p. 155.

25 Stephanie Pratt, 'From the margins: the native American personage in the cartouche and decorative borders of maps', *Word and Image*, xii, 4 (1996), pp. 349–65.

26 R.V. Tooley, *Some portraits of geographers and other persons associated with maps*, Map Collectors' Series, cv (London, 1975), unpaginated; Rodney W. Shirley, *The mapping of the world: early printed world maps, 1472–1700* (London, 1983), plate 127, p. 174; plate 256, p. 360.

27 Arnold Horner, 'Cartouches and vignettes on the Kildare estate maps of John Rocque', *Quarterly Bulletin of the Irish Georgian Society*, xiv, 4 (1971), p. 62; David Smith, 'The cartographic illustration of land surveying instruments and methods', *Bulletin of the Society of Cartographers*, xxvi, 1 (1992), pp. 11–20.

28 Delano-Smith and Kain, *English maps, a history*, p. 247.

29 J.B. Harley, 'Meaning and ambiguity in Tudor cartography' in Sarah Tyacke (ed.), *English map-making 1500–1650* (London, 1983), p. 31.

30 R.V. Tooley, editorial, *The Map Collector*, xv (1981), p. 2, differently rendered in Leo Bagrow (ed. R.A. Skelton), *History of cartography* (London, 1964), p. 220 and in Ronald Rees, 'Historical links between cartography and art', *Geographical Review*, lxx (1980), p. 65.

31 National Archives (Public Record Office), London, MPF 1/89: east Ulster, 1568; MPF 1/73: Munster, 1571.

32 E.G.R. Taylor, *The haven finding art* (London, 1956), p. 100.

33 François de Dainville, *Le langage des géographes* (Paris, 1964), pp. 286, 299; Monique Pelletier, 'La symbolique royale française: des globes et des rois' in *Tours et contours de la terre* (Paris, 1999), pp. 64–6.

34 G.S. Ritchie, '500 years of graphical and symbolical representation on marine charts', History of Cartography Conference, Greenwich, 1975, typescript, p. 3; W.E. May, *A history of marine navigation* (Henley-on-Thames, 1973), p. 55.

35 Arthur Charles Fox-Davies, *A complete guide to heraldry* (London, 1993), pp. 272–6; Michel Mollat du Jourdin and Monique de la Roncière with Marie-Madeleine Azard, Isabelle Raynaud-Nguyen and Marie-Antoinette Vamereau, *Sea charts of the early explorers, 13th to 17th century* (New York, 1984), pp. 202, 217.

36 Armando Cortesao and Avelino Teixeira da Mota, *Portugaliae monumenta cartographica*, v (Lisbon, 1960), plates 546–8, 550.

37 Derek Howse and Michael Sanderson, *The sea chart: an historical survey based on the collections in the National Maritime Museum* (Newton Abbot, 1973), p. 101.

38 Robert Morden and William Berry, *A new map of the English plantations in America* (London, *c*.1673), reproduced in *Imago Mundi*, liv (2002), plate 13. See above, Fig. 2.3.

39 David Buisseret, *The mapmakers' quest: depicting new worlds in renaissance Europe* (Oxford, 2003), p. 74.

40 Reproduced in Goss, *The mapmaker's art*, p. 76.

41 J.H. Andrews, 'Map and language: a metaphor extended', *Cartographica*, xxvii, 1 (1990), pp. 1–19.

42 Wallis and Robinson, *Cartographical innovations*, pp. 247–8.

43 Anon., 'Cartouches', *Imago Mundi*, xxvii (1975), p. 8.

44 Vicomte de Santarem, *Atlas composé de mappemondes, de portulans et de cartes hydrographiques et historiques depuis le xie jusqu'au xvie siècle* (Paris, 1849), plate 49: Giovanni Leardo, Venice, 1448; http://www.maphist@geo.uu.nl, 11–19 September 2006, 'Cartouche'; 9–10 January 2008, 'Book on cartouches'.

45 Joan Blaeu, *Le grand atlas*, i (Amsterdam, 1663, 1967), p. 21: Jan Mayen.

46 Trinity College, Dublin, MS 1209/15; note 54 below.

47 *A new map of the island of Barbadoes*, c.1675, reproduced in Peter Barber and Christopher Board (eds), *Tales from the Map Room: fact and fiction about maps and their makers* (London, 1993), p. 30; Tony Campbell, *The printed maps of Barbados*, Map Collectors Series, xxi (London, 1965), pp. 11–12; Jeanette D. Black, *The Blathwayt atlas, ii: commentary* (Providence, 1975), pp. 180–3.

48 Günter Schilder, 'Jodocus Hondius, creator of the decorative map border', *The Map Collector*, xxxii (1985), pp. 40–3.

49 Pieter van der Aa, *Galérie agréable du monde* (Leiden, 1729).

50 J.B. Harley and Yolande O'Donoghue, *The old series Ordnance Survey maps of England and Wales, scale 1 inch to 1 mile*, ii (Lympne, 1977), p. xlii.

51 Valerie A. Kivelson, 'Cartography, autocracy and state powerlessness: the uses of maps in early modern Russia', *Imago Mundi*, li (1999), pp. 88–9, 91.

52 Aaron Rathborne, *The surveyor in foure bookes* (London, 1616), p. 174.

53 James A. Welu, 'The sources and development of cartographic ornamentation in the Netherlands' in David Woodward (ed.), *Art and cartography: six historical essays* (Chicago, 1987), p. 149; Robin Halwas Ltd, *John Rocque's survey of the Kildare estates: manor of Kilkea, 1760. A rediscovered atlas ornamented by Hugh Douglas Hamilton* (London [2005]), pp. 22–3.

54 Trinity College, Dublin, MS 1209/15; National Maritime Museum, Greenwich, MS P.49/18; Saxton, Warwickshire-Leicestershire and Derbyshire.

55 H. George Fordham, 'Ships on maps', *The History Teachers' Miscellany*, v, 3 (1927), pp. 40–2.

56 Edward Lynam, 'Period ornament, writing, and symbols on maps, 1250–1800' in *The mapmaker's art* (London, 1953), pp. 48–54.

57 J.H. Andrews, 'Science and cartography in the Ireland of William and Samuel Molyneux', *Proceedings of the Royal Irish Academy*, lxxx C (1980), plate II; original Holkham Hall, Norfolk, reproduced by permission of Viscount Coke and the Trustees of the Holkham Estate. For other examples see J.H. Andrews, *Irish maps* (Dublin, 1978), no. 7 (c.1640); J.H. Andrews, *Shapes of Ireland: maps and their makers, 1564–1839* (Dublin, 1997), p. 126 (1659).

58 Marcel Watelet, *Gérard Mercator cosmographe: le temps et l'espace* (Antwerp, 1994), p. 24: 1584.

59 William Ravenhill, *John Norden's manuscript maps of Cornwall* (Exeter, 1972), pp. 34–7.

60 Roger J.P. Kain and Elizabeth Baigent, *The cadastral map in the service of the state: a history of property mapping* (Chicago, 1992), p. 148: Niederzwehren, Hessen, Germany, 1625.

61 R.A. Skelton, introduction to Gerard Mercator, Henry Hondius and Joannes Janssonius (ed. R.A. Skelton), *Atlas or a geographicke description of the world*, i (Amsterdam, 1636, 1968), pp. xii, xxv.

62 David Smith, 'Jansson versus Blaeu: a study in competitive response in the production of English county maps', *Cartographic Journal*, xxiii, 2 (1986), pp. 106–14.

63 François de Dainville, *Le langage des géographes* (Paris, 1964), pp. 64–5; Mary Sponberg Pedley, *Bel et utile: the work of the Robert de Vaugondy family of mapmakers* (Tring, 1992), pp. 64–7; Peter Barber, 'Maps and monarchs in Europe 1550–1800' in Robert Oresko, G.C. Gibbs and H.M. Scott (eds), *Royal and republican sovereignty in early modern Europe: essays in memory of Ragnhild Hatton* (Cambridge, 1997), p. 120; Laurence Worms, 'Thomas Kitchin's "journey of life": part two', *The Map Collector*, lxiii (1993), pp. 16–17; Pieter Mortier, *Le duché de Milan* (Amsterdam, 1701), title reproduced in Marco van Egmond, 'The secrets of a long life: the Dutch firm of Covens and Mortier (1685–1866) and their copper plates', *Imago Mundi*, liv (2002), p. 72.

64 J.H. Andrews, *Plantation acres: an historical study of the Irish land surveyor and his maps* (Belfast, 1985), p. 119 (plate 4), pp. 153–4; Andrews, *Irish maps*, plates 16, 17; Raymond Refaussé and Mary Clark, *A catalogue of the maps of the estates of the archbishops of Dublin* (Dublin, 2000), passim.

65 Essex County Council, *The art of the map-maker in Essex, 1566–1860* (Chelmsford, 1947), plate XVII: Danbury etc., 1758; Bendall, *Maps, land and society*, plate 11, p. 63: East Hatley, 1750.

66 Horner, 'Cartouches and vignettes on the Kildare estate maps of John Rocque', pp. 57–76; Bendall, *Dictionary of land surveyors*, i, plate 11: John Rocque, Woodstock, Ireland, 1756.

67 Delano-Smith and Kain, *English maps*, p. 86: Joel Gascoyne, Cornwall, 1699, cartouche.

68 Essex County Council, *Art of the map-maker in Essex*, plate IV: Mistley Thorn, 1778; Vladimir E. Bulatov, 'Eighteenth-century Russian charts of the Straits (Bosporus and Dardanelles)', *Imago Mundi*, lii (2000), plates 9 and 10: c.1782.

69 Kain and Baigent, *Cadastral map*, p. 230: Noisy-le-Roi, France, 1819.

70 Ibid., p. 64: Svanholm estate, Sweden, 1785; p. 316: Adelaide, Australia, 1841; p. 321: Port Nicholson, New Zealand, 1843.

71 J. Pigot & Co., *British atlas, comprising the counties of England* (London, [1840], 1990); Ashley Baynton-Williams, *Town and city maps of the British Isles, 1800–1855* (London, 1992), pp. 14–55.

72 Thomas Moule (ed. Roderick Barron), *The county maps of old England* (London, 1990); Baynton-Williams, *Town and city maps of the British Isles*, pp. 62–127.

73 Samuel Wyld, *The practical surveyor, or the art of land-measuring, made easy* (London, 1725), p. 115; J.B. Harley and Kees Zandvliet, 'Art, science and power in sixteenth-century Dutch cartography', *Cartographica*, xxix, 2 (1992), pp. 12–14.

74 Welu, 'Cartographic ornamentation in the Netherlands', pp. 166–7.

75 David Fletcher, 'The Careswell atlas: working tool and work of art', *The Map Collector*, lxxiii (1995), p. 35.

76 Andrews, *Plantation acres*, p. 152.

77 Donald L. McGuirk Jr, 'The mystery of Cuba on the Ruysch map', *The Map Collector*, xxxvi (1986), pp. 40–1; W.P. Cumming, R.A. Skelton and D.B. Quinn, *The discovery of North America* (London, 1971), p. 36: Juan de la Cosa, world map, c.1500 (central American mainland).

78 Reproduced in Goss, *The mapmaker's art*, plate 3.13, pp. 69–70.

79 Tony Campbell, *Weinreb + Douwma Ltd catalogue 7: maps* (London, 1971), p. 13: Jean Baptiste Nolin and Vincenzo Coronelli; Helen Wallis, '"So geographers in Afric-maps"', *The Map Collector*, xxxv (1986), p. 32; Nicholas Crane, *Mercator: the man who mapped the planet* (London, 2002), p. 248.

80 R.A. Skelton, *Maps: a historical survey of their study and collecting* (Chicago, 1972), p. 61.

81 David B. Quinn, 'Artists and illustrators in the early mapping of North America', *Mariner's Mirror*, lxxii (1986), pp. 244–73.

Chapter 19: Maps and society

1 J.B. Harley (ed. Paul Laxton), *The new nature of maps: essays in the history of cartography* (Baltimore, 2001).

2 Michel Mollat du Jourdin and Monique de la Roncière with Marie-Madeleine Azard, Isabelle Raynaud-Nguyen and Marie-Antoinette Vamereau, *Seacharts of the early explorers, 13th to 17th century* (New York, 1984), p. 234.

3 Adam Smith (ed. Edwin R.A. Seligman), *The wealth of nations*, i (London, 1910), p. 117.

4 C. Koeman, 'Levels of historical evidence in early maps (with examples)', *Imago Mundi*, xxii (1968), p. 79.

5 Robert Dunlop, 'Sixteenth-century maps of Ireland', *English Historical Review*, xx (1905), p. 334.

6 Helen Wallis (ed.), *The maps and text of the Boke of Hydrography presented by Jean Rotz to Henry VIII now in the British Library* (Oxford, 1981).

7 Mary Sponberg Pedley, *Bel et utile: the work of the Robert de Vaugondy family of mapmakers* (Tring, 1992), p. 64.

8 Matthew H. Edney, *Mapping an empire: the geographical construction of British India, 1765–1843* (Chicago, 1997), p. 32. See also Matthew Edney, 'Mathematical cosmography and the social ideology of British cartography, 1780–1820', *Imago Mundi*, xlvi (1994), pp. 101–16.

9 W.R. Taylor, *An outline of the re-triangulation of Northern Ireland* (Belfast, 1967), p. 23.

10 National Archives of Ireland: Ordnance Survey, letter registers (inwards), 8983 (13 January 1831).

11 J.B. Harley and David Woodward (eds), *The history of cartography, volume one: cartography in prehistoric, ancient, and medieval Europe and the Mediterranean* (Chicago, 1987); J.B. Harley and David Woodward (eds), *The history of cartography, volume two, book one: cartography in the traditional Islamic and south Asian societies* (Chicago, 1992); J.B. Harley and David Woodward (eds), *The history of cartography, volume two, book two: cartography in the traditional east and southeast Asian societies* (Chicago, 1994); David Woodward and G. Malcolm Lewis (eds), *The history of cartography, volume two, book three: cartography in the traditional African, American, Arctic, Australian, and Pacific societies* (Chicago, 1998), passim.

Glossary

IN THE FOLLOWING DEFINITIONS, alternative significations irrelevant to the present book are ignored. The list does not include words that are both (a) used only in a single passage of the text and (b) defined at the point where they first occur.

Achromatic lens: two lenses juxtaposed to counteract the decomposition of refracted light in optical instruments

Acre, statute: 4840 square yards

Acute angle: an angle of less than 90 degrees

Alidade: a straight ruler carrying a pair of sights

Alphanumeric grid: a rectangular grid whose squares are identified by numbers in one direction and letters in the other

Azimuth: the horizontal angle between any great circle and a meridian

Azimuthal (or zenithal or central) projection: a projection in which the directions of all the lines radiating from a central point are the same as the directions of the corresponding lines on the earth's surface

Back staff: an adaptation of the cross staff which allowed the observer to turn his back to the sun

Beam compass: an instrument incorporating a heavy horizontal beam, used for drawing larger circles than can be described by an ordinary pair of compasses

Binnacle or bittacle: a box holding the magnetic compass used on a ship

Cable: as a unit of distance, 720 feet

Cadastral map: a map which, by virtue of its precision and large scale, is capable of showing property boundaries with a high degree of accuracy

Camera lucida: an aid to drawing in which light rays from the object to be represented are reflected by a prism to produce an image on a surface incorporated within the instrument

Camera obscura: a darkened chamber or box, into which light is admitted through a double convex lens, forming an image of external objects on a surface placed at the focus of the lens

Cartometry: the taking of measurements on a map as a means of elucidating its author's methods

Cartouche: a decorative border in the margin of a map, enclosing the title or other supplementary matter

Centimetre: 0.3937 inches

Centuriation: the laying out of agricultural land by Roman surveyors in an extensive grid of large squares

Chain: as a unit of distance, 66 feet

Chorochromatic map: a map distinguishing different areas by colour or shading

Circumferentor: a surveying instrument comprising a flat bar with sights at the ends and a circular box in the middle, containing a magnetic needle that rotates over a graduated circle

Collimation, line of: line of sight or optical axis in a levelling instrument

Command: the vertical height of a viewpoint above the object under observation

Comparative costs, law of: the tendency in a free market economy for producers to specialise in goods or services for which their competitive advantage is greatest or their competitive disadvantage least

Compass point: a horizontal direction defined by its angular relationship to a north-south (geographical or magnetic) alignment

Complementary angles: angles which together make up one right angle

Conformality: the property of a map projection by which the horizontal scale at any point is the same in all directions, though different from the scale at other points on the same map

Coordinates: Linear or angular values that define the position of a point with reference to an origin

Cosine: for an acute angle in a right-angled triangle, the ratio to the hypotenuse of the other side adjoining that angle

Cotangent: for an acute angle in a right-angled triangle, the tangent of the complementary angle

Cross staff: (1) an instrument for measuring vertical angles in which a crosspiece slides along a graduated shaft, one end of the cross being aligned with the horizon, the other end with the object under observation; (2) an instrument for setting out offsets in a chain survey, in which two pairs of sights are permanently set at an angle of 90 degrees

Diopter, dioptra: an invention of the Greek scientist Hero (first century AD) similar in principle to a modern theodolite

Eidograph: an instrument for enlarging or reducing drawings

Electrotyping: a process of duplication employing the electrolytic deposition of copper first as a matrix on an original printing plate and then on the matrix to form a second plate identical with the first

Ephemerides: an astronomical calendar or almanac in which the positions of heavenly bodies are tabulated in advance for each day of a certain period

Equivalence: a property of a map projection by which the area of any enclosed figure on the map is equal to the area of the corresponding figure on a globe of the same scale

Fathom: 6 feet

Foot (English): as a unit of distance, 12 inches or 30.48 centimetres

Forme: type composed in a frame ready for printing

French curve: a curved ruler

Furlong: 660 feet

Geodesy, geodetic: the scientific study of the exact size and shape of the earth as a whole

Geoid: the shape that the earth would assume, under existing gravitational forces, if its surface were entirely covered by sea

Gimbals: a mounting that keeps a mariner's compass in a horizontal position on a moving ship

Globular projection: a map projection (other than a transverse case of an azimuthal projection) in which hemispheres are represented by circles, and both central meridian and equator by straight lines

Gore: a map of a strip of the earth's surface between two closely-spaced meridians running from one pole to another, usually designed to form part of a terrestrial globe

Graphometer: a semi-circular surveying instrument for measuring horizontal angles

Graticule: a network of lines on a map representing meridians and parallels. Not to be used as a synonym for 'grid'

Great circle: the largest circle that can be drawn on a sphere

Hachures: lines drawn in the direction of steepest slope as a means of representing relief

Impression: in cartobibliography, a single printed sheet

Inch: 2.54 centimetres

Isoceles triangle: a triangle with two sides the same length and the third side of a different length

Isoline, isopleth: a line on a map connecting points assumed to have the same numerical value

Kilometre: 1000 metres, 0.6214 miles

Knot: as a measure of speed for ships, 1 nautical mile per hour

Latitude: angular distance on a meridian measured outwards from the equator

League: English, 3 statute miles; marine, 3 nautical miles

Link: as a unit of distance, 0.66 feet

Lithography: planographic printing on paper from an image drawn in greasy ink on a flat surface formed of stone or (in more recent usage) some other substance

Log: a device for measuring the speed of a ship in which a floating piece of wood is attached to a line wound on a reel

Longitude: the angle measured in a plane parallel to that of the equator between the plane of a given meridian and the plane of some other meridian

Magnetic pole: the place on the earth's surface to which an accurate magnetic compass needle appears to point when unaffected by magnetic materials in its immediate vicinity

Mappamundi: a medieval world map, not based on strict geometrical principles

Meridian: a semi-great-circle terminated by the earth's geographical poles

Meridian, central: the meridian about which the geometric properties of a map projection are symmetrically disposed

Meridian, prime: the meridian from which terrestrial longitude is measured

Metre: 39.3701 inches

Metrology: the scientific study of weights and measures

Mile: statute, 5280 feet; nautical, formerly 6080 feet, currently 6076.1 feet

Millimetre: 0.1 centimetres

Neat line: a single line, usually part of a rectangle, delimiting the geographical content of a map

Oblique projection: a projection in which geometrical properties normally unique to the equator or a meridian are vested in some other great circle

Obtuse angle: an angle of more than 90 degrees

Octant: an angle-measuring instrument in the form of a graduated eighth of a circle

Offset: a short horizontal measurement at right angles to a longer survey line, used to fix the position of geographical detail

Optical square: a reflecting instrument used for laying out lines at right angles to each other

Orthographic projection: a projection in which part of the surface of a sphere is projected on to a tangential plane by perpendicular rays from a light-source an infinite distance away

Pace, geometrical: as a unit of distance, 5 feet

Pantagraph or pentagraph: an instrument for copying a drawing with or without a change of scale, composed of a framework of movable arms, a tracing point and a drawing point

Parallax: the angle formed by straight lines connecting a heavenly body with (a) the centre of the earth and (b) a point on the earth's surface

Perch, statute: 16.5 feet

Planiform: simulating, either exactly or schematically, the appearance of a landscape feature seen from above

Planimetry: the precise representation of horizontal angles and distances

Plat carrée: a map projection with straight meridians and parallels intersecting at right angles, and with all the meridians and one parallel (the equator) drawn true to scale

Poles, celestial: points where the earth's axis notionally meets the celestial sphere and about which the heavens appear to rotate

Poles, geographical: extremities on the terrestrial surface of the earth's axis of rotation, the north being the pole from which the star Polaris can be seen

Polyconic projection: a map projection in which parallels of latitude are represented by non-concentric circular arcs with their centres on a straight line representing the central meridian

Projection: any systematic arrangement of meridians and parallels portraying the curved surface of a sphere or spheroid on a plane

Proportional circle: a circle whose area or apparent area on a statistical thematic map is exactly or approximately proportional to the quantity it represents

Proportional dividers: adjustable dividers which with a single setting can lay off two distances in a known proportion

Protract: to draw or otherwise graphically to define the points, lines and angles of a measured survey at a prescribed scale as the framework for a map

Pythagoras's theorem: a proof that the square on the hypotenuse of a right-angled triangle is equal to the sum of the squares on the other two sides

Register: the correct matching, on paper, of superimposed images (usually in contrasting colours) printed from different plates

Representative fraction: an arithmetical statement of map-scale, with paper distances represented by unity as numerator and real-world distances by an appropriate multiple as denominator

Resection: the fixing of a point in a plane-table survey by taking angles from it to points already established

Rhombus: a quadrilateral with all four sides of equal length and none of the angles right angles

Rhumb line: (1) a line crossing successive meridians at a constant angle; (2) more generally, one of a system of radiating straight lines drawn on a map

Rood: as a measure of area, 0.25 acres

Scale: the numerical ratio of distances on a map to the terrestrial distances that they represent

Scale of chords: a scale-line graduated in degrees, in which each length (measured from the origin) defines the angle subtended at the centre of a circle by a chord of the same length

Sector, zenith: an astronomical instrument for the exact measurement of small angles in which a telescope turns about the centre of a graduated vertical arc

Sextant: a surveying or navigational instrument for measuring angles, with an arc of 60 degrees

Sine: for an acute angle in a right-angled triangle, the ratio of the side opposite to that angle, to the hypotenuse

Spheroid: a solid figure differing only slightly from a sphere; usually, the shape of a rotating sphere as affected by gravitational forces

Spline: a flexible strip of wood or other material used by draughtsmen in setting out curves

State: what is common to all the impressions printed from an engraved plate between one deliberate alteration of that plate and the next

Station pointer: an instrument with three movable arms for locating the observer's position on a chart from angles taken to three objects

Tangent: (1) a straight line that touches a curved line without cutting through it; (2) for an acute angle in a right-angled triangle, the ratio of the side opposite that angle to the adjacent side other than the hypotenuse

Thematic map: a non-topographical map that lays particular emphasis on a small number (usually one) of geographical variables

Theodolite: an instrument for measuring angles in the field, incorporating an alidade with sights that rotates horizontally on a graduated circular plate and, in later designs, vertically against a graduated arc

Topographical map: a map whose principal purpose is to portray the more permanent visible features of the earth's surface as faithfully as its scale allows

Topology: the study of those spatial properties (independent of measured lengths and angles) that are not changed by continuous deformation such as stretching or twisting

Transverse projection: a map projection whose axis or line of minimum distortion has been rotated through 90 degrees from what is considered its normal position

Trapezium: a quadrilateral having only one pair of its opposite sides parallel

Traverse: a survey of the lengths and bearings of a connected series of lines in which there are no triangles

Triangulation: a survey method in which angles are measured in a network of triangles

Trilateration: a survey method in which distances are measured or estimated in a network of triangles

Well-conditioned triangle: a triangle in which small angular changes at the apices have little effect on the lengths of the sides

Zenith: a point vertically above a terrestrial observer

Zincography: the printing of maps from zinc plates

Index